Birley/Haworth/Batchelor

Physics of Plastics

Front cover illustrations

centre	CAD CAM Analysis of polymer melt flow in injection moulding (courtesy of MOLDFLOW (Europe) Ltd.)
Top-left	Transmission electron micrograph, showing the morphology of acrylic-elastomer modified PVC (courtesy of Rohm & Haas (France) S.A.)
Top-centre	Orientation of short glass fibres in injection moulded PP, examined by contact microradiography
Top-right	Unmelted granules of PA6 in an injection moulded component; section viewed in an optical microscope, through crossed polars
Bottom-right	Scanning electron micrograph of an etched surface of a HDPE extrusion blow moulded container
Bottom-centre	Fibrillar morphology of highly drawn PVC, showing ductile yielding at the root of a notched impact test specimen (reproduced by permission of Elsevier Applied Science Publishers, Ltd.)
Bottom-left	Optical micrograph showing the development of spherulites of different sizes, in a semi-crystalline thermoplastic material

Arthur W. Birley
Barry Haworth
Jim Batchelor

Physics of Plastics

Processing, Properties and Materials Engineering

Hanser Publishers Munich Vienna New York Barcelona

Distributed in the United States of America and Canada
by Oxford University Press, New York

The Authors:

Arthur W. Birley, M.A., D.Phil., C.Eng., F.P.R.I.
Barry Haworth, B.Sc., M.Sc., C.Eng., M.P.R.I.
*Jim Batchelor**, Ph.D., C.Eng., F.P.R.I., A.N.C.R.T.

Institute of Polymer Technology and Materials Engineering (I.P.T.M.R.)
Loughborough University of Technology
Loughborough
Leicestershire LR11 3TU, United Kingdom

*Jim Batchelor is currently employed as Technical Director at:

Stirling Lloyd Polychem Ltd.
New Mills
Stockport
Cheshire SK12 4AE, United Kingdom

Distributed in USA and in Canada by
Oxford University Press
200 Madison Avenue, New York, N.Y. 10016

Distributed in all other countries by
Carl Hanser Verlag
Kolbergerstrasse 22
D-8000 München 80

Library of Congress Cataloging-in-Publication Data

Birley, Arthur W.
Physics of plastics : processing, properties, and materials engineering / Arthur W. Birley, Barry Haworth, Jim Batchelor.
p. cm.
Includes bibliographical references and index.
ISBN 0-19-520782-3. --ISBN 0-19-520918-4 (pbk.)
1. Plastics--Mechanical Properties. 2. Plastics--Molding.
3. Plastics--Extrusion. I. Haworth, Barry. II. Batchelor, Jim.
III. Title.
TA455.P5B513 1991
620.1'92392--dc20 91-73381 CIP

CIP-Titelaufnahme der Deutschen Bibliothek

Birley, Arthur W.:
Physics of plastics : processing, properties and materials engineering / Arthur W. Birley ; Barry Haworth ; Jim Batchelor. - Munich ; Vienna ; New York ; Barcelona : Hanser ; New York : Oxford Univ. Press. 1991
ISBN 3-446-16274-7 brosch.
ISBN 3-446-15098-6 Pp.
NE: Haworth, Barry:; Batchelor, Jim:

ISBN 3-446-15098-6 (Pp.) / ISBN 3-446-16274-7 (brosch.) Carl Hanser Verlag Munich Vienna New York Barcelona
ISBN 0-19-520782-3 (hardcover) / ISBN 0-19-520918-4 (pbk.) Oxford University Press

TeX-typesetting: Formelsatz Steffenhagen, Königsfeld, Germany
Printed in Germany

Dedication

to
Joy, Helen and Shirley

Foreword

The Plastics and Rubber Institute (PRI) is delighted to sponsor 'The Physics of Plastics – Processing, Properties & Materials Engineering', by Prof. A.W. Birley, Mr. B. Haworth and Dr. J. Batchelor of Loughborough University. It is hoped that this book will represent the first in a long line of co-operative partnerships between Carl Hanser Verlag, who are renowned for the high quality of their polymer titles, and the PRI.
The PRI is a professional body whose aim is to promote excellence in the polymer field generally, and to advance standards in the education, research and training in polymers, in particular. A vital part of doing this is to bring together the worlds of Industry and Academic Research; it is strongly felt that 'Physics of Plastics' achieves all of these objectives.

SERENA AITCHISON
Publications Officer
PRI - The Plastics and Rubber Institute

Preface

The Objectives

It has been commented that plastics are chemicals which are applied on the basis of their physics. Whilst an all-embracing statement of this type is obviously not strictly true (witness the importance of oxidation or hydrolysis, which limit the usefulness of many polymers), an understanding of the physics of plastics is essential if optimum performance is to be achieved in polymer products. Our general intention is to provide a modern treatment of the physical properties of plastics materials, in a manner which defines and develops first principles, and which emphasizes all factors which influence the behaviour of plastics products in service.

The versatility of polymers with respect to shaping allows a variety of techniques to be employed. This results in a plentiful supply of problems for the technologist, particularly concerning polymer-process interactions. This is a unique field which does not confront other materials technologists to the same extent; it can be defined by the four *p*'s: *processing, properties, product and performance.* In this context, characterization takes on a new significance: product characterization must also reflect upon the thermomechanical history of the material during processing, the consequential effects on properties and the likely effect on performance. It is not sufficient that a correctly-specified polymer has been used for an application; the product quality also depends on the degree of optimization at the manufacturing stage.

This book is written to extend the outlook of physicists to polymers, not from a purely scientific point of view, but relevant to the practical world of industry, paying particular attention to the problems which arise. It is not the intention to invade the domain of the industrial chemists, nor of their colleagues in research laboratories: it is the objective to provide a physics foundation for those who wish to learn about the downstream end of the plastics industry, particularly students of polymer technology, and industrialists who are concerned with commodity products. *Quality* is a term and a creed which will intrude increasingly into our working lives; the attainment of high and consistent quality is possible only with mature understanding of the processes and materials involved.

The Approach

It would be pertinent to suggest that the most competent industrial engineers and technologists are those who have commendable levels of experience and expertise at their disposal. Since experience must be accumulated and cannot therefore, be substituted, we must concentrate on how this text can best contribute to an individual's knowledge. We have attempted to do this with reference to some theoretical concepts of polymer physics; however we hope that the deliberately-adopted "theory-in-practice" approach can be justified for this subject area.

Following an introductory Chapter written especially to provide a scientific basis for those new to polymer physics and technology, plastics processing, and the physical properties which influence the choice and optimum conditions of any manufacturing technique, are considered in detail (Chapters 2–5). A range of mechanical, and other physical properties are then discussed (Chapters 6–10), yet as the text progresses, we have attempted to include a degree of overlap with the early Chapters in order to demonstrate how the context of fundamental theory is determined, or modified by the manufacturing process. This approach is especially evident in Chapter 7, where the mechanical principles of product failure are studied in detail.

Wherever appropriate, we have attempted to emphasize key points and principles with reference to real examples of material characteristics and behaviour, many of which have been determined within our own laboratories. In general, we follow the unwritten philosophy which has always been a prominent feature of Polymer Research at Loughborough University of Technology: only if the effects of the processing stage are fully characterized and understood can a theoretical concept be realized in practice.

Although the physical characteristics of plastics materials are described and documented throughout the textbook, it has become our intention to emphasize the factors which enhance, modify or destroy a predicted level of in-service performance; these overtones are present throughout the presentation. We have not dedicated a Chapter to *"Process-Property Interactions"*, since we do not feel able to do justice to the importance of these relationships in an autonomous piece of text; instead, it forms the overall theme of the book.

Within the constraints of a single text, it is impossible to address all the relevant factors and principles associated with such a vast and important field. We acknowledge therefore, that our contribution contains, or refers directly to but a fraction of the total range of important issues in the world of manufacturing with plastics. However, we trust that what we feel to be a realistic approach to this subject helps, in its way, to complement the many notable attempts which have previously been made to describe the behaviour of these modern, unique, and increasingly successful and established materials.

Acknowledgements

We have compiled this textbook in a manner which we hope has reflected the diverse range of work with which we have become associated, at the Institute of Polymer Technology and Materials Engineering (IPTME), at Loughborough University of Technology (LUT). Such projects cross several scientific and engineering disciplines, and these activities inevitably have a practical undercurrent associated with them. During the compilation of this textbook, it has therefore been one of our main intentions to illustrate specific physical principles by reference to some real, practical examples of the behaviour of plastics materials and products, thereby providing an approach which, if not unique, should nevertheless provide a useful link between theoretical concepts and in-service behaviour. We hope we have succeeded, in this respect.

Reference to this type of information, much of which has been derived by Staff and Postgraduate Students from within IPTME, would not have been possible without the efforts, skills and achievements of many personnel based presently, or previously at Loughborough, and contact with our much valued colleagues from the industrial polymer community. We would like to express our gratitude to all those who have offered contributions of many different kinds, towards the stimulating environment in which much of the work reported has been carried out. To acknowledge all these individuals would be impossible; instead, we would like to offer some special gratitude to those colleagues whose work appears in this manuscript:

F.H. Axtell; R.S. Blacker; A. Blinkhorn; J. Brewer; G. Buchalter; D.J. Calvert; G.M. Chapman; X.Y. Chen; F.P. Chen; C.F. Chin; P.R. Christodoulides; J.A. Cook; P.J. Costigan; J.A. Covas; N.K. Datta; J.V. Dawkins; M.F. Edwards; B.C. Fisher; P.K. Freakley; S.K. Garg; M. Gilbert; R.J. Heath; D.A. Hemsley; C. Hepburn; R.P. Higgs; K.M. Jones; M. Kanagendra; C. Kistrup; P.L. Koh; T.C. Law; D.A.N. Luck; L. Mascia; D.E. Marshall; A. Miadonye; M.I. Mulla; A.O. Ola; S.R. Patel; T.M. Robinson; N.A. Sandy; R.M. Shelley; M. Singh; D.G. Smoker; D. Srdic; R.C. Stephenson; H.E. Strauss; R. Streamer; R.A. Sykes; G.M. Walsh; R.E. Wetton; J.R. White; D.M. Wilkinson; J. Wood; P.L. Yeh ...

...and all the Technical Staff at Loughborough whose skills are always required to make the 'impossible' practical work possible.

We would also wish to acknowledge the following companies whose products have been used to generate data, or who have given us permission to reproduce their intellectual property: APV Chemical Machinery; Ian Barrie Consultancy; BASF Plastics; Battenfeld Fischer; Bekum Maschinenfabriken; British Plastics Federation; British Standards Institution; Carters Packaging; European Vinyls Corporation; Hoechst Celanese Plastics; Husky Injection Moulding Systems; ICI Chemicals & Polymers; Moldflow (Europe) Ltd.; Orme Polymer Engineering; Polymer Laboratories; Polymer Microscopy Services; RB Blowmoulders; Solvay-Laporte; Transmet Corporation; WES Plastics.

Finally, we would like to thank Dr. Wolfgang Glenz and colleagues at Carl Hanser Verlag, for all their help in the production of the text.

Table of Contents

Chapter 1
Introduction

1.1 The Nature of Plastics

Plastics are now universally available and recognized as materials in their own right, yet a matter of only three decades ago, the materials produced and applied by the newly developing plastics industry were regarded as special, with properties which were not rational, or understood by the average engineer. Two factors have contributed to the very rapid growth in plastics usage: first the facility of being able to shape plastics under mild temperature conditions; indeed, the term "plastic" implies the capability of being easily forged, or moulded into shape. The second factor is the relatively high resistance of these materials to chemical attack, which has obvious advantages in a variety of applications: plastics pipes replacing copper, and the replacement of a number of metal components in the transport industry.

Whilst the number of possible plastics is almost limitless, the economics of their manufacture, and the properties they possess, restricts the number of commercially significant materials to about thirty-five, although, as we shall see, there are frequently many variants for any one polymer, which have been developed for particular end uses. Plastics are but one group of many in which large molecules are involved; others are fibres, adhesives, rubbers and surface coatings. These are built from relatively simple substances of low molecular weight, sometimes called *"mers"*, or more usually, *monomers*, and the reaction by which they are joined together is *"polymerization"*. The product of such a reaction is a polymer, and can in principle be a very pure material if care is taken over the polymerization conditions. Commercial polymers generally contain initiator residues and other impurities. However, polymers which are relatively pure may not give optimum performance, and a wide range of additives has been developed to match the properties of a plastics compound to its application requirements. These will be considered later in this chapter.

It is inherent in polymers that their responses to external stimuli, that is, their reactions to applied forces etc., differ from those of non-polymeric materials. This difference is shown, however, only in certain property regions; in others, polymers respond in a manner similar to the behaviour of "small molecule" materials, and have properties associated with their organic structures, that is their *repeat units*, including:

(a) density,
(b) refractive index and relative permittivity,
(c) enthalpy, (heat content), thermal conductivity and diffusivity,
(d) thermal expansion,
(e) chemical behaviour, (including oxidation),
(f) melting and softening temperatures, and
(g) solubility, (which may be restricted by long chain nature and by crystallinity, and may be confined to swelling for network structures).

Properties which are consequent on the long chain nature of polymers include:

(h) high viscosity of polymer melts and solutions, (even when dilute),
(i) elasticity of polymer melts and solutions,
(j) capacity to accommodate very large strains without fracture, (the strain may be recovered when the stress is removed),
(k) marked time dependence in many properties, (including creep and stress relaxation),and
(l) frequency-dependence of permittivity and loss factor.

It is the purpose of the later chapters of this book to examine in detail the properties of plastics in both categories, and to discuss and illustrate the effect on them of factors which contribute to the structure of the final product. First, however, it is necessary to set down some of the basic truths of polymer science which are relevant to the main theme of this book.

Elements of the molecular structure and organization of polymers, including reference to the methods of synthesis, will be covered in the remainder of this section, followed by a review of the more important methods of polymer characterization, including structural examination. The final section of the chapter introduces the subject of product characterization, the cornerstone both of the quality control of plastics products and of failure analysis.

1.2 Basic Concepts of Polymer Science

1.2.1 Nomenclature

Polymers have defied the attempts of classical chemists to impose a rigorous scheme of nomenclature; in some cases familiar, but non-informative terms like "nylon" have been retained, in others, more descriptive but not unambiguous names have been adopted, such as poly(ethylene terephthalate), which is the polymer derived from ethylene glycol and terephthalic acid. At this time, it seems appropriate to adopt the common names, particularly since there is an International Standard being evolved for the standardization of such names and the relevant abbreviations[1.1]. A Table based on this document is reproduced in Table 1.1.

Table 1.1 Symbols for Homopolymers

CA	Cellulose acetate	PMMA	Poly(methyl methacrylate)
CAB	Cellulose acetate butyrate	PMP	Poly-4-methylpentene-1
CAP	Cellulose acetate propionate	POM	Polyoxymethylene:
CF	Cresol-formaldehyde		polyformaldehyde
CMC	Carboxymethyl cellulose	PP	Polypropylene
CN	Cellulose nitrate	PPOX	Poly(propylene oxide)
CP	Cellulose propionate	PPS	Poly(phenylene sulphide)
CS	Casein	PPSU	Poly(phenylene sulphone) (Sic)
EC	Ethyl cellulose	PS	Polystyrene
EP	Epoxide, epoxy	PTFE	Polytetrafluoroethylene
MF	Melamine-formaldehyde	PUR	Polyurethane
PA	Polyamide	PVAC	Poly(vinyl acetate)
PB	Polybutene-1	PVAL	Poly(vinyl alcohol)
PBA	Poly(butylene acrylate)	PVB	Poly(vinyl butyral)
PBTP	Poly(butylene terephthalate)	PVC	Poly(vinyl chloride)
PC	Polycarbonate	PVDC	Poly(vinylidene chloride)
PCTFE	Polychlorotrifluoroethylene	PVDF	Poly(vinylidene fluoride)
PDAP	Poly(diallylphthalate)	PVFM	Poly(vinyl formal)
PE	Polyethylene	PVK	Polyvinylcarbazole
PEOX	Poly(ethylene oxide)	PVP	Polyvinylpyrrolidone
PETP	Poly(ethylene terephthalate)	Si	Silicone
PF	Phenol-formaldehyde	UF	Urea-formaldehyde
PIB	Polyisobutylene	UP	Unsaturated polyester

Table 1.1 (cont.) Symbols for Copolymeric Materials

ABS	Acrylonitrile/butadiene/styrene
A/MMA	Acrylonitrile/methyl methacrylate
ASA	Acrylonitrile/styrene/acrylate
A/PE-C/S	Acrylonitrile/chlorinated PE/styrene
A/EPDM/S	Acrylonitrile/ethylene-propylene-diene/styrene
E/EA	Ethylene/ethyl acrylate
E/P	Ethylene/propylene
EPDM	Ethylene/propylene/diene
E/VAC	Ethylene/vinyl acetate
E/VAL	Ethylene/vinyl alcohol (or E/VOH)
FEP	Perfluoro (ethylene/propylene): tetrafluoroethylene/hexafluoropropylene
MPF	Melamine/phenol formaldehyde
S/B	Styrene/butadiene
S/MS	Styrene/α-methyl styrene
VC/E	Vinyl chloride/ethylene
VC/E/MA	Vinyl chloride/ethylene/methyl acrylate
VC/E/VAC	Vinyl chloride/ethylene/vinyl acetate
VC/MA	Vinyl chloride/methyl acrylate
VC/MMA	Vinyl chloride/methyl methacrylate
VC/OA	Vinyl chloride/octyl acrylate
VC/VAC	Vinyl chloride/vinyl acetate
VC/VDC	Vinyl chloride/vinylidene chloride

1.2.2 Polymers and Polymerization

A polymer molecule contains a large number of similar repeat units, linked to each other by *covalent bonds,* in which the electrons are shared equally by the two atoms involved. Thus the simplest type of structure is a chain, for which the only non-identical units are at the ends; these end-groups are usually so dilute that they do not affect the polymer properties, although in some cases their nature is significant. If all the repeat units are identical, the polymer is termed a *homopolymer,* and the number of repeat units is the *degree of polymerization,* n.

$$\text{Polymer chain} = \text{End group} - (\text{Repeat unit})_n - \text{End group}$$

Polymer molecules can be formed by a sequence of classical chemical reactions, starting with small molecules called *monomers.* For a linear polymer to be formed, the *functionality,* (linking capability) of the monomer units must be two. Functional groups are sites for reaction, for example -OH in methanol, CH_3OH, (one functional group): ethylene glycol, $HO{-}CH_2CH_2{-}OH$, is difunctional, as is a double bond in $CH_2 = CH_2$. Polymerization can also give non-linear chains and network structures: both require branch points, which are units with a functionality greater than two. Further, with extensive branching, there is increasing opportunity for the branches to become interconnected, resulting in a network structure, which can also be formed by deliberately *cross-linking* existing polymer chains.

Two distinct reaction mechanisms are found in polymerization : in *addition polymerization* the polymer chain grows by the successive addition of monomer units, and this is the route followed by monomers where the functionality is in a double bond, (or sometimes when a cyclic monomer is polymerized). An example of addition polymerization is the reaction of vinyl chloride, $CH_2{=}CHCl$:

$$H(CH_2{-}CHCl)_{n'} + CH_2{=}CHCl \rightarrow H(CH_2{-}CHCl)_{n+1'}$$

Growing chain + monomer → chain, active, longer by 1 unit. By contrast, *step growth polymerization* is the reaction of monomers to produce dimers and trimers initially, which then combine progressively to produce longer chains. This process continues until acceptable degrees of polymerization have been attained. Step growth polymerizations frequently involve the elimination of a small molecule, often water, during the reaction, leading to the older term *condensation polymerization.*

An example of such a step-growth reaction is the condensation of ω-amino carboxylic acids; amide links are formed by the different functional groups of two molecules, and a dimer is formed:

$$\begin{aligned} &H_2N{-}(CH_2)_a{-}COOH + H_2N{-}(CH_2)_a{-}COOH \\ &\rightarrow H_2N{-}(CH_2)_a{-}CO{-}NH{-}(CH_2)_a{-}COOH + H_2O \end{aligned}$$

"a" is an integer, (a = 5 for PA6).

This dimer retains the same two functional groups as the monomer, and can therefore react with other monomers, dimers, or larger molecules, to build a polymer. Step growth polymerization can also take place between two different types of difunctional molecules, for example a diacid and a diamine; repeat units from each of the components are then included in the polymer chain. (The true repeat unit contains both a diacid residue and a diamine residue.)

All the materials considered so far have been *homopolymers.* In *copolymerization,* two or more different monomers are polymerized together; the product, a *copolymer,* incorporates repeat units from both monomers. These may be distributed irregularly in *random copolymers,* (e.g. ethylene/vinyl acetate) whilst certain pairs of monomers, which have particular reactivities towards each other, may form *alternating copolymers.* The molecules of a *block copolymer* contain long uninterrupted sequences of each type of repeat unit; examples are the thermoplastic rubbers, SBS and SIS. Finally, *graft copolymers* consist of branches of one species attached to a main chain of a different repeat unit. The various types of copolymer are illustrated below.

```
Random       —A–A–B–A–B–B–A–A–B–A–B–B–B–A–B–A–A–A—
Block        —(A... A)–(B... B)–(A... A)–(B... B)—
Alternating  —A–B–A–B–A–B–A–B–A–B–A–B–A–B–A–B–A–B—
Graft        —A–A–A–A–A–A–A–A–A–A–A–A–A–A—
                             |
                             B
                             |
                             B
                             |
                             B
                             |
                             B
```

1.2.3 Chain Length, Molecular Weight (MW), and Molecular Weight Distribution (MWD).

A polymer generally contains molecules of different chain length, since its formation is the result of a succession of random events, and the chain length reflects the growth history. Such a polymer with a distribution of chain lengths is described as *polydisperse;* in contradistinction, a polymer in which all the chains are of similar length is *monodisperse.* The molecular weight of the latter is a unique quantity, whilst there are many ways of expressing the distribution of chain lengths, and consequently molecular weights, for a polydisperse polymer. If the reaction mechanisms are known, the *chain length distribution* can, in principle, be derived theoretically. Whilst homopolymers may be polydisperse, and in this sense not defined as unique chemical compounds, they are also likely to have at least one chain end derived from the initiator. Further, they may be adulterated with the products of rogue reactions, or the remnants of other ingredients of the polymerization. With copolymers, however, there is opportunity for variety in structure of a more far-reaching kind; the individual molecules of a copolymer may differ in chemical composition, since the reactivity of the two monomers with the active growth point may well be, and usually is, different. Thus, in a batch copolymerization, one monomer is consumed preferentially, thereby changing the ratio of monomers available for reaction, and thus the constitution of the copolymer being formed.

1.2.4 Molecular Structure and Physical Properties

As we have seen earlier in the chapter, many of the properties of polymers can be ascribed to their organic nature or to the chain structure; it is now appropriate to examine these structure - property relationships in more detail. In particular, the nature of the individual molecules and the forces they exert on each other require closer examination. The primary bonds which hold together the atoms in a polymer chain are strong: polymer degradation only occurs at relatively high temperatures. The forces between the polymer chains comprise dispersion (Heitler - London), polar and induced dipolar forces. These are weak, although the fact that they operate over the whole chain length increases their effectiveness: even so, the softening points and the strengths of polymers are both comparatively low.

The properties of polymers are mainly dependent on two factors:

- the flexibility of the polymer chain, and
- the interaction of the chain with its neighbours.

The first factor is very much a function of the repeat unit; bulky groups stiffen the chain, as illustrated by comparing polybutadiene rubber (softening point −80 °C) with polystyrene (softening point +100 °C). The rubber is acknowledged to have a very flexible chain. A further example of stiffening is found in replacing all the hydrogen atoms in linear polyethylene by fluorine, yielding poly(tetrafluoroethylene):

$$-CH_2-CH_2- \qquad \text{and} \qquad -CF_2-CF_2-$$

The PTFE chain is notoriously stiff, because of the packing of the chain by the fluorine atoms, resulting in the melt viscosity being about a million times higher than for most other polymers. Further examples will be cited after consideration of crystalline melting point. Another effective way of increasing chain stiffness is by introducing very stiff units into the polymer chain, shown by comparing the melting point of poly(ethylene terephthalate), PETP with that of poly(ethylene adipate). The aromatic ring in the former gives it a 200 °C advantage in melting temperature: the respective values are 265 and 60 °C. The interaction of a chain with its neighbours is governed by the force and by the distance over which it acts, the chain separation. Interchain forces are highest for polar groups, illustrated by the effect of the amide group -CO-NH-, whilst a polyethylene chain $-CH_2-CH_2-$ does not interact strongly with adjacent chains. This is reflected in the much higher melting temperature of polyamide 66, (265 °C), compared with that of polyethylene (135 °C). The introduction of side chains of increasing length pushes the main chains further apart, decreasing the effectiveness of any attractive centres along the chains. To decrease the softening temperature, it is important that the side chains be flexible, otherwise the stiffening effect on the main chain may more than compensate for the increased chain separation. The influence of increasing length of flexible side chains is exemplified in the melting temperatures of the poly α-olefins (see Table 1.2), general structure of the repeat unit, CH_2-CHR:

Table 1.2 Effect of Side-Chains on Crystalline Melting Point

Polymer	Length of side chain (R)	Melting point, T_m (°C)
Polyethylene	$-H$	135
Polypropylene	$-CH_3$	170
Polybutene	$-CH_2-CH_3$	128
Polypentene	$-CH_2-CH_2-CH_3$	80
Polyhexene	$-CH_2-CH_2-CH_2-CH_3$	Below room temperature

The increase in melting temperature from polyethylene to polypropylene is a *chain stiffening* effect; the decrease thereafter is a consequence of increasing *chain separation.*

An example of increasing softening point with increasing *polarity* is furnished by the series of α-substituted acrylic esters: poly(methyl methacrylate), poly(methyl chloroacrylate) and poly(methyl cyanoacrylate). These polymers are of increasing softening temperature, which is correlated with increasing dipole moments of the $C-CH_3$; $C-Cl$ and $C-CN$ bonds. The argument that the changes are the result of stiffening the main chain is refuted by the fact that the sizes of the substituents are all very similar.

1.2.5 The Glass Transition Temperature

Amorphous polymers are glassy solids at low temperatures, since molecular motion is severely restricted. The temperature at which a polymer softens (the *free volume* increases suddenly to a point at which the required longer-range segmental motion becomes feasible[1.2]) is known as the *glass transition temperature,* T_g. Immediately above T_g, the physical properties are *"rubber-like"*, where large, elastic deformations can be accommodated by linear, or lightly-crosslinked systems; this remarkable behaviour is only seen in high polymers. As temperature increases further, all types of molecular chain-segment motion become feasible; the molecules then assume a state of continuous, random motion, which defines the *melt state* of glassy polymers.

The glass transition is therefore an important characteristic, which can be found by a number of tests, which are detailed in Section 1.3.4. Some of these tests however, lead to a range of values of T_g, the differences arising from the different timescales of the observations, lower frequencies (longer timescales) leading to lower T_g values. At T_g the solid amorphous polymer changes to the rubber state and, with further increase in temperature, to a viscous liquid.

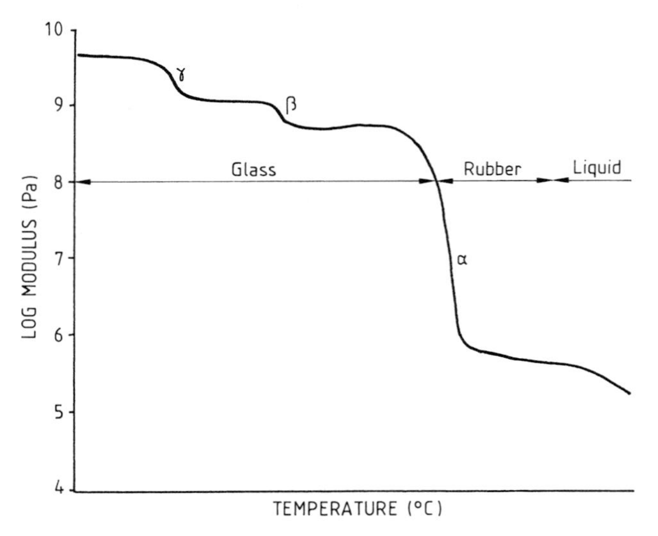

Figure 1.1 Modulus (log scale) vs. temperature for plastics materials.

This is illustrated by plotting modulus versus temperature, Figure 1.1. In addition to the glass transition temperature, a polymer can have several subsidiary transitions at lower temperatures, associated with particular molecular groups becoming mobile: starting with T_g, they are designated $\alpha, \beta, \ldots$ etc., for transitions at decreasing temperatures.

These points are taken further in Section 6.3.2, where the factors affecting secondary relaxation temperatures are discussed in relation to the temperature dependence of mechanical properties.

1.2.6 Crystallinity, and Crystalline Melting Point

It has been assumed so far that the structure of all linear polymers is similar, a random array of the chains; this is so for amorphous polymers. If, however, the polymer chains possess sufficient chemical and geometrical regularity, small regions of local order are developed on cooling the melt; these are known as *crystallites* ("small crystals") and their existence can be demonstrated most commonly by X-ray diffraction, and inferred from the enthalpy (heat content) vs. temperature behaviour. Unlike low molecular weight materials, the process of crystallization never reaches completion, since chain entanglements preclude the unlimited growth of a crystallite, hence the term *"semi-crystalline"* is often applied to such polymers. A single chain may participate in several different crystallites at different parts of its length.

Both the *rate of nucleation* and the subsequent *crystallite growth rate* are strongly dependent on temperature; it is possible, therefore, to change the *degree of crystallinity* and the *texture* or *morphology* by thermal treatment. This has important consequences, since many properties differ for crystalline or non-crystalline states, and linear polymers are necessarily subjected to thermal treatment during their processing. The crystallites are, almost by definition, *anisotropic*, although this is only of major significance if the system is oriented, (see Section 1.3.8).

Crystallinity is characterized at the *crystalline melting point* (T_m), which is usually independent of molecular weight, but is modified by the chemical structure of the chains. This point is exemplified by comparing two thermoplastic polyesters, PBTP and PETP, which are polymerized by reacting terephthalic acid with glycols of different structure: the additional methylene groups in the former polymer enhance chain flexibility, resulting in a depression of T_m (and a lower glass transition temperature).

PETP (From Ethylene glycol):
Repeat Unit --[-0-C0-⟨benzene ring⟩-C0-0-$(CH_2)_2$-]--
$T_m = 260\,°C$

PBTP (From Butylene glycol):
Repeat Unit --[-0-C0-⟨benzene ring⟩-C0-0-$(CH_2)_4$-]--
$T_m = 230\,°C$

In general, it is true to suggest therefore, that the same aspects of chemical structure in plastics (chain flexibility, side-chains, copolymerization) tend to modify T_g and T_m in a similar manner[1.3].

Only a fraction of known polymers are partially crystalline, indeed, there are many well-known polymers which cannot be induced to show any sign of crystallinity. One of the criteria for crystallization is regularity of the repeat unit, so that it is unusual for random copolymers, which have at least two kinds of repeat unit distributed along the chain, to crystallize.

An additional requirement for the appearance of crystallinity is geometric regularity; this is important in the case of polypropylene, where only the configurationally regular forms can crystallize (*isotactic* and *syndiotactic* forms), illustrated schematically in Figure 1.2: those of lesser regularity are amorphous. Standard Polymer Chemistry texts should be consulted for further detail[1.4].

```
Isotactic            — CH2 — CH — CH2 — CH — CH2 — CH — CH2 — CH —
                             |           |           |           |
                            CH3         CH3         CH3         CH3

Syndiotactic                CH3                     CH3                     CH3
                             |                       |                       |
                     — CH2 — CH — CH2 — CH — CH2 — CH — CH2 — CH — CH2 — CH —
                                         |                       |
                                        CH3                     CH3

Atactic                                             CH3                     CH3
(or Heterotactic)                                    |                       |
                     — CH2 — CH — CH2 — CH — CH2 — CH — CH2 — CH — CH2 — CH —
                             |           |                       |
                            CH3         CH3                     CH3
```

Figure 1.2 Isotactic, syndiotactic and atactic (heterotactic) forms of polypropylene.

The significance of crystallinity in a polymer is that it affects both processing and properties and, in its turn, it is influenced by processing, especially by thermal history. The effects on processing include:

(a) the melting point has to be attained before shaping;
(b) in cooling to stabilize the shape, the temperature of crystallization may be tens of degrees Celsius below the fusion temperature, a phenomenon known as *supercooling;*
(c) additional thermal energy is needed to melt the crystallites, the latent heat of fusion; perhaps more important, this latent heat is evolved during crystallization, thus retarding the cooling; and
(d) since the density of the crystalline regions is generally higher than that of amorphous material, there is additional shrinkage during the cooling process.

Further, the following properties are affected by increasing crystallinity:

(1) increased strength and stiffness, but possibly increased brittleness;
(2) lower solubility and lower permeability;
(3) increased density; and
(4) transparency may be adversely affected.

In its turn, crystallinity is influenced by:

- the melt temperature reached during the process;
- the mould temperature, and the rate of cooling;
- molecular alignment induced during processing;
- the presence of a crystal-nucleating additive; and
- chain length; polymers of high molecular weight (MW) are more reluctant to crystallize than those of lower MW.

1.2.6.1 Crystalline Superstructure

So far there has been little consideration of the crystalline texture, a subject which has intrigued polymer physicists for many years. The appreciation that the dimensions of a crystallite are considerably smaller than the average length of a polymer chain came early, with the *fringed micelle* offering a solution to the dilemma (Figure 1.3a). However, the study of crystals grown from polymer solution, notably by KELLER and coworkers at Bristol[1.5], indicated a laminar structure with the polymer chain folding back and forth, perpendicular to the plane of the lamina, Figure 1.3b.

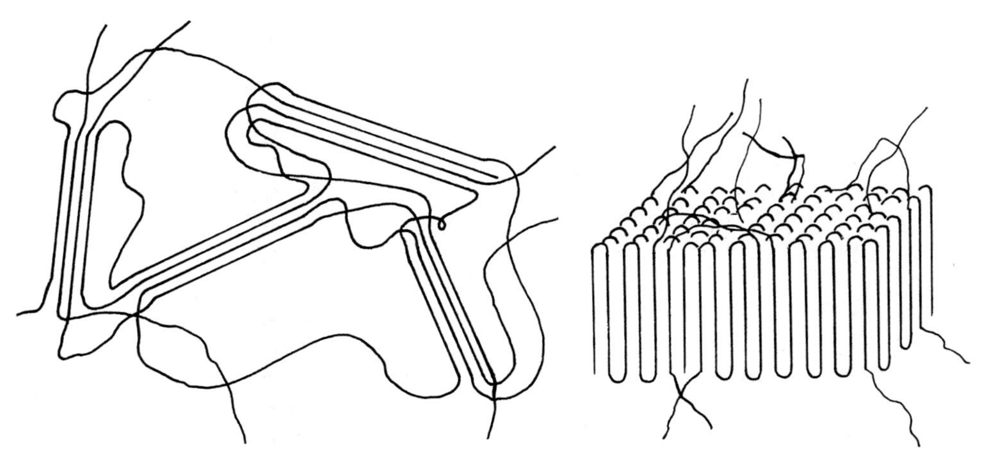

Figure 1.3a (left) Fringed micelle model of polymer crystallinity;
Figure 1.3b (right) Folded chain model of polymer crystallinity.

There is strong evidence that the *folded chain structure* is present in polymers crystallized from the melt but the structure is not universal, for crystals along the length of a polymer chain can be found, as can mixed structures, the so-called *shish-kebabs*[1.6]. Polymers which are crystallized from the melt under quiescent conditions frequently form spherically symmetrical structures, known as *spherulites,* which are of such size as to be readily visible in a light microscope; sometimes with the naked eye. The birefringence of the crystallites, when aggregated into a spherulite, leads to a Maltese Cross extinction pattern when viewed through crossed polars. A spherulite, which dominates the texture of most crystalline polymers, is illustrated in the micrograph in Figure 1.4; the fine detail confirms the growth pattern of the structure. Spherulites are similarly important with respect to properties such as transparency and toughness.

Figure 1.4 Spherulite, showing detail of the structure

1.2.7 Branched Polymers and Network Structures

With increasing departure from linearity, the precise description of a polymer chain becomes more difficult; a branched polymer chain, for example, requires description of the distribution of chain lengths with, additionally, the distribution of branch lengths and the number of branches per molecule. These quantities are mainly accessible by measurement of the end-groups.

For network structures, it is even more difficult to determine characterizing parameters, since such materials are insoluble, although it is possible to assess the molecular weight between cross-links. This difficulty of obtaining information concerning network structures has retarded understanding, particularly of cross-linked plastics and, to a lesser extent, of rubbers, which are frequently cross-linked by a reaction involving sulphur.

Both branching and the formation of network structures interfere with the capability of a polymer to crystallize; indeed, the presence of a dense array of cross-links effectively eliminates crystallinity. It should be noted, however, that a few cross-links may improve the crystallinity obtained on stretching a material, since orientation is increased markedly by restricting the flow of the polymer. This is of considerable significance for orientation of plastics films, sheet and thermoformings.

The polymers and copolymers of ethylene afford examples of the effect of chain branching on crystallinity and morphology. High density polyethylene (HDPE) is an almost perfectly linear polymer and can, therefore, be obtained in a highly crystalline state with high melting temperature (80-85 % crystalline, relative density 0.96, melting point 133 – 135 °C). The original polyethylene, made at high pressure and high temperature, now classified as low density polyethylene (LDPE) has substantial numbers of ethyl and butyl branches and long chain branching, Figure 1.5; crystallinity levels are 55 %, relative density 0.92 and melting temperatures 110 – 115 °C. A more recent development, linear low density polyethylene (LLDPE), is manufactured by copolymerizing ethylene with butene, hexene, octene, or 4-methylpentene. They can be produced over a range of densities, are generally similar to LDPE, but do not have long chain branching.

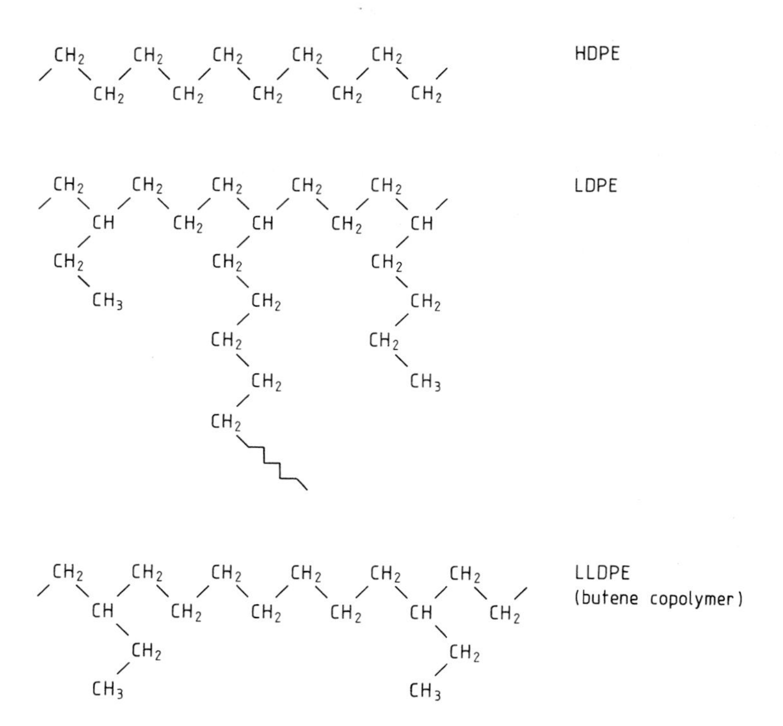

Figure 1.5 Structure of polyethylene: HD, LD and LLD types.

The interplay of molecular structure and crystallinity is illustrated by the tensile properties of the three types of polyethylene, Figure 1.6[1.7]. Properties up to, and including yield are determined by the level of crystallinity: thus, the yield stresses of LDPE and LLDPE are similar, whilst that of HDPE is considerably higher. After yield, the strain accommodation depends on the available amorphous material and on the "shape" of the polymer chain. Despite the higher crystallinity, the strain at failure of HDPE is higher than that of LDPE, due to the absence of long chain branching in HDPE. The linearity, and the greater amorphous content, gives LLDPE the very high elongation, typically of 1400 %.

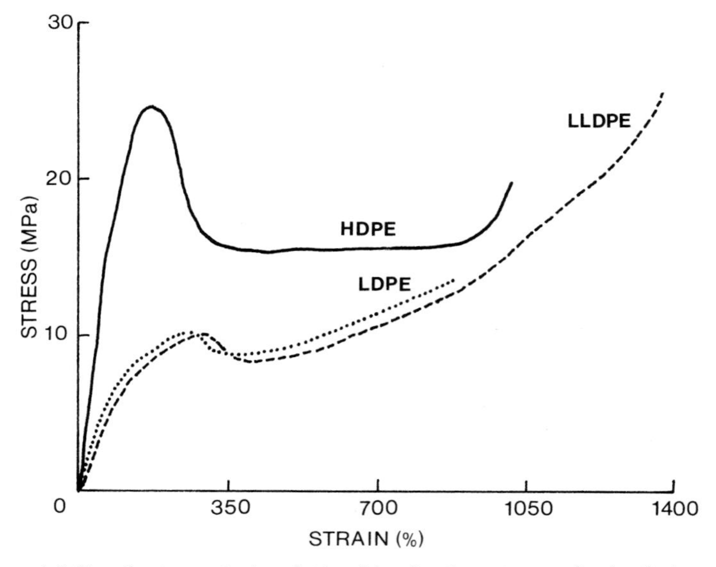

Figure 1.6 Tensile stress-strain relationships for three types of polyethylene.

1.3 The Characterization of Polymers

1.3.1 Determination of Chain Structure

The development of physical methods of analysis, especially *infra-red, (IR), spectroscopy* and *nuclear magnetic resonance, (NMR), spectroscopy,* has enabled the structures of small molecules to be determined precisely. Although the application of these techniques to polymers does not lead to such determinate analysis, they, together with the older chemical methods, have enabled the structure of many polymers to be established in considerable detail. However, chemical methods based on *pyrolysis,* followed by examination of the degradation products by *gas chromatography, mass spectrometry,* or *infra-red spectroscopy* are still much used in the analysis of polymers. *Elemental analysis* is limited in the information it yields, since many polymers can have the same composition, for example high density polyethylene, low density polyethylene, linear low density polyethylene, polypropylene, polybutene and ethylene-propylene rubber, and blends and copolymers all have the empirical formula $H-(CH_2)_n-H$.

1.3.1.1 Pyrolysis of Polymers

Polymers may be broken down by dry distillation, or heating with acid or alkali; products obtained can be collected, separated and identified by a variety of methods. Of particular importance is *gas chromatography* carried out on the pyrolysis products, which not only identifies polymers, but also allows the quantitative determination of the monomer ratio of copolymers. It is possible by this method to distinguish between random or block copolymers and mechanical mixtures of the same composition. A further important method of separating and identifying the products of pyrolysis is *mass spectrometry,* whilst IR and NMR techniques can be employed to identify functional groups of degradation products which have been separated by gas chromatography.

1.3.1.2 Infra-red Spectroscopy

This technique is, without doubt, the most widely used for polymer identification, and for determining, qualitatively and quantitatively, the presence of specific groups in a material. This is possible since each group has characteristic frequencies at which absorption of radiation occurs, usually in the infra-red region, and the frequency of this absorption is, to a first approximation, independent of the environment of the group. By reference to a collection of infra-red spectra[1.8], it is possible to assign specific absorptions to particular groups, confirming the assignment by obtaining the spectra of model compounds.

There have been several advances in instrumentation and in technique which have increased the versatility and sensitivity of measurement. A more fundamental change has been to *Fourier Transform* methods, in which the response of the material to a single excitation is transformed by digital computer into a spectral scan, the process being repeated many times, and the results averaged, to attain high accuracy.

The use of polarized infra-red radiation allows comment to be made concerning the direction of particular bonds in an organized polymer structure.

1.3.1.3 Nuclear Magnetic Resonance Spectroscopy, (NMR)

Although introduced much more recently than infra-red spectroscopy, this method, capable of being applied in a number of variants, is possibly the most powerful technique available for determining chain structures. It depends on the very small magnetic moment resulting from nuclear spin, which is affected by the environment. Hydrogen nuclei (protons) and ^{13}C nuclei are generally important, whilst nuclei other than these can be employed in special cases, for example, the response of fluorine has been studied in considerable detail in elucidating the structures of fluoropolymers. A library of NMR spectra has been assembled, from which it is usually possible to identify unknown species. A further factor contributing to the usefulness of NMR is that the responses are related directly to the number of nuclei involved.

As with infra-red spectroscopy, scanning methods were introduced first, but these have been overtaken in accuracy and speed by *Fourier Transform* methods.

1.3.2 Determination of Chain Length

There are many methods of determining *chain length,* (or *molecular weight);* some are direct measurements, others rely on the measurement of properties which are considerably affected by molecular weight. As we have already seen in Section 1.2.3, in a practical polymer the chains are usually of differing lengths, so that the description of the *degree of polymerization* can only be in terms of averages. The *number average molecular weight* (M_n) is perhaps easier to obtain than the *weight average molecular weight* (M_w): both are defined below.

$$M_n = \frac{\sum n_i M_i}{\sum n_i} \tag{1.1}$$

$$M_w = \frac{\sum w_i M_i}{\sum w_i} \tag{1.2}$$

The number average molecular weight is the summation of the product of each molecular weight value, M_i, and the number of such molecules present, n_i, divided by the total number of all molecules present. In contrast, the weight average molecular weight sums the weight fraction of each species, (w_i is the weight of each component).

Molecular weights may be averaged with various other kinds of weighting which are of value in some applications: the *viscosity average molecular weight* (M_v) is that derived from solution viscosity measurements. It is usually closer to weight average, than to number average molecular weight.

The following methods give the number average molecular weight:

- end group analysis;
- increase in boiling point of solvent by solution of polymer;
- decrease in melting point of solvent by solution of polymer; and
- osmometry.

A disadvantage of the last three methods is that the polymer must be in true solution, a state which is difficult to achieve with crystalline polymers, except at high temperatures. To this end, appropriate instrumentation has been developed and is available commercially.

The weight average MW is generally determined by measurement of the light-scattering of polymer solutions; less often, by sedimentation in an ultracentrifuge. Both methods again depend on the availability of polymer solutions, and so can only be applied to crystalline polymers, such as polyethylene and polypropylene, by working at elevated temperatures.

A useful practical monitor of chain length and especially of changes in chain length, is afforded by the measurement of solution or melt viscosity. There are many designs of solution viscometer which compare the viscosity of a dilute polymer solution to that of the pure solvent, generally by comparing flow times, as in the Ubbelhode instrument[1.9].

For polymer melts rotational viscometers, parallel plate plastometers and capillary rheometers find considerable application in the everyday business of polymer technology (see Chapter 3; Section 3.6). The *Melt Flow Indexer*[1.10] is a particularly widely used type of capillary rheometer, extensively applied in the quality control of thermoplastics.

1.3.3 Chain Length Distribution (Molecular Weight Distribution MWD)

The distribution of chain lengths has been referred to in a previous section (1.2.3) and is a factor of considerable practical significance, since many properties are affected by it. Both the flow behaviour of the melt and the strength of the solid material depend on MWD.

Characterizing the MWD can be achieved by *fractionating* the polymer, making use of the fact that some properties of polymers depend on molecular weight. Solubility is one such property and its dependence on MW can be used to effect a separation of a polydisperse polymer into a number of *fractions,* differing in MW. Fractionation can be carried out either by changing the temperature, or the solvent power of the liquid by varying the solvent/non-solvent ratio.

An elegant way of determining the MWD is by *Gel Permeation Chromatography (GPC),* a technique which has been developed comparatively recently[1.11], the principle of which is that chains of differing length behave differently when a solution passes through a column packed with a gel with a uniform and appropriate pore size. Whilst the smaller molecules can diffuse freely into the pores of the gel, and are thereby retarded in their passage through the column, the larger chains are not so delayed and emerge first from the column. A detector, sensitive to the presence of polymer, monitors the polymer content of the effluent; the classical detector was a differential refractometer, but infra-red detectors, operating at a fixed frequency, have also been widely employed.

A *chromatogram* obtained on a "Novolak" (phenol-formaldehyde prepolymer) resin, monitored at a frequency of 3400 cm^{-1}, is shown in Figure 1.7: the gel was cross-linked polystyrene (with a molecular weight exclusion limit of 2000), and the solvent was tetrahydrofurane[1.12]. The various fractions can be identified by comparison with model compounds, on the basis that molecules of similar size will respond similarly during passage through the gel. Applying the technique to high MW polymers requires high pressure pumping to achieve fractionation in an acceptable time, and, since this is again a solution method, high temperature GPC is necessary for the fractionation of crystalline polymers.

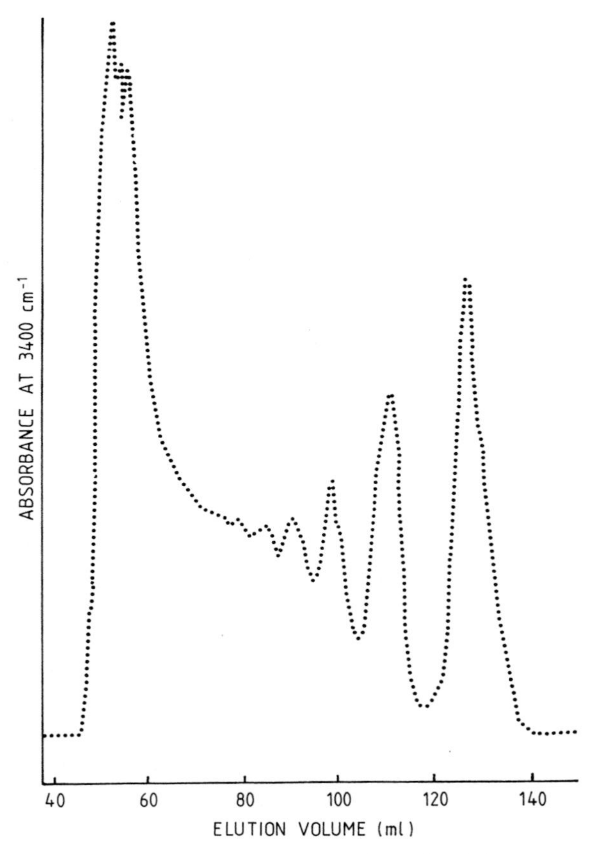

Figure 1.7 Gel permeation chromatogram of Novolak (phenol-formaldehyde prepolymer).

1.3.4 Determination of T_g

If a plastics specimen is heated, the *glass transition temperature* (T_g) represents the point at which the material first acquires significant molecular motion. It is a characteristic of the *amorphous phase;* although it also carries fundamental relevance to semi-crystalline plastics (for example, highly-crystalline materials often suffer embrittlement effects on passing below T_g; witness PP homopolymer at −20 °C), it becomes more difficult to measure for highly crystalline materials, as the amorphous content, and its effects, become dominated by the crystalline phase.

The molecular mobility of a polymer depends upon the thermal energy available, so that T_g varies according to the chemical constituents and structure of the polymer chain. An important characteristic of polymers is the change in the rate at which *specific volume* increases with temperature, about T_g. This effect arises due to the onset of enhanced chain mobility, and provides a convenient principle of measurement for T_g, using *dilatometric techniques.* Other techniques which can be used to characterize T_g are based upon physical properties which change markedly at the point of the transition; these include *thermal analysis* (heat content), *thermomechanical analysis (TMA)* (thermal expansion), *optical methods* (refractive index), NMR, and temperature scans of mechanical (dynamic modulus) or electrical (permittivity) properties.

1.3.5 Determination of Crystallinity

The *fractional crystallinity* of a semi-crystalline polymer is an important characteristic, since it determines the level of a number of properties which are significant in diverse applications. Determining this is, therefore, a matter of some importance, and a variety of methods have been developed. The most fundamental technique is *X-ray diffraction,* the crystal planes leading to discrete diffraction angles in the polycrystalline material. By making assumptions about the level of X-ray scattering from the amorphous polymer, the fraction of crystalline polymer can be determined. There is a degree of arbitrariness in assigning the level of amorphous scattering and, therefore, in the final result, a comment which applies to all the methods of monitoring crystallinity.

Since the *density* of a crystalline polymer is usually higher than that of the amorphous material, the density of a sample is directly related to its fractional crystallinity. Whilst the density of the crystalline polymer can be calculated from the unit cell by X-ray diffraction, the density of the amorphous phase can generally only be estimated, and the further assumption made that the crystalline and amorphous phases do not interfere with one another.

Differential scanning calorimetry (DSC) is a recently developed technique which can be adapted to the assessment of crystallinity; the latent heat of fusion is measured to monitor the crystalline content of a sample. The method depends on knowing the heat of fusion of pure crystalline polymer, a state which cannot be achieved; therefore the method requires calibration against a sample of the same polymer of known crystallinity. Further, there is always the danger that the crystallinity level is increased by heating the polymer during the measurement. Such a change is observed in poly(ethylene terephthalate)[1.13], for example.

1.3.6 Melting Temperature and Crystallization Temperature

In Section 1.2.6, it was stated that the *crystalline melting temperature* (T_m) is an important property, in respect both of processing and applications. It is this temperature which must be exceeded for processing a material as a thermoplastic, and it also represents the limit of solid state properties. The *crystallization temperature* (T_c) is equally significant in processing, since it defines the solidification of a crystallizing polymer, and is, therefore, the temperature to which the polymer has to be cooled, to regain form stability. The crystallization temperature is always lower than the fusion temperature, the difference ranging from some 10 °C in polyamides to 75 °C for polypropylene: as noted earlier, the phenomenon is known as *supercooling*.

Any property which is dependent on crystallinity can be utilized for determining the melting temperature: there is a variety of standard methods involving modulus etc. However, two of the most convenient methods depend on intrinsic properties of the crystals: the birefringence of crystalline aggregates and the latent heat absorbed or evolved during melting and crystallization, respectively.

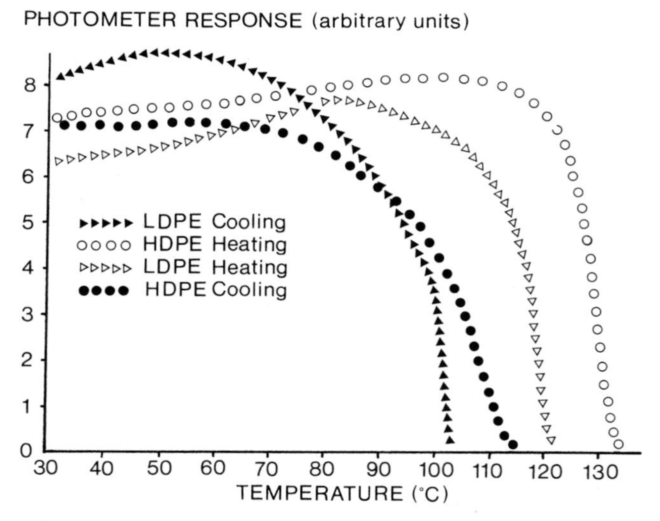

Figure 1.8 Measurement of fusion and crystallization temperatures by crystallite depolarization of polarized light.

When a polymer sample is interposed in a crossed polars system of illumination, the spherulites interfere with the radiation and cause *light scattering,* allowing some of the radiation to pass through the system. Monitoring the intensity of this light as the temperature is increased at a linear rate identifies the melting point as the temperature at which light no longer passes through the optical system, since the crystallites have melted[1.14]. The same technique can be used in the cooling mode.

As an example, the behaviour of HDPE and LDPE is shown in Figure 1.8; the heating rate was 10 K/min, and the cooling rate 3 K/min.

Differential thermal methods are particularly convenient for the measurement of these two quantities and depend on the latent heat affecting the programmed heating or cooling of the sample. In *differential thermal analysis (DTA)* the temperature difference between the sample and a reference material is monitored as both are heated equally, whilst in *differential scanning calorimetry (DSC)* the differential energy input to maintain sample and reference at the same temperature during programmed heating or cooling defines the melting point and the crystallization temperature.

Other methods have been employed, the most fundamental being *variable temperature X-ray diffraction.* Less fundamental is the monitoring of mechanical properties which change appreciably with crystallinity, for example, *softening point* tests.

1.3.7 Rate of Crystallization

The rate of crystallization determines the speed at which a polymer solidifies, if shape stabilization is achieved by this means. Polymers crystallize at very different rates, examples of fast and slow crystallization are polyethylene and poly(ethylene terephthalate). Crystallization is a two-stage process; the first step is the formation of a stable nucleus, which occurs by the local ordering of polymer chains. This is followed by the growth stage, which is governed by the rate of addition of other chains to the nucleus. The rate is increased by increased supercooling, but retarded by the increasing viscosity as the temperature is reduced. There is, therefore, a maximum in the curve relating rate and temperature; this occurs about midway between the crystal melting temperature and the glass transition temperature. In favourable cases, the growth of spherulites can be observed directly in an optical microscope, particularly when illuminated by a crossed polars system. A DSC instrument operated isothermally is a further method of investigating the crystallization behaviour, see Figure 1.9[1.15].

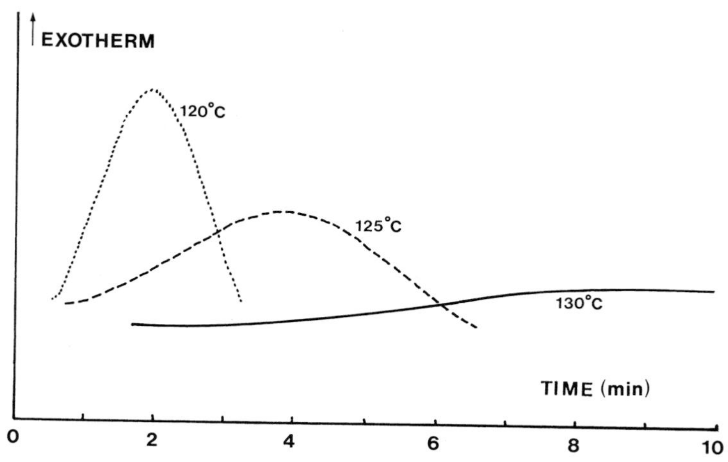

Figure 1.9 Investigation of crystallization behaviour of polypropylene by isothermal DSC measurements.

1.3.7.1 Heterogeneous Nucleation

The crystallization, and the associated latent heat, determine the shape stabilization stage in the processing of thermoplastics; typically, nuclei formed homogeneously from the melt have to be cooled below the fusion temperature by some tens of degrees before they are sufficiently stable to support subsequent growth. If a nucleating agent is added to the system, crystallization can be induced at higher temperatures; with more nuclei available and the growth rate unaffected, crystallization is completed sooner and the cycle time is reduced[1.16]. The availability of many nuclei also leads to smaller spherulite size, which will be discussed in a later section.

1.3.8 Polymer Texture

On first consideration, the texture of amorphous polymers is simple, consisting of a random array of polymer chains, the interactions of which determine the properties. Density fluctuations are minimal, so the light scattering is low and the polymers appear transparent. On a finer scale, however, there are slight density fluctuations, due to the uneven packing of the chains, and the consequent light scattering, although very low, precludes many polymers being used in communications fibres in competition with glass. The incorporation of insoluble additives, or a second polymer phase, generally leads to a loss in transparency, unless the refractive indices of matrix polymer and additive are closely matched.

Figure 1.10 (left) Section of a spherulitic crystalline polymer, viewed through crossed polars.
Figure 1.11 (right) Nucleation of crystallization at a shear plane.

Crystalline polymers, on the other hand, may have very complex *texture*, which is strongly dependent on the processing conditions, in particular, the thermal history. Conversely, examination of the texture allows comment to be made on the likely thermal treatment imposed on a product. *Polarized light microscopy*[1.14] is a technique much used in the examination of *microtomed* thin sections of crystalline polymers (see, for example, Figure 1.10). Other textural features which are investigated by this method include *nucleated crystallites,* where nuclei are provided by a "foreign" surface or by a shear plane; the latter is shown in Figure 1.11.

For more complex polymer systems, such as the sequential PP copolymer (in which ethylene is polymerized on to polypropylene), the texture can become very complicated. In Figure 1.12 there is clear evidence of a *globular texture,* which is superposed on the spherulitic structure of the polypropylene component[1.17]. This very complicated structure undoubtedly plays a part in determining the properties and performance of products made from these copolymers, but our understanding is far from perfect.

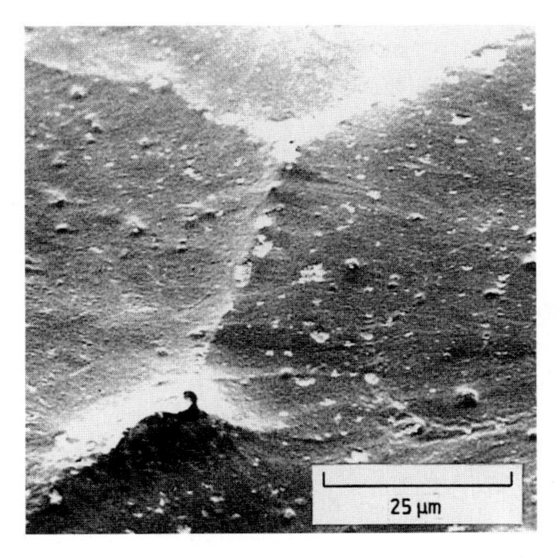

Figure 1.12 'Globules' superposed on polypropylene spherulites.

One relationship which has been established is between optical properties, especially transparency, and crystalline texture; striking improvements in transparency result from decreased spherulite size. An inverse relationship between toughness and spherulite size is more tenuous, since there are many other interacting variables.

For the examination of texture which is beyond the resolving power of the light microscope, the *Scanning Electron Microscope (SEM)* is a valuable tool; the characteristic large depth of focus of this instrument being an added virtue. However, specimen preparation is tedious: etching, followed by metal (gold) coating to prevent charge accumulation on the surface, which would distort the image. Further, the visual acceptability of the image may conceal the true nature of the interaction between the irradiating electrons and the surface. Figure 1.13 is a micrograph of polyethylene obtained by SEM.

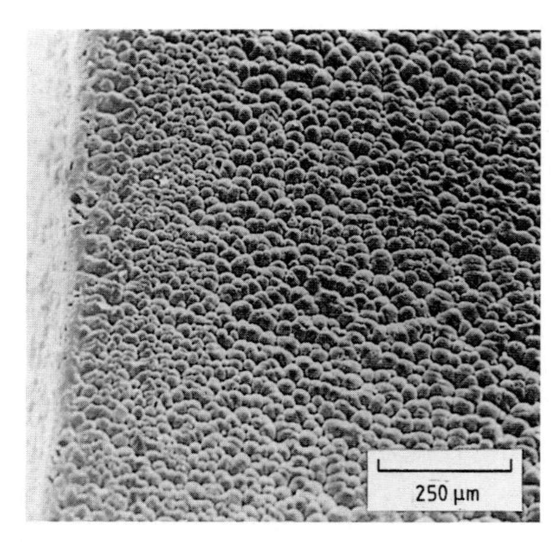

Figure 1.13 SEM micrograph of a section through the thickness of high density polyethylene fuel tank.

1.3.9 Orientation

This occurs in many polymer products; sometimes it is introduced deliberately to enhance the mechanical properties in one direction, *(uniaxial orientation),* as in fibres or strapping tapes, or in the plane of the surface of a film or bottle, *(biaxial* or *planar orientation).* Frequently, however, orientation appears adventitiously, occurring as a result of flow and rapid cooling during processing; this may lead to some unexpected and often unwanted characteristics. This point is considered further in Chapter 5.

Orientation occurs in both amorphous and crystalline polymers, and is a consequence of the ordering of polymer chains during flow with freezing of the morphology before relaxation occurs. Anisotropy in mechanical properties results from the directional character of the polymer chains, which in their length are sustained by covalent bonds (strong), whilst an assembly of chains is kept together only by much weaker forces. The usual method of assessing orientation is by measuring the *birefringence,* that is the difference in refractive index in the orientation and transverse directions. *Shrinkage* is also a satisfactory monitor of orientation, especially for amorphous polymers, when they are heated to a temperature marginally higher than the glass transition temperature.

For a crystalline polymer, the corresponding temperature is just above the melting point, but much greater care is necessary in interpreting the results, since orientation of crystallites also occurs, and is significant for many properties: it is usually monitored by X-ray diffraction.

1.4 Product Characterization: Preliminary Comments

In the preceding pages, we have been concerned with polymers and their characterization, that is determining the "finger-print" by which a particular polymer might be recognized: the combination of features which is unique to that polymer and defines it. However, only those employed in the industry are concerned about polymers; the world at large is more concerned with the performance of *plastics products,* which is two stages removed from the simple polymer concept. The performance of a plastics component will depend on the polymer involved, but the polymer may be modified by additives, or blended with other polymers; further, the processing or shaping operation will inevitably affect the subsequent behaviour of the product.

Clearly, it is very desirable that the effects of additives and of processing should be monitored, requiring that tests be carried out on the product, tests which would be meaningless and irrelevant if made on the initial polymer, or on specimens fabricated especially for a given test, (e.g. "dumb-bell" samples for tensile tests). Thus there is a need for *product characterization,* which is essentially a more technically demanding version of quality control. This topic will be discussed briefly in the following sections.

1.4.1 From Polymer to Plastics Compound

Polymers are, as a consequence of the chemistry of their formation, remarkably pure materials, with impurities present at a level of a few parts per million. However, they are not pure in the sense that each molecular chain is identical with the other chains, since we have seen earlier in this chapter that there is a distribution of chain lengths. Further, depending on the route of manufacture, some of the ingredients in the polymerization recipe may remain in the polymer; thus a polymer prepared by *emulsion polymerization* will almost certainly contain *residual surfactant,* which may affect the properties of the final product.

Pure polymers are not generally the optimum materials for the best performance of the final product; additives are frequently employed to enhance a particular property region. The blend of polymer and suitable additives is known as a *plastics compound* or often simply as a "plastic". Additives[1.18,1.19] involved include *stabilizers,* (antioxidants, antiozonants and photochemical stabilizers), *pigments* and *dyes, lubricants, antistatic agents, flame retardants, slip* and *anti-blocking additives, fillers* and *plasticizers:* more detail is given in Table 1.3.

An important feature of a plastics compound is that the *dispersion* of particulate additives must be very good, otherwise performance will deteriorate[1.20]. To obtain a satisfactory dispersion usually entails mixing in the molten state of the polymer, although powder blending prior to melt processing (with rigid PVC, for example) frequently gives a cheaper overall operation.

Table 1.3 Some Additives for Plastics

Additive	Purpose	Example of Plastics Use
Hindered phenol	Antioxidant	PP
2-hydroxybenzophenone	UV stabilizer	LDPE
Tribasic lead sulphate	Stabilizer	PVC
Calcium stearate	Stabilizer/Lubricant	PVC
Quaternary ammonium salts	Antistatic agent	Polyolefins
Oleamide	Slip additive	LDPE film
Silica, diatomaceous earth	Antiblocking additives	LDPE film
Stearic acid	Lubricant	PVC (rigid)
Dioctyl phthalate	Plasticizer	PVC (flexible)
Calcium carbonate	Filler	PP
Wood flour	Filler	UF/PF moulding powders
Glass fibres	Reinforcing filler	PA66; PBTP
Antimony oxide and chlorinated hydrocarbons	Flame retardant	PP
Ammonium phosphate	Flame retardant	ABS
Carbon black	Pigment	LDPE
Molybdenum disulphide	Solid-phase lubricant	POM

Another technique which facilitates mixing is the use of *masterbatches,* which are concentrates of the additives, generally up to 50 %, well dispersed in a polymer-carrier. It might be regarded as axiomatic that the masterbatch carrier polymer should be identical or compatible with the major polymer in the product, but this seeming common sense does not always prevail. That the masterbatch carrier should be of lower molecular weight – dispersion is facilitated in such polymers – is a reason for doubt, as is the choice of a higher molecular weight carrier, which is, of course, more difficult to disperse. However, the use of an alien polymer leads almost inevitably to deteriorated performance, and is a trend much to be deprocated.

Poor dispersion of additives is a frequent contributor to premature failure in a plastics product; even more disastrous is poor distribution of an alien polymer, as polymers are generally incompatible. An example of poor pigment dispersion is given in Figure 1.14, which shows pigment clearly defining the flow in the pinch-off region of a blow moulded part: the lines of pigment are stress concentrators of the worst type. This dispersion is particularly bad, although a masterbatch technique was used[1.20].

Figure 1.14 Poor pigment dispersion in a plastics blow moulding (pinch-off region).

1.4.2 Polymer-Polymer Miscibility

Immiscibility in polymers is the general rule[1.21], although there are some notable and commercially significant exceptions: the PS-PPO system is fully compatible and marketed under the trade name "Noryl", (General Electric Company of America). Lack of miscibility is even more pronounced in polymers which crystallize; indeed, there are remarkably few examples of *cocrystallization,* although one such blend is known[1.22,1.23]: HDPE-LLDPE, (high density and linear low density polyethylenes). The immiscibility of most polymers makes the concept of a "universal masterbatch" a very undesirable proposition; however, the convenience and economy of a cheap carrier polymer which accepts additives readily, and which would "service" the range of polymers used by a manufacturing company, is very attractive to processors. *Product characterization* highlights the danger of this practice: experience has shown the presence of an alien polymer to be a frequent cause of failure[1.20,1.24].

1.4.3 The Analysis of Plastics Compounds

The task of determining quantitatively the amount of a particular additive in a plastics compound is formidable; the problem of ascertaining whether a second polymer has been used in a masterbatch process is even more difficult. Frequently the additive is present in only small concentration, hopefully well dispersed in the matrix polymer; almost without exception, the analysis depends on isolating the additive, for example:

(a) by extracting the additive with a solvent which does not dissolve the matrix polymer, thereafter identifying the additive by normal chemical methods;
(b) for inorganic additives, the polymer, and any organic additives, can be removed by pyrolysis and the residue analyzed; and

(c) isolation of the polymer carrier from a plastics compound is frequently difficult, if not impossible, and the problem devolves into identifying a small amount of one polymer in the bulk of a different one. Infra-red spectrometry may help, whilst for crystalline polymers, differential thermal analysis, (DTA or DSC), might identify the polymer. For grossly incompatible systems, solvent extraction of the masterbatch carrier polymer may be possible.

Identifying the additives present in a plastics compound is only part of the product characterization; their distribution, including the masterbatch carrier polymer, is at least as important as the types of additive and the amounts present. Since interest centres on dispersion on a very small scale, the *optical,* or *light microscope* is a valuable tool. The modern microscope acts not only as a magnifier but enables other optical techniques to be used which render visible aspects of the morphology which cannot be seen in common light. As examples, whilst the poor pigment dispersion in Figure 1.15A was obtained in transmitted common light through a thin section of the product, the micrograph in Figure 1.15B was obtained in crossed polars illumination, based on the birefringence which is generally a characteristic of polymer spherulites.

Refractive index differences in the constituents of a plastics compound are the basis for analysis by *differential interference contrast (DIC) microscopy.* In favoured cases, accurate measurement of the refractive indices of the components is possible, allowing positive identification[1.25]. There are many other variants of microscopy, including *ultra-violet* and *fluorescence* methods, which contribute to an understanding of the structure of plastics, the latter is particularly useful in monitoring polymer degradation.

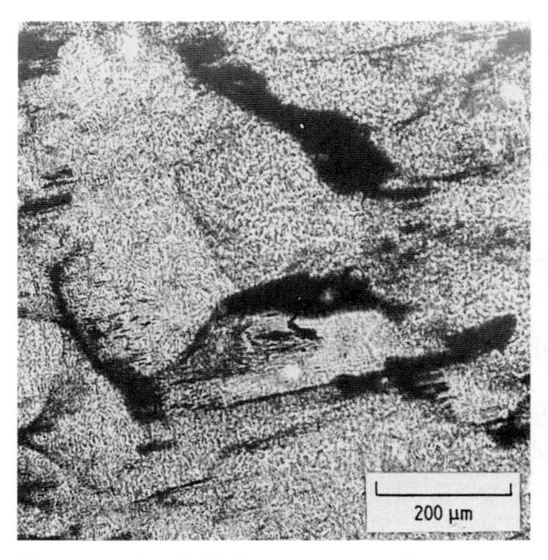

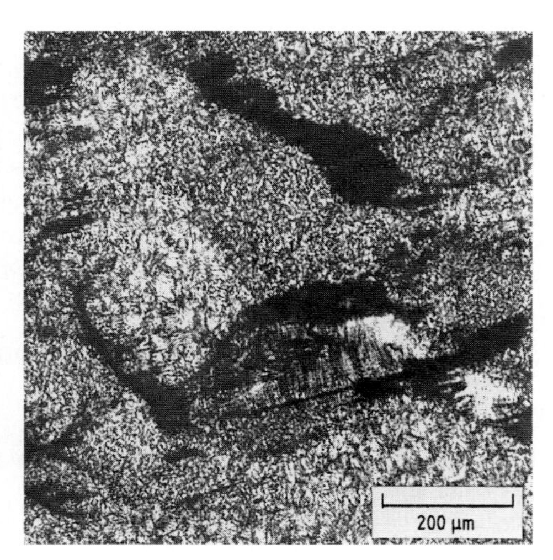

Figure 1.15a (left) Poor pigment dispersion viewed in common light;
Figure 1.15b (right) As 1.15a viewed between crossed polars.

1.4.4 The Effect of Processing on Product Performance

The light microscope and optical techniques enable many features of a product to be investigated; comment can be made about structural details which arise inevitably during the shaping process, with additional features which appear as a consequence of less-than-optimum processing conditions. It is helpful to consider the effects of processing in two categories: those which reasonably cannot be avoided in an economical shaping operation, and those which arise through ignorance or carelessness, or by trying to force the process to lower cycle times (or higher line speeds), in seeking greater profits. The intrusion of contaminants, often particulate, is detrimental to performance, and is frequently a consequence of poor housekeeping or working methods. Since thermoplastics constitute the majority of plastics usage, it is appropriate to start with discussion of these materials.

1.4.4.1 Processing under Optimum Conditions

The two most important means of shaping are the *die* and the *mould:* both require that the melt presented to them for shaping shall be homogeneous with respect to material distribution, temperature, (and possibly thermal history), and viscosity, (and possibly shear and tensile stress history). The shaped plastics product is stabilized by cooling, which sets up stresses, as the cooling cannot be homogeneous, but must occur by heat transfer through the bulk of the plastic, to be extracted at the boundaries of the product, ultimately being transferred to the cooling medium. For plastics products of thick section, there is obviously a difference in cooling rate for material adjacent to the cooling surface and material more remote from this boundary. The plastics material in contact with the cooling surface quickly attains the temperature of the surface, and solidifies. Thermal contraction of material in the interior is restrained, and mechanical stresses are generated (see also Chapter 5). Further, the different cooling rates at the surface and in the interior cause a textural mismatch which can contribute to failure[1.26]. This *morphological interface* is more pronounced for very cold surfaces and is thus most evident for injection or blow mouldings of thick section with crystallizing plastics.

Even when shaping is carried out under optimum conditions, there are a number of features which result from the flow of the polymer melt to take up the shape of the mould or die. This leads to *orientation,* especially at the high shear rates in injection moulding. With the high cooling rates employed to achieve economical cycle times, much of the orientation is retained in the moulding or extrudate, although due to lower shear rates, and slower cooling, the effects are less significant in the latter case. For filled thermoplastics, particularly with anisotropic fillers such as glass fibres, orientation effects may be even more serious.

The lining-up of the polymer chains causes the refractive index in the flow direction to differ from the indices in the two transverse directions. Whilst this is not visible in common light, viewing through crossed polars renders the phenomenon visible, and allows quantitative measurements of optical *birefringence* to be made.

The design of a mould for a plastics product often requires that the melt be "split" to circumvent a feature of the design, for example a hole or an insert, the melt thereafter recombining. The *internal weld* which results is a weakness, the effect of which should be minimized by attention to the processing conditions. The extrusion of pipe leads to internal welds as the melt recombines, after splitting as it passes the "spider" which supports the central mandrel. To monitor the severity of this problem, thin sections can be viewed through crossed polars. As with orientation, the inclusion of anisotropic fillers may enhance the effect: Figure 1.16 shows an internal weld in linear low density polyethylene.

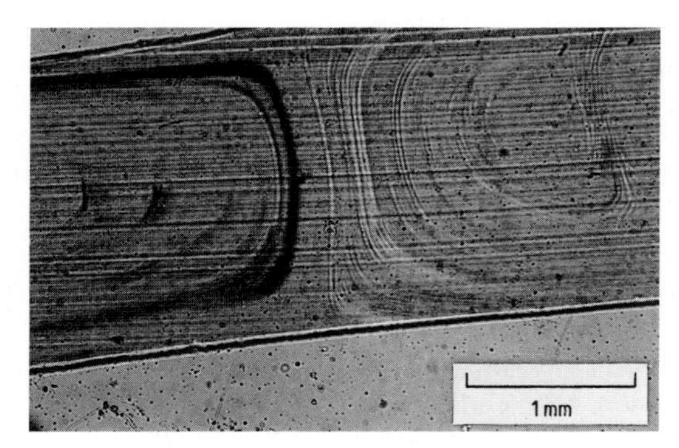

Figure 1.16 An internal weld in linear low density polyethylene.

The gate region where the plastics melt enters the mould is usually a region of confused structure, since injection moulding requires that more polymer be packed into the mould during cooling to compensate for shrinkage. For crystallizable plastics, this leads to textural hiatus, which can be viewed through crossed polars, and which must be carefully-controlled and minimized to achieve satisfactory products. An example of a defective gate microstructure can be seen in Figure 1.17.

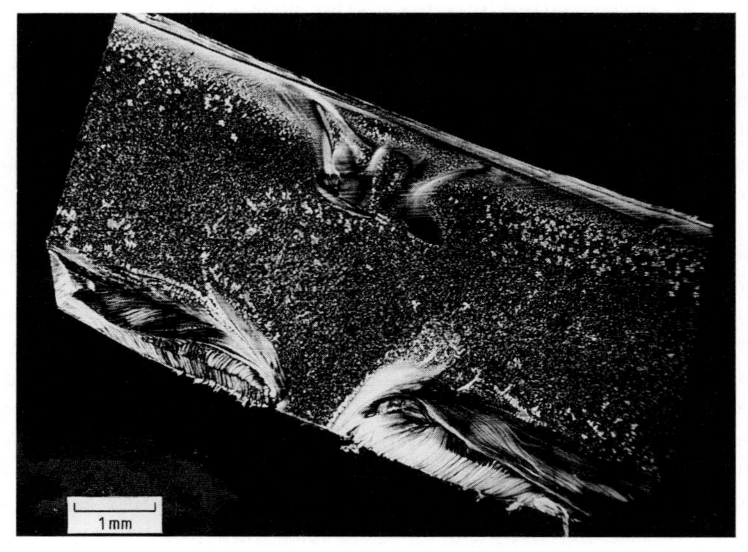

Figure 1.17 Gate region in plastics injection moulding, viewed through crossed polars.

1.4.4.2 Product Faults arising from Non-optimum Processing

Performance of a plastics product can be unsatisfactory for a variety of reasons: design of the product; design of the mould or die; and less-than-optimum processing. Design of the product to give acceptable service is governed by the rules of good design in any material, especially avoidance of stress concentrations, including sudden changes in section thickness. Stress analysis can now be carried out by finite element methods; the complexity of these programs poses no problems if sufficient computer power is available. Software is also available for the design of moulds, thereby taking much of the trial and error out of mould-manufacture.

As a feature of product design, there is often a need that two plastics subcomponents be joined. This may be accomplished by one of several methods: hot plate welding is a popular choice. Since a weld is a weakening feature, even if perfect, it is good design that the weld be located in a relatively unstressed region of the product. Further, the welding conditions must be optimized with respect to hot plate temperature, pressure of the subcomponents on the hot plate, time of heating, time of changeover, welding pressure and welding time. With so many variables, it is all too easy to make a bad weld, as exemplified in Figure 1.18. Both common light and crossed polars illumination are helpful in assessing the quality of a weld.

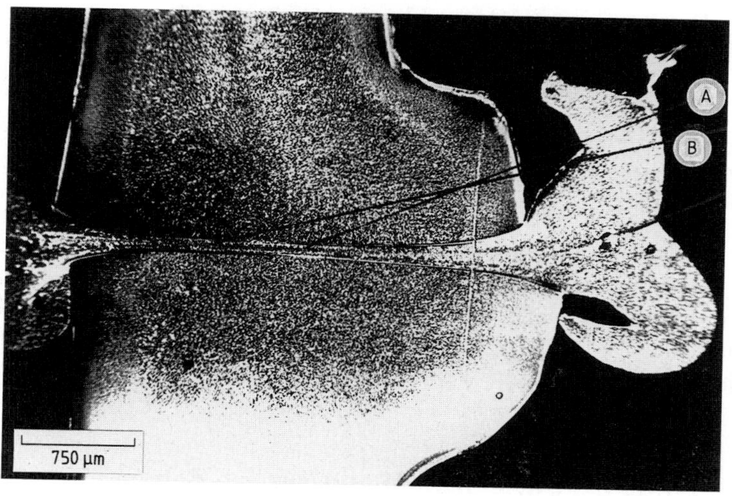

Figure 1.18 Example of a poor hot-plate weld: most of the molten polymer has been squeezed into the beads. 'A' is the original join, and has recrystallized, whilst 'B' was unaffected by welding.

Thermal effects give rise to many phenomena which determine product performance, many of which can be readily monitored. The temperature of the melt is the first consideration, to obtain a melt of optimum viscosity for shaping: too cold a melt often leads to inhomogeneity and in the limit, *unmelted granules,* which carry through to the product, resulting in very large stress concentrations. Conversely, too high a melt temperature, or a long residence time at a lower temperature, can give *degradation,* with consequent deterioration in properties and performance. In air, oxidation may occur, again giving degradation and loss in performance.

A slow cooling rate, which is almost inevitable for shapings of thick section due to the low thermal conductivity of plastics, results in high crystallinity and coarse texture, which favours *interspherulitic cracking*[1.27]. Once again, microscopic analysis using common light and crossed polars illumination is helpful in diagnosing this behaviour, which is especially prone to occur in blow mouldings of thick section, cooled from the outside only, see Figure 1.19. This same type of product of small dimensions may be subject to surface defects unless high blowing pressures are used, as the moulding may shrink away from the mould.

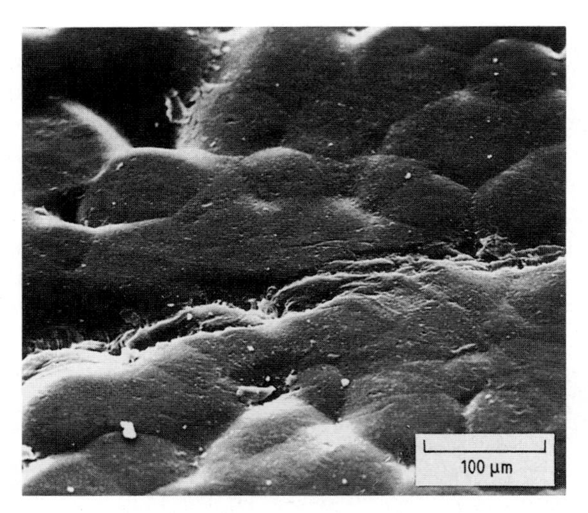

Figure 1.19 Interspherulitic cracking on the slowly-cooled inner surface of a PP car coolant reservoir tank.

This also results in much reduced heat transfer and an appreciable increase in cooling time. Another factor is the heat generated by mechanical working, as a consequence of which "melt" temperatures may be higher than those "set" by the equipment.

Further analyses of many of these aspects of plastics processing are covered in Chapters 2 – 5.

1.4.5 Product Characterization and Service Behaviour

The fact that a plastics product has been manufactured under apparently optimum conditions does not guarantee that it will not fail in service. It is unlikely that product characterization will be able to forecast failure exactly, and it is only by experience that we can estimate the likely response to unusually severe service conditions. However, it is possible to examine products which have proved to be less than satisfactory in service and comment on the cause and course of failure[1.20]. It is also possible if extreme conditions of a particular type are anticipated to choose a plastics material which is more resistant to that failure mechanism. Prolonged loading, rapid loading and cyclic loading all encourage brittle failure in an otherwise ductile polymer; if any of these conditions is likely to be encountered, then design should be on the basis of *fracture mechanics* (see Chapter 7), employing the relevant data for mechanical properties. Time dependent properties, which will be discussed in more detail in Chapter 6, represent a further element of complication.

Degradation of a plastics product, by thermal or photochemical means, and including oxidation, can occur in service (see Chapter 10). If such damaging conditions are likely to be encountered, the products should be protected by appropriate additives: stabilizers and antioxidants in the present case. If degradation does occur, it can be monitored by changes in melt or solution viscosity, whilst oxidation can be followed by infra-red spectrometry. Again, if extreme conditions are anticipated, the plastics compound should be chosen accordingly.

Environmental stress cracking (ESC), the failure of a plastics product under the combined effects of stress and an aggressive environment, (whereas it can withstand either element separately), is not yet susceptible to a general theoretical treatment, and so forecasting is unreliable, except from experience. As with many failure phenomena, however, it is possible by examination of the failed parts to ascertain the mechanism, and thus know in what direction to take remedial action. A typical environmental stress cracking failure is illustrated in Figure 1.20.

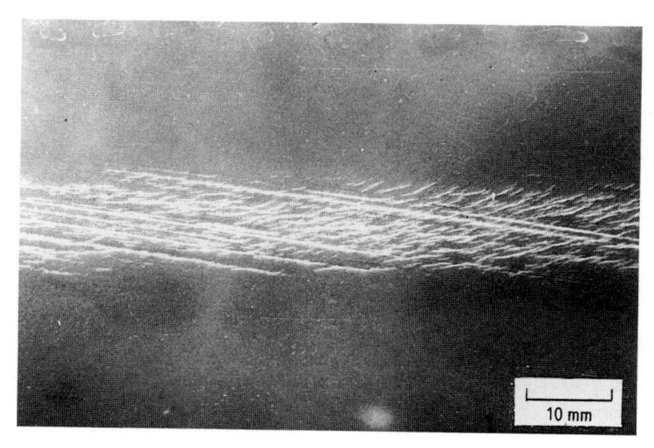

Figure 1.20 Typical environmental stress cracking on a car coolant tank, which is pressurized at elevated temperature, in the presence of ethylene glycol.

1.4.6 Products based on Cross-linked Plastics

For convenience, and because the majority of plastics products are based on thermoplastics, much of the discussion has been concerned with these polymers. Cross-linked plastics present somewhat different problems for product characterization, especially since fillers, which are often fibrous in nature, are prevalent in cross-linked systems. Thus, greater attention has to be paid to filler orientation and to processing techniques to minimize deleterious effects.

Orientation of fibres can be studied by common light microscopy, but *contact X-radiography* is a powerful technique for obtaining fibre orientation distribution, since samples of appreciable thickness can be examined without danger of disturbing the structure. Characterization of the polymer constituent is considerably more difficult than for a thermoplastic, since solution methods are no longer viable. X-ray diffraction and DSC are not very informative, and infra-red spectrometry either gives virtually no response, (the usual case), or a very complicated spectrum. Destructive chemical methods are generally the only routes of analysis available, and even these require gross assumptions about the reactions taking place.

Whilst the general use of fillers, and their often anisotropic nature complicate product characterization, the *prepolymers* involved (which are shaped and then cross-linked) are not as susceptible as thermoplastics to molecular orientation, since they are generally of smaller molecular size and lower viscosity. Furthermore, the presence of the filler overshadows other causes of stress concentration, so that failure mechanisms are more restricted, but potentially complex in many fibre-reinforced, cross-linked materials, where resin failure, fibre fracture and *interfacial debonding* are all common.

References

1.1 ISO DR 1043: International Standards Organization.

1.2 FOX, T. G., and FLORY, P. J.: *Second Order Transition Temperatures and Related Properties of Polystyrene,* J. Appl. Phys., **21,** (1950), 581.

1.3 BOYER, R. F.: *The Relation of Transition Temperatures to Chemical Structure in High Polymers,* Rubb. Chem. Tech., **36,** (1963), 1303.

1.4 RODRIGUEZ, F. H.: *The Principles of Polymer Systems,* 2nd Ed., Hemisphere Publishing Corporation; McGraw-Hill, New York, (1982).

1.5 KELLER, A.: *Polyethylene as a Paradigm of Polymer Crystal Morphology, Plast. Rub. Proc. & Appl.* **4,** (1984), 85.

1.6 BASSETT, D. C.: *Principles of Polymer Morphology,* Cambridge Solid State Science Series, Cambridge University Press, (1981), 189.

1.7 DATTA, N. K.: Ph.D. Thesis, *A Study of the Miscibility of Crystalline Polyolefins,* Loughborough University of Technology, (1982), 193.

1.8 HASLAM, J. A., WILLIS, H. A., and SQUIRRELL, D. C. M.: *The Analysis of Plastics,* 2nd Ed., Iliffe, London, (1972).

1.9 BILLMEYER, F.: *Textbook of Polymer Science, 3rd Ed.,* Wiley-Interscience, New York, (1984), 209.

1.10 BS 2782, Method 720A, (1979), *Determination of the Melt Flow Rate and the Melt Volume Rate of Thermoplastics,* (Corresponding International Standard ISO 1133), British Standards Institution, London.

1.11 DAWKINS, J. V., Ed.: *Developments in Polymer Characterization Vol. 1,* Applied Science, London, (1978).

1.12 BLINKHORN, A., and BIRLEY, A. W.: *The Technology of PF Moulding Powders – 1 Characterization of Novolak resin and derived Moulding Powder,* Brit. Poly. J., **18,** (1986), 151.

1.13 CHEN, X. Y., and BIRLEY, A. W.: *A Preliminary Study of Blends of Bisphenol-A Polycarbonate and Poly(ethylene terephthalate),* Brit. Poly. J., **17,** (1985), 347.

1.14 BUCHALTER, G.: *A Study of the Miscibility and Compatibility of Polybutene and LLDPE,* M.Sc. Project Report, (1986), Loughborough University of Technology.

1.15 Instruction Manual for DSC System: Du Pont, (1978)

1.16 HEMSLEY, D. A.: *The Light Microscopy of Synthetic Polymers,* RMS Microscopy Handbook 07, Oxford Science Publications, (1984), 48.

1.17 YEH, PO-LEN, HEMSLEY, D. A. and BIRLEY, A. W.: *The Structure of Propylene-Ethylene Sequential Copolymer,* Polymer, **26,** (1985), 1155.

1.18 MASCIA, L.: *The Role of Additives in Plastics,* Edward Arnold, (1974).

1.19 GAECHTER, R., and MUELLER, H.: *Plastics Additives,* Carl Hanser, Munich, Vienna, New York (1987).

1.20 HEMSLEY, D. A. and BIRLEY, A. W.: *Failure Analysis of Plastics,* in *Encyclopedia of Materials Science and Engineering,* BEVER, M. B., Ed., Pergamon Press, Oxford, (1987).

1.21 OLABISI, O., ROBESON, L. M. and SHAW, M. T.: *Polymer-Polymer Miscibility,* Academic Press, New York, (1979).

1.22 DATTA, N. K. and BIRLEY, A. W.: *Thermal Analysis of Polyethylene Blends,* Plast. & Rub. Proc. & Appl., **2,** (1982), 237.

1.23 EDWARD, G. H.: *Crystallinity of Linear Low Density Polyethylene and of Blends with High Density Polyethylene,* Brit. Poly. J., **18,** (1986), 88.

1.24 HEMSLEY, D. A. and BIRLEY, A. W.: *Case Studies of Failure in Plastics, Products,* Prog. Rubb. Plast. Technol. **2,** (1986).

1.25 HEMSLEY, D. A.: As Reference 1.16, pages 26 and 30.

1.26 GARG, S. K.: Ph.D. Thesis, *The Effect of Morphology and Microstructural Interfaces on the Properties of HDPE Moulded Containers,* Loughborough University of Technology, (1982).

1.27 YEH, PO-LEN: Ph.D. Thesis, *Material and Process Problems in the Manufacture of Coolant Reservoir Tanks,* Loughborough University of Technology, (1984).

Chapter 2
Thermal Properties

2.1 Introduction

Temperature has a considerable effect on many of the properties of plastics and these are sometimes classified as *thermal properties.* Tests for such properties are to be found in the National and International Standards compilations; examples include the Vicat Softening Point[2.1], the Deflection Temperature under Load[2.2], the Cold Flex Temperature for PVC[2.3] and many others. Strictly, however, such tests are concerned with the effect of temperature on other properties, in the cases cited, mechanical properties. The term *"thermal properties"* will be restricted to consideration of the following:

- Enthalpy, the relationship between heat energy and temperature (2.2);
- Melting and crystallization (2.3);
- Thermal conductivity, thermal diffusivity and heat transfer (2.4);
- Thermal expansion and contraction (2.5).

These are important in understanding the processing of plastics, for both thermoplastics and cross-linking systems. Further, the limiting temperature for solid state plastics is the *crystalline melting point* T_m, whilst thermal insulation and dimensional stability are sometimes critical factors in specific applications.

Some discussion of the *glass transition temperature,* T_g, is inevitable in comparing the softening behaviour and thermal properties of amorphous and semi-crystalline plastics.

2.2 Enthalpy: The Relationship between Heat Energy and Temperature

On supplying heat energy to a substance, the normal observation is that the temperature will increase, and that there is a linear relationship between energy and temperature. This defines the *specific heat capacity* as the amount of heat required to raise the temperature of unit mass, 1 kg, by 1 K; the usual units are kiloJoules (kJ).

The *enthalpy* or *heat capacity* is the heat energy required to raise the temperature of unit mass by a defined amount and is, therefore, the sum of the specific heat capacities over a temperature range. For reasons which will become evident, it is usual to present enthalpy data as a function of temperature; this is exemplified for naphthalene in Figure 2.1[2.4]. It is seen that the curve has three parts: the slope AB defines the specific heat capacity of the solid material, whilst slope CD similarly defines the same quantity, characteristic of the liquid state. The portion BC, when heat is supplied but there is no increase in temperature, represents the heat energy required to melt unit mass of the crystals to a liquid at the same temperature, the *latent heat of fusion.*

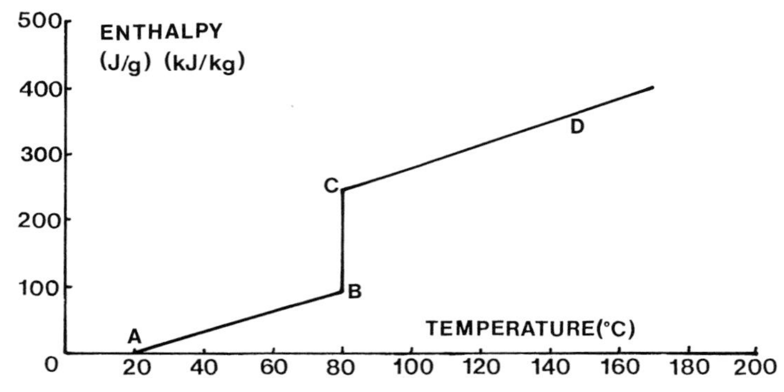

Figure 2.1 Enthalpy versus temperature relationship for naphthalene.

2.2.1 Supercooling in Semi-Crystalline Plastics

If the same material is cooled from the liquid state, the *enthalpy vs. temperature* relationship retraces the curve obtained on heating more or less exactly. However, Figure 2.2 shows a similar heating curve for a typical partially crystalline thermoplastic, HDPE, but the relationship is only linear at high temperatures, being appreciably curved at low temperatures. The construction A′B′C′D′ provides an indication of the origin of the curvature, in that the level of crystallinity gradually decreases with increasing temperature. Also, the construction identifies the latent heat of fusion, which is greater than that of naphthalene, although the polymer is only partially crystalline, (75 – 80 %), whilst the naphthalene is virtually 100 % crystalline. This highlights the considerable energy inherent in a polymer crystal, and provides explanation of the importance of crystallinity in plastics technology.

The enthalpy-temperature relationship for an amorphous plastic (PC) has also been included in Figure 2.2: in this case, the gradient changes in the approach to the glass transition temperature.

The curve in Figure 2.2 was obtained in the heating mode; if the liquid polymer is cooled, a different picture emerges. Figure 2.3 is similar to Figure 2.2, together with results for the same polymer in the cooling mode.

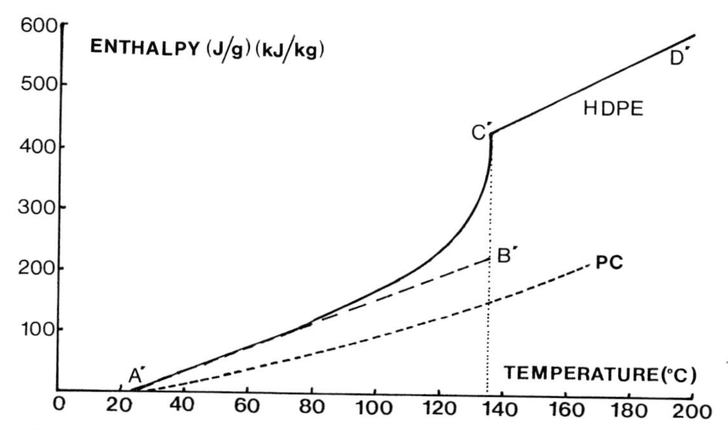

Figure 2.2 Enthalpy versus temperature curves for semi-crystalline high density polyethylene (HDPE), and amorphous polycarbonate (PC).

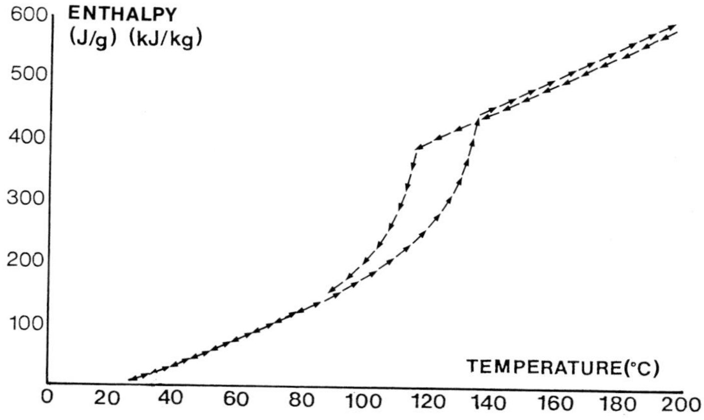

Figure 2.3 Enthalpy versus temperature curves for HDPE: heating and cooling modes.

It is immediately obvious that the response of the polymer during heating and cooling is different; in Figure 2.3, the crystallization temperature is lower than the melting or fusion temperature. This has been encountered in Chapter 1, and is termed *supercooling;* its implications in plastics technology will now be considered.

The extent of supercooling depends on the polymer, with polypropylene and polybutene showing large effects, and polyethylene and the polyamides much less supercooling. For any single polymer, cooling rate has a marked effect, higher rates of cooling being associated with increased supercooling. This is illustrated in Table 2.1, where cooling rates of 0.5, 5.0 and 50 K/min have been imposed on a variety of propylene polymers[2.5]. The crystallization temperature for the polypropylene is lowered from 118 – 122 °C for cooling at 0.5K per minute, to 108 – 109 °C for 5K per minute, and to 96 – 98 °C for 50K per minute. The *nucleating* effect of 0.25 % talc is perceived, whilst the fusion temperatures are but little affected by the changes in cooling rate. Other variations, to time and temperature in the melt had little effect on the crystallization or fusion temperatures.

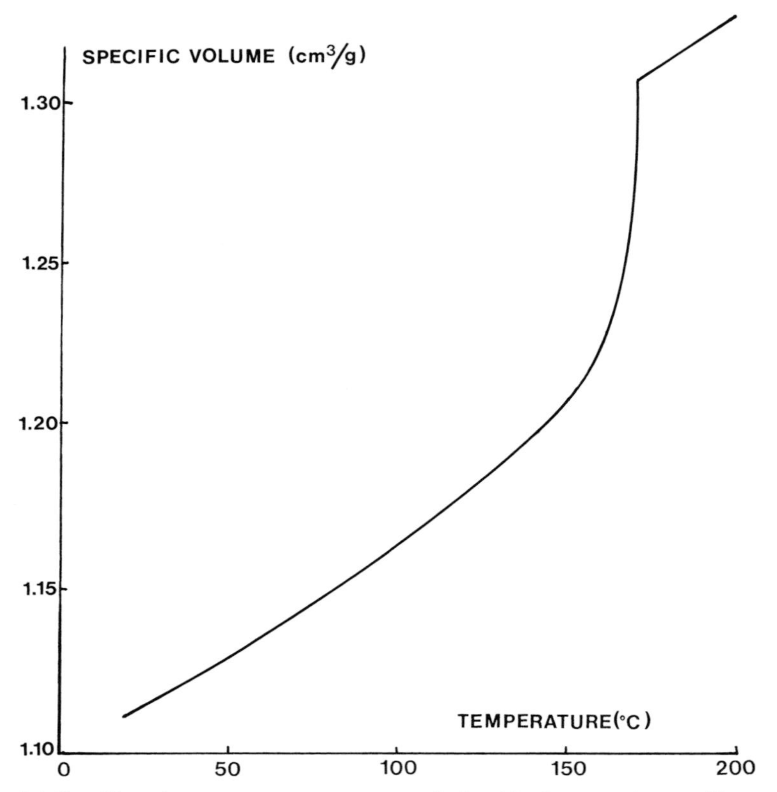

Figure 2.4 Specific volume versus temperature relationship for a semi-crystalline polymer, polypropylene (PP).

The data in this table are very relevant to processing, since it is usual to impose very fast cooling rates on thermoplastics melts to achieve short cycle times. Under these conditions, crystallization is likely to be delayed, and occurs at a much lower temperature than might have been anticipated. The persistence of the liquid state to low temperatures has the disadvantage of increased cycle times, but the compensation that the relaxation of stresses incurred in processing can take place during this time may be of greater practical significance. Thus, polypropylene is reported to be more tolerant in processing than high density polyethylene, suffering to lesser extent defects like *distortion* and *warping.* It must be stated, however, that

Table 2.1 Cooling Rate Effects on Melting and Crystallization Temperatures: Propylene Copolymers

Each sample was subjected to the following thermal treatment:

(1) Heat at 10K/minute, noting melting point, $T_{m(1)}$, of PP;
(2) Retain at maximum temperature, usually 200 °C, for 5 minutes;
(3) Cool at 0.5, 5.0 or 50 K/min, noting crystallizing temperatures, T_c;
(4) Reheat at 10K/min, noting melting points, $T_{m(2)}$, for PE and PP.

Polymer	PP transitions (°C)			PE transition (°C)
	Tm(1)	Tm(2)	Tc	Tm(2)
	Cooling rate 0.5 K/min.			
HSE 110(A)*	162.0	161.5	122.0	120.0
HSE 110(B)	162.5	161.0	119.0	—
Profax 7824**	162.0	161.5	118.0	119.0
	Cooling rate 5.0 K/min.			
HSE 110(A)	161.5	159.5	109.0	120.0
HSE 110(B)	164.0	162.0	108.0	120.0
Profax 7824	163.0	161.5	108.5	120.5
	Cooling rate 5.0 K/min: Nucleated with 0.25 % talc			
HSE 110(A)	163.0	161.5	119.0	122.0
	Cooling rate 50 K/min.			
HSE 110(A)	161.5	161.0	97.0	120.0
HSE 110(B)	162.5	161.0	97.5	120.0
Profax 7824	163.0	161.0	96.0	121.0
	Cooling rate 5.0 K/min: Effect of maximum temperature			
HSE 110(A)			109.0	(Maximum Temp. =190 °C)
HSE 110(A)			110.0	180
HSE 110(A)			110.3	170
	Cooling rate 5.0 K/min: Stage 2 extended to 20 mins.			
HSE 110(A)			109.5	

* HSE 110 is a grade of propylene-ethylene copolymer (Propathene™) from Imperial Chemical Industries PLC; A and B are two batches of the polymer;
** Profax 7824 is a similar polymer supplied by Hercules Powder Corporation.

the higher crystallinity, and the greater density differential between the crystalline and amorphous states, enhance the effect for polyethylene.

It is seen in Figure 2.2 that at the fusion temperature there is, in principle, a discontinuity in the enthalpy vs. temperature curve, although this is blurred for polymers, as crystallites under strain, or of very small size, melt at lower temperatures than the more perfect crystals. This discontinuity defines melting as a *first order transition.* A similar phenomenon can be seen in the volume vs. temperature relationship, Figure 2.4; indeed, not unexpectedly, volume and enthalpy vs. temperature curves have similar shapes.

2.2.2 Amorphous Plastics: The Effect of T_g

In the absence of crystallinity in a polymer, the enthalpy vs. temperature curve does not have any discontinuities, but there are temperatures at which a change of slope can be detected, i.e. there are discontinuities in the temperature derivatives of enthalpy (specific heat capacity) and of volume (coefficient of expansion). Such a transition is termed a *second order transition,* and that occurring at the highest temperature in the solid is arbitrarily defined as the *Glass Transition Temperature,* T_g.

The thermal properties of amorphous plastics are also modified by other factors. Above the glass transition temperature, the effect of changing *pressure* has a significant effect on the enthalpy-temperature, or specific volume-temperature relationships; due to the *compressible* nature of high polymers, an increase in pressure decreases the specific volume at any given temperature. This type of observation is very important in melt processing, especially where part consistency and minimum thermal shrinkage are important requirements; pressure-volume-temperature (PVT) relationships are used therefore to optimize the packing pressure cycle in injection moulding, in order to comply with the stated objectives for this process (see also Chapter 5; Section 5.2.5).

The glass transition temperature (and therefore the thermal properties of any compound, which depend upon the polymer chain mobility either side of it), is affected by thermal history and is therefore modified by cooling rate effects[2.6]. Rapid cooling appears to lower T_g, and is associated with greater specific volume (higher *free volume)* within the glassy phase below the transition[2.7]. As a result, amorphous plastics fabricated by techniques which are associated with severe changes in cooling rate often contain through-thickness variations in density, which reflect the inhomogeneous cooling which occurs in manufacture.

The lower density, higher specific volume surface layers can be modified by subsequent *ageing,* or *annealing* effects, in order to modify other physical properties. More generally, the concept of free volume has been used by WILLIAMS, LANDEL and FERRY[2.8] to relate the temperature dependence of mechanical (stiffness, viscosity) or electrical properties to the relaxation kinetics about the glass transition.

2.3 Melting and Crystallization

Inevitably, these phenomena have already been considered in the previous section on enthalpy, since they contribute considerably to the behaviour of crystalline polymers. Further, we have already seen in Figure 2.2 that the melting of polymers is not a single catastrophic process occurring at one temperature, as is the case for low molecular weight materials, (Figure 2.1). Melting is affected by many factors, which are now discussed under the general headings of materials, environmental conditions and processing.

2.3.1 Effect of Molecular Weight on Melting Temperature

Polyethylene has a repeat unit of $-(CH_2)-$, with terminal methyl groups, and is thus a high molecular weight member of the *paraffin series* of hydrocarbons. The low members of this series are gases, the somewhat higher members are liquids (e.g. octane), followed by crystalline solids, with the melting point increasing with molecular weight. This trend is continued with the *paraffin waxes,* with molecular weights of one to five thousand, which melt in the range 75 – 95 °C, and linear polyethylene, with molecular weight fifty thousand, melting at 133 – 136 °C. This is the *practical* equilibrium melting point for HDPE, although some writers have suggested values exceeding 140 °C.

In the range of plastics commercially available, increase in fusion temperature with molecular weight is not significant. At very high molecular weights, however, there is some tendency for the melting point to decrease, due to strains within the crystallites and their imperfections, caused by *chain entanglements* and the extremely *high melt viscosity,* which impedes crystal growth. It is often found for such high molecular weight polymers that neither the level of crystallinity nor the high melting temperature of the polymer "as manufactured" can be regained after melting. An example is polytetrafluoroethylene, (PTFE), which, as polymerized, has a crystallinity of 90 % and a fusion temperature of 340 °C; even after slow-cooling from the melt, the crystallinity rarely reaches 70 %, and the melting point is 328 °C.

2.3.2 Effect of Branching and Copolymerization on Melting Point

Both branching and copolymerization disrupt molecular order and chain-regularity, thereby reducing both crystallinity and melting point; see Section 1.2.7, where ethylene polymers are discussed. A further example is the random copolymerization of 2.4 % ethylene in propylene, which low level of copolymerization reduces the melting point from 165 to 150 °C.

Copolymers of industrial importance include ethylene-vinyl acetate (E/VAC) and vinyl chloride-vinyl acetate (VC/VAC) copolymers, both of which represent reasonably ideal copolymerization in that the two monomers are distributed randomly.

Block copolymers exhibit, to a first approximation, both the crystallinity level and melting temperature appropriate to the block unit, although both may be slightly reduced by dilution.

2.3.3 Effect of Polymer-Polymer Blending on Melting Point

Polymers are rarely miscible with one another; even more unusual is *cocrystallization* occurring when two partially crystalline polymers are mixed. The general observation is that the crystal phases remain separate, with the melting temperatures of each component degraded slightly. *Phase separation* is found even in blends of HDPE and LDPE, for which data obtained by differential scanning calorimetry are given in Table 2.2, and there is a further example of immiscibility in the observation of both PP and PE crystallinity in propylene-ethylene sequential copolymers, (Table 2.1). It is not possible to comment for these systems on possible changes in melting point by interaction, as the influence of molecular weight is not known.

The system HDPE-LLDPE does appear to be cocrystalline, although the analysis is complicated by the LLDPE, (Dowlex® 2045, an ethylene-octene copolymer), showing multiple transitions. In the data in Table 2.3, the increasing melting temperature as progressively more HDPE is present in the blend can be seen clearly; there is a corresponding, but smaller change in the crystallization temperatures at both cooling rates.

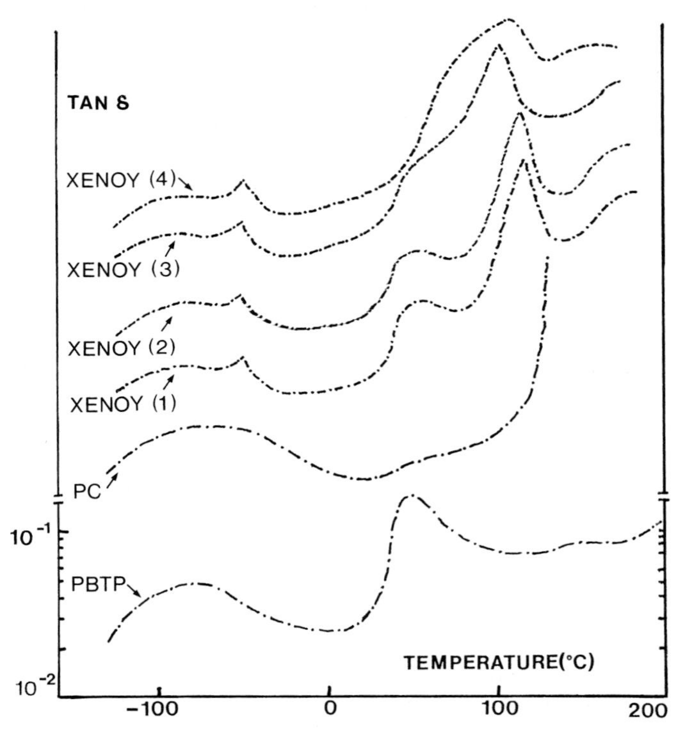

Figure 2.5 50/50 blend of PC and PBTP: dynamic loss curves after various heat treatments – (1) 3 minutes at 240 °C; (2) 3 minutes at 242 °C; (3) 10 minutes at 270 °C; (4) 30 minutes at 270 °C;

Blends of polycarbonates with saturated aromatic polyesters are of commercial significance, for example as marketed under the trade name "Xenoy"® by the General Electric Company of America. In such blends, the polycarbonate does not crystallize from the melt, but the polyesters, usually PETP or PBTP, are normally crystalline. The PETP-PC blend behaves differently to the PBPT-PC blend, the former being apparently immiscible in the amorphous regions, manifested by the transition temperatures of both components being observed. The latter blend seems to become miscible on vigorous mixing, possibly as a result of some *ester-interchange* occurring. In both cases, the terephthalate esters crystallize to a limited extent, but the melting temperatures are decreased by blending. Transition temperatures for PC, PBTP and the 50/50 PC/PBTP blend are given in Figure 2.5, and are seen to be separate initially, but to merge into a single transition at about 100 °C with heat treatment, which also results in a drastic deterioration in melting temperature[2.9].

2.3.4 Effect of Nucleating Agents on Crystallization Temperature

The use of nucleating agents with crystallizing polymers is increasing; their purpose and effect is to increase the temperature of crystallization by providing a very large number of stable nuclei available at high temperature. A further inevitable consequence of their inclusion is the development of very fine and controllable texture, which might be advantageous in respect of both mechanical properties and transparency. An example of *heterogeneous nucleation* is included in Table 2.1.

Table 2.2 Melting (Tm) and Crystallization Temperatures (Tc) of HDPE-LDPE Blends

Each sample was subjected to the following thermal treatment:

(1) Heat at 10 K/min to 200 °C;
(2) Cool immediately to 25 °C at 20 K/min ("fast") or 1 K/min ("slow");
(3) Reheat to 150 °C at 10 K/min, (second heating).

Composition	First melting	Fast cooled		Slow cooled	
LDPE/HDPE (%)	Tm (°C)	Tc (°C)	Tm (°C)	Tc (°C)	Tm (°C)
100/0	112.00	92.00	112.00	101.00	113.00
70/30	110.75	94.75	110.75	103.50	111.50
	127.00	109.75	127.00	118.25	130.25
50/50	112.25	96.00	112.25	104.50	112.50
	128.50	111.25	128.50	118.75	132.25
30/70	130.00	111.50	130.00	119.50	109.00
					133.50
0/100	133.25	111.75	133.25	120.00	133.50

Table 2.3 Melting (Tm) and Crystallization Temperatures (Tc) of HDPE-LLDPE Blends

Composition	First melting	Fast cooled		Slow cooled	
LLDPE/HDPE (%)	Tm (°C)	Tc (°C)	Tm (°C)	Tc (°C)	Tm (°C)
100/0	107.25	100.25	107.25	102.00	108.50
	119.75		119.75		
	124.50		124.50	111.75	124.75
70/30	127.25	109.00	127.25	117.75	129.50
50/50	128.50	110.50	128.50	119.25	131.25
30/70	130.25	111.25	130.25	119.75	133.75
0/100	133.25	111.75	133.25	120.00	133.50

(The results in Tables 2.2 and 2.3 are taken from Reference 1.22)

2.3.5 Effect of Cooling Rate on Crystallizing and Fusion Temperatures

Thermal Analysis data for a number of polyolefins have been given in Tables 2.1, 2.2 and 2.3; they show that the rate of cooling has a marked effect on the amount of *supercooling*, that is, on the *crystallization temperature*, with rather less effect on the subsequent melting temperature, a lower rate of cooling being associated with a higher melting point, although the effect is small.

2.3.6 Effect of Pressure on Melting and Crystallization Temperatures

As we have seen, the crystalline state usually has a higher density than amorphous polymer; thus, from the Clapeyron-Clausius equation, we might expect both the melting and crystallization temperatures to increase with pressure:

$$(T_m)_p = (T_m)_{p=1} \exp\left[(V_a - V_c)\frac{p-1}{(H_f)_{p=1}}\right] \tag{2.1}$$

where V_a and V_c are the specific volumes of the amorphous and crystalline phases, H_f is the heat of fusion, and p is the hydrostatic pressure (in bars).

Van der Wegt and Smit[2.10] gave an early account of the unexpected appearance of crystallinity in polypropylene at temperatures above the normal fusion temperature in a pressurized, flowing system, implicating both pressure and orientation.

2.3.7 Orientation and Crystallinity

It is a feature of polymer chains that they may become aligned during processing, either intentionally or accidentally; such alignment is known as *molecular orientation,* and affects many properties, making them dependent on direction. The polymer chains may be aligned in one direction only, *uniaxial fibre-type orientation,* or in a plane, *planar* or *biaxial orientation.* There is frequently confusion between orientation and crystallinity, although in the latter the chains are locally aligned and in crystalline register, they do not have necessarily any overall directionality, unless, of course, the crystallites themselves are oriented. Orientation does, however, have very large effects on both crystallite morphology and rates of crystallization.

Strain-induced crystallization was first reported for natural rubber some sixty years ago, the material crystallizing at room temperature when oriented, especially uniaxially. More recently, studies on crystallizing plastics have revealed that *nucleation* is facilitated by orientation, the nuclei being formed along the lines of flow, hence *row nucleation.* Growth then occurs on these nuclei, at a rate enhanced by a factor of a thousand by the orientation. By applying a standard thermodynamic approach:

$$\triangle G = \triangle H - T\triangle S$$

where G is the Gibbs free energy, H the enthalpy and S is the conformational entropy; the "$\triangle$" designations represent the change in these variables during the phase change. For an equilibrium process, such as fusion at the melting temperature, $\triangle$G = 0, and we have:

$$T_m = \frac{\triangle H}{\triangle S}$$

Chain alignment will decrease $\triangle$S significantly, but will have no effect on the enthalpy; we must expect, therefore, the melting point to increase with orientation, and at any temperature, the driving force to crystallization to increase, as supercooling increases.

Orientation is promoted by flow, that is, by shear, or by an elongational stress. For a thermoplastics melt, there is a relationship between shear rate and nucleation: at very low shear rates the chains have time to relax before they are drawn into stable nuclei, but at higher shear rates the induction time for crystal nucleation is progressively reduced.

2.4 Thermal Conductivity

It is generally appreciated that certain applications require high thermal conductivity, whilst others need insulation. Consider the example of a domestic iron; high conduction of heat from the heater element to the article being smoothed is important, whilst the user must be insulated from the heat source, so that the handle is an insulator, frequently a plastics material.

The *coefficient of thermal conductivity,* more usually known as *thermal conductivity,* is defined, for steady state conditions, by Fourier's law of heat conduction. For the flow of heat perpendicular to the faces of a flat slab:

$$q = K\,A\,\frac{\triangle T}{t} \tag{2.2}$$

where q is the rate of flow of heat (W or J/s), A is the surface area across which the heat flows (m^2), $\triangle T$ is the temperature difference between the two faces (K), t is the thickness of the slab (m) and K is the thermal conductivity (W/mK).

It may be observed that this expression is formally identical with that which describes mass transport through a permeable barrier (see also Chapter 10; Section 10.1.2.1).

Measurements of thermal conductivity are made by classical steady-state methods, but for plastics, care has to be taken over the experimental detail, since all materials are poor thermal conductors. For accurate data, it is usual to work with a 5K difference in temperature between the hot and cold faces.

The fact that steady-state methods are employed precludes making the measurements relevant to cooling, which are important in processing. There are two solutions to the problem: extrapolation of melt data to lower temperatures, or the measurement of transient thermal behaviour by determining *thermal diffusivity* directly.

2.4.1 Thermal Diffusivity

Thermal conductivity implies a system at equilibrium, but in many practical situations heat transfer is a transient phenomenon, as, for example, in the cooling of a moulding. In such cases both conductivity and enthalpy factors are involved: *thermal diffusivity (α)* is defined as the ratio of thermal conductivity to the heat capacity per unit volume:

$$\alpha = \frac{K}{\varrho \cdot C_P} \tag{2.3}$$

where C_P is specific heat (the heat capacity per unit mass, J/kgK), ϱ is the density at the appropriate temperature (kg/m^3) and α is the thermal diffusivity, expressed in m^2/s.

There are many ways of applying thermal diffusivity data to the solution of heat transfer problems; when no phase change occurs, it is sufficient to make use of relevant functions and average values for the thermal properties involved[2.11]. For heat transfer calculations involving phase changes, numerical methods may be used; such a technique has been developed by EDWARDS et al[2.12,2.13], and represents a good example of heat transfer theory applied to a commercial plastics processing operation. Both types of solution are illustrated in case studies, but first the effects of materials' variables on thermal conductivity and thermal diffusivity must be examined.

2.4.2 Effect of Crystallinity on Thermal Conductivity

The regular arrangement of a crystalline polymer leads to a higher thermal conductivity, in comparison to amorphous plastics. This is exemplified by considering HDPE and LDPE, with crystallinity levels of 80 % and 55 %, the thermal conductivities being 0.50 and 0.33 W/mK respectively. A consequence of the higher conductivity of crystalline polymers is that the conductivity decreases with increasing temperature, particularly through the melting region; this is illustrated for propylene-ethylene sequential copolymer in Figure 2.6[2.13].

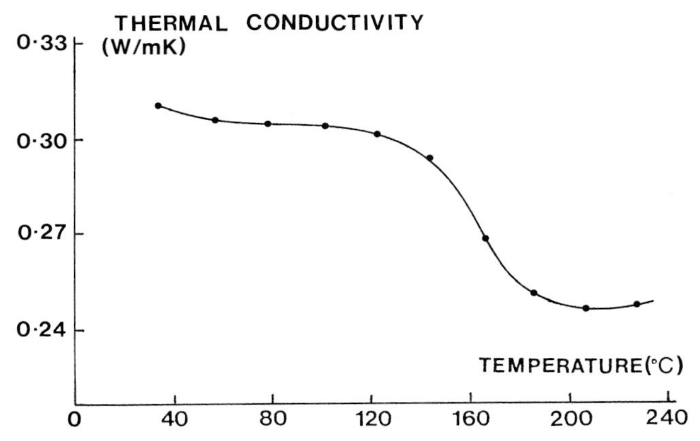

Figure 2.6 Temperature dependence of thermal conductivity for a partially crystalline polymer: behaviour of PP through the melting region.

The dependence of thermal conductivity on crystallinity may also be significant in the cooling of plastics products: it may be counter-productive to use a very low mould temperature in blow moulding a crystallizing polymer, since the quenched polymer adjacent to the mould surface may prove to be a thermal barrier, whilst a layer of more crystalline material, albeit with the mould at a higher temperature, may allow more heat to be transferred. Model experiments, the results of which are shown in Figure 2.7[1.26] have not disproved these ideas.

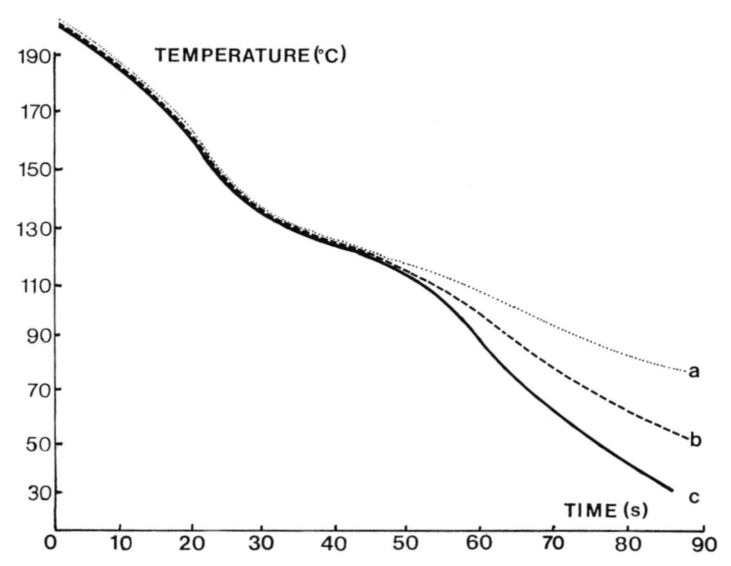

Figure 2.7 Cooling rate (as measured by temperature at the inner surface of a blow moulding) as a function of mould temperature: (a) 75 °C; (b) 40 °C; (c) 20 °C.

2.4.3 Effect of Fillers on Thermal Conductivity

Generally the thermal conductivity of a filled polymer is dependent on the characteristics of the filler, addition of a more conductive filler increasing the conductivity. Various formulae have been advanced to quantify this behaviour[2.14]: the rule of additivity is the simplest.

A particularly important "filler" is air, or another inert gas such as nitrogen; these are introduced into plastics to produce *foams* (*cellular polymers*); low-density plastics foams have thermal conductivities at least an order of magnitude lower than those of the solid polymers.

2.4.4 Effect of Orientation on Thermal Conductivity

Having noted the effect of crystallinity, it might be anticipated that the response of oriented polymers might be different in the direction of orientation, (transmission along primary bonds), compared with the transverse direction, where only secondary bonds link the chains together. This is found to be so, and the combined effects of orientation and crystallinity lead to very large differences being observed. This again has implications in processing technology, especially in injection moulding, for which it may be desirable to generate less orientation, and allow more efficient and/or consistent heat transfer.

2.4.5 Effect of Pressure on Thermal Conductivity

There are very few data available concerning the effect of pressure on conductivity; results indicate marginal effects only, some 1 % increase in conductivity for a 75 bars increase in pressure. If the pressure results in a change of phase, somewhat larger effects might be anticipated.

2.4.6 Heating a Plastics Product

An example of the application of thermal diffusivity in calculations involving transient heat transfer may be found in the manufacturing process for uniaxially oriented polypropylene tapes, which are being used increasingly, replacing metal, as strapping. The process is, briefly, the extrusion of tape into a quench bath, followed by reheating to a specified temperature, and subsequently *drawing* to orient the tape. The reheating stage is carried out by passing the tape through a liquid bath at high temperature for which we are interested in the minimum residence time:

Assume a drawing temperature of 95 °C, and that the tape is initially at 20 °C. The relevant thermal properties of PP in the temperature range 20 – 95 °C are:

Thermal conductivity,	K	= 0.31 W/mK
Specific heat,	C_p	= 2.5 kJ/kgK
Density,	ϱ	= 900 kg/m^3
whence Thermal diffusivity,	α	= 1.38×10^{-7} m^2/s

The Fourier Number (Fo) is a function of thermal diffusivity, time and the thickness of material ($2x$):

$$\text{Fourier Number} \quad (Fo) = \frac{\alpha t}{x^2} \tag{2.4}$$

For PP, with tape thicknesses 1.5, 2.0 and 3.0 mm, and a contact time of 5 seconds, the corresponding Fourier Numbers are 1.23, 0.69 and 0.31. Standard texts on heat transfer provide graphical solutions for the relationship between Fourier Number and temperature gradient for selected geometric forms (these procedures are described more completely, and discussed in the context of injection moulding, in Chapter 5; Section 5.2.4.2). Equating the tape to a flat sheet, the temperatures at the centre of the tape are 91.25, 80.0 and 50.0 °C, respectively.

An attempt to speed the process by limiting the residence time in the bath to 4 seconds will limit the centre temperature to 87.5 °C.

2.4.7 Cooling a Blow Moulding of Appreciable Thickness

A wide variety of containers are made by blow moulding, some of which are required to withstand pressure in service; for example, car coolant reservoir tanks. They have to operate at high temperatures and are, therefore, of considerable thickness. The conventional cooling of such blow mouldings is very time-consuming, as the heat is extracted via the surface of the mould, which is maintained chilled, typically at 10 °C.

The inside surface is cooled only by *natural convection* of the limited amount of air used for blowing. It will be shown that a considerable increase in cooling rate can be achieved by circulating air inside the moulding, and a further improvement effected by employing a low temperature coolant, e.g. liquid carbon dioxide.

A theoretical model of cooling and solidification for blow moulding is based on one-dimensional heat flow and a knowledge of the relevant thermal properties: conductivity, specific heat capacity and density, or directly, of thermal diffusivity. The temperature distribution through the thickness is calculated as a function of cooling time for various starting temperatures, cooling times and wall thicknesses[2.12]. The calculation involves the *cooling mode,* and it is vital that the data be appropriate to this mode; supercooling must be taken into account. It is similarly important that the rate of cooling employed should be realistic in respect of practical blow moulding.

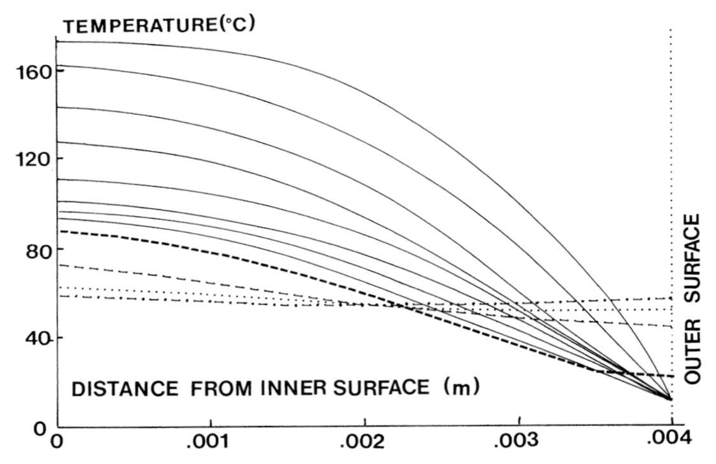

Figure 2.8 Natural convection cooling curve for a PP container of wall thickness 4 mm. The time interval between curves is 10 seconds; broken lines indicate behaviour when the mould is opened.

Results obtained for the model for "classical" blow moulding, for which the only cooling of the inside surface is by natural convection, are shown in Figure 2.8, for a wall thickness of 4 mm. It can be seen that the inner surface cools more slowly than the remainder of the moulding; for this case the cooling curve is simply the temperature versus time relationship for the inner surface. Predictions for a 4 mm thick moulding, with the interior cooled by static or circulating air are given in Figure 2.9. Here, the slowest cooling occurs within the moulding, near to the inner surface. Comparing these two sets of data, and assuming that the moulding can be ejected when its temperature reaches 90 °C, cooling times are 100 seconds for the moulding cooled by natural convection, and 60 seconds for that cooled by internally circulated air.

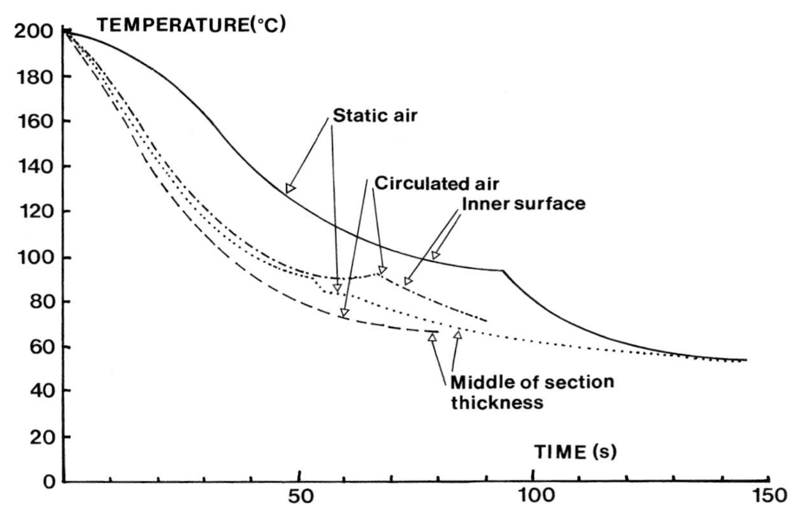

Figure 2.9 Improvement in cooling rate by circulating air.

An estimate may be made of the amount of air required for interior cooling, making the following assumptions:

(1) half the cooling is via the inner surface;
(2) the latent heat of crystallization is separable from the total enthalpy change, and is evolved at 90 °C;
(3) the remaining heat energy is evolved at an average temperature of 145 °C, midway between the initial parison temperature and the final temperature;
(4) the mass of the moulding is 500 g.

Employing the known density and specific heat capacity of air, and basing the calculation on a cooling time of 100 seconds and a blowing pressure of 6 bars, the volume flow rate required is 1.1 litre/seconds, which is well within the capacity of a normal compressor.

2.4.8 Liquid Carbon Dioxide as an Internal Coolant

The use of circulating air as an internal coolant is effective for two reasons: there is sufficient capacity to carry away the heat, but more important, the heat transfer coefficient is increased by a factor of about ten by causing the air to move. It has been found empirically that a similar further increase in heat transfer coefficient results when liquid carbon dioxide is injected into the moulding. This, together with the much lower temperature of the fluid, compared with air at ambient temperature, results in a much faster cooling rate overall. This has been investigated both with theoretical modelling, and also in practice[2.15].

Cooling times in the mould to reach a temperature of 90 °C in the hottest part, with an initial parison temperature of 200 °C, are given in Table 2.4.

At each thickness level, the possible reductions in cooling time are shown to be highly significant. The further improvement in cooling rate obtained by using

Table 2.4 Cooling Data for Polypropylene Blow Moulding

Internal Coolant	Part Thickness (mm)	Cooling Time (s)
Natural convection	4	90
	6	200
Circulating air	4	60
	6	100
Liquid carbon dioxide	4	30
	6	60

liquid carbon dioxide in place of circulating air is compromised by the cost of that commodity. The effectiveness of liquid CO_2 depends on the storage conditions and it has been further found that *pulsing* the ingress improves the economy of usage, so that a liquid carbon dioxide usage of 0.19 kg per 1 kg of HDPE can be achieved[2.15]. On the basis that similar data are applicable to polypropylene, and that only half the cooling is by the inner surface, the limiting use of liquid carbon dioxide is 0.1 kg per kg of polymer.

2.5 Thermal Expansion and Contraction

When heat energy is applied to a material, two phenomena are generally observed: there is an increase in temperature and the material usually expands, which may be seen in length, area or volume. Orientation in the sample, particularly for crystalline or reinforced polymers, results in *anisotropic expansion* (or contraction). It is then usual to quote data for the direction or plane of orientation, and perpendicular to this.

The relationship between expansion and temperature is best presented graphically, following BS 4618[2.16], although the convenience of giving the coefficient is sometimes adopted, in spite of the inaccuracies, such as when comparing families of plastics with metals, Table 2.5:

Table 2.5 Coefficients of Linear Thermal Expansion[2.17]

	Material	Coefficient of linear expansion at 20 °C(m/mK)
Metals	Mild steel	11×10^{-6}
	Aluminium	25
Plastics; amorphous	PS	45
	SAN	50
	PMMA	50
	PC	60
	HIPS	70
	PVC (Rigid)	70
	PVC (Plasticised)	200
Plastics; semi-crystalline	POM	80
	PETP	80
	PP	110
	HDPE	150
	LDPE	200
Plastics; crosslinked	PF (Cellulose-filled)	30
	UF (Glass-reinforced)	20

2.5.1 Factors Affecting Thermal Expansion: Crystallinity

Curves of thermal expansion versus temperature have been given in Figure 2.4; the significant effect of crystallinity has been noted. The level of crystallinity is thus an important variable affecting expansion and contraction. Indeed, contraction is probably more important than expansion in the processing of thermoplastics, since it is the principal cause of *shrinkage* which is encountered in plastics products in a number of guises. From Figure 2.4, the shrinkage of polypropylene from the processing temperature to ambient conditions is some 22 % by volume; some of this is eliminated by *packing* the mould after filling, but the compressibility of polymer melts is limited, and contraction leads to a shrunken surface *(sink marks)*, or to voids in the moulding interior, since this is the last material to solidify. Contraction voids are illustrated in Figure 2.10.

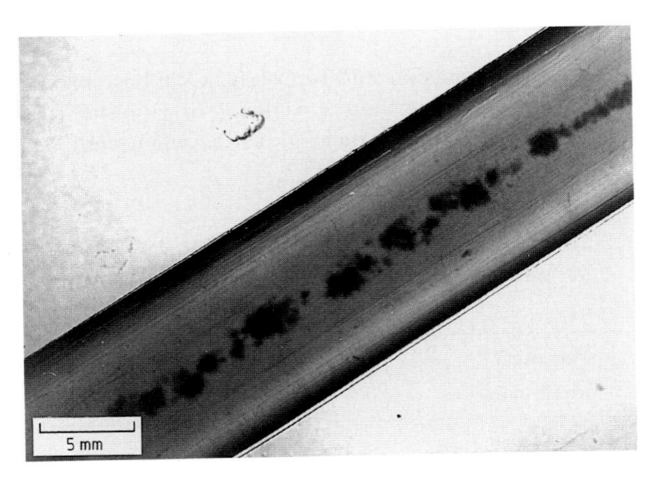

Figure 2.10 Contraction voids.

Expansion and contraction are similar to conductivity in being dependent on crystallinity, with higher crystallinity being associated with lower expansion. However, most crystalline plastics have higher expansions than glassy amorphous materials, since the amorphous regions in crystalline plastics are usually in the *rubbery state*, and these are more sensitive to temperature than glassy amorphous plastics (see Table 2.5).

2.5.2 Factors Affecting Thermal Expansion: Cross-linking

Whilst rubbers and crystalline thermoplastics generally show high expansion, at the other end of the plastics materials range cross-linked plastics have lower expansion coefficients, and these are reduced still further by the inevitable fillers employed in such systems. Illustrative data are given in Table 2.5.

2.5.3 Factors Affecting Thermal Expansion: Fillers

The high thermal expansion and contraction of plastics is reduced by the addition of fillers; particulate and fibrous fillers give significant improvements, but for products involving the latter, attention must be given to the effects of *fibre orientation*, which introduces anisotropy into the expansion and contraction behaviour. The following thermal expansion data were obtained on PA66 containing 30 % glass fibre by weight:

Orientation direction,	30×10^{-6} m/mK
Transverse direction,	70×10^{-6}

2.5.4 Factors Affecting Thermal Expansion: Orientation

Working from the premise that expansion is restricted by chemical bonds, which may be inferred from the low values for cross-linked plastics, oriented polymers would be expected to show anisotropy in their expansion behaviour, this being lower in the orientation direction. This is indeed found to be generally true; similar considerations apply to oriented crystalline plastics, and to the shrinkage observed on crystallization.

References

2.1 BS 2782, Method 120A, (1976, confirmed 1983), *Vicat Softening Temperature of Plastics,* (Corresponding International Standard ISO R306), British Standards Institution (BSI), London.

2.2 BS 2782, Method 121A, (1976, confirmed 1983), *Temperature of Deflection under a Bending Stress of 1.8 MPa,* (Corresponding International Standard ISO R75), BSI, London.

2.3 BS 2782, Method 122A, (1976, confirmed 1983), *Deformation under Heat of Flexible PVC Compounds,* BSI, London.

2.4 Spaght, M.E., Thomas, S.B., and Parks, G.S.: *Some Heat Capacity Data on Organic Compounds,* J. Phys. Chem., **36,** (1932), 882.

2.5 Streamer, R.: *Factors Affecting the Fusion and Recrystallisation Behaviour of Polypropylene,* Internal Project Report, Loughborough University of Technology, (1984).

2.6 Roberts, G.E., and White, E.F.T.: *Relaxation Processes in Amorphous Polymers,* in *The Physics of Glassy Polymers,* Haward, R.N., Ed., Applied Science, London, 1973.

2.7 Kovacs, A.J.: *La Contraction Isotherme du Volume des Polymeres Amorphes,* J. Polym. Sci., **30,** (1958), 131.

2.8 Williams, M.L., Landel, R.F., and Ferry, J.D.: *The Temperature Dependence of Relaxation Mechanisms in Amorphous Polymers and other Glass-Forming Liquids,* J. Amer. Chem. Soc., **77,** (1955), 3701.

2.9 Birley, A.W., and Chen, X.Y.: *Studies of Polycarbonate – Poly (Butylene Terephthalate) Blends,* Brit. Polym. J., **16,** (1984), 77.

2.10 Van der Wegt, A.K., and Smit, P.P.A.: *Crystallization Phenomena in Flowing Polymers,* SCI Monograph No 26, (1967), 313.

2.11 CRAWFORD, R.J.: *Plastics Engineering, 2nd Ed.,* Pergamon Press, Oxford, (1987).

2.12 EDWARDS, M.F., GEORGHIADES, S., and SUVANAPHEN, P.K.: *A Study of the Cooling of Blow Moulded Objects,* Plast. Rubb. Proc. Appl., **1,** (1981), 161.

2.13 BIRLEY, A.W., YEH, P-L., BARNES, C., EDWARDS, M.F., and SMITH, A.: *The Cooling of Polypropylene Blow Mouldings,* Plast. Rubb. Proc. Appl., **6,** (1986), 337.

2.14 ORR, C.: *Particulate Technology,* MacMillan, New York, (1966).

2.15 EDWARDS, M.F., ELLIS, D.I. and GEORGHIADES, S.: *Liquid Carbon Dioxide as an Internal Coolant in Blow Moulding,* Plast. Rubb. Proc. Appl., **5,** (1985), 143.

2.16 BS 4618, Part 3.1 (1970): *Presentation of Data on Thermal Expansion versus Temperature,* BSI, London.

2.17 Data Book for O.U. Course PT 614, *Polymer Engineering,* The Open University, Milton Keynes, (1984).

Chapter 3
Melt Flow Properties

3.1 Introduction

Plastics processing is primarily concerned with forming the raw compound into the required shape and fixing the shape by cooling or by chemical reaction. The shaping operation will involve the flow of the polymer in melt form at some time during the process. A study of the flow behaviour of the material in the melt-state is essential if we are to achieve a complete understanding of processing operations, and be able continually to improve the efficiency and accuracy of product manufacture.

Rheology is the science of the flow and deformation of real materials. In this chapter we give a brief introduction to the rheological properties of plastic melts and how they may be quantified. The application of rheological properties to design of processing operations is considered later. It should be appreciated that there are a number of non-rheological factors which also influence the shaping operation including:

- Polymer type and molecular characteristics;
- Heat transfer mechanism;
- Process control features;
- Any mixing of the material during processing;
- Mechanical design of the processing equipment.

3.2 Fundamental Concepts of Rheology

Rheology is concerned with the relationships between *stress* (force/unit area), *strain* (change of shape) and *time.* The flow of plastics in actual processes is a very complex mechanism, involving completely three dimensional flow under *non isothermal* and *non steady state* conditions. In many cases it is reasonable to approximate the behaviour by modelling the flow as a series of simplified idealized viscometric flows. The general objectives of rheology are to:

- Understand the behaviour of the plastic relevant to the conversion process;
- Quantify the response of the material;
- Use the data to predict the performance of the polymer in actual industrial processes;
- Relate the molecular structure to the rheological response of the polymer.

These are very optimistic objectives but significant advances have been made in recent years[3.1]. There is little doubt that more advanced rheological studies have led to better designed processes and have enabled the polymer chemist to "tailor make" polymers to match particular aspects of conversion processes.

3.3 Geometry of Flow

There are three types of deformation which we generally need to consider: simple shear, elongation and bulk compression[3.2]

3.3.1 Simple Shear

In shear deformation a force (F) is applied tangentially to an elemental volume as shown in Figure 3.1, so that the top layer is displaced by u. The relative displacement of the two layers is the simple *shear strain* (γ):

$$\gamma = \frac{u}{h} \tag{3.1}$$

The *shear stress* (force per unit area, τ) is given by:

$$\tau = \frac{F}{A} \tag{3.2}$$

For viscous flow it is more usual to consider the rate of strain, or *shear rate,* which is given by:

$$\dot{\gamma} = \frac{v}{h} \quad \text{or} \quad \frac{dv}{dx} \tag{3.3}$$

where v is the velocity in the z direction.

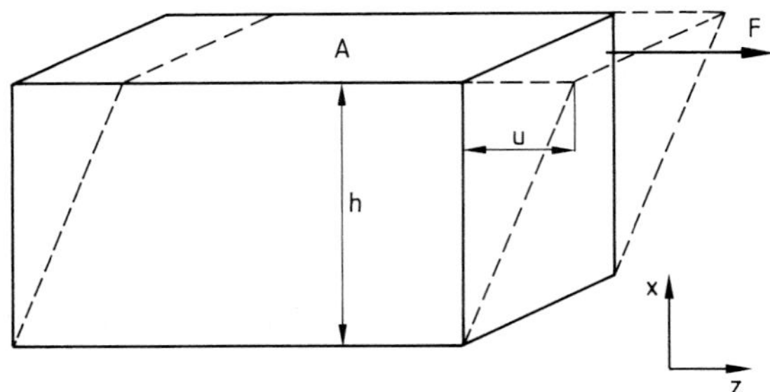

Figure 3.1 Geometry of shear flow.

3.3.2 Elongational Flow

This is achieved by applying a force normal to the opposite faces of the element, as indicated in Figure 3.2. The *tensile stress* (σ) and the conventional *engineering strain* (ε) are given by:

$$\sigma = \frac{F}{A} \tag{3.4}$$

$$\varepsilon = \frac{l - l_0}{l_0} \tag{3.5}$$

where l_0 is the original length of the sample and l is the length at time t. Similarly, we can define an *elongational strain rate* ($\dot{\varepsilon}$), which is given by:

$$\dot{\varepsilon} = \frac{dv}{dz} \tag{3.6}$$

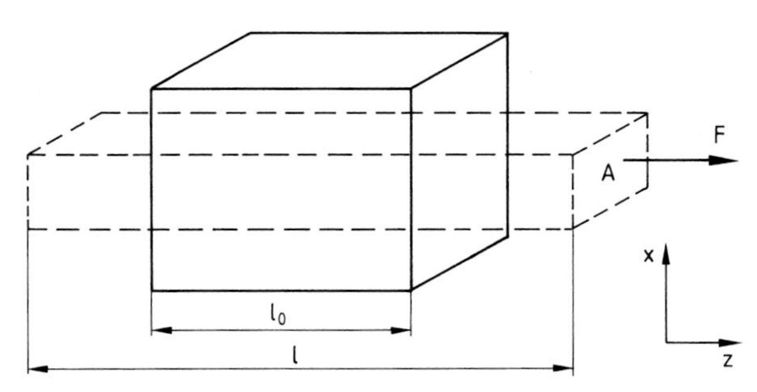

Figure 3.2 Geometry of elongational flow.

3.3.3 Bulk Deformation

A uniform pressure is applied to all faces and the strain is determined from the change of volume of the element. This type of deformation is of less significance in polymer processing and it is usual to assume that polymer melts are incompressible. However, it may be important in processes where high, near-hydrostatic pressures are developed, such as the packing phase of injection moulding.

In practice, most processes can be classified as either simple shear or elongational, or a combination of both. Simple shear flow is a good approximation for many melt processing operations such as extrusion or injection moulding. Elongational flow is important in fibre spinning, film extrusion processes, in various phases of blow moulding, and in flows through convergent sections.

3.4 Rheological Behaviour in Simple Shear

Polymer melts are *viscoelastic* in their response to an applied stress. Thus under certain conditions they will behave like a liquid, and will continually deform while the stress is applied. Under other conditions the material behaves like an elastic solid, and on the removal of the applied stress there will be some recovery of the deformation. This viscoelastic response is illustrated in Figure 3.3.

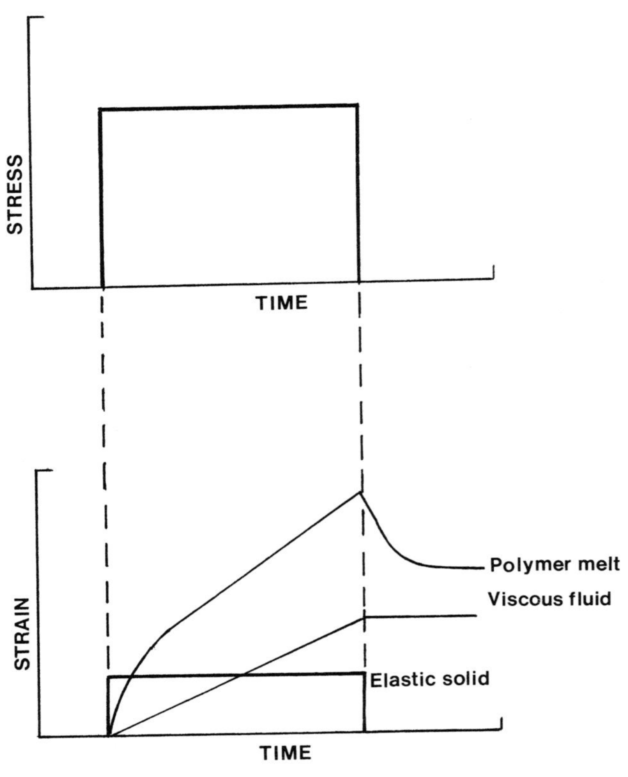

Figure 3.3 Viscoelastic behaviour of polymer melts.

Conversely, if the strain is held constant at the end of an experiment, the stress will not immediately return to zero, but will relax with time. Thus to characterize the rheological behaviour of polymer melts completely it is necessary to measure both the elastic and viscous response to the applied stresses. The elastic properties are important in many actual processing operations, for example when the melt flows out of an extrusion die, or through abrupt changes in cross section. Elastic properties will be discussed later and we will restrict the discussion to viscous behaviour at this stage.

3.4.1 Viscous Behaviour in Simple Shear

It is necessary to define *viscosity*, which is a fundamental rheological parameter of fluids:

$$\text{Viscosity}\ (\eta) = \frac{\text{shear stress}}{\text{shear rate}} = \frac{\tau}{\dot{\gamma}} \tag{3.7}$$

The SI units of viscosity are Nsm^{-2}.

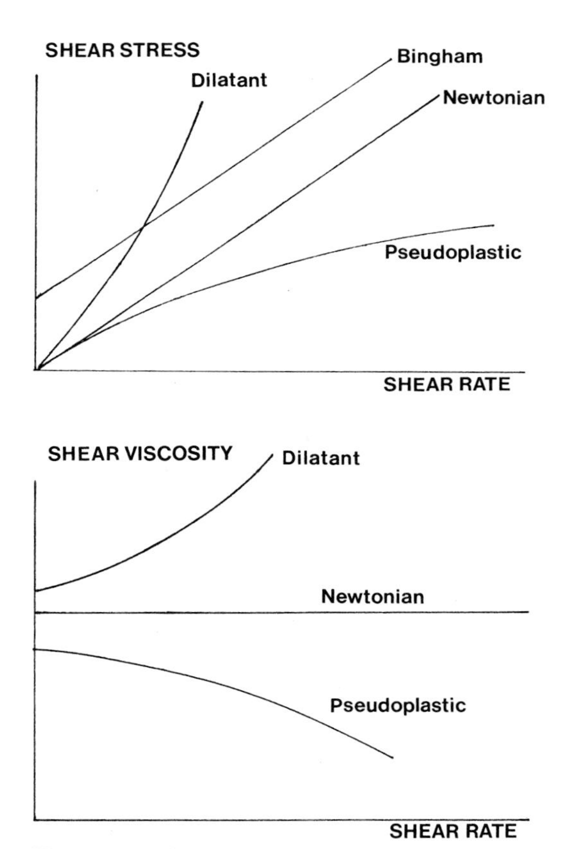

Figure 3.4 Flow curves of idealized fluids.

Newtonian fluid. This is the simplest model which describes the behaviour of real fluids. They are characterized by a constant viscosity, which is independent of shear rate. This model is an accurate description of many low molecular weight fluids such as water, benzene and most organic liquids.

Bingham body. Materials of this class do not deform below a certain yield stress. Above the yield stress, the model behaves as a simple Newtonian fluid. Plasticine and some clay slurries are examples of materials which can be described as Bingham bodies. Some highly filled polymers may also exhibit a yield stress before flow occurs[3.3].

Dilatancy. For this class of fluids viscosity increases as shear rate increases. It is found to occur in certain PVC plastisols over a limited range of shear rates[3.4].

Pseudoplastic. For these materials, viscosity decreases as shear rate increases. This is the most common response for polymer melts and we cannot specify the viscosity of the material without specifying the shear rate.

The flow curves of these idealized materials are shown in Figure 3.4, but they do not take account of time effects. There is a certain class of materials where the viscosity will change during an experiment even at constant shear rate. These effects are not due to elasticity but are caused by structural changes which can occur throughout the duration of the experiment. The structural changes are reversible on resting the fluid and there is no elastic recovery when the stress is removed. A reduction in viscosity during an experiment is known as *thixotropy* and is exhibited by some filled polymer systems and paints. A time-dependent increase in viscosity is referred to as *rheopexy,* or negative thixotropy.

3.5 Viscous Properties of Plastic Melts in Simple Shear

3.5.1 General Characteristics

The flow curves of plastic melts have a number of typical characteristics which are common and there are few exceptions to this general behaviour, which can be summarized as follows:

(A) The materials have high viscosities, as is demonstrated in Table 3.1:

Table 3.1 Typical values for Viscosities (Nsm^{-2}) of Materials

Water	10^{-3}	Polymer melts	10^2 to 10^7
Olive oil	10^{-1}	Pitch	10^9
Glycerine	10^0	Glass	10^{21}
Golden syrup	10^2		

(B) In general, viscosity decreases with increasing shear rate, i.e. polymers are *pseudoplastic*. The actual form of the *flow curve* will depend on the type of polymer and its molecular characteristics such as molecular weight, molecular weight distribution and degree of branching. There are exceptions however, and some polymer systems exhibit *dilatant* behaviour under certain circumstances[3.4]. However, the large majority of plastic melts behave in a pseudoplastic manner.

(C) The viscosity tends to a constant value as shear rate tends towards zero. This constant value is often referred to as the "Newtonian viscosity", and is denoted by μ or by η_0 .

(D) There is some evidence that viscosity also tends to a constant value at very high shear rates, but this is much less well established.

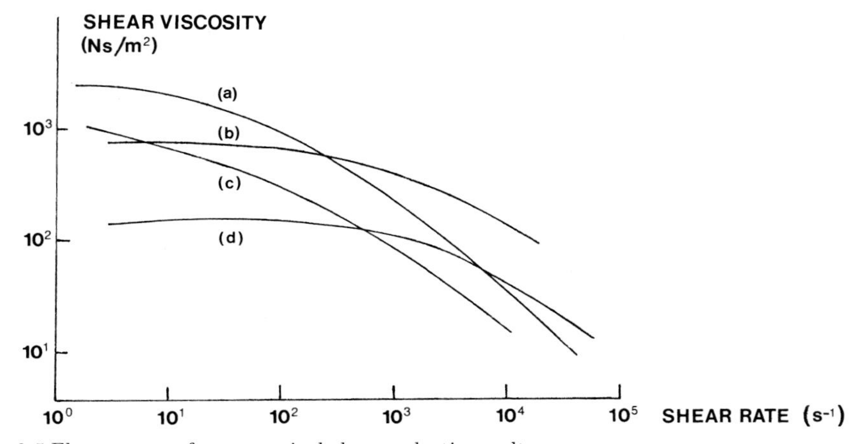

Figure 3.5 Flow curves of some typical thermoplastics melts:
(a) Poly(methyl methacrylate) at 240 °C; (b) Polyethersulphone at 350 °C;
(c) LDPE (MFI 20 g/10 min) at 170 °C; (d) Polyamide 66 at 285 °C.

Typical flow curves for a number of plastics melts are given in Figure 3.5.

This graph immediately indicates the pitfalls associated with single point measurements and the difficulty of specifying a single polymer melt 'viscosity'. There is a large variation in the shear rates which exist in manufacturing processes, as shown in Table 3.2. It is essential to relate the rheological properties to the shear rates which exist in practice, particularly if the data are to be used for purposes of process design.

Table 3.2 Typical Shear Rates (s^{-1}) for Polymer Processing

Compression moulding	1	to	10
Calendering	10	to	10^3
Extrusion	10^3	to	10^3
Injection moulding	10^3	to	10^5

It must be emphasized that the values quoted in Table 3.2 are only approximate, but they do give some indication of the shear rates which occur in particular manufacturing processes.

3.5.2 Empirical Representation of Experimental Data

It is convenient to have some mathematical models to describe viscosity data. These are usually empirical models and there have been numerous forms proposed for representing polymer melt behaviour[3.2,3.5]. The purpose of such models is to simplify the analysis of real flow problems and provide a convenient basis for comparing materials. The most commonly used equations for describing polymer melt behaviour are described below:

Power law model. In practice we find that a plot of log τ versus log $\dot{\gamma}$ is linear over a limited range of shear rates, which suggests a model of the form:

$$\tau = k \cdot (\dot{\gamma})^n \tag{3.8}$$

where k and n are material parameters. k is the *consistency index* and is a measure of the fluidity of the material; a high value of k indicates a very viscous material.

The *power law index* (n) is a measure of the non Newtonian behaviour of the fluid. For a Newtonian fluid, $n = 1$ and k is the viscosity. There are a number of comments about the power law model which should be noted:

- It is only valid over a limited range of shear rates, as we have seen in Figure 3.5. This may not be a serious limitation in practice, since different values for n and k can be used, depending on the conditions under investigation.
- The model predicts an infinite viscosity at low shear rates, which is clearly not true. Thus it does not predict the Newtonian region where the viscosity is constant, which could present problems in predicting flow behaviour in the low shear rate region.
- Only two parameters are needed to fit the model; these can be readily obtained experimentally.

– The model simplifies the presentation of material properties and makes it mathematically easier to analyse many flow problems.

Polynomial model. This represents a modification of the power law, to allow for some non-linearity in the flow curve:

$$\log \tau = A_0 + A_1 \cdot \log(\dot{\gamma}) + A_2.(\log \dot{\gamma})^2 \tag{3.9}$$

Thus there are three material parameters: coefficients A_0, A_1 and A_2. The model is valid over a wide range of shear rates and the constants in the model can be evaluated from experimental data by regression analysis. Extrapolation outside the range of experimental shear rates is unreliable and can give nonsensical results. This model has been used successfully by BOWERS[3.6] for the analysis of rubber injection moulding. It is possible to extend the model to include higher terms in the polynomial, but this is very rarely carried out in practice.

Ellis equation. This model has been developed to allow for a constant viscosity region at low shear rates:

$$\frac{\eta_0}{\eta} = 1 + \left(\frac{\tau}{\tau_*}\right)^{\alpha-1} \tag{3.10}$$

Thus there are three constants: η_0, τ_* and α. The model has some of the features of the power law, since it predicts that the log τ versus log $\dot{\gamma}$ is linear at high shear rates, and therefore α is related to n. Although there are only three constants, and the model generally satisfies some of the objections concerning the power law, it has not found favour with polymer rheologists because it is not easy to use in analyzing practical flow problems.

Carreau model. This is a development of the various integral rheological equations of state which have been proposed for modelling polymer melts[3.5]:

$$\frac{\eta - \eta_\infty}{\eta_0 - \eta_\infty} = [1 + (\lambda\dot{\gamma})^2]^{(n-1)/2} \tag{3.11}$$

where η_0, η_∞, λ and n are constants.

The main attraction of Equation 3.11 is that it predicts a constant viscosity at both high and low shear rates. In its more generalized (tensor) form it will predict elastic behaviour.

Models of this type seem to have been developed from the simplified molecular theories of LODGE[3.7]. Good representations of practical polymer melt flow data have been obtained by SINGH[3.3]. The model has not been extensively used in studying processing behaviour, because of the difficulties associated with greater mathematical complexity.

There are various other models, of varying degrees of complexity and accuracy, which have been proposed[3.5]. However, they have generally found limited application in practice and are outside the scope of the present treatment. The power law model and its derivatives have been most extensively used to represent the shear flow properties of polymer melts.

3.6 Measurement of Shear Properties

There are two basic types of *viscometer* (rheometer) used for these measurements: rotational and extrusion. The detailed mathematical analyses of the methods are discussed by WHORLOW[3.2] and we will only outline the main principles here. The assumptions common to both types are:

- Steady state flow conditions are present, which implies that the fluid velocity at any point is independent of time;
- The fluid is at constant temperature;
- There is no slip at the solid/fluid interface;
- The polymer melt is incompressible.

3.6.1 Cone and Plate

The geometry for this instrument is shown diagrammatically in Figure 3.6. A cone of radius R rotates about its axis with a constant angular velocity Ω. The apex of the cone is in contact with a horizontal stationary plate and the polymer fills the gap between the cone and the plate. If the angle (θ) between the cone and plate is small, this configuration is analogous to the situation depicted in Figure 3.1. The velocity (v) at any radius r is $r\Omega$, and the shear rate will be given by:

$$\dot{\gamma} = \frac{r\Omega}{h} = \frac{\Omega}{\theta} \tag{3.12}$$

where h is the gap separation at radius r; θ is in radians.

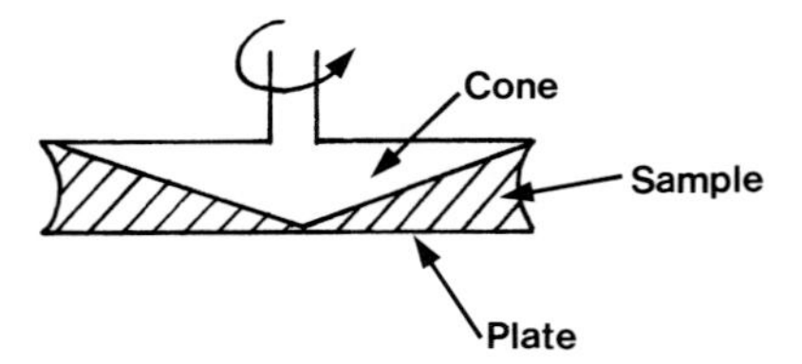

Figure 3.6 Schematic diagram of a cone and plate viscometer.

We see that the shear rate in this geometry is constant and therefore the shear stress will be independent of r. The total torque (M) developed is:

$$M = \int_0^R 2\pi r^2 \cdot \tau \, dr = \frac{2\pi R^3 \tau}{3} \tag{3.13}$$

More rigorous analyses have shown that for small angles ($\theta < 6\,°$), the error in using Equation 3.12 to calculate shear rate is less than 0.5 %[3.8].

Thus we have τ and $\dot{\gamma}$ in terms of measurable quantities, torque and angular velocity, and the flow curve for the material can therefore be obtained. The apparatus can be operated at either constant speed or constant torque; the latter method, using a dead weight and a lamp and scale to measure rotation, is probably the cheapest and simplest rheometer available for plastic melts[3.1].

The main advantage of this technique is that the flow conditions are precisely defined. Only a small amount of material is required and good temperature control can be achieved. Also, it is possible to measure time dependent effects such as thixotropy and elastic properties. However, for highly viscous fluids such as polymer melts, the shear rate range is very limited. As the shear rate is increased, the flow pattern breaks up at the free boundary, giving rise to non steady state conditions and anomalous readings. For most plastics melts, the maximum shear rate for reliable data is about 10 s^{-1}, which is much lower than practical shear rates in commercial processes. Some improvement can be achieved by using a *biconical* geometry (Figure 3.7), where the cavity is under pressure. Any friction between the tip of the cone and the plate will interfere with the torque mechanism and is avoided by truncating the cone very slightly. If the radius of the truncated section is less than (0.2).R, then from Equation 3.13, the error in the torque will be less than 1 %. Truncation of the cone is a feature of most commercial viscometers of this type.

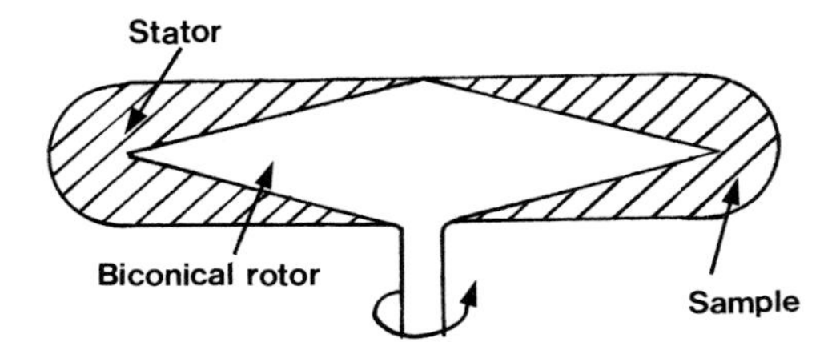

Figure 3.7 Schematic diagram of a biconical viscometer.

Viscous flow is a dissipative process and energy will be converted into heat which will produce a temperature rise in the flowing polymer. The magnitude of viscous heating can be estimated by assuming there is no heat lost from the sample. Clearly, this calculation will overestimate the effect, but it will indicate if shear heating is likely to be significant. The temperature rise ($\triangle T$) per unit time is given by:

$$\frac{d\triangle T}{dt} = \frac{\tau\dot{\gamma}}{\varrho C_P} \tag{3.14}$$

where C_P and ϱ are the specific heat and density of the polymer melt, respectively. If it is found that shear heating may be significant, more exact computations can be carried out[3.2].

There is a large surface to volume ratio with this type of viscometer, which may induce *slip* at the boundary; this effect can be studied using cones of different angles or by using grooved surfaces.

The cone and plate rheometer is a useful technique for measuring the flow properties of polymer melts, but the shear rate range is limited. An added advantage of the method is that elastic effects can evaluated directly.

3.6.2 Concentric Cylinder

In this type of viscometer, the sample is sheared between two concentric cylinders (Figure 3.8), one of which is rotating at a constant angular velocity Ω. The shear rate and shear stress at any radius r are given by[3.2]:

$$\dot{\gamma} = r \cdot \frac{d\omega}{dr} \tag{3.15}$$

$$\tau = \frac{M}{2\pi r^2} \tag{3.16}$$

where M is the torque per unit length of the cylinder and ω is the angular velocity at radius r.

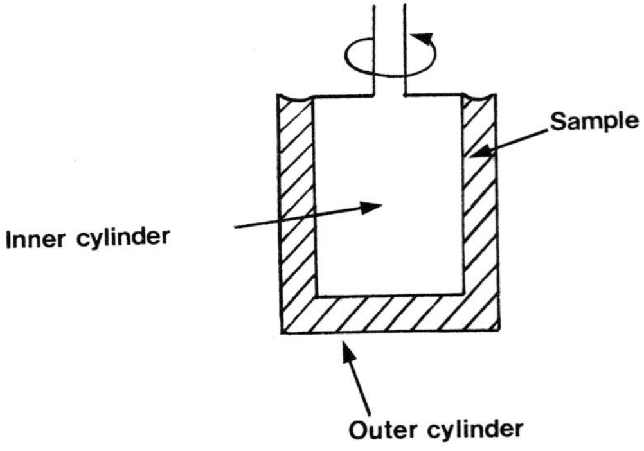

Figure 3.8 Schematic diagram of a concentric cylinder viscometer.

A feature of this geometry is that the shear rate in the sample is not constant but varies with radius, in a manner which can only be determined by assuming some behaviour model for the material. However, if the gap is small, it is reasonable to approximate Equations 3.15 and 3.16 as follows:

$$\dot{\gamma} = \frac{R_a \Omega}{R_0 - R_i} \tag{3.17}$$

$$\tau = \frac{M}{2\pi R_a^2} \tag{3.18}$$

where R_i, R_0 and R_a represent the inner, outer and average radii.

If the ratio R_0/R_i is 1.1, then the error in using Equations 3.17 and 3.18 is less than 1 % for Newtonian and many power law fluids. In practice this is a reasonable gap size and the simplified approach is justified. WHORLOW[3.2] reviews more rigorous methods of analyzing concentric cylinder data.

Concentric cylinder viscometers are suitable for suspensions and other low viscosity liquids and are used extensively for these materials. For high viscosity polymer melts, it is difficult to fill the gap and alignment may be a problem. The ill-defined flow conditions at the ends can introduce errors, which may only be eliminated by repeating experiments using cylinders of different length. Heat generation can also be a problem[3.2] and for polymer melts the apparatus has a limited shear range. Applications to polymer melts for this type of viscometer have been limited.

There are a number of other geometries which have been developed for rotational viscometers, but they are all basically variants of the concentric cylinder and cone and plate instruments. They have been introduced to minimize end corrections and extend the experimental ranges of shear rate. The main advantage of rotational viscometers is their ability to measure time dependent effects, and elastic properties.

3.6.3 Capillary (Extrusion) Viscometer

The operating principle of this viscometer is shown in Figure 3.9. The melt is forced through a capillary die and the flow curve can be evaluated from data relating pressure drop to the volume rate of flow. The shear rate and shear stress at any radius (r), are given by[3.5]:

$$\dot{\gamma} = \frac{dv}{dr} \tag{3.19}$$

$$\tau = \frac{Pr}{2L} \tag{3.20}$$

where:

v = axial velocity at r;

P = total pressure drop across the capillary;

L = capillary length.

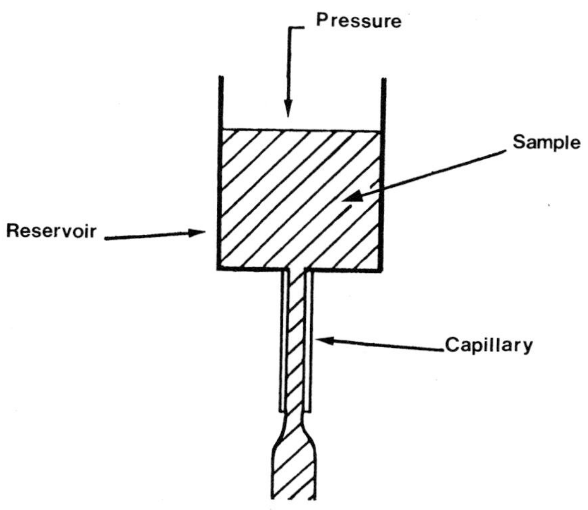

Figure 3.9 Schematic diagram of a capillary viscometer.

Thus the shear stress is a maximum (τ_R) at the wall:

$$\tau_R = \frac{PR}{2L} \tag{3.21}$$

where R is the radius of the capillary.

The volume rate of flow (Q) is:

$$Q = \int_0^R 2\pi r^2 \cdot v\, dr \tag{3.22}$$

To analyse the flow it is necessary to assume that there is a unique function between $\dot{\gamma}$ and τ: i.e.

$$\dot{\gamma} = f(\tau) \tag{3.23}$$

The boundary conditions are:

$$v = 0 \qquad \text{at} \qquad r = R \tag{3.24}$$

Thus from 3.19 to 3.24 we find[3.5]:

$$\frac{4Q}{\pi R^3} = 4(\tau_R)^{-3} \int_0^{\tau_R} \tau^2 \cdot f(\tau)\, dr \tag{3.25}$$

Since the right hand side of Equation 3.25 is a function of τ_R only, a plot of $(4Q/\pi R^3)$ is unique. We can assume some model for $f(\tau)$ and integrate this expression. Thus for a power law (Equation 3.8) we find:

$$\tau_R = k \cdot \left[\frac{3n+1}{4n}\right]^n \cdot \left[\frac{4Q}{\pi R^3}\right]^n \tag{3.26}$$

where the shear rate at the wall, $\dot{\gamma}_R$, is given by:

$$\dot{\gamma}_R = \left[\frac{3n+1}{4n}\right] \cdot \left[\frac{4Q}{\pi R^3}\right] \tag{3.27}$$

We can evaluate the fluid parameters "n" and "k" from a plot of $\log(PR/2L)$ against $\log(4Q/\pi R^3)$. It is not necessary to assume an analytical form for $f(\tau)$, since 3.25 can be rearranged and differentiated to give the *Rabinowitsch equation*[3.5]:

$$\dot{\gamma}_R = \frac{(3n'+1)}{4n'}\dot{\gamma}_a \qquad n' = \frac{d(\log \tau_R)}{d(\log \dot{\gamma}_a)} \tag{3.28}$$

where

$$\dot{\gamma}_a = \frac{4Q}{\pi R^3} \tag{3.29}$$

$\dot{\gamma}_a$ is often referred to as the "apparent" or "Newtonian" shear rate.

3.6.3.1 Analysis of Experimental Results from Capillary Flow Measurements

There are a number of possible sources of error which have to be considered:

(A) End effects. The shear stress is calculated from the total pressure drop across the capillary (P), but there are a number of energy losses and P must be corrected to take these into account. They arise from the kinetic energy in the issuing stream, leakage flow past the piston and as a consequence of the convergent flow at the die entrance. These effects can be accommodated by including an *effective die length* (eR) in Equation 3.21, where "e" is an end correction which depends on shear rate (Figure 3.10); R is then calculated from:

$$\tau_R = \frac{PR}{2(L + eR)} \tag{3.30}$$

BAGLEY[3.9] has applied Equation 3.30 to polymer melts and calculated values of τ_R which are geometry-independent. Therefore, τ_R is obtained from measurements of pressure drop (P) using capillaries of different length at constant output rate $4Q/\pi R^3$. There is some evidence that the Bagley plot becomes non linear at very high values of L/R, which has been associated with the effect of pressure on viscosity[3.10]. In general, the Bagley approach gives consistent results and is the preferred method for analyzing end effects from experimental data.

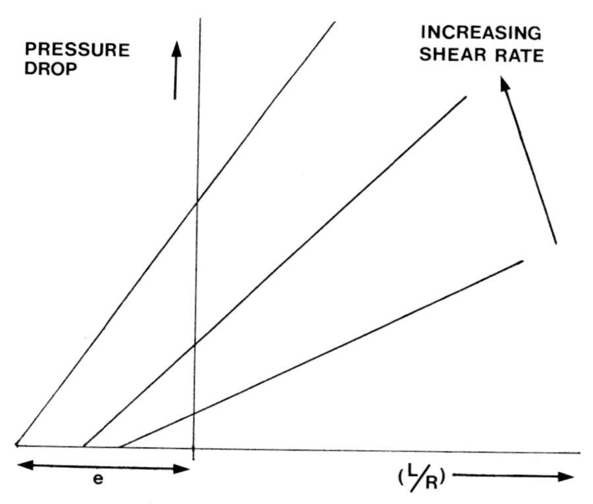

Figure 3.10 End corrections in capillary flow: the Bagley plot[3.9].

(B) Slip at the wall. It has been assumed that there is no slip at the wall of the capillary (Equation 3.24), which seems to be a reasonable assumption for most materials. However, some lubricating additives may migrate to the surface and induce slip at the solid boundary. The presence of wall slip can be established by experiments with capillaries of different radii. If we assume that there is a constant slip velocity v_s, the boundary conditions become:

$$v = v_s \qquad \text{at} \qquad r = R \tag{3.31}$$

From 3.17 to 3.27 we find:

$$\frac{4Q}{\pi R^3} = 4(\tau_R)^{-3} \int_0^{\tau_R} \tau^2 \cdot f(\tau)\, d\tau + 4\frac{v_s}{R} \tag{3.32}$$

Thus the slip velocity can be obtained from a plot of $1/R$ against $4Q/\pi R^3$, at constant τ_R.

(C) Reservoir effects. There can be a pressure drop across the material in the barrel due to some viscous or elastic effect. The magnitude will depend on the precise geometry but the apparatus is effectively functioning as two viscometers in series. In a constant speed instrument the pressure on the piston will decay, as shown in Figure 3.11. Similarly, in a constant pressure device the volume rate of flow will increase as the volume of material in the reservoir decreases. The results must be extrapolated to the point where the reservoir is empty, in order to obtain reliable results. Errors due to this effect do not arise in constant speed viscometers where the pressure is measured by a transducer close to the capillary entrance.

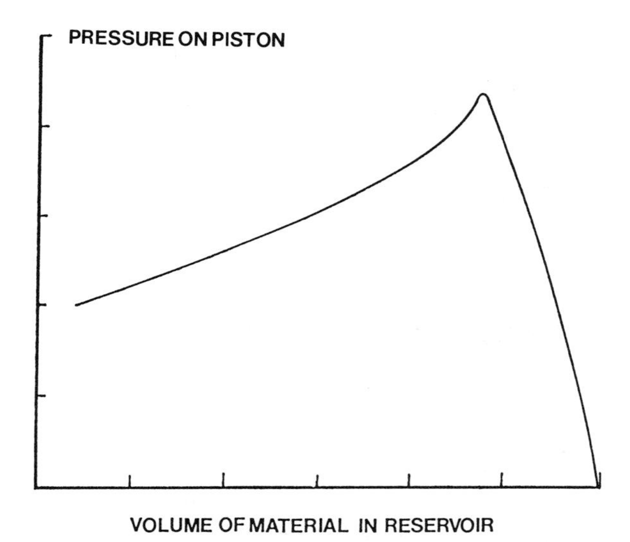

Figure 3.11 Reservoir effect in a constant speed capillary viscometer.

(D) Shear heating. The effect of shear heating is less severe in extrusion viscometers. It can be estimated by assuming that there is no heat loss from the sample, which will overestimate the effect. Under these circumstances the temperature rise ($\triangle T$) will be given by:

$$\triangle T = \frac{P}{\varrho C_P} \tag{3.33}$$

If the calculated rise is significant, a more detailed analysis is required[3.2].

(E) Pressure effects. In the analysis we have assumed that the polymer melt is incompressible, an assumption which is generally invalid. However, the error is small and can be detected from experiments with very long capillaries. In practice the increase of viscosity associated with high pressures tends to cancel out the shear heating effects[3.1]. Thus, it is reasonable to assume that the material is incompressible.

3.6.3.2 Types of Capillary Viscometer

(a) Melt indexer. This is the simplest type of instrument and is found extensively in industrial laboratories. A dead load is applied to a piston and the mass flow rate, (the *Melt Flow Index,* (MFI)) is measured. The melt indexer is therefore a *pressure-imposed* capillary rheometer.

The technique was originally developed for polyolefins and is the basis of many national and international standards e.g. ISO 1133, DIN 53735, BS 3412. Many raw material producers' quote the MFI of their polymers, which is the mass of material extruded in a given time under constant load and temperature conditions. The equipment is relatively crude and inexpensive, but is suitable for quality control purposes and for comparisons between materials of the same family; the technique is not suitable for fundamental rheological studies.

(b) Gas driven viscometer. The melt is forced through a die by constant pressure from a gas cylinder. A ball bearing or plug may be inserted between the melt and the pressure line to prevent the gas from channelling through the sample. The maximum pressure available is usually below 35 MPa, which limits the highest shear rate that can be achieved. The measurement of flow rate is not ideal, since we must convert mass to volume and an independent measurement of density at the melt temperature is required. However, the viscometer is simple to use and suitable for measurements over a limited range of shear rates.

(c) Constant speed rheometer. Pressure is applied by a piston which is driven at constant speed to achieve a constant volume rate of flow (a *rate-imposed rheometer).* The pressure drop is measured by some suitable device, such as a transducer situated immediately above the die. This is a convenient method and several commercial models are available for polymer melts. Pressures up to 170MPa can be accommodated without too much difficulty.

The great advantage of capillary viscometers is that the data can be obtained over a wide range of conditions on highly viscous materials. It is possible to measure the rheological properties under conditions relevant to processing and in consequence, this is the most common type of instrument used for polymer melts. The main disadvantages of the method are that elastic and time dependent effects cannot be measured directly.

There are a number of variations of the method where the sample is forced through a rectangular slit and the pressure drop obtained from pressure transducers mounted flush with the die wall. This can eliminate the need for end corrections.

3.7 Factors Affecting Shear Flow

3.7.1 Temperature

Flow occurs as a result of polymer chains sliding over each other. The ease of flow will depend on the mobility of the chains and the *entanglement forces* holding them together. An increase in temperature will increase the mobility, and hence reduce the viscosity. For simple low molecular weight Newtonian fluids, the viscosity is related to temperature (T) by an *Arrhenius equation* of the form:

$$\eta_0 = A \exp\left(-\frac{B}{T}\right) \tag{3.34}$$

where A and B are constants.

For polymer melts at low shear rates, this is a reasonable representation but is restricted to temperature spans of about 50 to 60 °C. A simpler form of the equation may also be used:

$$\eta_0 = a \exp[-b(T - T_0)] \tag{3.35}$$

where a and b are constants, T_0 is a reference temperature.

The variation of viscosity with temperature depends on the polymer type and varies widely[3.1]. It is important in the design of processes and to the polymer scientist who wishes to relate structure to the fundamental physical properties of the material. Typical flow curves at various temperatures are shown in Figure 3.12 for LDPE.

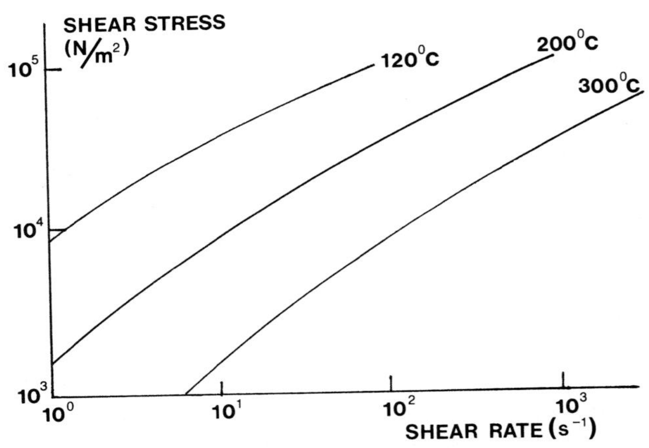

Figure 3.12 Typical flow curves for LDPE at various temperatures.

The measurement of melt viscosity as a function of both shear rate and temperature is a time consuming procedure and there is a need for a generalized method of predicting flow at any temperature. The power law (Equation 3.8) can be modified to take account of temperature by using Equation 3.35[3.5]. Thus we obtain:

$$k = k_0 \exp[-c(T - T_0)] \tag{3.36}$$

where k_0 is the value of k at T_0 and c is an empirical parameter.

Equation 3.36 will only be suitable for fluids which can be represented by a power law. MENDELSON[3.11] has described a more generalized technique for producing a master curve which takes account of temperature. It is based on a shear rate - temperature *superposition* (which is similar to the well established approach used for representing linear viscoelastic behaviour which will be referred to in Chapter 6). A shift factor at constant $\tau(a_\tau)$ is defined:

$$a_\tau = \frac{\dot{\gamma}_{\text{ref}}}{\dot{\gamma}(T)} \tag{3.37}$$

where $\dot{\gamma}_{\text{ref}}$ is a reference shear rate. A mastercurve is then obtained by plotting τ versus the product of $\dot{\gamma}$ and a_τ. MENDELSON has shown that the method is valid for a wide range of polymers, and has found that a_τ can be expressed by an Arrhenius relationship similar to Equation 3.35. The shift factor appears to be independent of molecular weight and only depends on polymer type. This is an extremely useful approach which also appears to have some basis in viscoelastic theory.

3.7.2 Pressure

In previous discussions we have assumed that polymer melts are *incompressible,* which is a reasonable first approximation for most applications. However, high pressures are evident in injection moulding where values up to 150 MPa can be developed. An increase in pressure will restrict the mobility of the polymer chains and increase viscosity. There has been little research carried out in this area and results tend to be variable because of experimental difficulties.

Data have been reported using a pressurized capillary rheometer[3.12] and two capillary rheometers in series. Since the effect of pressure is determined from small differences between two large pressure recordings, there tends to be a problem with accuracy and reliability, which is reflected in the published results. COGSWELL[3.1] used a pressurized concentric cylinder which separates the effect of pressure from the measurement of viscosity. The use of a non-linear Bagley plot has also been suggested as a promising method of approach[3.13].

3.7.3 Effect of Molecular Structure

The importance of *molecular weight, molecular weight distribution* and *chain branching* on physical properties of polymers has been discussed earlier in Chapter 1. Polymer chains are highly coiled in the equilibrium state and there is a high degree of both inter- and intra-chain *entanglement.* When a stress is applied, the chains tend to become aligned and disentangled. There will also be some slippage of the chains over each other, and an individual molecule (or molecular segment) is able to drag others along, to comply with the effect of external stress. Thus, we would expect the resistance to flow and hence viscosity to increase as the chain length (Z) increases, which occurs in practice. The relationship between η_0 and Z is shown in Figure 3.13, where there are two distinct regions. The behaviour can be represented by:

$$\eta_0 = K_1 Z \qquad \text{for} \quad Z < Z_c \tag{3.38}$$

$$\eta_0 = K_2 Z^{3.4} \qquad \text{for} \quad Z > Z_c \tag{3.39}$$

where K_1 and K_2 are temperature-dependent constants. The abrupt change in the slope occurs at a critical chain length Z_c where the effect of entanglements becomes significant. As chain length increases, the number of entanglements increases sharply and we obtain a very strong dependence on Z. The above relationships are valid for a wide range of linear polymers, but Z_c depends on the polymer type and is lower for *polar* polymers.

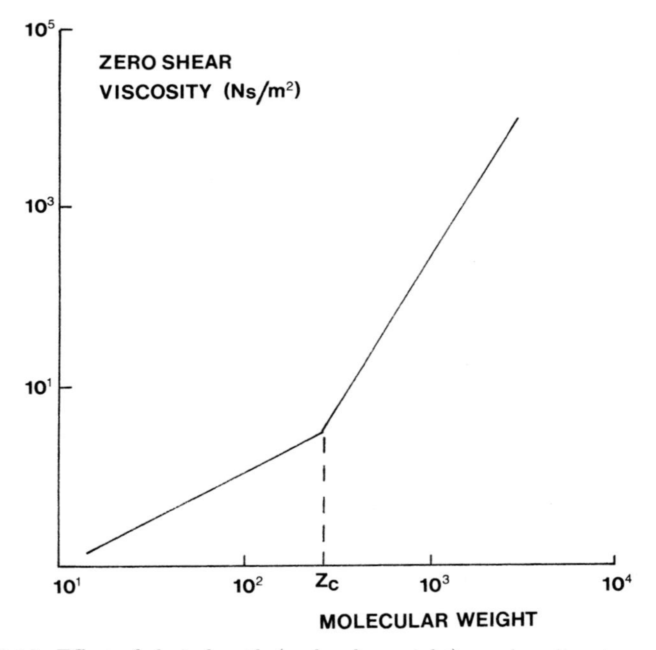

Figure 3.13 Effect of chain length (molecular weight) on viscosity at zero shear rate.

At high stresses there will be greater disentanglement and the chains will become more aligned in the direction of flow. This will create less resistance to flow and the viscosity will decrease as we have seen (Figure 3.5). Since there will be less entanglements at higher stresses, there will be less dependence on molecular weight. A detailed theoretical basis for this argument is given by GRAESSLEY[3.14].

The effect of molecular weight distribution is shown in Figure 3.14. Deviation from Newtonian behaviour occurs at a much lower shear rate in polymers with a broad distribution. At high shear rates the viscosity appears to depend only on an average molecular weight.

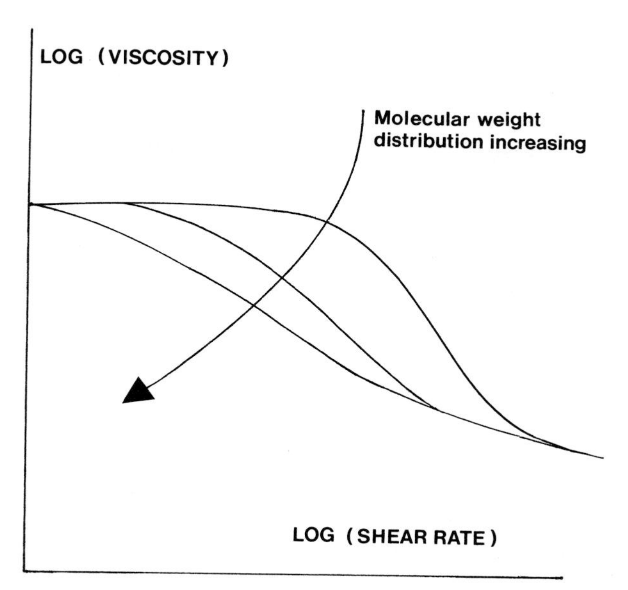

Figure 3.14 Effect of shear rate and molecular weight distribution on flow curves.

The effect of branching is not as well understood and some of the evidence seems to be conflicting. It is clear that the the length of the branch is significant. Polymers with long side chains have a lower η_0 than linear polymers of the same molecular weight. This is physically reasonable as a branched polymer will be more compact and less entangled[3.15].

3.8 Elongational Flow

There is an increasing awareness of the significance of elongational flow but it has not been studied in the same detail as shear flow. This is mainly due to some associated experimental difficulties, and a general lack of awareness of its importance in practice. The *Elongational Viscosity* (η_e) of fluids is much higher than the shear viscosity and for a simple Newtonian fluid $\eta_e = 3.\ \eta_0$. The dependence of η_e on elongational rate is a function of the polymer[3.1,3.16].

3.8.1 Methods of Measurement

Elongational deformation is usually apparent as *free surface flows;* i.e. there is no boundary between the melt and a wall of a viscometer. In essence, these flows are simulated by a simple tensile experiment, but it must be appreciated that the cross section of the test specimen is changing rapidly during the experiment. We need to establish what dimensional changes must be provided to ensure that the elongational rate is constant throughout an experiment. From 3.6, the velocity v in the z direction (see Figure 3.2), is given in terms of the specimen length (1) at time t by:

$$v = \frac{dl}{dt} = \dot{\varepsilon} \cdot l(t) \tag{3.40}$$

where it is required that the *elongational strain rate* ($\dot{\varepsilon}$) is constant. Integrating with the boundary condition:

$$t = 0 \qquad \text{at} \qquad l = l_0 \tag{3.41}$$

where l_0 is the original length of the sample, we find:

$$l(t) = l_0 \cdot \exp(\dot{\varepsilon} \cdot t) \tag{3.42}$$

Therefore the length of the sample must increase exponentially to achieve steady-state conditions where the elongation rate is constant. The *elongational stress* is given by:

$$\sigma = \frac{F}{A} \tag{3.43}$$

where F and A are the force and cross sectional area of the specimen at time t. We can assume that the melt is incompressible, thus:

$$A(t) \cdot l(t) = A_0 \cdot l_0 \tag{3.44}$$

where A_0 is the original cross sectional area of the sample. We obtain from 3.43 and 3.44:

$$\sigma = F(t) \cdot \frac{\exp(\dot{\varepsilon} \cdot t)}{A_0} \tag{3.45}$$

Equation 3.42 defines an experiment at constant elongation rate. It is also possible to specify the requirements for a constant stress experiment; from 3.43 and 3.44:

$$\frac{dF}{dl}\, \alpha\, l^{-2} \tag{3.46}$$

Equation 3.46 defines the rate of variation of F with l which must be attained for a constant stress experiment.

3.8.2 Types of Elongation Viscometer

A conventional tensile test principle can be adopted for these experiments. The constant elongation rate is achieved by using a servo-drive system so that the cross head speed satisfies Equation 3.42. A load cell is fixed to one end of the specimen to measure the force (Figure 3.15). The test specimen can be in the form of a strip or a ring[3.16]; since the melt tends to slip in the grips, it is preferable to use a ring which is looped over two pulleys. The sample is immersed in an oil bath to control the temperature and minimize the sag in the sample under its own weight. This type of apparatus can also function in the constant stress mode by appropriate programming of the servo-drive system.

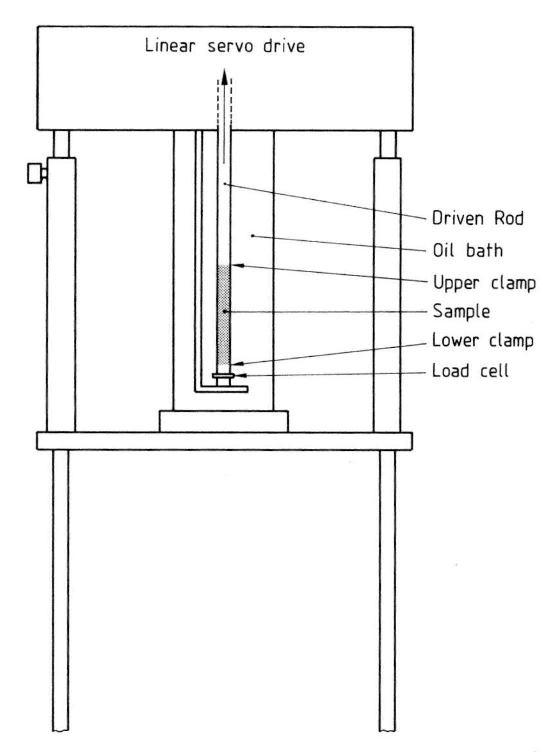

Figure 3.15 Schematic diagram of equipment for measuring elongational viscosity.

Constant stress devices have been used[3.2,3.17], where the constant load has been applied through a profiled cam such that 3.46 is satisfied (Figure 3.16). These types of viscometer are simple and inexpensive but can only be used with one size of specimen, which must be prepared in the appropriate form.

MEISSNER[3.18] described a novel method where the sample of constant length was held between a pair of rollers rotating at constant speed (Figure 3.17). In this way a constant elongation rate was attained. This apparatus is very flexible and can handle a wide range of materials.

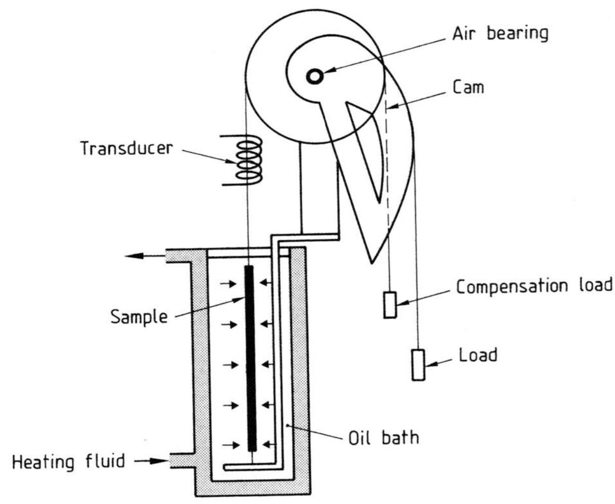

Figure 3.16 Constant stress elongational viscometer (after MUNSTEDT[3.17]).

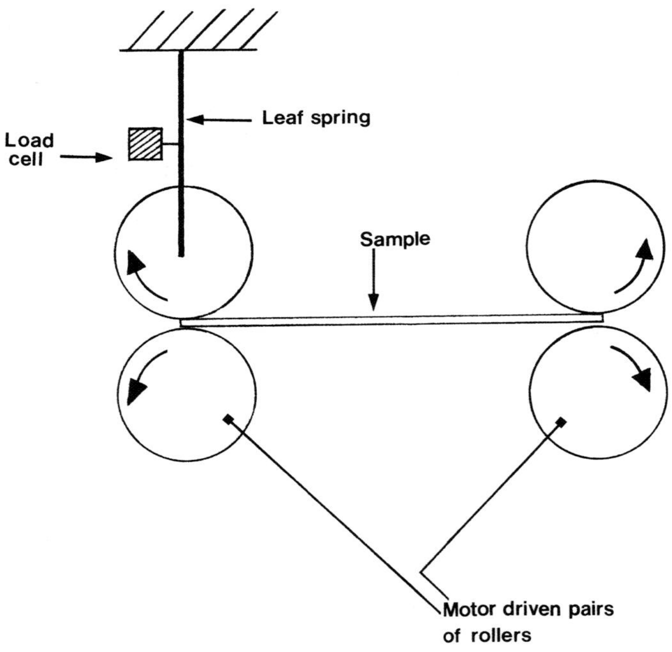

Figure 3.17 Constant strain rate elongational viscometer (after MEISSNER[3.18]).

All the above methods are based on free surface flows. COGSWELL[3.1] has pointed out that there is an element of elongational deformation when a melt flows into a capillary (Figure 3.18) and has proposed a method of analyzing the data to obtain elongational viscosity. The relevant equations are as follows:

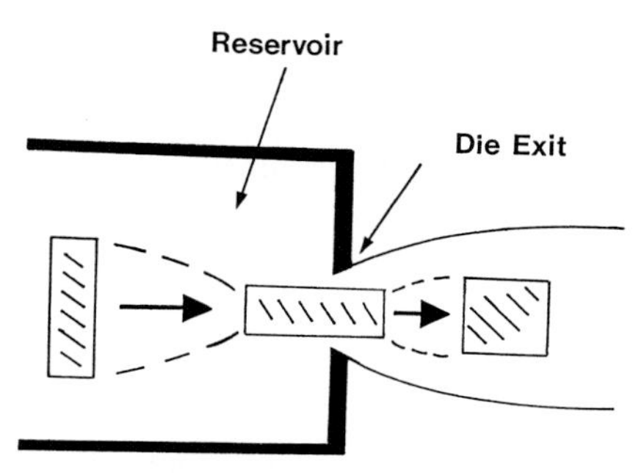

Figure 3.18 Elongational deformation in convergent flow.

$$\sigma = \frac{3(n+1)}{8} P_0 \quad (3.47)$$

$$\eta_e = \frac{9(n+1)^2 P_0^2}{32 \eta \dot{\gamma}_a^2} \quad (3.48)$$

where P_0 is the pressure drop through a zero land length capillary, for a material which has a shear viscosity η at an apparent shear rate $\dot{\gamma}_a$.

The theory is not rigorous but results obtained by this method agree qualitatively with data from more precisely defined techniques. The main attraction of the convergent flow method is that the measurements can be obtained on the same viscometer which is used for shear flow experiments, without any changes in procedure.

3.9 Factors Affecting Elongational Viscosity

The study of elongational flow in polymer melts is fairly recent and there is considerably less published information compared with shear effects. Thus the various factors affecting elongational flow are not so well established.

3.9.1 Elongation Rate

This is very dependent on the polymer, as is indicated in Figure 3.19. The viscosity maximum in the LDPE data has been noted by several workers using different methods. There is no obvious explanation for this behaviour but the *tension-stiffening* effect is probably related to *chain branching* in the polymer[3.1]. For linear polymers, elongational viscosity is either substantially constant or decreases (*"tension-thinning"* behaviour), as elongation rate increases. As $\dot{\varepsilon}$ tends to zero, then η_e tends to approach a constant value of 3. η_0, i.e. to respond like a Newtonian fluid.

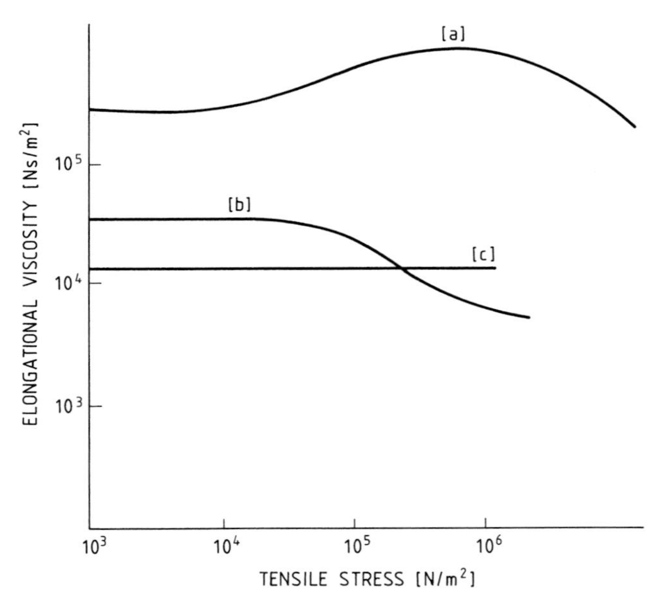

Figure 3.19 Elongational viscosity data for some thermoplastics:
(a) Extrusion grade LDPE at 170 °C;
(b) Extrusion grade propylene-ethylene sequential copolymer at 230 °C;
(c) Moulding grade poly(methyl methacrylate) at 230 °C.

3.9.2 Molecular Parameters

Elongational viscosity increases with molecular weight and at low stresses depends on chain length in a similar manner to shear viscosity (Equation 3.39). For

LDPE, there does not appear to be a strong dependence on molecular weight distribution[3.16].

3.9.3 Temperature

Elongational viscosity decreases with increase in temperature and appears to follow an Arrhenius type dependency. SMOKER[3.16] has shown that the Mendelson superposition procedure discussed earlier for shear properties can also be applied to elongational viscosity.

3.10 Melt Elasticity

Polymer melts are *viscoelastic;* the elastic properties can be very significant in shaping operations. The elasticity arises as a result of entanglements in the polymer chain. In the unstressed state the chains will be randomly orientated; when the polymer is deformed there will be some disentanglement and orientation in the direction of the applied stress. In systems which are not chemically cross-linked, there will be some slippage of chains over each other. On removal of the stress, chains will tend to return to the equilibrium random-coil state and thus there will be a component of elastic recovery. The recovery is not instantaneous because of the entanglements still present in the system. Furthermore, because of the slippage of chains during flow, recovery will not be complete (Figure 3.3), unlike the behaviour of cross-linked polymers. There are several manifestations of elasticity, as summarized below.

3.10.1 Stress Relaxation

If a polymer melt is subjected to a certain flow history and then held at constant deformation, the stress does not become zero instantaneously, but decreases to zero with time. This behaviour is referred to as *stress relaxation.* The rate of relaxation depends on the strain rate during deformation, the magnitude of strain and molecular characteristics. Relaxation is much more rapid as the strain rate increases, since the chains are less entangled. Similarly, a lower molecular weight polymer will relax more quickly. The effect of molecular weight distribution is shown in Figure 3.20. The faster response at short times (for the broad molecular weight distribution polymer) is due to the dominating influence of the low molecular species; at long times the high molecular weight chains dominate the process. PETICOLAS[3.19] developed a theory to explain this mechanism and suggested stress relaxation as a method with which to characterize molecular weight distribution.

3.10.2 Stress Growth

Similarly, when a constant strain is applied to a material at rest, the stresses do not immediately attain a steady value. Under certain conditions stress overshoot can occur (Figure 3.21). The chains are forced to disentangle faster than their natural response before they relax to a steady state. This overshoot region may be commensurate with, for example, the filling times in injection moulding and may be the dominating factor. The amount of overshoot depends on the resting time between successive straining periods.

3.10.3 Elastic Recovery

This occurs when the stress is removed following a period of steady flow. The polymer recovers some of its original shape (Figure 3.3). The amount of recovery increases with strain rate, molecular weight and molecular weight distribution.

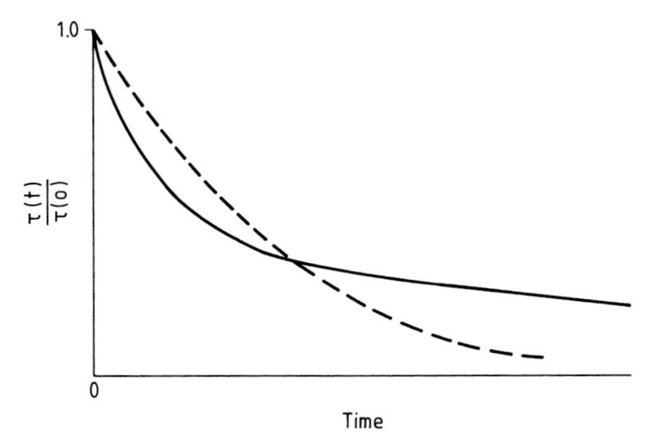

Figure 3.20 Effect of molecular weight distribution on normalized stress relaxation at time t, following cessation of steady shear flow.
______ Broad distribution; – – – – Narrow distribution)

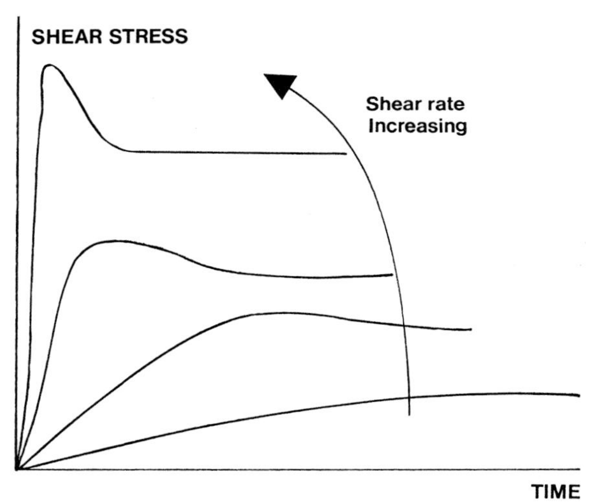

Figure 3.21 Stress growth on commencement of shear flow.

3.10.4 Die Swell

In capillary flow the diameter of the extrudate is greater than the capillary diameter, a phenomenon known as *die swell.* It is associated with *elasticity* in the polymer[3.7], but Newtonian fluids also exhibit a small amount of die swell[3.20] under certain conditions. Die swell is important in extrusion and must be taken into account by the die designer; its effects are briefly discussed within this context in Chapter 4.

The amount of die swell depends on the applied stress and capillary length (Figure 3.22). At low values of L/R the swell is dominated by elastic recovery from elongational stresses set up in the entry region[3.1]. These stresses relax during flow down the capillary and thus die swell decreases with increasing die aspect ratio. As

L/R tends to infinity die swell reaches a constant value, where recovery from shear deformation becomes the dominating influence.

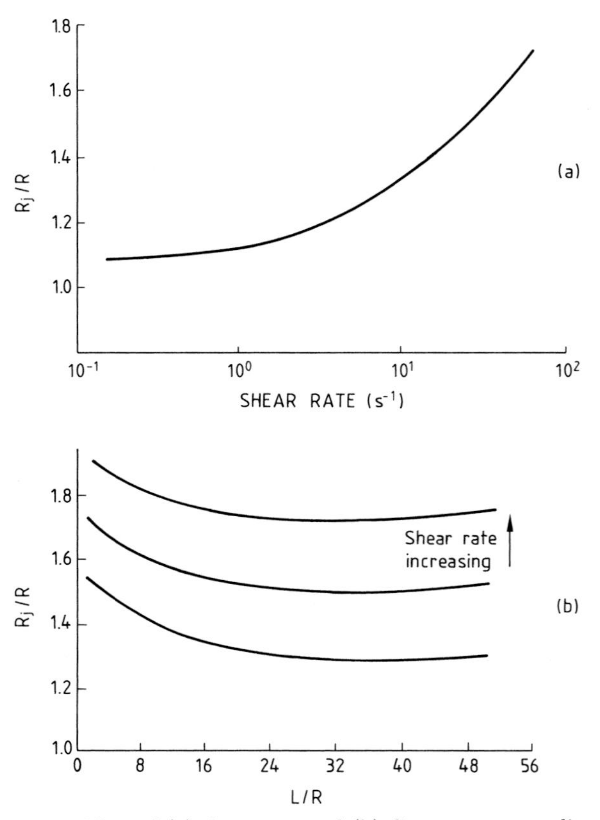

Figure 3.22 Effect of (a) shear rate and (b) die geometry on die swell.

TANNER[3.21] has applied Lodge's entanglement theory[3.7] to capillary flow and has obtained the following relationship between shear recovery and die swell:

$$\left(\frac{R_j}{R}\right)^2 = 0.1 + (1 + \gamma_R^2)^{1/6} \tag{3.49}$$

where γ_R is the recoverable strain at shear stress τ_R, and R_j is the radius of the extrudate.

Using a different approach, CRAWFORD[3.22] has developed the following expression:

$$\left(\frac{R_j}{R}\right)^2 = 0.67(\gamma_R)\left[(1 + \gamma_R^{-2})^{3/2} - \gamma_R^{-3}\right] \tag{3.50}$$

Predictions from the two theories are very similar but neither accounts for the observed die swell in Newtonian fluids. For short dies, CRAWFORD has calculated the swell from the elastic recovery obtained from elongational flow:

$$\left(\frac{R_j}{R}\right)^2 = \exp(\varepsilon_R) \tag{3.51}$$

where ε_R is the *recoverable strain* from elongational flow under a tensile stress σ.

The die swell theories are not always accurate (especially when studying flow through complex dies during processing), but are in qualitative agreement with experimental data. There is little doubt that die swell is related to elastic recovery and is a useful method for comparing the elastic properties of polymer melts. Die swell increases as the molecular weight distribution of the polymer increases[3.23], which is in qualitative agreement with stress relaxation observations discussed earlier (Figure 3.21).

3.10.5 Melt Fracture

In capillary flow the extrudate becomes irregular and distorted at high flow rates. This phenomenon takes a variety of forms and is associated with elastic behaviour, and is referred to as *melt fracture.* The shear rate at which this occurs increases with capillary length and decreases with die entry angle[3.1,3.5]. It is suggested that fracture occurs as a result of high tensile stresses set up in the entry to the die. If the stress exceeds the tensile strength of the melt, then melt fracture will occur. This explanation is consistent with Figure 3.23, where the tensile stress calculated from convergent flow analysis is constant at the onset of fracture[3.24]. Further confirmation has been reported by SHAW[3.25] in experiments with lubricated dies.

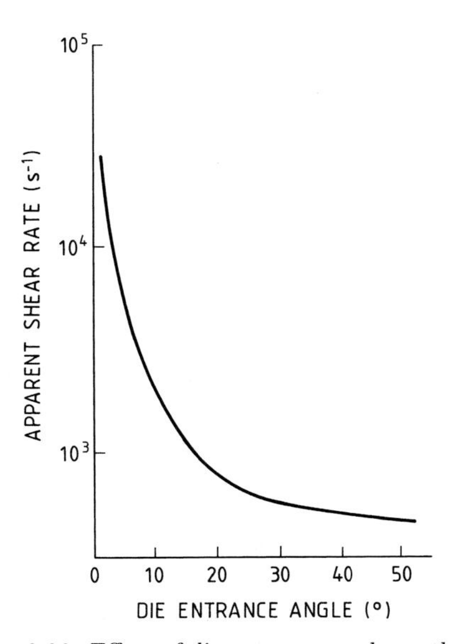

Figure 3.23 Effect of die entrance angle on the onset of melt fracture.

Melt fracture limits the maximum output which can be achieved in extrusion but, as Figure 3.23 indicates, modification to the die entry geometry can improve the situation. The extrusion rheometer is a convenient method for studying melt fracture in the laboratory.

References

3.1 COGSWELL, F.N.: *Polymer Melt Rheology,* George Godwin, London, (1981).

3.2 WHORLOW, R.W.: *Rheological Techniques,* Ellis Horwood, Chichester, (1980).

3.3 SINGH, M.: Ph.D. Thesis, *Modelling the Rheological Behaviour of Elastomers and Elastomer Compounds,* Loughborough University of Technology, (1990).

3.4 COOK, J.A., and GILBERT, M.: *Effect of Particle Size Distribution on PVC Plastisols,* Presented at Third International PRI Conference PVC 87, Brighton, (1987).

3.5 TADMOR, Z., and GOGOS, G.C.: *Principles of Polymer Processing,* John Wiley, New York, (1979).

3.6 BOWERS, S.: *Prediction of Elastomer Flow in Multi-cavity Injection Moulding,* Plast. Rubb. Proc. & Appl., **7,** (1987), 101.

3.7 LODGE, A.S.: *Elastic Liquids,* Academic Press, London (1964).

3.8 WALTERS, K., and WATERS, N.D. in: *Polymer Systems,* R.E. WETTON, and R.W. WHORLOW, Eds., Macmillan, London, (1968).

3.9 BAGLEY, E.B.:*End corrections in Capillary Flow of Polyethylene,* J. Appl. Phys., **28,** (1957), 624.

3.10 GOLDBLATT, P.H., and PORTER, R.S.: *A Comparison of Equations for the effect of Pressure on the Viscosity of Amorphous Polymers,* J. Appl. Poly. Sci., **20,** (1976), 1199.

3.11 MENDELSON, R.A.: *A Generalized Melt Viscosity-Temperature Dependence for Styrene and Styrene-Acrylonitrile based Polymers,* Poly. Eng. Sci., **16,** (1976), 690.

3.12 WESTOVER, R.F.: *Effect of Hydrostatic Pressure on Polyethylene Melt Rheology,* SPE Trans., **1,** (1962), 14.

3.13 CHOI, S.Y.: *Determination of Melt Viscosity as a function of Hydrostatic Pressure in an Extrusion Rheometer,* J. Polym. Sci., **6,** (1968), 2043.

3.14 GRAESSLEY, W.W.: *The Entanglement Concept in Polymer Rheology,* Adv. Polym. Sci., **16,** (1974), 1.

3.15 BUECHE, F.: *Viscosity of Molten Branched Polymers and their Concentrated Solutions,* J. Chem. Phys., **40,** (1964), 484.

3.16 SMOKER, D.G.: Ph.D. Thesis, *A Study of the Extensional Flow Behaviour of Low-Density Polyethylenes,* Loughborough University of Technology, (1984).

3.17 MUNSTEDT, H.: *Viscoelasticity of Polystyrene Melts in Creep Experiments,* Rheol. Acta., **14,** (1975), 1077.

3.18 MEISSNER, J.: *Dehnungsverhalten von Polyathylene-Schmelzen,* Rheol. Acta., **10,** (1971), 230.

3.19 PETICOLAS, W.L.: *Molecular Viscoelastic Theory of Polymers and its Connotations,* Rubb. Chem. & Tech., **36,** (1963), 1422.

3.20 BATCHELOR, J., BERRY, J.P., and HORSFALL, F.: *Die swell in Elastic and Viscous Fluids,* Polymer, **14,** (1973), 297.

3.21 TANNER, R.I.: *A Theory of Die Swell,* J. Polym. Sci., **A – 28,** (1970), 2067.

3.22 CRAWFORD, R.J.: *Plastics Engineering, 2nd Ed.,* Pergamon Press, Oxford, (1987).

3.23 GRAESSLEY, W.W., GLASSCOCK, R.D., and CRAWLEY, R. L.: *Die swell in Molten Polymers,* Trans. Soc. Rheol., **14,** (1970), 519.

3.24 EVERAGE, A.E., and BALLMAN, R.L.: *A Mechanism for Polymer Melt or Solution Fracture,* J. Appl. Polym. Sci., **18,** (1974), 933.

3.25 SHAW, M.T.: *Flow of Polymer Melts through a well-lubricated Conical Die,* J. Appl. Polym. Sci., **19,** (1975), 2811.

Chapter 4
Processing I – Continuous Methods

4.1 Introduction

The ease with which compounds may be shaped at relatively low temperatures represents an enormous cost advantage for plastics, within the materials processing sector. Those features of modern process technology listed below have also contributed to the economic advantages generally associated with plastics-based products:

- the ability to mould, or otherwise manufacture complex shapes in a single part;
- ease of incorporation of a vast range of compound constituents to modify specific properties (eg. self-colouring afforded by pigmentation; reinforcement by fibrous fillers);
- excellent quality of finish, making ancillary phases of manufacture unnecessary;
- application of computer-based process control methods and associated automation facilities.

Although many plastics processing methods were initially developed from existing technologies, the last 20-25 years have witnessed significant advances to define a more exact processing science. Recent growth has been more associated with an understanding of processing-physical property interactions; biaxially-oriented containers represent a good example of the range of plastics products, whose market was created by innovative processing.

At present, certain processes have been researched to an extent where a basic understanding has been unquestionably achieved. Other newer, or more complex techniques are still to receive such attention. To completely fulfil present levels of potential however, processing technologies must expand to become more thorough and precise in nature. To this end, it is the intention in Chapters 4 and 5 to present a "state-of-art" overview of the most important plastics manufacturing methods. Brief indications of hardware, working parts and modes of operation will be given, but special emphasis will be placed upon the role played by the processing characteristics of plastics relevant to each phase of production. Establishment of the initial but fundamentally vital link between polymer structure and thermo-rheological history defines the basis from which the properties of fabricated products can be predicted with an enhanced degree of confidence.

The route "from polymer to plastics commodity" involves several distinct, yet equally important stages. To an extent governed by specific choice of polymer grade, and also by the degree of overlap between successive phases of manufacture, the appearance, general quality and often - of prime importance - fulfilment of mechanical property specifications of the product are dictated (or at least modified) by the response of the polymer system to the thermal and mechanical demands of the process. A simplified schematic diagram of a total sequence of manufacturing operations for plastics parts is presented in Figure 4.1. If the processing elements in level (3) of this chart are considered in isolation from any mixing or compounding processes which may precede them, each constitutive sub-process may be directly associated with some key physical properties.

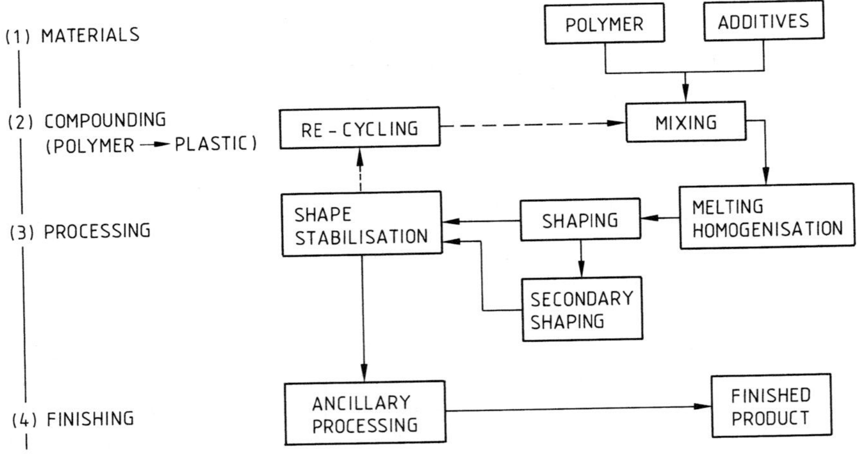

Figure 4.1 Processing flow chart for the general manufacture of plastics parts.

4.1.1 Melting

Based upon temperature alone, a process viability window is defined by a lower-bound crystalline *melting point* (or a "maximum viscosity" constraint) and an uppermost temperature which is usually associated with the onset of *thermal degradation.* The thermal energy requirement is specified by the *enthalpy* of the compound between the feedstock temperature (a lower reference point) and the maximum shaping temperature. For an economically-feasible process where output rates dictate machine design and overall viability, the *thermal diffusivity* (commensurate with the melting requirement) can often prove to be an important process rate-controlling factor. An additional characteristic – the tendency to dissipate "mechanical" heat during viscous flow – assumes an increasingly important role as the extent of melting (for example, the volumetric size of the melt pool in the compression zone of a single screw extruder; see Section 4.2.3.1) becomes significant. The rate of application of deformation (*shear rate),* in combination with the material's own resistance to flow (its *shear viscosity)* under the imposed conditions, contrives to create the generation of heat by mechanical means.

4.1.2 Shaping

Molten plastics materials are shaped relatively easily, under the influence of an externally applied stress; this is the most fundamental and important of all processing characteristics of long chain polymer systems. Fluidity or compliance functions are usually extremely sensitive to changes in temperature, strain rate and the magnitude/mode of stress. Taking the common example of melt flow under shear (eg. in an extrusion die or an injection mould) *shear viscosity* data are required to model the processing behaviour, and should be determined precisely, with specific reference to the temperatures, geometric constraints and processing rates likely to be encountered in practice.

To reiterate an underlying message from Chapter 3, however, it should be noted that an exact appreciation of shaping behaviour can be approached only if a more extensive and exacting flow simulation is available. Included within this category are viscosity data for *elongational flows* (where tapered dies or post-extrusion shaping processes are important), the magnitude and likely effects of *melt elasticity* and also the practical limitations in process rate imposed by *melt instabilities.* It is also desirable that deformation and flow properties be available to simulate the behaviour of plastics in secondary processes such as welding.

4.1.3 Shape Stabilization

Consolidation of shape by hardening is achieved by *cooling,* or by chemical *crosslinking,* for thermoplastics and thermosetting polymers respectively. Whilst the former route is a purely physical effect, the chemical crosslinking reaction in the latter is usually also associated with a heat exotherm. Each mechanism of shape-consolidation is associated with a degree of physical *shrinkage,* which must be pre-determined in order that equipment be designed to yield parts of optimum dimensional tolerance.

The time factors associated with shape stabilization are often process rate-controlling, simply because organic polymers are poor conductors of heat. Such situations, often further exaggerated by geometric shape constraints (eg. blow mouldings or hollow extruded profiles, where significant cooling occurs predominantly from one surface only) are more problematic, where products of thick section are concerned. Semi-crystalline plastics provide additional complexities: *latent heat* of crystallization and *supercooling* characteristics (see Chapter 2) each add to heat extraction difficulties.

Detailed discussions of the salient thermal and polymer melt characteristics, together with an indication of those factors which determine the magnitude of physical constants, have been presented in Chapters 2 and 3. Having established a brief overview of their relevance to practical commercial processes, it is necessary to discuss individual fabrication methods in more detail.

Continuous processes are considered first in the present Chapter; of particular note are processes based upon screw extrusion, since many of the more important principles may be extended or modified, for other processing methods. Chapter 5 is concerned with cyclic or intermittent *moulding processes,* where the complete 3-dimensional shape of the product is formed by the tooling and shaping technique. Although partly based upon the principles of screw extrusion, extrusion blow moulding and in-line thermoforming are considered in the later chapter, since the final form of the product is determined by the shape and dimensions of a mould.

4.2 Extrusion Processes

Defined broadly, the term *"extrusion process"* may represent any manufacturing operation in which a fluid is pumped through an orifice *die* to produce an article of constant cross-section. However, the initial shaping phase, where the plastics material is forced under pressure through an appropriately-shaped die, often represents only a convenient initiation point for manufacture, since there exists a multitude of products which only assume their final form following a secondary, *"post-extrusion"* processing phase. Such articles include components stretched uniaxially (fibres, tapes) or biaxially (films, sheets), plastics foams, laminates, and many other coated and/or composite structures[4.1 – 4.3].

Processes based upon *screw extrusion* of *polymer melts* dominate the modern extrusion industry; a vast diversity of plastics products (many groups of which are discussed in Section 4.2.5) is fabricated using extrusion principles. Although the largest proportion of the modern extrusion industry is associated with screw viscosity pumping of "dry" polymer melts, an expanded context of this process could also include a range of less-familiar shaping methods.

Alternative *pumping modes:*

(1) Intermittent extrusion, by direct ram or plunger methods, is still apparent in certain application areas, such as PTFE powder compaction (PTFE has an abnormally high "melt" viscosity and cannot be processed by conventional means), and rubber-blank preparation.
(2) Planetary gear pumps are utilized in series with screw extruders, in order to induce rapid, but more even and controllable flow characteristics in low viscosity melts (eg. PA, PETP monofilament extrusion: apparent viscosity in the range 10 – 100 Ns/m^2) in multi-orifice, spinneret dies.
(3)] More modern developments in solid-phase processing have been used to produce high-strength, oriented polymer sections by hydrostatic extrusion, and by die-drawing[4.4 – 4.6].

Alternative *feedstock forms:*

(1) Processing of solvated ("wet") polymer systems, where ultra-high viscosity or thermal instability precludes conventional "dry" extrusion.
(2) Solid-phase forming of thermoplastics in the thermoelastic or pre-melting range.

Process optimization is achieved only in situations where a degree of design compatibility exists between the mechanical operation of the extruder and the thermal/rheological characteristics of individual polymer grades. However, extremely high levels of compatibility are only achieved at the expense of machine flexibility. In practice, a compromise situation is usually sought, according to the expected length and diversity of production runs.

4.2.1 Machine and Material Requirements – An Overview

Machinery

The adopted specification for plastics extruders is primarily associated with maximum practical levels of mass output for candidate polymers. Particular design features are also specified, and power requirements must be sufficient to generate and continuously maintain high pressures required for shaping high viscosity materials at correspondingly high output levels, allowing also for the occurrence of secondary modes of power dissipation (e.g. adiabatic heating due to internal shear). It follows that the barrel and die assemblies should be subject to rigorous pressure-engineering design procedures and must offer good resistance to corrosion and wear during service life. Particular problems of this type for thermoplastics extrusion include the release of hydrogen chloride vapour from PVC dehydrochlorination, and barrel wear due to the highly abrasive nature of mineral-filled or fibre-reinforced compounds.

Materials

Specification of candidate extrusion-grade polymers is centred upon two important selection criteria: melt flow properties (at intermediate/high shear rates) and thermal stability.

(A) Melt Flow Properties

(1) Melt Viscosity. Although high melt-index (low shear viscosity) polymers offer advantages by way of lower working pressures and reduced running costs, extruders are sufficiently powerful to process high molecular weight grades so that an additional degree of melt strength can be realized, thereby assisting shape stability in the extrudate prior to sizing. Product properties (notably mechanical strength) are also enhanced by an appropriate choice of high molecular weight resin.

(2) Melt Strength. The tensile or shear rupture strength of polymer melts imposes an effective working limitation upon throughput rate in, for example, tapered die sections. It is the combination of these two parameters – output rate and taper angle – which determines the elongational strain rate within the section and thus the corresponding tensile stress acting on the melt. If this is excessive, the rupture stress of the melt may be exceeded, and melt fracture (see Chapter 3) may occur. Similar arguments can also be applied to other types of elongational flows, such as bubble fracture during film blowing, or parison rupture in extrusion blow moulding.

(3) Melt Elasticity. A precise knowledge of the recoverable shear strain characteristics of melts allows the die designer to account for the discrepancy between the cross-section of extruded products (under any given operating conditions) and the die exit dimensions. In this way, die swell characteristics can be used at an early stage by equipment designers. Swelling ratios lie within a typical range 1.0 – 2.5, but are inevitably sensitive to changes made to processing temperature, strain rate, die land geometry and molecular weight.

(B) Heat Stability

Thermal instability is a commonly encountered facet of melt processing and occurs not only because of high temperature, but also due to excessive residence time spent at lower temperatures which may otherwise be considered safe. In consequence, the *residence time distribution,* for a given process, assumes a high level of importance when measures are taken to prevent any physical hang-up of material in the extruder, for example by reducing taper angles, or by applying generous radii to sharp changes in section.

4.2.2 Extruder Hardware and Machine Component Functions

A brief review of extruder hardware is presented in this section, which provides a basis from which the physical and dynamic flow characteristics of polymers can be related to some different, but equally important phases of extrusion.

4.2.2.1 Machine Components

In order to compete in specific areas of the market, screw-plasticating extruders are subject to many radically different mechanical designs. However, the general schematic shown in Figure 4.2 typifies what could be considered a "standard" single-screw machine assembly.

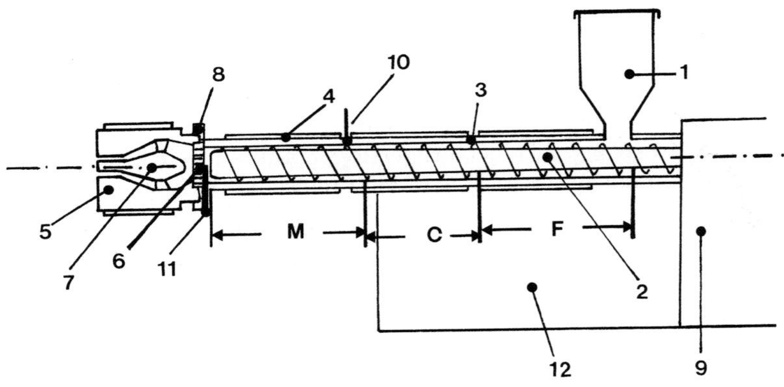

Figure 4.2 Functional parts of a typical single screw plastics extruder: (1) feed hopper (gravimetric discharge); (2) screw; (3) barrel; (4) electrical heaters; (5) die body; (6) breaker plate and screen pack; (7) die mandrel; (8) die-head clamp; (9) motor and reduction gearbox; (10) thermocouples; (11) pressure transducer; (12) control panel. (F), (C) and (M) refer to the feed, compression and metering zones of the extruder.

Feed System. Simple vertical (gravimetric) feeding of powdered or granular feedstock is the dominant feeding mode in single screw extrusion. Any minor fluctuations in feed rate are generally absorbed and nullified by the compaction which occurs in the downstream screw channel. However, the use of controlled-rate *dosing units* (usually consisting of a shallow, rotating, ribbon-bladed feed-screw assembly above the main hopper) is recommended for most twin-screw processes (to avoid excessive loading on the thrust bearings; or, to provide accurate metering rates in starve-fed processes) and during situations in which uniformity of composition is important. Some additional "bolt-on" features include the following:

- pre-heating facilities (circulating air-heaters or desiccant-type dehumidifying driers) to remove moisture from hygroscopic plastics (notably polyamides and polyesters);
- vibrating agitator units and/or hopper throat cooling (water circulation) each help to prevent the possibility of material "bridging" at the feed port, due to premature surface melting of feedstock;
- automatic feedstock loading devices, operated by an air-pressure differential.

Barrel and Screw Assembly. The rotating screw must form a precise concentric fit within the cylindrical steel barrel; otherwise, premature wear of metallic surfaces occurs with significant losses in machine performance (see Section 4.2.3). It is possible to reduce wear rates in tool steels by nitriding, or by specification of case-hardened steel alloys.

The *screw* is housed in a robust *thrust-bearing* assembly and is driven by a variable-speed *electric motor* (speeds up to 100 RPM (1.66 s^{-1}) are typical) via a reducing gear train. Externally applied heat energy is derived from electrical cuff-heater bands and is conducted to the polymer through the barrel wall; deep-set thermocouples are an integral part of closed-loop temperature control systems. It is clear from Figure 4.2 that the extruder screw is sub-divided into various zones according to the geometry of the helical screw channel. In the "general-purpose" screw design, the screw pitch and flight land width, together with the helix angle and outside diameter, are each held constant (see also Figure 4.3): the *compression* required to accompany the solid-melt phase change is imposed by decreasing the depth of the screw channel (hence decreasing the channel volume per turn, usually linearly as a function of distance along the screw) in an intermediate *compression zone.* Preceding this transition section is the *feed zone* which functions as a conveying device for polymer granules, but of greater interest with respect to process dynamics is the *metering zone,* where fully-homogenized melt is pumped to the die at a controllable, but precise and consistent rate.

The flow mechanisms which underlie these aspects of single screw extrusion, in conjunction with the factors which influence output rates, represent the fundamental principles of the extrusion process: they are closely scrutinized in Section 4.2.3.

The Die. The bilateral form of the extruded product is determined by the extrusion die. However, the final cross-sectional dimensions of the part are modified by *thermal contraction,* by *drawdown* effects and perhaps of greatest importance, by time dependent relaxation of elastic shear strain. The last effect is known as *die swell;* the magnitude of elastic recovery is strongly dependent upon process temperature, shear rate, melt viscosity and molecular weight. These molecular and dynamic interrelationships must be determined at the drawing board stage, in order to fulfil any die design specification for a particular product.

Instrumentation is provided by independently controlled heaters to various zones of the die, in addition to a pressure monitoring device (transducer) between the die head and screw tip.

Breaker Plate. Located on a purpose designed seat (under the connection flanges of the barrel and die) is a circular breaker plate which is machined to allow the passage of polymer melt through a multitude of small, axially-bored, horizontal parallel channels. (Typical bore dimensions are length 10 – 20 mm and diameter 2 – 3 mm). Incorporated in flutes at the upstream side of the breaker plate is a steel-mesh *screen pack,* whose functions are twofold:

(1) to filter contaminating particles of significant size (e.g. metal slivers, oxides, unmelted granules) to diminish the likelihood of stress concentration sites being promoted within the product;
(2) to provide a means by which a degree of fine-tuning can be imposed upon the die head pressure, thus enabling an additional degree of mixing to be imposed on the flowing melt.

More recent developments in breaker plate design have allowed *continuous melt filtration* to become feasible on a commercial scale, thereby increasing throughput rates by reduction of working pressures and down time.

4.2.2.2 Machine Specification

Having listed the major constituent parts of the extruder, it is useful to summarize the elements of a typical machine specification.

Barrel Diameter (D). As will become evident in a subsequent section, the theoretical output capability of an extruder is proportional to the square of the barrel diameter.

Length-to-Diameter Ratio (L/D). For a given screw diameter, the capacity available to melt, mix and homogenize at a given screw rotation speed increases with increasing length of screw. High L/D ratio extruders are also preferred in the following situations:

- to maintain a given residence time distribution in instances where line speed increases are envisaged;
- to allow a precise level of temperature control to be exercised;
- to help generate additional pressure required for long or narrow die sections;
- to allow for the provision of additional screw zones, such as decompression (venting), double-compression, or pupose-designed mixing sections.

Typical L/D ratios for thermoplastics are in the range $20:1 - 30:1$.

Screw Zone Configuration. Ultimate choice of the number and geometric design of screw zones is a complex and iterative procedure. Such a decision depends not only upon the envisaged die design and throughput rates, but also upon the softening/melting characteristics of the polymer, together with an awareness of how the melt viscosity changes as a function of process temperature and output rate. A simple, three zone, single-start screw is usually expressed in terms of the number of turns (diameters) in the respective feed, compression and metering zones.

Compression Ratio. An exact definition of compression ratio is "the volumetric ratio of single, helical turns of the extruder, in the respective feed and metering zones". It is more commonly expressed in terms of the equivalent *channel depth ratio,* an approximation which is, of course, only valid if the helix angle, pitch and flight width are constant throughout. Typical compression ratios lie between 2.0 and 4.0.

Rather than focus on the compression ratio itself, it is often more convenient or meaningful to consider the metering zone channel depth as the parameter of greater relevance, especially if we consider that the feed zone channel depth to diameter ratio is usually held constant (0.15 - 0.2, typically). A shallow metering zone (higher compression screw) imposes a greater degree of shear strain on the melt (for a given screw speed), and is also associated with a much steeper pressure gradient.

4.2.3 Plastics Flow Analysis in the Screw Extruder

Highlighted within this section, in the context of machine performance, is the special importance of the melt pumping (metering) zone of the extruder.

In consequence, the development of the flow equation (the *extruder characteristic)* assumes a special degree of significance in understanding the functions of processing variables, and their influence on machine efficiency and product quality. Simplified flow analyses of both single screw and intermeshing twin screw systems are reviewed, together with the effects of die restriction.

4.2.3.1 Single Screw Extrusion

First, consideration is given to the successive polymer transportation modes which occur within each zone of the extruder:

Feed Zone

Feedstock is delivered along the feed zone to a position where significant melting and compression can be developed. It follows that there is usually an element of "over design" in the solids conveying zone, if a starve fed situation is to be avoided. In consequence, the characteristics of polymers quoted below are often seen as holding only secondary significance: the output capability of the machine is likely to be determined by the metering zone and its interaction with the extrusion die.

Transportation of granular or free-flowing powdered feedstock occurs by a predominantly *plug-like flow,* represented by relatively flat velocity profiles across the screw channel. The "zero wall velocity" criterion assumed for shear flows is generally unfulfilled. Instead, the relevant characteristics of polymers with respect to solids conveying are the respective *coefficients of friction* between polymer granule surfaces and the barrel (μ_B), and the screw (μ_S). For example, if μ_B is increased (by increasing the temperature or by introducing an element of surface roughness on the internal barrel surface), the ratio $\mu_B : \mu_S$ is therefore increased, and transportation rates are then enhanced when polymer-barrel adhesion is significant.

Any pressure generation in the feed zone is slight and relatively unimportant, although a degree of compaction is beneficial to subsequent softening and melting, since it enhances the extent of inter-particulate thermal contact. The progressive nature of compaction often negates the necessity for a venting zone, because air and gaseous volatiles are able to migrate back to low-pressure regions through the compacted yet porous solid-bed.

Compression Zone

Parallel developments of compression, melting and viscous flow are complex and near-impossible to analyse quantitatively. Melting mechanisms specifically are discussed in Section 4.2.4; with respect to "flow", it may be helpful to summarize as follows:

- the degree of viscous flow is proportional to the volume of the melt pool, which is itself developed by the combined influences of compaction pressure and external/internal heat energy sources;
- in common with the comments made earlier, the influence of the compression zone on overall machine throughput is secondary.

It is required that the polymer melt is completely developed at the downstream end of the compression zone.

Metering Zone

The geometry of a typical metering zone screw channel, for a single-start unit, is illustrated in Figure 4.3; the geometric variables are defined as follows:

Barrel diameter, $D = 2R$	Channel depth, $H = R - R_i$
Screw helix angle $= \theta$	Screw clearance, $h = R - R_0$
Screw pitch $= (B + b)$	Channel/flight widths $= W; w$
Screw rotational speed $= N$	Axial channel/flight widths $= B; b$

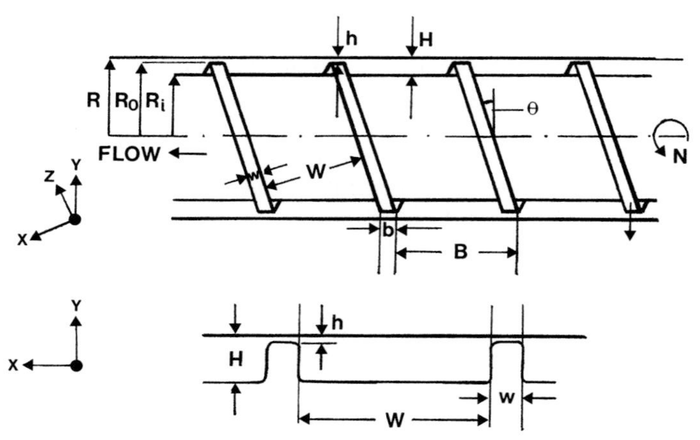

Figure 4.3 Metering zone of a single screw plastics extruder; definition of screw channel geometry and cartesian co-ordinate system.

When developing the extruder flow equation, it is usual to introduce some further assumptions to simplify the analysis:

(1) the element of curvature along the barrel is ignored i.e. gravitational forces are considered negligible and the screw channel is assumed to be "unrolled" to lie in a single x-y plane;
(2) the barrel (as opposed to the screw) is assumed to move linearly with velocity V;
(3) the lateral dimensions of the rectangular channel are constant; width (W) is considered to be much greater than depth (H); any flow perturbations due to geometric irregularities (i.e. finite flight-root radii) are ignored.

Having defined a starting point, the most convenient way in which the flow equation can be developed is by considering individually each of the different flow mechanisms which occur:

(A) Screw Rotation – Drag Flow

The velocity differential between adjacent screw and barrel surfaces induces the primary driving force for melt transportation (i.e. by shear) in the positive, down-channel direction. In a planar simulation, the flat barrel surface is propagated at a steady rate (velocity = V) giving a down-channel velocity component V_z, where:

$$V_z = V \cdot \cos\theta \tag{4.1}$$

This concept is illustrated in Figure 4.4, where the downstream (z-axis) velocity profiles are shown. The general form of the equation relating volumetric flow rate to screw channel geometry is expressed as:

$$Q_D = \int_O^H W \cdot v(y)\, dy \tag{4.2}$$

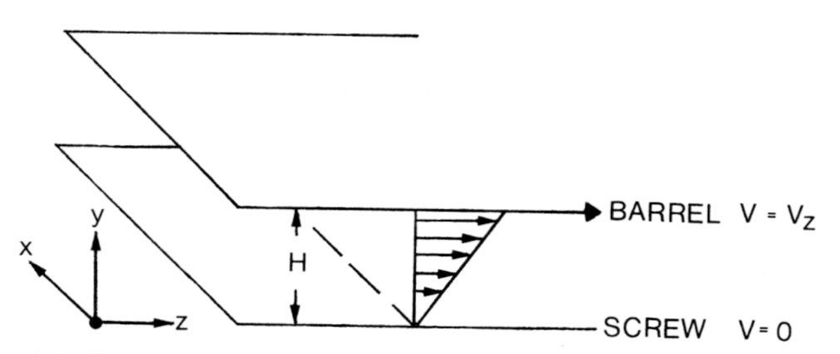

Figure 4.4 Drag flow mechanism in a single screw extruder.

The boundary conditions are:

$$v = V_z \quad \text{at} \quad y = H; \qquad v = 0 \quad \text{at} \quad y = 0.$$

However, in the simple case where the velocity profile is linear, the mean velocity may be used: i.e.

$$v(y) = V_z \cdot \frac{y}{H}$$

The integral in (4.2) is expanded thus:

$$Q_D = \frac{WV_z}{H} \int_0^H y\, dy = \frac{WV_z}{H} \cdot \frac{H^2}{2}$$

$$Q_D = \frac{WV_zH}{2} \tag{4.3}$$

Equation (4.3) relates volumetric throughput (due only to the shearing forces imposed by screw rotation – *Drag Flow)* to the product of "mean velocity" ($V = V_z/2$) and cross-sectional area of the slit-like channel. It is of equal relevance to situations where the screw is not a "single-helix" type, so long as the relevant multiple for the number of thread starts (per screw) is inserted into the expression.

The tangential velocity (V) at the barrel surface is given by:

$$V = \pi DN \tag{4.4}$$

Combining (4.1) and (4.4):

$$V_z = \pi DN \cdot \cos\theta \tag{4.5}$$

Therefore:

$$Q_D = \frac{1}{2} \cdot (WH\pi DN \cdot \cos\theta) \tag{4.6}$$

Immediately, it is noted that drag flow is directly proportional to the rotational speed of the screw, to an extent dictated only by the geometry of the extruder. As long as the boundary conditions are fulfilled (for example, barrel-adhesion must be apparent), melt properties play no direct part in determining the magnitude of drag flow.

(B) Effect of Flow Restriction – Pressure Flows

Overall output is equal to drag flow (Equation 4.6) in *open-discharge* situations ie. where no downstream restriction is imposed on the flowing melt. Whenever a die or any additional component (such as a breaker plate assembly) which suppresses flow is attached to the extruder, the pressure created (often over-simplistically termed *back-pressure)* acts against the drag forces to create a *Pressure Flow* (Q_P) on the material within the screw channel. This can be considered to be superimposed upon (and counteracting) drag flow, as illustrated in Figure 4.5. In such circumstances, Q_P is simply a reaction to the restriction imposed upon Q_D.

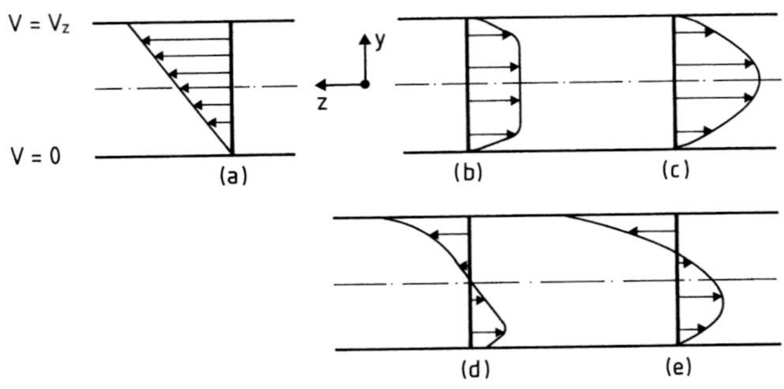

Figure 4.5 Pressure flows in single screw extrusion (b,c), and the superimposition with drag flow (a), for melts of low pseudoplasticity index (b and d) and for near-Newtonian plastics (c and e).

The velocity profiles in Figure 4.5 (d) and (e) simply represent convenient analogies to illustrate the effect of increasing die restriction on fluids of different pseudoplasticity indices. Although pressure flow detracts from overall pumping capacity, flow suppression also tends to enhance the degree of mixing within the screw channel, as a result of:

(1) a local increase in shear stress (dispersive mixing);
(2) increase in the total imposed shear strain (distributive mixing).

Pressure flows created in extrusion dies are identical in nature to those for which the derivations made in Chapter 3 were evaluated. For a slit-section of width W and depth H, volumetric output due only to pressure flow is expressed in the form:

$$Q_P = -\frac{WH^3}{12\mu} \cdot \frac{dp}{dz} \tag{4.7}$$

(Note that the negative sign, by adopted convention, indicates flow away from the die head, towards the feed end of the extruder.)

A particular type of pressure flow is the backflow which occurs through a shallow, "twisted annular slit" channel, defined by the gap between barrel surface and screw flight land (see Figure 4.6). *Leakage Flow* is apparent whenever extruders are old or become worn and may be included within the overall extruder flow equation. However, unlike the direct pressure flow Q_P, it represents a real backflow of melt, towards the upstream end of the machine.

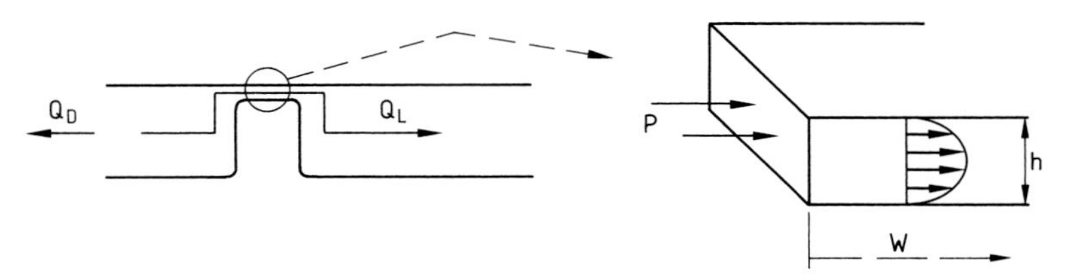

Figure 4.6 Leakage flow mechanism in a single screw extruder: pressure flow acting across the screw-flight gap.

(C) Total Output

Total machine output is assumed to be a simple algebraic sum of the individual flow components described:

$$Q = Q_D + Q_P + Q_L \tag{4.8}$$

(Q_P, Q_L are negative)

It is possible to restructure the individual terms of Equation 4.8 to evaluate a more meaningful expression for output. From Figure 4.7:

$W = (B - b) \cdot \cos\theta$, and since $\tan\theta = B/\pi D$, it follows that $W = \pi D \cdot \frac{(B-b)\sin\theta}{B}$.

If $B \gg b$, then $W \approx \pi D \cdot \sin\theta$, and the term for drag flow (Equation 4.6) reduces to:

$$Q_D = \pi^2 D^2 NH \cdot \frac{\sin\theta\cos\theta}{2} \tag{4.9}$$

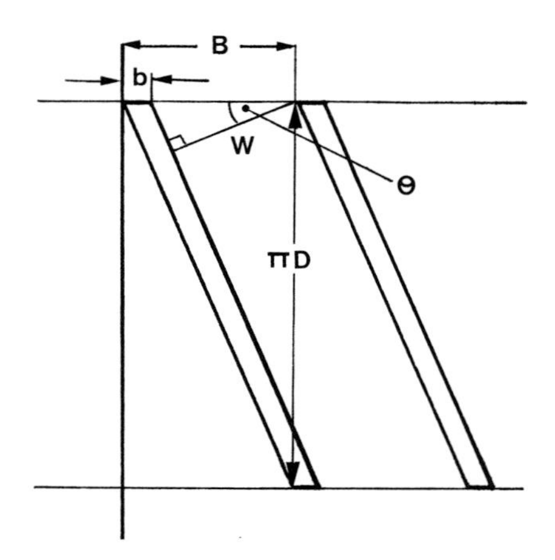

Figure 4.7 'Unrolled' single turn of an extruder screw helix.

A similar substitution for W can be made to the pressure flow term (Equation 4.7), and rearranging in terms of the axial pressure gradient, it is noted that:

$$\frac{dp}{dz} = \frac{dp}{dl} \cdot \sin\theta$$

Assuming a linear pressure gradient:

$$Q_P = -\pi D H^3 \sin^2\theta \cdot \frac{\triangle P}{12\mu L} \tag{4.10}$$

Leakage flow is a particular type of pressure flow, so that the general expression for Q_L is of the form quoted earlier in Equation 4.7. However, the shape of the helical slit channel through which leakage flow occurs can be described by:

$$\text{"width"},\quad W = \frac{\pi D}{\cos\theta};\qquad \text{"height"},\quad H = h;\qquad \text{"length"},\quad L = w = b\cdot\cos\theta$$

Using a similar procedure to that above, and translating to an axial pressure gradient, we obtain:

$$Q_L = -\pi^2 D^2 E h^3 \tan\theta \cdot \frac{\triangle P}{12\mu b L} \tag{4.11}$$

where E is an eccentricity factor relating the axial alignment between screw and barrel (equal to unity for a perfectly concentric housing).

Substitution from Equations 4.9, 4.10 and 4.11 can be made into 4.8 to give the *single-screw extruder characteristic:*

$$Q = \pi^2 D^2 N H \frac{\sin\theta\cos\theta}{2} - \pi D H^3 \sin^2\theta \cdot \frac{\triangle P}{12\mu L} - \pi^2 D^2 E h^3 \sin^2\theta \cdot \frac{\triangle P}{12\mu L} \tag{4.12}$$

This has the general form:

$$Q = \alpha N - \frac{\beta}{\mu} \cdot \triangle P \tag{4.13}$$

(α and β are constants of screw geometry).

Equation 4.12 is referred to as the *Extruder Characteristic* and assumes fundamental importance in allowing predictions of extruder output rates (for plastics compounds of pre-determined viscosity) to be made. The foregoing analysis, based on an array of assumptions, is inevitably subject to error, the magnitude of which is best determined by experimental verification. Nevertheless, it does allow an immediate semi-quantitative assessment of the likely effects of input variable alterations. The most convenient way of doing this is by using a graphical form of the output-pressure extruder characteristic (Figure 4.8). Two practically-relevant situations are immediately apparent:

(1) Open Discharge ($\triangle P = 0$); this situation is encountered in the absence of flow restriction; that is, zero pressure flow results in the condition $Q = Q_D$. For a given screw speed, Q is a function only of the barrel diameter, channel depth (compression ratio) and helix angle. Melt flow characteristics play no direct role in establishing the maximum theoretical output for a given machine design.
(2) Finite Working Pressure ($\triangle P = \triangle P_x$); if the die-head pressure ($\triangle P_x$) can be predicted, it is a simple matter to anticipate the resultant machine output Q_x (Figure 4.8), under such conditions.

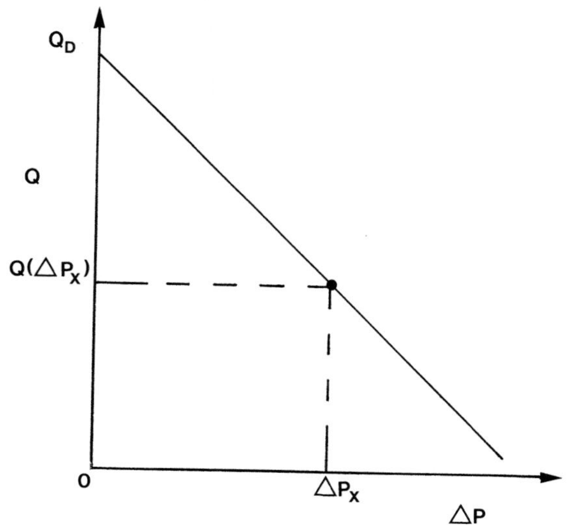

Figure 4.8 Pressure-dependence of extruder output: the *extruder characteristic* (linear, Newtonian solution).

The apparent sensitivity of the output capability of single screw extruders to head pressure effects is well-documented and often imposes severe limitations upon the likely profitability of manufacturing lines. An expression to relate a *process inefficiency factor* Q_P/Q_D, to several important parameters is obtained by combining Equations 4.9 and 4.10:

$$\frac{Q_P}{Q_D} = H^2 \tan\theta \cdot \frac{\triangle P}{6\pi N \mu L D} \tag{4.14}$$

High Q_P/Q_D ratios emphasize the inefficiency of single screw extruders; the sensitivity of output to head pressure contributes to the necessity for the development of more positive pumping mechanisms, such as the twin screw systems described in a following section.

Influence of polymer melt properties

With respect to the flow analysis, only a single term for 'viscosity coefficient' assumes a direct relevance to throughput rates. However, it must be noted that the compound must fulfil the usual assumptions demanded for shear flow analysis, which underlie the basis of the pressure flow derivation. Although the simple mathematical treatment is a direct consequence of the "Newtonian Fluid" assumption in Equation 4.7 (hence also Equations 4.10 and 4.11), it must be appreciated that since most thermoplastics melts are pseudoplastic, the analysis is subject to error whenever the shear strain rate is changed; the magnitude of the discrepancy is determined by the deviation of the power law index from unity.

Nevertheless, this approach is sufficient to illustrate trends; the degree of quantitative error is often sufficiently small to alleviate the need for a more rigorous approach.

4.2.3.2 Twin Screw Extrusion

The most common example of the family of multi-screw extruders is the parallel *twin-screw,* the development of which was partly stimulated by the inability of single screw systems to process materials which exhibit tendencies to degrade, or to slip excessively on the barrel surface.

In order to describe the principles of melt transportation and mixing, reference is given to two different types of twin screw extruders: the intermeshing, *counter-rotating* machine is used almost exclusively for the extrusion of UPVC dry blends, whilst the *co-rotating* extruder is designed to operate as a *continuous compounding* device.

(A) Counter-Rotating Twin-Screw: Forward Pumping Mechanisms

Whereas the pressure-sensitive pumping action of the single-screw machine is based on a flow mechanism determined by the wetting characteristics of the polymer to the respective surfaces of screw and barrel, the more *positive conveying* characteristics of a typical intermeshing twin-screw machine are realized by a completely different and discrete transportation system.

Consider Figures 4.9 and 4.10, which illustrate how the plastics compound occupies individual, twisted C-shaped chambers; these convey the compound along the extruder as the screws rotate. These cavities are effectively bounded by leading and trailing edges of adjacent flights, and by the rotating root surface of the opposite screw. Once the polymer is held within these chambers, the maximum pumping capability of the machine can be readily characterized:

$$Q = 2mNV_c \tag{4.15}$$

where m = number of thread-starts per screw; N = screw rotation speed; V_c = trapped volume of a C-cavity.

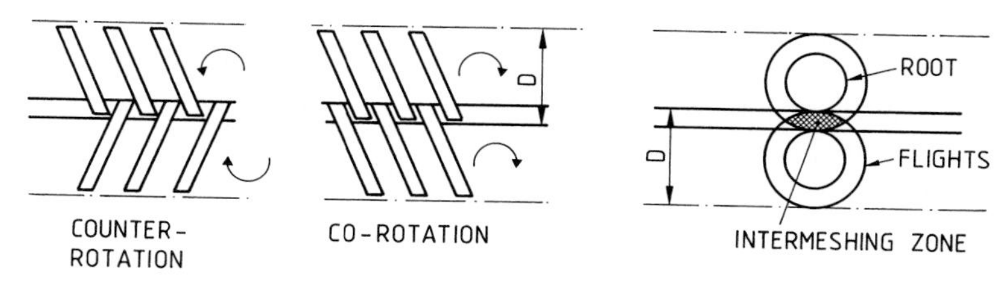

Figure 4.9 Principle of intermeshing in twin screw extruders.

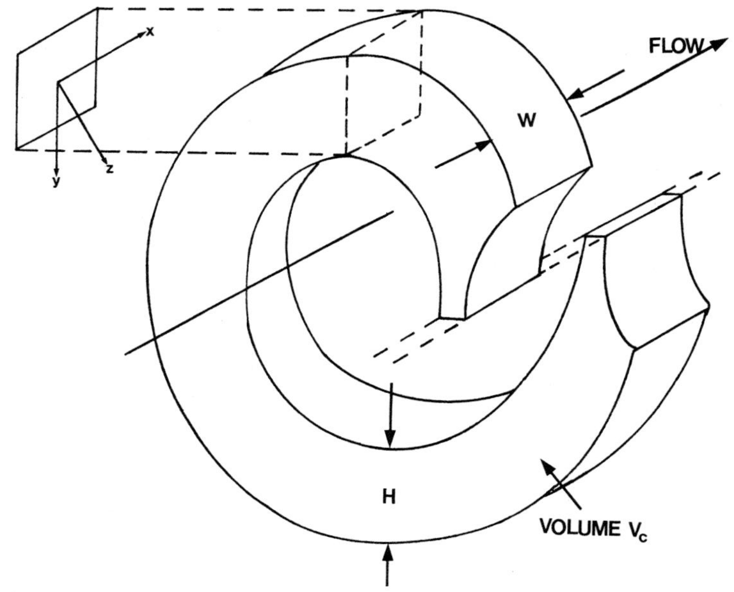

Figure 4.10 Geometric shape of the discrete "C-shaped" chambers, in an intermeshing, counter-rotating twin screw extruder.

The factor of two, in Equation 4.15, denotes the existence of two parallel screws. Geometric computations (refer to Figure 4.10) allow V_c to be estimated:

$$V_c = \pi WH \cdot \frac{(D - H)q}{\cos\theta} \tag{4.16}$$

(q is a dimensionless factor representing the proportion of cavity which is unobscured by meshing.)

Direct substitution yields a simple form of the *Twin-Screw Extruder Characteristic*[4.7]:

$$Q = 2\pi mNWH \cdot (D - H) \tag{4.17}$$

- both q and Cos θ usually lie in the range 0.9 – 0.95, and can often be considered to be self-cancelling;
- geometric variables are relevant only to the discharge end of the machine.

Equation 4.17 is analogous to the expression for drag flow presented earlier (Equation 4.6), although it must be appreciated that the underlying polymer flow mechanisms are entirely different, in either case. As with the earlier derivation for single screw machines, only the overall downstream component of velocity has been considered. It is possible, in turn, to assess the cross-flows within an enclosed cavity; the contributions from these to mixing characteristics are described elsewhere[4.7].

Flow Restriction – Pressure Effects. Intermeshing twin-screw extruders deliver polymer melts to the die in an efficient, positive manner; since "flow" is discontinuous, it is not feasible to consider a direct pressure flow effect (ie. analogous to Equation 4.10). In consequence, the gradient ($-dQ/d\triangle P$) of the twin-screw characteristic is very shallow (Figure 4.11) in comparison to the single-screw machine. This important observation emphasizes the advantages which twin-screw machines assume: an ability to melt and pump plastics compounds in a much more *energy efficient* manner, which is itself relatively insensitive to changes in polymer viscosity, surface melting characteristics and screw speed.

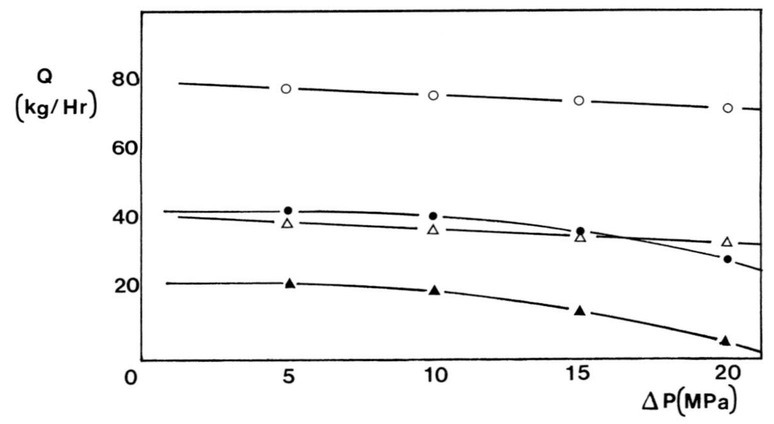

Figure 4.11 A comparison between the pressure-dependent output characteristics of twin-screw (△,○) and single-screw (▲, ●,) extruders. The screw speed depicted by the open symbols can be considered to be twice the speed represented by the closed symbols.

A glance at Figure 4.11 reveals that, despite a relatively low gradient, $dQ/d\triangle P$ for the twin screw machine is indeed negative, indicating that die head pressure – albeit to a lesser extent than in single screw extrusion – still in part determines the overall extruder throughput. A number of *leakage flows* are responsible for this effect[4.7]; these are illustrated in Figure 4.12 and are summarized below:

Q_F – *flight-gap* leakage flow, which occurs between screw flight tip and barrel surface and is analogous to Q_L in single screw extrusion;

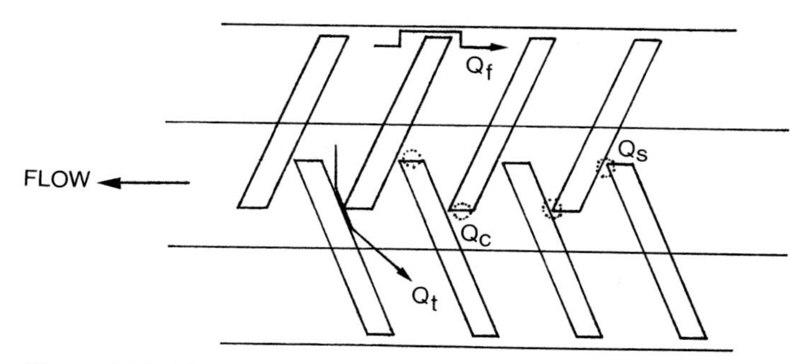

Figure 4.12 Mechanisms of leakage flow in a closely-intermeshing, counter-rotating twin-screw extrusion system.

Q_C – *calender-gap* leakage flow, between flight-tip and adjacent screw root;

Q_S *(side-gap)* and Q_T *(tetrahedron-gap)* leakage flows each occur between adjacent (rotating) screw flight flanks. Q_T differs from each other mode, since the path over which back flow occurs is not confined to any one of the screws; this occurrence, despite being associated with pumping inefficiency, makes a positive contribution to the mixing performance of the extruder.

The magnitude of individual leakage flows is a function of screw geometry, the degree of intermeshing and the relative rotation direction of the screws. Since these reverse flows are induced by both drag and head-pressure forces, it is supposed that high viscosity polymer melts will suppress the driving forces for leakage flow. However, analysis of each individual mechanism is complex, since variable pressure distributions exist (within chambers), screw clearances are complex in shape and are, of course, also subject to changes in dimensions as wear increases. For a thorough analysis of these aspects of twin screw extrusion, the reader is referred to more specialist monographs[4.7,4.8].

The total contribution to leakage flow is given by:

$$Q_L = 2Q_F + 2m(Q_C + Q_S) + 2Q_T \tag{4.18}$$

Therefore, the twin screw discharge characteristic (combining 4.17 and 4.18) is expressed as:

$$Q = 2\pi mNWH \cdot (D - H) - 2(Q_F + Q_T + m \cdot [Q_C + Q_S)]$$

i.e.

$$Q = 2m\left\{\pi NWH \cdot (D - H) - \left[Q_C + Q_S + \frac{1}{m} \cdot (Q_F + Q_T)\right]\right\} \tag{4.19}$$

where Q_C, Q_S, Q_F and Q_T are each positively related to the pressure-viscosity quotient $\triangle P/\mu$.

(B) Co-Rotating Twin-Screw: Some Aspects of Machine Output

The primary function of most commercial co-rotating twin screw extruders is to act as an effective *continuous mixing* device; not surprisingly, it is the design of the kneading zones, and the agitator segments which make up the mixing sections (rather than the pumping, or discharge zones), which are most critical to the performance of the extruder. Moreover, for a typical compounding process, the extruder is often *"starve-fed"* (by a constant-rate feed-screw unit) so that output cannot be affected by the material flow characteristics, or by the machine variables.

However, it is worth noting that the pumping mechanism in the discharge zones of co-rotating designs is fundamentally different from the situation described above (in (A), for counter-rotating machines), because the material cannot exist in closed, discrete cavities. Instead, by a "figure-of-eight" type drag flow mechanism (see 4.2.4.2; Figure 4.22), the material is able to flow along these passages, in pseudo-spiral fashion, to move towards the downstream end of the screws.

MARTELLI[4.8] has described how forward pumping in co-rotating extruders can be understood in terms of an "equivalent screw", in order to come to terms with the complex nature of the drag-induced discharge mechanism. The maximum output capability is evaluated as the product of screw channel area, the "figure-of-eight" length and the rotational speed of the equivalent screw; the solution is given by:

$$Q = \frac{1}{2} \cdot (\pi^2 D^2 N H \cdot \tan\theta) \tag{4.20}$$

This expression is the closest analogy to Equation 4.17, relating theoretical output to machine variables and screw design. However, in common with other pumping mechanisms, the effects of downstream pressure and associated reverse flows detract from practical attainment of these output values. The co-rotation design is thought to represent an effective means of forward melt transportation; however, this must remain only of partial relevance to practical processes based upon starve-feeding.

(C) Summarized Comparison between Single-Screw and Twin-Screw Designs

A review of the characteristics of both single-screw and twin-screw machines is presented in Table 4.1; this is perhaps especially useful when studying UPVC extrusion, which is feasible by each of these methods.

Table 4.1 Summary of Extrusion Characteristics: Single and Twin Screw

	Single Screw	Twin Screw (Counter-rot.)	Twin Screw (Co-rot.)
Flow Mechanism	Continuous Shear	Discrete C-sections	Figure-of-8 (Continuous)
Pumping Efficiency	Variable	Good; Positive	
Die Restriction	Often Severe	Smaller Effects (Leakage Flows Only)	
L/D Ratios	> 20	$16 - 20$	Various
Compression	Decrease Channel Depth	Various Designs	
Screw Speeds	20 – 100 RPM	50 RPM (Max.)	Up to 500 RPM
Heating Mode	High Proportion by Shear	Controllable; Low Shear	Near-Adiabatic
Residence Time	Large Spread; Wide Distribution	Narrow Distribution Often Easy to Control	

4.2.3.3 Extrusion Dies and Effects of Die Restriction

However well-designed the extruder and its operating conditions are, it is the *extrusion die* which often plays a dominating role in determining part quality and correct dimensional consistency. Mechanical die design techniques are usually based on an appropriate balance of experience and technological expertise; since the latter is developed from a knowledge of the viscous and elastic characteristics of molten plastics, and their interaction with die geometry, this section highlights the effects of extrusion dies on the performance of the complete extrusion process. Before advancing to discuss machine-die interaction, the overall construction of dies is briefly reviewed.

Elementary Concepts of Die Design. Depicted in Figure 4.13 is a simplified diagram of a plastics extrusion die for circular-section profile.

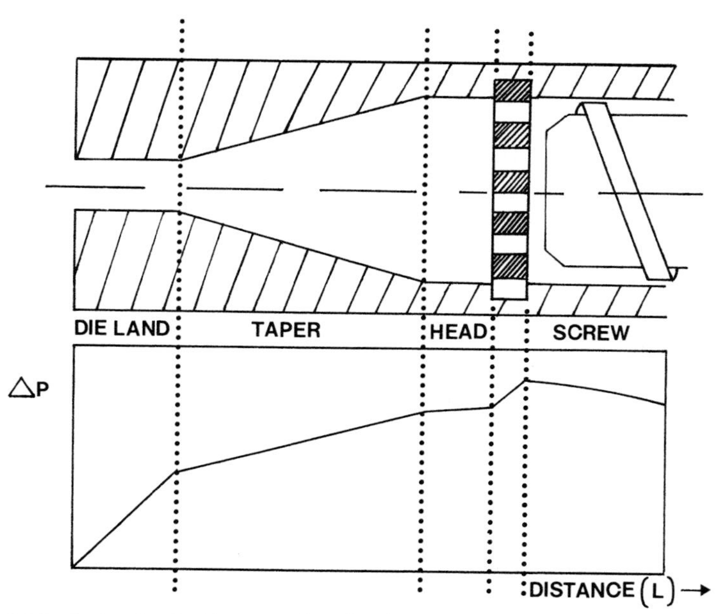

Figure 4.13 A simple plastics extrusion die; die construction and a steady-state pressure profile.

Die design methods to evaluate the optimum dimensions of each constituent zone are inevitably iterative in nature, and such procedures are usually initiated by considering what are often the only pre-constrained input variables: the required product diameter (related, but not equal to the die exit dimension), together with the anticipated or desired processing rates.

Successive requirements for land and taper length, taper angle and other 'upstream' dimensions are evaluated by working from plastics flow curves to determine the magnitude of pressure drop for steady flows throughout each die sub-section, using techniques described in Chapter 3. The total pressure drop is computed by adding a series of individual pressure losses; these include:

(1) Flow in parallel sections (breaker plate, pre-taper and die-land): pressure losses due to shear flows;
(2) Flow in converging tapers: the total pressure drop is approximated by the algebraic sum of individual contributions from shear and elongational flows. The magnitude and importance of elongational-flow pressure losses is related to the degree of taper;
(3) Entrance pressure losses due to severe changes in section.

Outlined in Figure 4.13 is a die-section pressure profile; the largest single contributor to the overall pressure drop is the "depth" dimension (in this case, the radius) of any section (for example, see Equation 4.21).

Since pressure losses also vary directly with orifice length, an immediate reaction is sometimes to minimize die section length, by increasing the severity of taper angles. Due to some of the special properties of polymer melts, there are limitations on the extent to which this is feasible[4.1]:

- Polymer *melt elasticity* demands that die land sections must be sufficiently long to erase *memory effects,* which would otherwise cause distortion of, or dimensional instability in, the extrudate.
- Similar arguments can be applied to situations where melts are split and reformed, as occurs during the extrusion of tubular films or pipes, where the melt streams are re-welded after flowing over the supporting "spider" legs of the die mandrel. Weld reformation zones (*split-lines,* or *spider-lines)* can be erased by systematic use of high shear stress and increased residence time; that

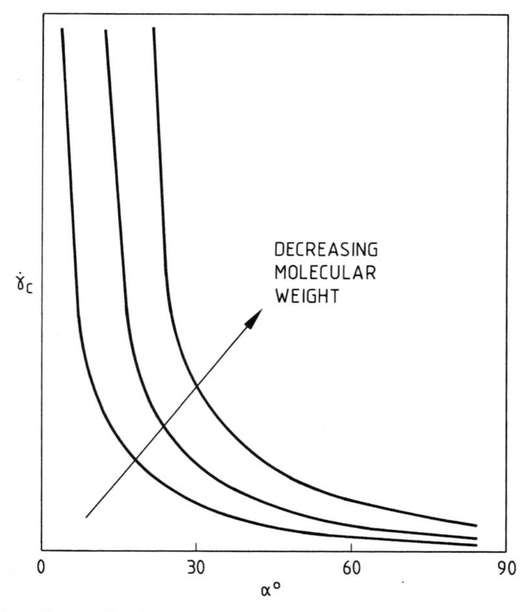

Figure 4.14 General relationship between critical shear rate ($\dot{\gamma}_c$) and taper angle (α) for a polymer melt, showing the general effect of changes in molecular weight.

is, by increasing the degree of *flow restriction,* for example, by increasing the die land-length.

- To avoid the possibility of *melt fracture* due to excessive elongational stress in the tapered zone, the taper angle itself should be controlled; this point was discussed in Chapter 3, and is also illustrated in Figure 4.14. Since melt rupture is dependent upon stress, the *critical shear rate* at which practical problems are encountered is a function of polymer melt viscosity, and hence molecular weight. Typical melt rupture stress is in the range $10^5 - 10^6$ N/m^2.
- Whenever wide *residence time distributions* are evident, or in extreme cases where significant taper angles exist (resulting in the promotion of circulating flows at the entrance to the tapered zone), severe melt degradation may become apparent if the polymer is inadequately stabilized.

Die Exit Dimensions. Disparity between die exit and product dimensions is an inevitable consequence of the elastic properties of polymers. It is further exaggerated by the dependence of die swell upon a range of material and process operating variables, and is influenced by the continuous commercial quest to seek higher output rates for any given product. There are three contributory groups of factors which must be considered, in order to link the extrudate, and die dimensions:

(A) Melt Elasticity – Die Swell

On the removal of an external stress, re-randomization of uncoiled long-chain polymers occurs, causing lateral swelling in the extrudate as it emerges from the die lips. This phenomenon is thought to originate from the relaxation of normal stress, as the melt is released from the die[4.9], and is undoubtedly influenced by the *elastic* properties of non-Newtonian liquids. Although it is a time dependent phenomenon, the greatest proportion of recovery occurs before cooling is able to suppress the relaxation which is responsible for swelling. Concepts of die swell have already been introduced in Chapter 3 (Section 3.10); to summarize, in the context of practical extrusion, die swell usually increases if:

(1) output rate increases, or lateral die dimensions decrease; ie. shear rate increases. This effect occurs up to a point where melt fracture and instability first become evident.

(2) melt temperature is decreased, since the shear stress generated in response to an imposed shear rate will be diminished when melt viscosity is reduced. However, rigid PVC is an exception in this respect: an increase in "swelling" with increasing temperature is often noted, since the pure elastic effect is accompanied by PVC gelation, and melting of a few percent of crystalline order, which occurs over a typical processing range 180 – 220 °C.

(3) average molecular weight increases. Increases in molecular weight distribution and the degree of chain branching also appear to increase die swell; these effects however, have not been universally established for complete groups of plastics.

(4) die L/D ratio is decreased, to an extent where steady flows are not established in the die-land; if relaxation times exceed die residence time significantly, the extent of swelling can be exceptionally high.

At any temperature, the ratio of viscosity coefficient (μ) to elastic modulus (E), often referred to as the *"relaxation time"*, or the *"natural time"* (λ), is sometimes used to determine the likely extent of swelling. Elastic effects will become important when the residence time is significantly less than λ. In real terms, this means that the molecules are still in the "fully-stressed" configuration at the die exit. Since much of the shear strain is recoverable, this mechanism helps to explain why die swell ratio increases dramatically (for short dies) with decreasing L/D ratio.

For non-symmetrical or more complex shapes, the function of the die designer is made more difficult since swelling occurs to a different extent around the periphery of the extrudate section. For example, Han has undertaken an analysis of swelling in a rectangular slot die[4.10], and has concluded that the observation of greatest swelling at the centre of the long-dimension of the extrudate can be correlated with the distribution of wall *normal stress* in the melt at the die exit. Dies of more complex shape will inevitably be associated with more serious problems of this type.

It is obvious that these effects lead to serious constraints at the design stage, because the correct choice of die shape and dimensions (to yield the desired product section, after swelling has occurred) depend not only upon the grade of compound chosen, but also upon the processing rates and specific machine settings.

Further references to the interrelationships between die swell, flow patterns and rheological characteristics, for more complex dies, are made in the text written by HAN[4.11], and in other references quoted from that source. More recently, swelling in annular dies (an occurrence of great practical relevance to a major proportion of the plastics extrusion industry), and its dependence upon material and processing variables, has been studied in detail[4.12].

(B) Haul-Off Rates – Drawdown Effects

The origin of *drawdown* – defined loosely as a positive and finite longitudinal strain imposed upon an extruded, partially-solidified melt, due to stress created by the haul-off velocity (v_h) exceeding the die output velocity (v_d) – is illustrated schematically in Figure 4.15. Wherever drawdown occurs, the more compliant melt-phase deforms preferentially, under a longitudinal tensile stress (σ).

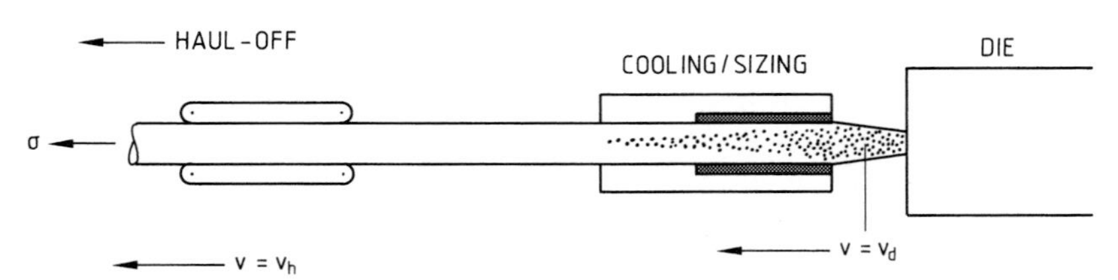

Figure 4.15 Origin of drawdown in plastics extrusion, induced when $v_h > v_d$. (The location of unsolidified material, in the extruded section, is illustrated by shading).

Although the changes in dimensions effectively oppose those due only to elastic recovery and die swell, it is sometimes inadvisable to attempt to balance these effects in practice, since the cooling/sizing facilities would need to be elaborately designed and also, excessive drawdown inevitably leads to poor surface characteristics and the introduction of molecular orientation in the "machine" direction of an extruded product. However, the principle of drawdown is put to effective commercial use in the production of highly-oriented films and fibres; see Section 4.2.5.

(C) Shape Stabilization – Thermal Effects

Superimposed upon die swell and drawdown phenomena is the *thermal contraction* which accompanies extrudate cooling. Specific volume differences between melt and ambient temperatures dictate the extent of thermal shrinkage, which will generally be greater for higher-crystallinity thermoplastics (see also Section 2.5). Inhomogeneous or constrained cooling of thick sections may lead to the formation of severe *residual stress* distributions (perhaps leading to warping if the extrudate section is not axisymmetric), or to the formation of internal voids.

A thorough review of extrusion die construction and design, showing how the specific rheological properties of plastics materials influence these rather complex procedures, has been presented by MICHAELI[4.13].

Effects of Die Restriction. In previous sections, the pumping modes and sensitivity to die head pressure of both single screw and twin screw extruders, have been examined. Since it is the presence of the extrusion die which induces pressure flow, and its modification of machine output, it is necessary to examine the complete process in order to optimize a choice of material grade, running conditions and process throughput. The pressure which contributes to inefficiencies in pumping capacity also acts as the driving force for polymer melt flow through the die; a general expression for shear flow through an orifice of constant dimensions is:

$$Q = K \cdot \frac{\triangle P}{\mu} \tag{4.21}$$

where K is a function only of die geometry; for example:

Circular Die $$K = \frac{\pi R^4}{8L}$$

Slot Die $$K = \frac{WH^3}{12L}$$

(The latter expression can also be used for thin-wall tubing dies, if "W" is assumed to represent the average circumference, and "H" the die gap.)

The gradient K/μ of an output-pressure *Die Characteristic* is positive. It is therefore convenient to illustrate output-pressure relationships, for both the extruder and the die, on common axes (see Figure 4.16): the pressure-output co-ordinates at which intersection occurs define the process *Operating Point* (for any given set of working variables), whose position can be verified by experimental observation.

One of the main objectives in running an extrusion line, as far as possible, is to design the process to run under "high-output, low-pressure" conditions. However, when the number of factors which determine the location of the operating point is considered, it becomes apparent that a great deal of skill is involved in optimizing equipment design and process dynamics on this basis. The effects of a number of

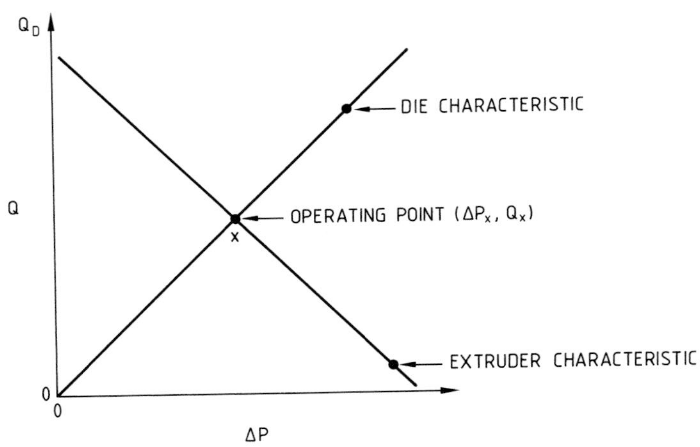

Figure 4.16 Output-pressure relationships for a plastics extrusion process, showing the extruder and the die characteristics, and the definition of the process operating point.

practically-relevant factors are discussed below, with reference to the operation of single screw machines.

Although Figure 4.16 shows linear extrusion and die characteristics (based on the Newtonian solutions given in Equations 4.13 and 4.21, respectively), the pseudoplastic nature of most polymers is exemplified by curved characteristics in practice, caused by the tendency for shear viscosity to decrease at high processing rates; it is for this reason that the families of characteristics in Figure 4.17, showing the effects of processing conditions, have been depicted in non-linear form.

Extruder Screw Speed (Figure 4.17a)

Increasing screw speed (see Equations 4.9, 4.12) increases the drag flow output component; ie. the open-discharge (zero pressure) intercept on the output axis. For a series of such changes, the locus of the die characteristic can be traced. As screw speed increases, both volumetric output and die head pressure each increase. Any departure from output-pressure linearity is also due to the generation of shear heat during processing; as the local temperature rises, melt viscosity will start to fall as dynamic equilibrium is approached (see also Figure 4.17d).

Extruder Screw Geometry (Figure 4.17b)

If the feeding and melting characteristics allow an increase in the metering zone length, the resulting pressure gradient decreases accordingly. The variable of greater importance is the channel depth (H). Since $-\frac{dQ}{d\triangle P}$ is proportional to the third power of H (Equation 4.10), a small decrease in H is very effective in diminishing the effect of die restriction and the magnitude of pressure flow, although it must also be appreciated that the open discharge component is also decreased if $H_2 < H_1$.

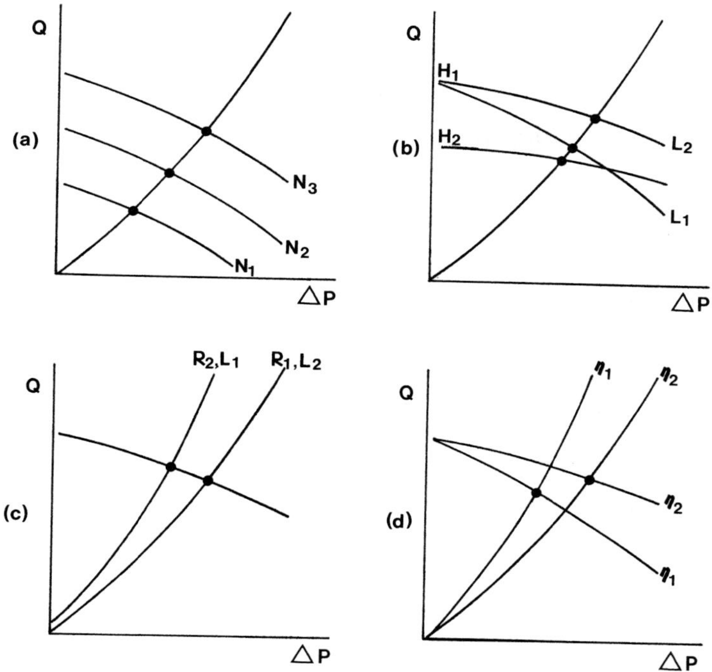

Figure 4.17 Practical output-pressure relationships, illustrating the effects of process-variable alterations: (a) screw speed, ($N_3 > N_2 > N_1$); (b) screw channel depth, ($H_1 > H_2$) and metering zone length, ($L_2 > L_1$); (c) die radius, ($R_2 > R_1$), and die land length ($L_2 > L_1$); (d) apparent viscosity, ($\eta_2 > \eta_1$).

Die Geometry (Figure 4.17c)

The operating point is shifted along the extruder characteristic, if the die length or lateral dimensions are adjusted. In order to comply with the "low pressure-high output" manufacturing criterion, the die restriction effect is diminished by decreasing orifice length, or by a tuning procedure associated with increasing lateral dimensions (Equation 4.21). Practical viability does, of course, limit the exploitation of this effect.

Melt Viscosity (Figure 4.17d)

Equations 4.13 and 4.21 each suggest that an increase in viscosity will decrease the gradient $-dQ/d\triangle P$ of each characteristic, resulting in increased die head pressure for an output rate which remains relatively unchanged. Since the opposite trend is beneficial, such decreases in viscosity may be achieved by:

- increasing the die and/or extruder temperature profiles;
- choosing a lower molecular weight (higher MFI) compound;
- judicious use of additives such as internal lubricants, plasticizers.

Any of these apparent advantages is associated with a corresponding trade-off, such as higher running costs or increased cooling times, or inferior product mechanical properties.

Accuracy of Output-Pressure Relationship. This approach represents an introductory aid towards the understanding of how the physical properties of plastics influence practical extrusion technology. It is an approximate, but potentially useful design tool. Extension of its application in quantitative terms is partially limited by the assumptions which have been made during the derivation of the flow equations, which are summarized in Table 4.2:

Table 4.2 The Extrusion Process - Validity of Rheological Assumptions

Polymer Melt Flow (General)		Comments/Validity
(A)	Isothermal flow	Poor assumption (Shear heat)
(B)	Incompressible fluids	Often self-cancelling with (A)
(C)	Steady flow	Good assumption, with the exception of short-land dies
(D)	Zero wall slip	Material and rate dependent (Lubricated PVC compounds are sensitive to wall-slip)
(E)	End effects	Can be calculated (Chapter 3) or accounted-for.
Extruder Constraints		
(F)	Non-Newtonian shear flow	Validity depends on sensitivity to shear heat, and magnitude of power law index
(G)	Screw curvature and perturbations	Reasonable approximations
(H)	Leakage flows ignored	Valid for new, unworn surfaces (single screw machines only)
(I)	Assume $\frac{dP}{dl} = \frac{\triangle P}{L}$ in the metering zone	Most pressure profiles are thought to be linear; departure from linearity depends upon thermal environment and correct screw design.

4.2.4 Plastics Materials in Extrusion: Melting, Mixing and Shaping

Most processing techniques based upon extrusion share common principles prior to and during the shaping stage; it is only during the cooling or post-extrusion phases that individual techniques diverge, appropriate to the extrudate shape and properties. To complete the section on extrusion machines, melting mechanisms and mixing performance are now discussed.

4.2.4.1 Melting Mechanisms

In order to develop a fully-homogenized melt in the metering zone, it is imperative that melting, together with wetting/coating of fillers (or other infusible additives) and expulsion of volatiles, be effectively completed within the compression zone. The heat transfer modes by which melting occurs may be summarized as follows:

(1) Conduction Melting (Electrical energy): Heat is transferred to the polymer from electrical resistance heaters, via the barrel wall;
(2) Conduction Melting (Electrical/Mechanical energy): Melting of the feedstock by solid-liquid contact, within the screw channel;
(3) Dissipative Melting (Mechanical energy): "Shear heat" generated within resident melt pools, in the screw channel.

Single Screw Extrusion. Figure 4.18 illustrates the generalized melting mechanism in a plastics extruder, for which the following notes represent important reference points:

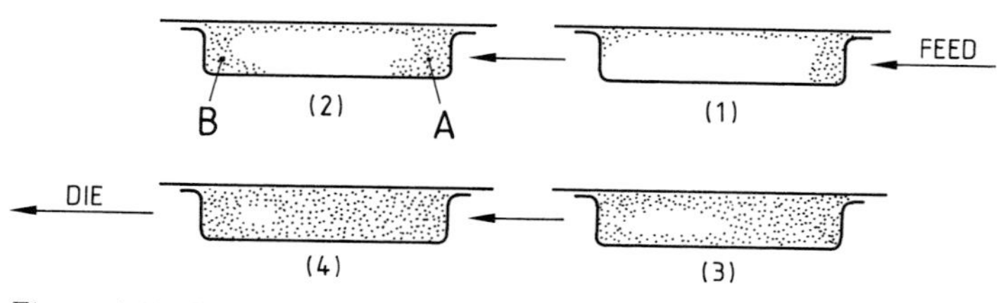

Figure 4.18 Melting mechanism in screw-channel sections normal to the downstream flow axis of a single screw plastics extruder; the shaded zones represent molten polymer. Sections (1) – (4) are presented in order of increasing compression.

(1) The conduction effect from the heated barrel occurs universally; the pushing flight tends to scrape the melt film from the barrel surface to form a melt pool (position A in Figure 4.18) which undergoes continuous and steady growth. Low shear viscosity promotes inter-particle flow, and accelerates melt pool development.
(2) Wherever wall-slip or leakage effects occur, melting is accelerated in locations behind the trailing flight of the screw (position B).
(3) The significance of the dissipative melting mode depends upon the respective volumetric sizes of melt pool and solid bed. Melt shearing assumes a greater level of importance, since shear strain contributes to distributive mixing of additives.

There has been a considerable amount of research carried out to establish the precise modes of melting in various plastics extrusion processes; a universal consensus of opinion has yet to be reached. The reader is referred to a recent paper by LINDT[4.14] for a more thorough review of previous work.

The melting characteristics of several groups of plastics are summarized below:

Polyolefins. High crystallinity resins (PE, PP) are able to endure a more severe compression gradient, in comparison to amorphous materials of equivalent melt viscosity. Equivalent numbers of screw turns are usually dedicated to feeding, compression and pumping, for most commodity polyolefins.

Crystalline Engineering Resins. As typified by polyamides, a narrow melting range for a low viscosity fluid is synonymous with a short compression zone and rapid completion of melting.

Rigid PVC. Unplasticized PVC powder compositions are capable of generating high levels of mechanical heat. Since these compounds are also prone to thermal degradation, it follows that compression (in parallel with gelation and "melting") must be gentle and gradual.

Rigid PVC is given special attention in this context, since it represents an unusual case in respect of the fact that the destruction of the inherent particle nature and sub-grain order (*PVC fusion* or *gelation)* is a precursor to the "melting" of a small proportion of crystalline order. The development of molten, fully-fused PVC is achieved only after a metastable, "sintered powder phase" has been destroyed. Direct reference is often made to the *CDFE gelation mechanism*[4.15], whereby the PVC powder grains are compacted, densified, fused and elongated (hence "CDFE") under the applied shear stress. In reality, however, a degree of *comminution* (mechanical breakdown of powder grains into primary particles before fusion) is thought to accompany the CDFE route, in most extrusion processes.

The progressive nature of melting in the screw is clearly accelerated at higher barrel temperatures (Figure 4.19)[4.16]; the maximum rate of melting appears to occur at the transition between the compression and metering zones. Increasing die resistance has been shown to have a similar effect[4.16]. Increasing screw speed induces more complex effects: although the shear rate in the screw channel increases, the average residence time diminishes. When this occurs, it may be anticipated that the onset of significant melting is delayed, but this would be compensated by an increased rate of melt pool growth.

PVC gelation is initially more apparent in the "melt" pool (Figure 4.20)[4.16] and develops more rapidly at higher barrel temperatures. Generally, the dependence of fusion and melting on extruder variables can be explained by similar arguments, since gelation is also accelerated by increases in shear stress and temperature.

Twin Screw Extrusion. Early melting of PVC dry blends is also evident close to the barrel surface in twin screw extrusion. The initial "tumbling" action of powders induces a more complex mechanism of fusion and melt pool growth, with the shape of the "molten" PVC rapidly becoming quite distorted[4.16]. On the basis of experimental observation, an extension of Allsopp's CDFE fusion mechanism has been proposed for the process of counter-rotating, twin screw extrusion[4.17].

4.2.4.2 Mixing Characteristics

Both *distributive mixing* and *dispersive mixing* occur in practical screw extrusion systems: the former is a function of the total shear strain imposed upon the melt, whilst the latter (which is perhaps more difficult to achieve yet possibly more relevant to final properties), depends upon the local shear stress distributions in the extruder.

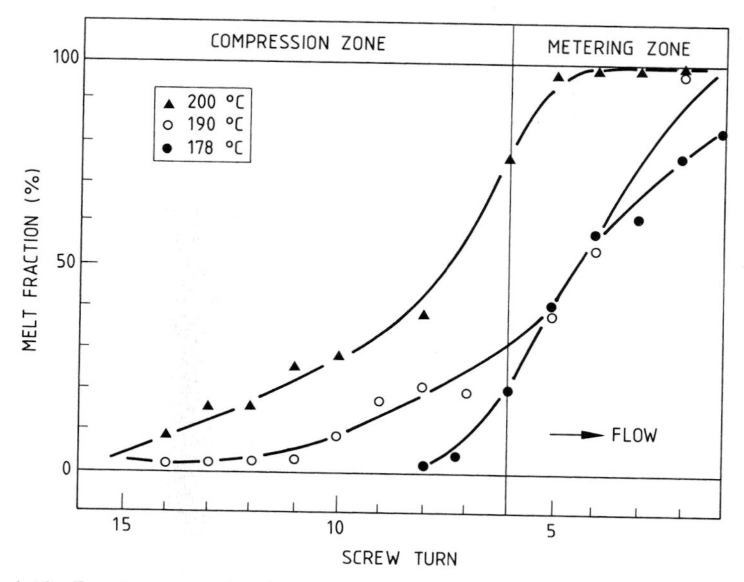

Figure 4.19 Development of melting for single screw PVC extrusion, illustrating the effect of changing barrel temperature. (After COVAS[4.16])

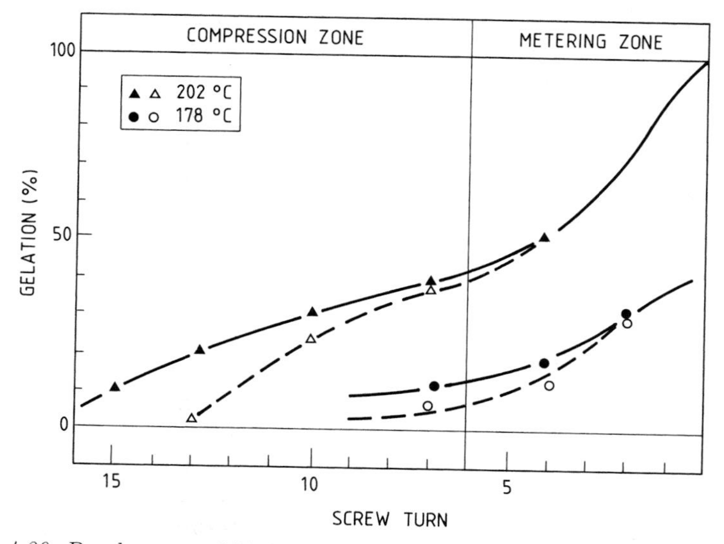

Figure 4.20 Development of PVC gelation in the single screw extrusion process; solid lines and dotted lines represent the "melt", and compacted solid-bed, respectively. (After COVAS[4.16])

Each of these mixing modes is important, in order to achieve a fully-homogenized compound, thereby making most effective use of the incorporated additives. The attraction of extruders as continuous mixing devices is therefore apparent, although it must be stated that overall mixing performance is most strongly influenced by the design of the screw.

(1) Single Screw Machines

Little effective blending occurs in the feed zone: on the contrary, segregation may start to occur if the surface characteristics (size, friction or electrostatic attraction) of separate constituents are incompatible. Distributive mixing occurs during the shearing of polymer melts, an effect which is promoted by the circulatory motion associated with cross-channel flow and the elements of pressure-flow due to die resistance.

The shear stresses required for dispersive mixing can be enhanced by increasing the degree of die restriction, or by including a fine-mesh breaker plate assembly to increase head pressure. However, the incorporation of specific elements (e.g. *mixing pins,* or channeled *dispersion heads),* to improve dispersion, is usually far more effective[4.18].

(2) Twin Screw Machines: Counter-Rotating

Given the constraints defined by the mechanism of flow within discrete chambers (Section 4.2.3.2), only a relatively poor mixing performance (promoted mainly by leakage flows) is usually observed. This statement is, however, of less consequence if PVC preblends, for example, are adequately mixed before the extrusion process.

(3) Twin Screw Machines: Co-Rotating Compounders

Intermeshing co-rotating twin-screw machines occupy important and expanding market-application areas, which are defined by the *high output* capability and *continuous mixing* characteristics of these extruders. A comparison between the effectiveness of each of numerous commercial designs is extremely difficult[4.18], so that the discussion here features only a single example of a modern commercial compounder, whose design specification includes the requirements to:

- optimize output rates (thereby increasing the economic advantages over batch mixers);
- optimize the energy-efficiency and quality of mixing, whilst avoiding the possibility of thermal degradation;
- fulfil these criteria for a vast range of combinations of plastics materials and additives.

The last requirement holds the key to extruder screw design for co-rotating twin screw machines; since the optimum design for blending pololefins of different melting and viscosity characteristics is likely to be very different to that required for incorporating mineral fillers and short glass fibres in nylon (for example), it is extremely beneficial to use a *modular screw design,* so that individual, custom-designed screw geometries can be attained.

Illustrated in Figure 4.21 are typical *agitator configurations* for single-stage, two-stage and three-stage mixing. Effective dispersive mixing is usually evident immediately after melting; since this is associated with high melt viscosity, it is important that sufficient motor power is available to run the extruder to achieve the desired quality of mixing at a predetermined output rate.

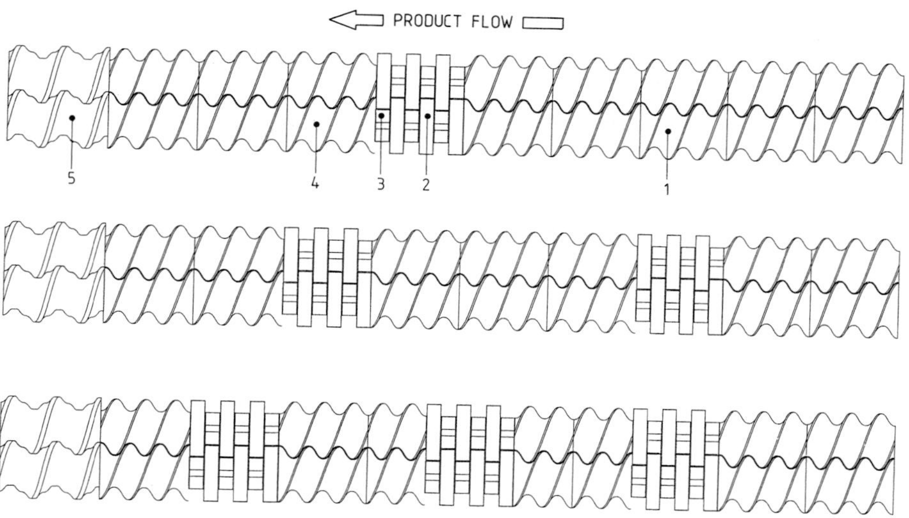

Figure 4.21 Typical agitator configurations for single-stage, two-stage and three-stage mixing, in a co-rotating twin-screw plastics extruder. Individual elements of the screw include feed/solids conveying zone (1), mixing paddles (2), restrictive "orifice plugs" (3), decompression and melt phase feeding (4) and "camelback" compound discharge (5). (Reproduced by permission of APV Chemical Machinery Ltd., UK.)

Figure 4.22 shows successive cross-sections of opposite agitators in the high-shear mixing zone. The lenticular screw paddles rotate together, and these define three available occupation volumes within the "figure-8" shaped barrel. This defines a *fully-intermeshing, self-wiping* action, which can be precisely controlled according to the desired mixing characteristics and residence time distributions. Material hang-up and thermal decomposition can also be controlled, by this mechanism. Co-rotating mixers are usually starve-fed, so that the degree of restriction imposed by the orifice plugs controls the effective mixing zone length, whilst the severity of shearing is determined by screw geometry and by the rotational speed of the screws. Forward conveying tendency is determined by the angle between successive mixing paddles. Some fundamental research on this design of extrusion compounding unit has been reported by TODD and co-workers[4.19,4.20].

When mixing filled or fibre reinforced plastics compounds, the overall objectives are usually:

(1) to achieve a good distribution of additives, whilst completely wetting-out the minor phases within the polymer melt matrix. This objective is fulfilled by the *"division and recombination"* effect. As the screws rotate (at speeds up to 500 RPM), material is transferred from one 'screw volume' to another, twice per revolution.
(2) to disperse solid additives to primary size and to avoid agglomeration or "bundling" of fibres.
(3) to avoid excessive fibre fracture.
(4) to avoid polymer degradation by controlling residence time, and by maintaining a narrow residence time distribution.

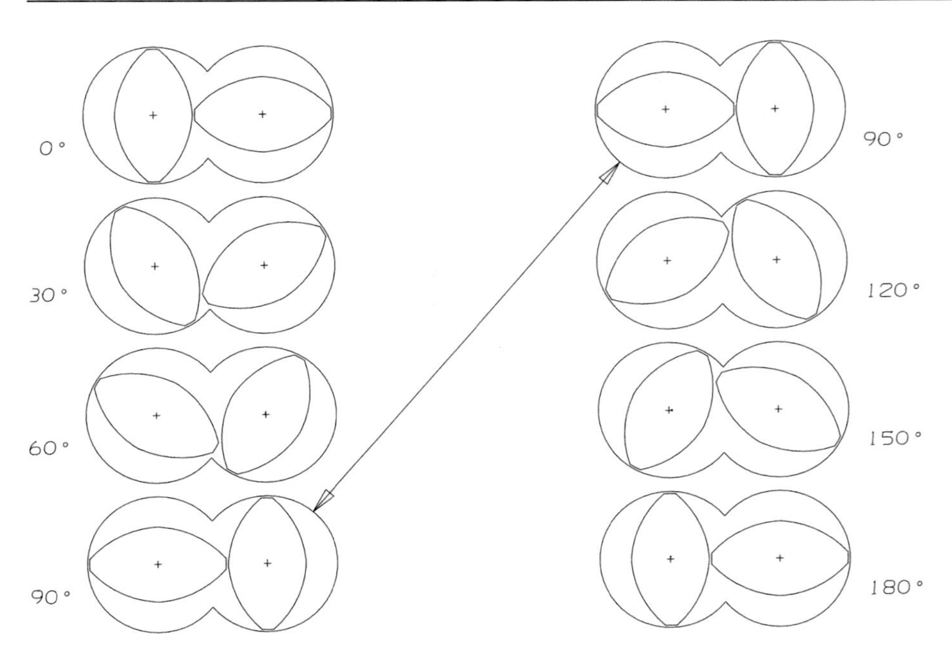

Figure 4.22 Mechanism of melt-chamber shape and volume changes, in a co-rotating plastics extruder, as the twin-lobed agitator sections rotate. (Reproduced by permission of APV Chemical Machinery Ltd., UK).

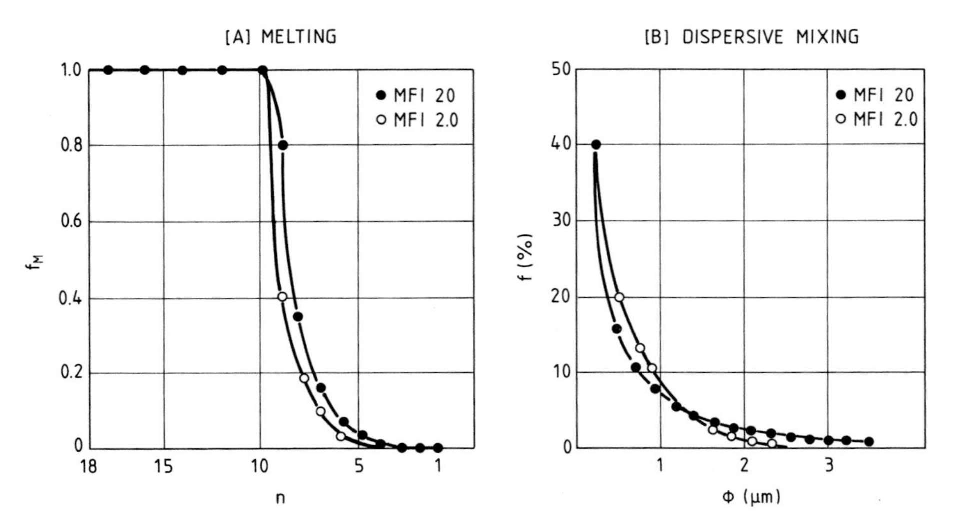

Figure 4.23 Melting and dispersive mixing of TiO_2 in a LLPDE masterbatch system, on a co-rotating twin screw extruder, and the influence of melt flow index (MFI); (A) Melt pool fraction (f_M) as a function of position within the mixing zone (the "extrusion direction" is right-to-left; there are 18 paris of agitator elements in total); (B) Image analysis of a 10 % TiO_2 masterbatch, expressed as cumulative size fraction (f) versus particle size (see text).

A high-shear environment is created by the high paddle speed shearing polymer in the narrow gap between the barrel surface and agitator element tips (Figure 4.22). Heating is almost totally *adiabatic,* yet excessive temperature rises are avoided since molten compound is rapidly recombined with solid, cooler material in the adjacent chamber.

It has been observed that the degree of dispersive mixing within a developing melt pocket increases rapidly, once the initiation of melting becomes significant; the data in Figure 4.23 refer to a white LLDPE masterbatch system, based upon titanium dioxide (TiO_2) pigment. In 4.23(A), the increase in melt fraction along the mixing zone is complete at the location of the 10th pair of mixing paddles. Figure 4.23(B) shows *dispersive mixing* within a fully-developed melt pool; for this particular compound, the mixing characteristics were shown to be critically dependent on screw configuration, with molecular weight of the carrier polymer assuming only secondary importance. The work was carried out on microtomed sections cut from the compound following a "dead-stop", and rapid-cooling experiment.

Comprehensive treatments on theoretical, and practical aspects of mixing in extrusion can be found in some dedicated texts[4.18,4.21 – 4.22].

4.2.4.3 Extrusion Materials and Shaping Variables

Summarized below are some of the most important variables which determine the viability of an extrusion process, and the quality of extruded products.

(A) Machine Variables

Screw Speed. An increase in screw speed will increase the volumetric output in all flood-fed extrusion processes; the effective shear rates (in both the screw channel and in the die) increase as a result. Screw speed, shear rate and output are in direct proportion under low-shear conditions: however, the pseudoplasticity of most thermoplastics, in conjunction with the development of shear heating, will detract from linear screw speed-throughput correlations. The mean residence time usually decreases with increasing screw speed, but the wider residence time distribution is often expected to contribute to an increase in laminar mixing efficiency.

Temperature. Since the temperature of several individual zones can be closely controlled on most commercial extruders, a great deal of fine tuning is possible. In brief, qualitative terms, increases in temperature induce the following effects, at the locations stated:

- **Feed Zone.** The polymer-barrel friction coefficient is likely to increase; solids transportation rates may therefore be enhanced.
- **Compression Zone.** Acceleration of the rate of melt formation (by the mechanisms given in 4.2.4.1) is usually apparent.
- **Metering Zone.** The extent to which shear viscosity is temperature-dependent determines the shift in process operating point (Section 4.2.3.3). High temperature is beneficial from the output-pressure (economic) standpoint, yet there

exist the possibilities of thermal degradation (wider residence time distribution), and rate limitations due to the additional cooling requirements.

- **Extrusion Die.** The effect on the process operating point has been described in Section 4.2.3.3. In addition, an increase in molecular mobility at higher temperature helps to overcome the problems associated with die *"spider" lines,* in hollow sections. Since temperature can often be varied and controlled precisely at various positions, the respective effects on swelling and viscosity can be used as a precise means of control on profile thickness uniformity. Modern computer-aided manufacturing methods make use of polymer melt properties in this manner.

Die Restriction and Head Pressure. The extent of restriction and pressure drop through a given die section can be varied by suitable adjustment of a *restrictor bar* or *mandrel.* Since the rheological properties are geometry-dependent, local output rates can be controlled to yield optimum thickness distributions in the extrudate. Subsequent increases in head pressure are synonymous with higher shear stress, which generally contributes positively to mixing, and also assists molecular flow across any longitudinal imperfections.

The following points are of additional significance to twin-screw machines:

Feed Rate. Unlike most gravity-fed single screw machines, twin screw extruders may not be flood-fed. If the machine is *starve-fed,* the mass feed rate controls productivity. Whenever feed rates are increased, the specific energy requirement and motor torque are raised accordingly; indeed, for many counter-rotating extruders, the limitation on feed rate is the possibility of overloading the thrust bearings.

Screw/Barrel Cooling. To control any possibility of a runaway adiabatic effect in co-rotating compounders (polymer decomposition being the likely result) barrel cooling facilities (eg. circulating water or low viscosity oil) are normally provided.

On counter rotating machines for UPVC extrusion, degradation is an obvious possibility. In this case, circulating oil is pumped along axial channels bored through the screw shafts (typically at 130 – 140 °C) as an additional mode of temperature control at the screw surface.

(B) Plastics Compound Characteristics

Polymer characteristics which influence melt viscosity will inevitably modify the performance of a given material in the extrusion process. Although these have been more fully reviewed throughout Chapter 3, a summary of some material variables relating to commercial plastics compounds is also presented here.

Polymer Molecular Weight. Chain length has a most profound effect on polymer melt viscosity; this is undoubtedly related to the presence of molecular *entanglements,* in a flowing liquid under high shear.

Post extrusion die swell is also positively related to molecular weight, since entanglements influence the number of sites about which chain uncoiling occurs in a capillary die.

Molecular Weight Distribution. As molecular size distribution (*"polydispersity"*) increases, the departure from classical Newtonian behaviour occurs more readily (at a lower shear rate), and the viscosity of broad-distribution polymers is known to be more shear-sensitive, as characterized by a lower Power Law Index. Also, viscosity relationships with temperature and pressure are usually more sensitive, probably due to the high molecular weight fraction.

Chain Branching. For a given M_w, chain branching yields an effect analogous to an increase in polydispersity. In addition, it is accepted that significant chain branching leads to the *tension-stiffening* effect seen in LDPE, which assists bubble stability in tubular film extrusion.

Polymer Blends. The formulation dependence of processability is a complex function of the characteristics of the individual constituents which make up the polyblend, together with the degree of compatibility which exists between them in the melt phase. Inevitably, rate dependence may differ, according to blend composition. In addition, any chemical changes (for example, *transesterification* in blends of thermoplastic polyesters, or more general thermo-oxidative degradation) may significantly alter processing behaviour from anticipated, or idealized trends.

(C) Effect of Additives

Fillers/Fibres. Over a range of concentrations, proportional increases in shear viscosity are usually apparent as the concentration of fillers or low aspect-ratio fibres is increased. The rate of increase depends upon the *surface wetting* characteristics of the filler, the interaction with the matrix polymer, and the tendency of heterogeneous additives to resist dispersion in the shear field.

For many groups of short-fibre reinforced thermoplastics, the pseudoplastic character of the base polymer tends to nullify the effect of fibre-addition, for typical fibre loadings of up to 40 weight percent[4.23]. Significant viscosity increases may not, therefore, be seen at shear strain rates relevant to extrusion. Melt elasticity appears to be sensitive to fibre addition: die swell diminishes as fibre concentration increases.

Plasticizers. Due to chemical compatibility, plasticizers are able to interpenetrate long chain polymer molecules. The enhanced chain mobility accounts for the decreases of viscosity observed as plasticizer concentration increases. Once again, different types of trends may be anticipated if the chemical nature of the plasticizer is altered.

Stabilizers. Unless stabilizers also possess a lubricating effect (eg. lead stearate in rigid PVC compounds) they have little direct influence on melt characteristics, but contribute strongly to the possibility of successfully increasing stock temperature, or practical residence time in the barrel.

Lubricants. *Internal lubricants* act in a similar way to plasticizers (ie. they are able to penetrate between polymer molecules) and can control shear heat generation quite effectively by maintaining shear viscosity (for a given processing rate) at an acceptably low level. *External lubricants,* whilst fulfilling a similar overall function, reside between the melt and equipment surfaces; wall-slip velocity tends to increase and the velocity profile becomes less severe.

Processing Aids. Extensive use of processing aids is made in the rigid PVC extrusion industry. Whilst the compatibility of long chain acrylics (for example) with PVC does not significantly modify viscosity, the onset of PVC gelation is greatly accelerated by process aids, and once fused, PVC "melt" extensibility, strength and elasticity are substantially enhanced.

4.2.5 Downstream Extrusion Processing - Cooling, Shape Stabilization and Development of Orientation and Microstructure

The approach adopted so far in Section 4.2 has been to consider the common principles of plastics extrusion up to the point of shaping through the extrusion die. However, since the properties of the extrudate are determined by *microstructural features* which are induced or modified by cooling effects, it follows that the mode, rate and homogeneity of cooling, combined with the effects of any subsequent deformation imposed downstream of the die (such as *molecular orientation* in extruded films or fibres), must be considered separately.

The number of different classes of extruded products reflects the diversity of downstream processing technologies which are feasible. Our approach is to concentrate on summarizing the process-material interface, appropriate to a number of downstream processes. Before doing so, it is necessary to consider the ways in which thermal energy is extracted from extrudates, following the shaping phase.

4.2.5.1 Heat Transfer During Extrudate Cooling

The major proportion of energy which is put into thermoplastics extrusion processes by mechanical (screw rotation) and electrical (barrel heaters) means, is extracted by direct cooling. The chosen mode of cooling depends upon the output rates required, extrudate thickness and shape-complexity, and also by the cooling characteristics of the polymer itself.

Output Rate. For a given polymer, the output rate determines the overall required rate of heat transfer; the length of the cooling zone and plant space requirements are determined accordingly.

Extrudate Thickness. Practical cooling rates rarely pose problems for relatively thin-section products; linear speeds are inevitably high and process stability is often of greater concern.

For section thicknesses of (say) 3mm or more, cooling requirements often determine maximum practical line-speeds: for example, on equivalent pipe production processes, limiting take-off rates of 13 m/min. and 7.5 m/min are observed for respective pipe wall thicknesses of 2.5 and 3.0 mm. Furthermore, if the shape complexity increases (for a given wall thickness) cooling constraints will further defy attempts to maintain production rates; this is especially true for hollow sections, since conductive heat transfer is severely retarded.

Thermal Properties. The total heat extraction requirement depends upon the difference between extrudate and safe handling temperatures. Total heat content of partially crystalline plastics (typified by PP, and other plastics which *supercool* over a wide temperature span; see Chapter 2) is especially significant.

Some possible means of achieving the desired objectives are summarized below:

Air Cooling

(Heat transfer coefficient, $h = 5\,W/m^2K$; natural convection

$h = 10 - 30\,W/m^2K$; forced convection)

Blown films are cooled by forced convection, and are simultaneously under biaxial tensile stress, in order to induce the required magnitude and distribution of orientation.

Water Bath $(h = 1000\,W/m^2K)$

Total immersion in water is a popular and effective means of heat transfer, for cooling a wide range of profile shapes. The conductive contact area is optimized and so long as the coolant temperature is prevented from rising, process stability and product microstructure are generally maintained.

Water Sprays $(h = 1500\,W/m^2K)$

Spraying induces turbulence and rapid evaporative cooling, each of which contributes to more effective heat transfer.

Cooling/Sizing Units $(h$, for steel $= 500 - 550\,W/m^2K)$

Profiles containing sharp section changes and arms which require support during cooling are usually sized simultaneously, by use of metallic *calibrators* which are themselves cooled by a circulating fluid. Direct conduction to the heat sink is effective, as long as the internal dimensions of the calibrator are extremely precise. If not, thermal contraction of the plastic may induce shrinkage away from the calibrator surface; such an air gap would detract from cooling efficiency. The link between cooling and sizing is critical to achieving given profile dimensions.

Chill-roll cooling of extruded sheet is based upon similar principles; since the intimacy of contact and total contact area determines the rate of heat transfer, roll diameter and separation can be designed accordingly. For thick sheets, higher temperature oil is the preferred coolant: "quench" surface cooling may otherwise induce unacceptably-high residual stresses into the section.

For any of the sub-processes listed below, a one-dimensional heat energy balance is applicable:

$$(T_{Melt} - T_{Take-off}) \cdot Q_P \cdot \varrho_P \cdot C_P = -Q_W \cdot \varrho_W \cdot C_W \cdot (T_{Out} - T_{In}) \qquad (4.22)$$

Q, ϱ and C are the volumetric flow rates, densities and specific heat contents of the polymer (subscript P) and coolant (subscript W).

This approach allows approximate calculations to be made, relating to the coolant flow rate requirements for products manufactured from prespecified materials at anticipated output rates. In order to estimate cooling time (and therefore plant space) factors, heat transfer modes must also be considered:

Conduction

Heat flux, $q = K \cdot A \cdot \frac{dT}{dx}$

Convection

Heat flux, $q = h \cdot A \cdot (T_M - T_C)$

(K = thermal conductivity; h = surface heat transfer coefficient; A = area across which heat transfer is taking place; T_M and T_C are the initial temperatures of the melt and the coolant).

Heat transfer rates in plastics are severely suppressed by the low thermal conductivities which these materials possess: typically, $K = 0.1 - 0.4$ W/mK. The obvious temptation when cooling thick sections is to use as low a coolant temperature as is practically-feasible. However, severe thermal gradients may then be generated through the extrudate thickness, resulting in microstuctural inhomogeneity, and the development of residual stress. A model of unsteady-state cooling of plastics pipe sections has been developed by EDWARDS et al[4.24], which predicts temperature distributions within the pipe, as a function of time and position. This type of approach provides useful information to allow the process engineer to optimize production conditions, whilst maintaining part-quality.

It is common practice in heat transfer theory[4.25 – 4.26] to refer to dimensionless groups which describe solutions to individual problems according to thermal gradients, heat flow properties and geometric/shape effects. These principles are applicable to extrudate cooling in plastics processing[4.27], where it should be noted that due to poor thermal conductivity, temperature gradients are relatively high; for example, the severity of thermal gradients in an extruded profile of thickness l, undergoing convective cooling, can be examined by the Biot Number:

$$Bi = h \cdot \frac{l}{K} \tag{4.23}$$

For tubular film extrusion of polyolefins, we find that Bi values are significantly less than unity, indicating that significant thermal gradients do not exist in thin-wall packaging films. However, when h and/or part-thickness are increased (eg. profile extrusion, involving contact with water or metallic surfaces), significant thermal gradients can be assumed when $Bi > 1$. Product microstructure is a function of cooling rate, and will itself be location dependent when this occurs. Further reference will be given to process heat transfer in Chapter 5 (injection moulding).

4.2.5.2 Pipe Extrusion

Polymers and Products. UPVC, PP, HDPE and ABS (including mineral-filled and grades containing regrind) are used for general purpose and non-pressure applications.

UPVC and MDPE are used for pressure pipes for water and gas distribution; they offer excellent, long-term creep rupture properties at relatively low material cost. PB, XLPE and CPVC are chosen for pressure applications at elevated temperatures, such as in domestic central heating, or undersoil heating. Small-bore tubes are extruded from a range of thermoplastics (depending upon application) including unmodified and reinforced grades of plasticised PVC, polyolefins and nylon resins.

Sizing. Traditional methods of tube-sizing by an internal mandrel have now been superseded by techniques based upon calibrator plates or sleeves, in conjunction with an external vacuum (Figure 4.15). Overall, the refinement and accuracy of the sizing mechanism depends upon the application and whether the internal, or the external dimensions are more critical. The diameter of the sizing unit, (and how it changes throughout the length of the system), must be chosen to allow for die swell and drawdown effects; in addition, the *surface lubricity* must be sufficient to allow the pipe to "slide" freely (as it cools) to produce a pipe of acceptable surface quality. Diametric size varies from 1mm (medical tubing) to large bore water pipes of 1 metre or greater; corresponding wall-thicknesses are sub-millimetre, to 100mm.

Cooling. For pipes of significant thickness, linear production rates are quite modest and the heat transfer is predominantly unidirectional. Water sprays are therefore preferred, having a higher surface heat transfer coefficient than water baths of equivalent (nominal) temperature. As pipe thickness decreases and line speeds increase, the practical attraction of cooling by total immersion in a water bath becomes more evident.

4.2.5.3 Profile Extrusion

Polymers and Products. "Profile" is a generic extrusion term which represents articles of constant and continuous cross-section that do not fit into any of the specific product areas referred to elsewhere in Section 4.2.5. Some of the more important commercial product groups can be summarized as follows:

UPVC – unmodified, for guttering, irrigation and other building/plumbing applications, together with domestic/industrial usage for curtain track, guides, protective covers etc.

UPVC – impact modified, for window and door frames and exterior cladding.

PC, PMMA – for transparent applications in building, construction and in lighting; PC is preferred where high toughness and durability are important.

Polyolefins; engineering thermoplastics – extruded into bar/rod stock for subsequent machining.

PPVC; thermoplastic elastomers – extruded for applications such as flexible seals and gaskets.

Some newer developments include *foamed profiles* (for greater longitudinal bending stiffness per unit weight), and hollow sections subsequently reinforced with metallic or reinforced plastics inserts.

Cooling/Sizing. The diversity of products precludes most attempts to generalize methods of cooling. However, although cooling/sizing modes are inevitably shape-dependent, the most common means of shape stabilization occurs by suitable use of a calibration box (whose surfaces are lubricated and whose dimensions gradually change appropriate to the product), with internal water cooling. Vacuum-assisted sizing is common for profiles of complex or non-uniform section.

The overall size range of extruded profiles is vast: profiled sheet extrusion dies of up to 1.5m are not uncommon.

4.2.5.4 Sheet, Flat Film and Coated Laminates

Polymers and Products. UPVC is used extensively as a general purpose sheet material for roofing and cladding, in the construction industry. Another important group application area is sheet for vacuum forming, where the traditional styrenic plastics (PS foam, HIPS and ABS) are complemented by other amorphous plastics such as PVC, PC, PMMA, PETP and more recently, by other combinations of plastics in *multi-layer* form, some of which are subsequently used for thermoforming. Although in-line thermoforming is becoming more important, there are still many 'trade' sheet-extrusion companies manufacturing successfully.

Chill-roll casting. This (see below) is an effective means of sizing and cooling extruded sheets; also, a variety of non-polymeric substrates can be bonded to sheet by this technique. Products include laminated paper, board, textiles and metallic foils, with polyethylene being the dominant thermoplastic material used. Higher temperatures than in conventional extrusion tend to be used, firstly to allow a sufficiently-low melt viscosity for impregnation to be achieved, and secondly, to induce *surface oxidation* of non-polar polyethylene, which assists interlaminar bonding. Products corrugated longitudinally or in the transverse direction are each manufactured on sheeting lines.

Sizing/Cooling. Sizing by roll-stacks (*"chill-roll casting"*) dominates sheet extrusion. Successively reducing nip-separation is used to control sheet thickness, whilst the sheet cools simultaneously by direct thermal contact with the roll surfaces, which are held at precisely-controlled temperatures. Water and oil are common coolants, the latter is preferred at higher roll temperatures, which are employed for thicker-gauge products to prevent quench-cooling effects and the formation of residual, through-thickness stresses. Roll temperature is also controlled to allow drawdown to be exercised to manufacture oriented film products.

4.2.5.5 Wire Covering

Polymers and Products. Some examples are:

PPVC, for flexible electrical insulation of wires and metallic twine;
PE (usually low-density resins, sometimes crosslinked), for sheathing/insulation in high-voltage cables and telecommunications products.

Sizing/Cooling. Water baths are widely used, although assistance from forced convection for thicker sections is also practiced. Cooling trains are often extremely long wherever high extrusion rates are operative (eg. up to 1000 m/min. for general purpose wires). For a given wire velocity, the extruder output rate and drawdown characteristics determine the cooling rate and the subsequent deposited thickness of the insulating polymeric layer. This is chosen according to operating voltage specification, and varies from sub-millimetre to several centimetres.

4.2.5.6 Co-Extrusion and Multi-Layer Products

Processing Technology. Different extrusion units are used to feed a common adapter and die (the *feed block);* for laminated sheets, separate dies may be used with sheet lamination downstream. Shear viscosities of the constituent polymers must be similar to maintain an even, laminar flow after melt combination, yet the die block and running conditions should be optimized such that interlaminar flow occurs to only a very small extent (ie. to achieve subsequent *adhesion* of adjacent layers, without destroying the discrete layer structure of the sheet). Adhesion-promoting *tie layers* are used whenever the combination of plastics to be co-extruded may be prone to delamination in service.

Polymers and Products. Co-extrusion is a relatively modern development; as a positive growth sector, new developments are continuously becoming apparent. The co-extrusion principle is used for reasons of increasing performance/cost or strength/weight ratios, improving aesthetic quality and especially, to improve the resistance of extruded films and sections to gas permeation. Product examples include:

- LLDPE/EVAL/LLDPE - 3-layer high barrier films, or 5-layer systems if adhesive tie layers are deemed necessary; other polyolefin films are also coextruded and other barrier materials include PAN "nitrile" resins, PVDC and polyamides.
- Coloured/Uncoloured combinations (including patterned effects such as spiralling).
- Solid/Foam/Solid - cellular profiles have been extruded from stryrene plastics and UPVC, to enhance the flexural stiffness, per unit weight of extrudate; the state-of-art of some variant processes associated with this principle, including "free-foaming" and the inward-foaming Celuka techniques, has been described by BECKMANN[4.28]. In addition, a process has been developed to produce 3-layer, internally-foamed profiles by coextrusion[4.29].

Sizing/Cooling. Methods are as for profile, sheet or film, according to product type and foaming process selected. Provision of longer calibration boxes may need to be considered for foamed co-extruded sections, to alleviate any possibility of post-swelling due to incomplete release of blowing agent.

4.2.5.7 Oriented Extrusion Products

(A) Uniaxial Orientation

Monofilament and Tapes. Polymers which are highly extensible and stable in the oriented state have formed the basis of the synthetic fibres industry for some considerable time; these include PP, cellulose esters, PA66 and PETP. Melts are extruded through multi-orifice dies, cooled, then subsequently re-heated to a temperature at which a drawdown process can most effectively induce high degrees of uniaxial *molecular orientation.* Polymers like PETP which *strain-crystallize* are ideal in these applications since the developed crystallinity (15-25 %) improves an array of physical properties, such as mechanical strength and, under some circumstances, enhanced stability at elevated temperatures. More recent applications, notably for PP and PETP, include tapes for strapping, audio and video applications, where the anisotropic strength characteristic is fully-utilised.

Cooling is rapidly achieved in water baths, to quench and leave a low-crystallinity state. Re-heating at an appropriate temperature between T_g and T_m allows stretching to be imposed at a pre-determined elongational strain rate; typical draw ratios are in the range 5:1-12:1, with the degree of longitudinal orientation (and/or chain-extended crystallinity) being positively related (but not necessarily proportional) to the stretch ratio.

Drawn product thickness varies from about $50\mu m$ to $2-3$mm.

(B) Biaxial Orientation

Tubular Films. The importance of the biaxially-oriented tubular (or layflat) film process is reflected by the magnitude of the polyolefins-dominated plastics packaging industry. LLDPE/LDPE still dominate the general purpose packaging market, but HDPE is used for stiffer, "crisper" films, EVAC for surface adhesion in "cling-films" and many co-extrusions and polyolefin blends are developed to occupy specific positions in the market. These include LLDPE/HDPE/LLDPE (heavy-duty sacks) and LLDPE/EVA adhesive/PP for heat-shrinkable film; LLDPE(or PP)/Tie layer/EVAL/Tie layer/LLDPE is known as "Hy-Bar"® film, manufactured for packaging of foodstuffs which require a particularly long shelf life against oxygen ingress.

Molten polyolefins are extruded vertically through a thin wall annular die. *Longitudinal orientation* is induced by drawdown: ie. a velocity differential between the take-off rolls and the film at the die exit; *transverse orientation* occurs as the annulus is inflated by an airstream which expands the extrudate to a "bubble" of increased diameter. Blow ratios and drawdown effects can be chosen according to production, gauge and film-property requirements. Cooling is by forced convection from another airstream; the rate of cooling can also be chosen to desired effect, and since high stretching rates induce higher levels of crystallinity, a great deal of processing flexibility is usually apparent.

Crystalline morphology of polyethylene is strongly influenced by film extrusion conditions. In LDPE, it has been shown that molecular orientation is mainly confined to the crystalline phase[4.30]. However, there are several processing parameters which strongly influence the direction and magnitude of crystallite orientation in HDPE; these include *blow ratio* , longitudinal *draw ratio, cooling rate* and *freeze-line height* , the effects of which have been described comprehensively in separate research studies[4.31 – 4.32].

Flat Films. *Biaxial orientation* of conventionally-extruded flat sheets is an alternative production method for those plastics whose thermal and rheological properties are less suitable to the tubular film process; notable examples are PP and PETP.

Extruded sheet is cooled by roll-stacks to $500 - 600\mu m$, then reheated to the correct temperature for subsequent stretching (80 – 110 °C for PETP and 100 – 110 °C for PP). A series of heated rolls rotating at gradually-increasing speeds induces longitudinal orientation, whilst transverse orientation is induced by gripping and lateral expansion on a gradually diverging *tentering frame;* gauge reduction ratios are in the range 10 – 50. *Heat-setting* can be achieved in PETP (eg. Melinex® film) by holding the film stable whilst raising the temperature in a subsequent heat tunnel; this tends to increase the degree and perfection of chain-extended crystallites.

Polymer Meshes and Oriented Grids. Extrusion technology has, more recently, been developed to allow direct continuous production of biaxially oriented polymer grids for packaging and civil engineering applications. "Netlon"® is produced by extruding through slots cut into two counter-rotating annular die rings; wherever slots in the respective rings coincide, a junction point for the mesh is obtained, and the melt streams are bonded together. Linear extrusion speed and relative rotational velocities of the die rings can be adjusted to yield meshes of different thickness, shape and size. Orientation is induced by biaxial stretching at a temperature some 20 – 30 °C below the crystalline melting temperature of PE and PP, the widely used thermoplastics in this process.

An alternative heavy-duty product, "Tensar"®, is manufactured by stamping a regular (meshed) pattern of holes in extruded polyolefin sheet, then reheating and stretching to achieve the required biaxial orientation. All the material within the sheet, including the "junction points", becomes oriented in the latter technique, giving the mesh much improved properties of strength and creep resistance, per unit weight of material used[4.33].

4.2.6 Some Common Faults and Structural Defects in Extruded Products

Any defective products (whether imperfect on a microstructural basis, or by non-fulfilment of selected performance criteria) may have deteriorated as a result of defects occurring from any of three main sources:

(1) polymer feedstock;
(2) non-optimum heat-shaping;
(3) non-optimum cooling.

Immediate suspicion is often aimed at the first of these possibilities and moreover, greater technological skill is often required to investigate possibilities (2) and (3) in a fault analysis situation. Presented below is a selection of common extrusion problems, which is subdivided according to defect origin.

4.2.6.1 Shaping Phase

1. Incomplete Melting. If all solid-phase crystalline order is not completely erased as the melt-state develops, the extruder head pressure rises, screen pack filters may become clogged and the boundary between "unmelted" and "remelted" material becomes a specific point of weakness. Microscopic examination in polarized light is a ready means of detecting partially-melted material. Remedial action might include increasing mechanical work by raising the screw speed, or increasing the rate of plastication by initiating melting at a point further upstream in the extruder barrel; these problems may be more effectively overcome, however, by a thorough re-appraisal of the screw design.

2. Inadequate Mixing. Such problems are often easy to diagnose in products such as coloured films (eg. *poor pigment dispersion* in LDPE; Figure 4.24), but are inevitably more complex to analyse and separate from other effects in thicker extrudates. Wherever a product appears to be of imperfect visual quality, poor mixing is always a possible contributary factor.

Figure 4.24 An example of poor pigment distribution, exemplified by excessive flow marks, in tubular LDPE film; the inhomogeneous flow pattern illustrates "spider lines", associated with the presence of the mandrel-support legs in the die.

These effects can be readily checked on thin sections, by use of a light microscope. Inadequate dispersion is the effect which yields most problems and contributes towards embrittlement: agglomerates themselves are stress concentrating features but often during processing, shear fields tend to align pigment agglomerates, or filler-rich regions, to produce planes of weakness in the extruded part.

Many mixing problems can be corrected by increasing the degree of work imposed on the melt during the compounding or extrusion processes. Special care must be exercised when extruding low viscosity resins with concentrated colour masterbatches of higher molecular weight, where additive dispersion is particularly difficult to achieve.

3. Degradation. Polymer degradation during processing, which can be detected by discolouration, or by various spectroscopic or other indirect methods, is induced by high temperature, or by excessive residence time in the melt state. Since many physical properties are related to molecular weight, it follows that a significant reduction in strength usually follows chemical degradation.

A successful solution to this type of fault can only be derived when the origin of decomposition is established: excess temperature settings, shear heat effects, residence time or inadequate stabilization are all common sources of degradative effects.

Figure 4.25 shows a bright band of thermally-degraded UPVC (which is *fluorescent* in near-UV radiation[4.34]; see also Chapter 10), at the joint interface of an extruded profile, hot-plate welded at 270 °C. The tear-shaped pit is formed by the evolution of CO_2 gas, which is itself formed by reaction of the degradation product (HCl) with $CaCO_3$ filler. Direct loss of molecular weight is therefore not the only means by which embrittlement occurs, as a consequence of polymer degradation.

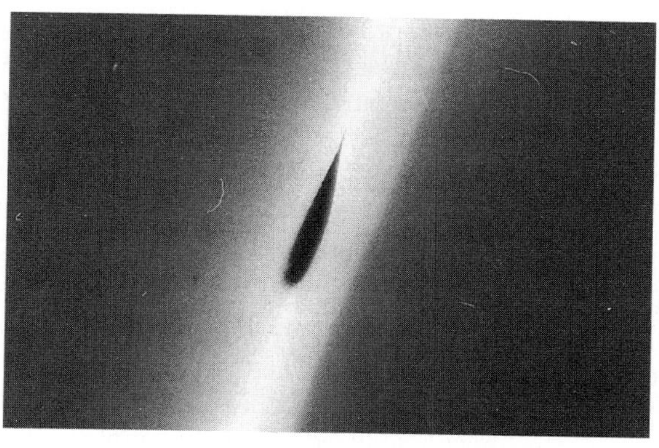

Figure 4.25 Fluorescence micrograph of an extruded UPVC profile, subsequently hot-plate welded at 270 °C, showing polymer degradation and interfacial voiding. (Reproduced from Ref. 10.23, with permission from Elsevier Applied Science Publishers)

4. Imperfect Re-Weld Zones. In hollow or complex extrudates, weld lines are formed when the melt recombines after passing through the mandrel-support ("spider") legs; the incomplete mixing evident in Figure 4.24 occurred as a result of this effect.

Although an increase in temperature would assist molecular flow across the weld lines, the most appropriate and effective remedy is to increase the shear stress on the melt by incorporating a mechanism of restricting melt flow, or by increasing the die land length. Enhanced mixing and homogeneity of weld zones may be achieved only at the expense of reduced output and increased die head pressure.

5. Contamination. Contaminants can be sub-divided into two main groups: undesired polymeric materials (originating from improperly-purged machinery, feed-line faults or imperfect supply), and other species including metallic slivers from processing equipment, and traces of paper/card from packaging cartons. Each type may contribute to unexpected defects in products. Figure 4.26 shows an array of irregular purging-reagent crystals, distributed in a plastics extrudate.

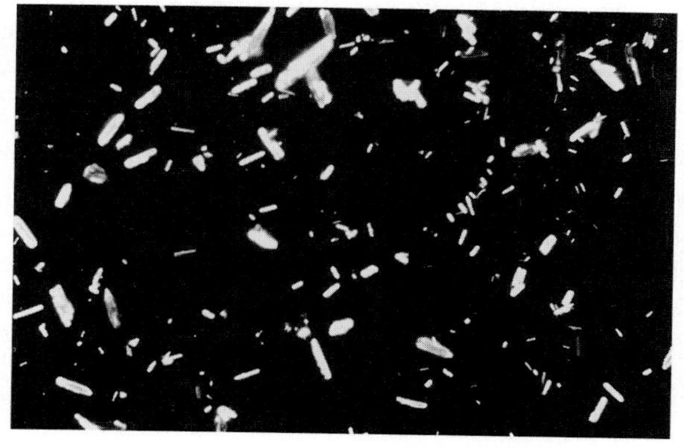

Figure 4.26 Birefringent crystals of a "contaminant" purging reagent, distributed in a matrix of amorphous PETP.

6. Melt Fracture and Sharkskin. Melt rupture was described in Chapter 3. Such defects are less likely to cause problems in service, since the occurrence of macro-defects is often detected readily during manufacture, and can be corrected during production. Adverse combinations of excess screw speed, high die taper angles and insufficient land length (exaggerated at low process temperature), are generally responsible. Solutions by way of material formulation might include choice of a lower molecular weight grade, or incorporation of a lubricant/processing aid.

7. Non-Optimum Gelation (PVC). It is evident that for any specific extruded PVC product, there appears to exist an optimum level of *gelation,* commensurate with the required properties. Under-processing leaves evidence of PVC's original particulate nature, whilst the achievement of very high levels of fusion can lead to a reduction in toughness, particularly if the morphology required for impact modification is destroyed[4.35].

An appropriate means of optimizing gelation for individual products must be sought. Increasing melt temperature, shear rate and residence time are all possible modes of accelerating the development of fusion. Great care is needed to avoid indiscriminate changes, however, since all these particular effects are likely to increase the possibility of degradation: a gradual, controlled-rate fusion process is the key to success.

4.2.6.2 Cooling Phase

Apart from some rather obvious imperfections such as melt fracture, gross distortion and voids, extrudate faults are usually investigated using more elaborate microscopic methods; it is the inhomogeneity of through-thickness cooling rate which leads to the defects given below.

1. Microstructural Effects. Both amorphous and semi-crystalline plastics are sensitive to cooling effects; crystalline texture, orientation and the total degree of crystallinity are likely to vary from point-to-point, according to local cooling conditions. *Morphological boundaries* are often formed as a result; it is the different physical characteristics (modulus, thermal expansion) of microstructures either side of such boundaries which add to the complexity of stress distributions, often leading to premature service failure. Similar and often more severe effects occur during moulding processes; more attention will therefore be given to this problem in the next Chapter.

2. Imperfect Surfaces. Cooling of extrudates is often accompanied by simultaneous sizing. The rate of cooling must therefore be balanced with the progression of die swell, and the onset of strain during the calibration of profile dimensions. Adequate surface lubricity and cooling uniformity are important elements of extrudate calibration, especially for filled or over-lubricated formulations where surface migration and *plate-out* would otherwise be likely to detract from surface quality.

4.3 Calendering of Plastics

Calendering is an alternative continuous processing method for the production of sheets and films from thermoformable thermoplastics feedstocks. A calender itself consists of a number (typically 3 or 4) of counter-rotating, cylindrical steel rolls which are internally heated. Raw compounds are squeezed between successive, inter-roll clearances (*"nips"*), and are metered and drawn off the calender to the required gauge, usually by suitably designed take-off rolls.

Calendering processes tend to be capital-intensive, for two main groups of reasons:

(1) Rolls are precisely machined and polished, and are very large and heavy; extremely robust bearing assemblies are therefore required;
(2) The calender itself usually represents only a constituent part of the total manufacturing operation.

It follows that calendering is preferable to extrusion only in situations where significant technical advantages are envisaged. Consequently, the development and refinement of calendering equipment tends to be very specific, towards individual combinations of polymer formulations and product groups.

Section 4.3 features only thermoplastics-based calendering processes (calendering has, of course, been extremely important in the manufacture of semi-finished rubber products for some considerable time), and examines the relationships between equipment layout and design, processing characteristics and the quality of calendered sheetstock.

4.3.1 Process Description

A calender may be considered as "a series of 2-roll mills" with gradually decreasing but adjustable nips, suitably arranged to allow continuous production of sheet or films. However, unlike the situation which occurs during milling, the total *residence time* which any element of feedstock spends under high shear is relatively short, since the material passes through each nip only once. Consequently, calenders are associated with only a modest mixing performance; the use of a premixed (and pre-gelled, for PVC compounds) feedstock is usually required. Although the combination of heat and shear in the calender is often considered insufficient to achieve complete homogeneity, this fact is considered to be beneficial when processing heat-sensitive plastics, since the thermomechanical history of calendered compounds is modest and controllable.

Rigid and flexible PVC compositions assume a large proportion of the plastics materials market for calendered products[4.36]; such compounds are therefore featured in an example of a typical operational sequence in a calendering plant, as described below:

(1) Pre-Mixing. Incorporation and dispersion of additives must be effective before gelation and shaping occurs. Rigid PVC is usually dry-blended in a high-speed mixer (typical discharge temperatures are 80 – 110 °C), whereas highly-plasticised compositions (being paste-like, rather than free-flowing powders) can be adequately homogenized by heated, ribbon-type blenders.

(2) PVC Gelation. A judicious combination of heat and shear deformation is required to break-down the *particulate structure* of as-polymerized PVC grains, although there are various mechanisms by which fusion can occur, according to the processing technique used[4.15]. Successive passes through the nips of a calender would be totally inadequate to achieve optimum or complete gelation; this is usually induced to produce a more homogeneous, mechanically worked "melt", in a separate stage.

The preferred method is based upon a high-shear internal mixer ("Banbury"® or "Intermix"® types), although continuous mixers are proving popular for extended production runs, and for mixing of high K-value (high molecular weight) rigid compositions which would otherwise require very high specific energies to be adequately dispersed by an internal mixer. Continuous extrusion-type mixers include the Buss kneaders, Farrel continuous mixers and more conventional twin-screw compounders (such as the APV, Liestritz or Werner-Pfleiderer systems), with relatively mild mixing screw configurations. Each commercial mixer usually has its own particular principle of compounding[4.18].

(3) Shaping. Once an acceptable, controlled-fusion PVC compound is developed, the feedstock is fed to the calender, usually in the form of strip. It is wise to maintain a positive bank of material at the first nip, in order to ensure that the sheet width is not too narrow. In counterbalance however, the residence time within the bank should be carefully controlled.

At the process design stage, the total number of rolls (and therefore nips) required for transmission of sufficient thermal and mechanical energy for shaping, together with the layout within the calender frame, must be balanced against machine size and capital expenditure. To achieve adequate flow and sizing characteristics, the number of calender rolls is usually between 3 and 5; see Figure 4.27. Equipment specification is concerned with the following points:

(1) Adequate feeding location – the 3-roll offset (design 2) was developed from the vertical stack system for this reason; this is especially crucial when processing low viscosity polymers;
(2) Direct conductive heating (at a given line speed), is determined by the contact area between polymer and roll surface; this must be optimized by layout enhancement, and by suitable use of take-off rolls;
(3) To avoid excessive shear stress and roll pressure, high speed processes demand that there be a greater number of nips, to induce a more gradual gauge reduction;
(4) The size and robustness of side-frames determines machine costs to a large extent; the development of Z-type calenders (designs 5 and 6) helps to reduce the overall frame size.

(4) Sizing and Cooling. Small take-off, *stripper rolls* act in conjunction with water-circulated cooling rolls, to act as a haul-off device for the calender. Embossing and printing units are often sited between these rolls, so that the PVC is relatively

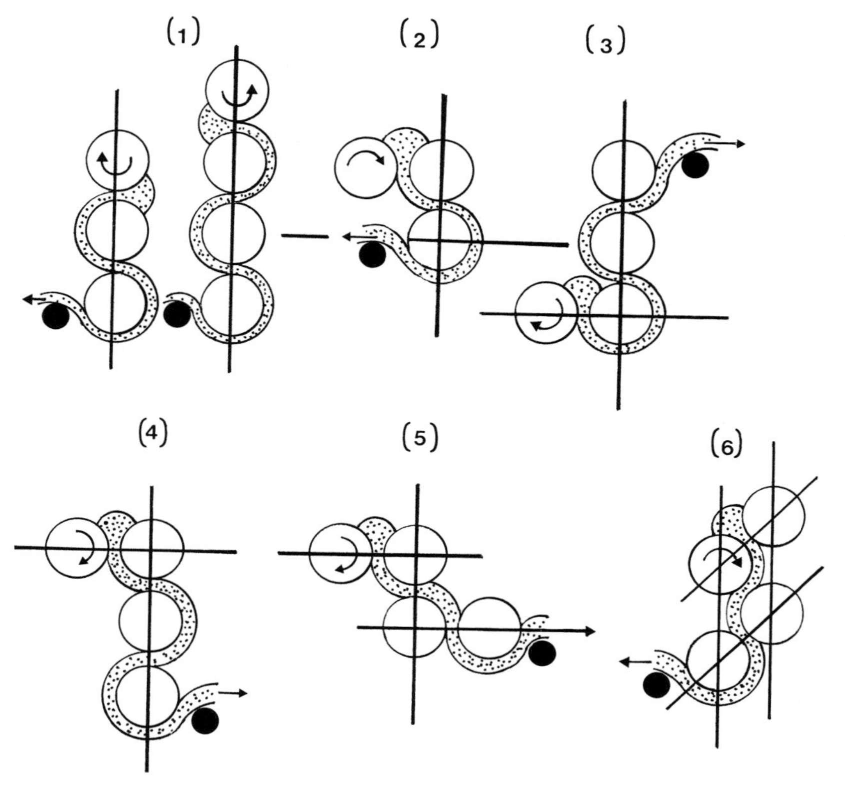

Figure 4.27 Calender roll configurations, presented in chronological order of development; (1) 3-roll and 4-roll vertical stacks; (2) 3-roll offset; (3) 4-roll "L"-type; (4) 4-roll inverted "L"-type; (5) 4-roll "Z"-type; (6) 4-roll inclined "Z"-type.

shape-stable, but sufficiently warm to allow relatively easy deformation to occur under external pressure.

4.3.2 Polymers, Melt Properties and Calendered Products

Calendering is traditionally associated with elastomer-based products and semi-finished goods, and generally has limited scope for thermoplastics due to the extreme versatility of screw extruders to manufacture sheet and films. Wherever calendering is considered a viable alternative to extrusion, the following range of properties should be considered:

Processing Temperature. A wide softening temperature range is required for processing; amorphous plastics are therefore more desirable for calendering techniques.

Viscosity. High viscosity resins are preferred if self-supporting characteristics are deemed important. However, low viscosity compositions allow more rapid throughput rates for equivalent roll pressure and in addition, low or modest viscosity is usually essential (eg. highly plasticised variants of PVC) where laminated or other composite sheet products (requiring precise but complete impregnation) are to be fabricated.

Shear and Mixing Characteristics. Shear rates in calender nips are moderate ($10^0 - 10^2$ s^{-1}); furthermore, since residence time under shear is inevitably short, mixing characteristics are poor and some degree of pre-mixing is usually desirable, especially for filled and heavily pigmented sheetstock.

Elasticity. Elastic characteristics of calendered plastics should be consistent and preferably insensitive to changes in output, deformation rate and temperature. This should ensure that the extent of elastic recovery is consistent, thereby keeping sheet thickness distribution under tight control.

Thermal Stability. Polymers sensitive to thermal or mechanically-accelerated degradation may be acceptably processed, provided that shear rate and temperature, together with the residence time within a material bank, are not excessive and are easily controllable.

Calendered product ranges for plastics are synonymous with the constraints listed above and are dictated by polymer selection in combination with process efficiency ie. production rate per unit input energy. *Reinforcement materials* (woven textiles, glass or carbon-fibre matting) can also be successfully processed by calendering. Typical product ranges are summarized below:

Flexible Goods (PPVC) – Sheeting for water-resistant products (rainwear, babywear, toys), and laminations for flooring, wallcoverings and book-binding.

Rigid Products (UPVC, Styrenics) – Clear film for packaging foodstuffs and pharmaceuticals, and for chemical applications; sheet for credit cards, tapes and general purpose cladding and panelling.

Product thickness may vary from about 50 μm (drawn thin films) to 5 mm, and calenders can accomodate sheet widths of up to 2 metres.

4.3.3 Operation and Analysis of Calenders

It is often convenient to consider the calender as a sequence of shaping nips of gradually reducing thickness; flow analysis can be carried out at each nip in order to study the effects of formulation, material flow properties, machine settings and geometric variables on the efficiency of the total manufacturing line.

During the passage through the calender, a limited degree of *roll-adhesion* is desirable. This usually occurs to the hotter or faster of the highly-polished pair of rolls which constitute the nip. Dynamic variables include roll speed (this includes *relative* rotational speeds of adjacent rolls: if these are unequal, a *"frictioning"* process is operative – this type of process is used to increase shear rate and/or assist fabric impregnation), roll temperature gradient, pressure gradient at the nip and the nip separations. (These variables are treated in detail in Section 4.3.4.)

It is useful to lead towards an analysis of pressure distribution and flow at a calender nip by first considering a 4-roll calender in terms of the individual functions of each successive part of the overall shaping system:

1st nip – rotation speed, in conjunction with feedstock temperature, determines the feed rate to initiate the process, together with the volume of material in the first feed-bank. The residence time under shear can be controlled in this way.

2nd nip – this acts as a metering device by decreasing the separation between rolls 2 and 3. The final elements of lateral spreading flow (ie. along the roll-axis) should occur here. This effect should be both controlled and minimized, since any unbalanced pressure distributions result in uneven flow and deformation history, perhaps leading to uneven thickness distribution in the calendered product.

3rd nip – this position represents the final shaping stage in the calender stack, where the design and commercial viability of the production unit must be proven by the manufacturer's ability to achieve uniformity and consistency. The functions of the final nip, and the flow mechanism which occurs within it, are equivalent to the analogous situation in an extrusion die. With this in mind, more detail is now given to the flow mechanism at a calender nip (see Figure 4.28).

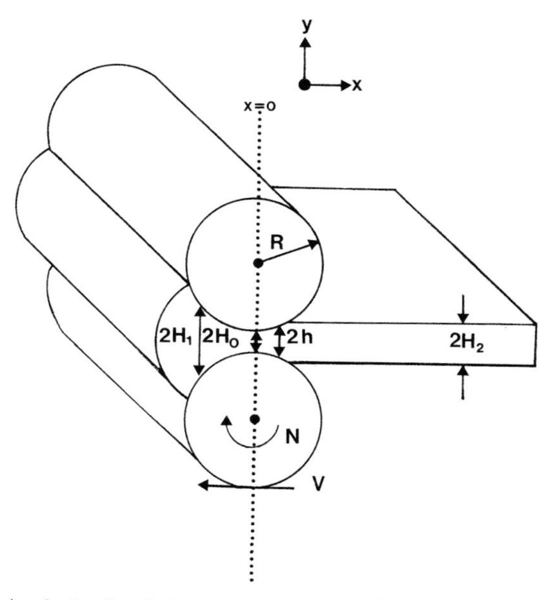

Figure 4.28 Analysis of melt flow in a calender nip; definition of geometry.

4.3.3.1 Flow and Pressure Generation at a Calender Nip

In order to analyse the flow of molten polymers through the nip of a calender or a 2-roll mill (as defined by the layout and geometric parameters in Figure 4.28), the qualitative aspect of flow (flow pattern and *velocity distribution)* must first be specified. As demonstrated previously[4.37], (see also Figure 4.29), the flow of melt in the calender bank is not symmetrical; nevertheless, most attempts to model this process are based upon a convenient axis of symmetry (at y=0; ie. along the

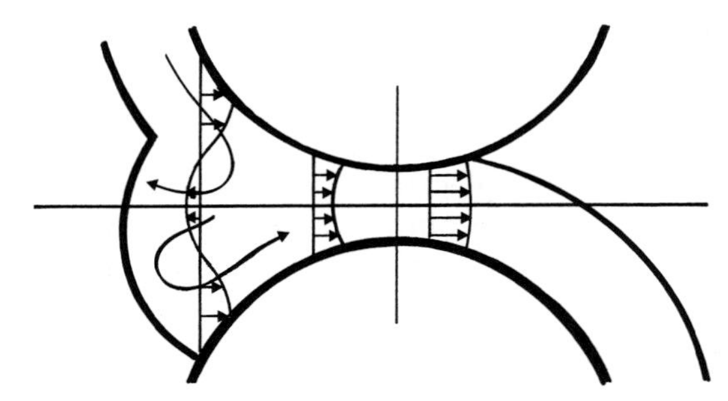

Figure 4.29 Material flow and velocity distributions in a calender nip.

horizontal axis midway between adjacent roll faces at the nips), to illustrate the melt velocity profiles.

It can be seen that the rotation of the rolls (and therefore the tangential speed at the calender nip) induces drag forces upon the melt. Also, the maximum pressure occurs not at the nip, but at a short distance upstream from it. Because of this occurrence, a *pressure flow* is superimposed upon the *drag flow* created by roll rotation. This accounts for the general shape of the velocity contours in Figure 4.29, and adds to the complexity of melt flow analysis and calender stack design.

Newtonian Analysis. Most analyses of calendering have been developed from the pioneering studies of GASKELL[4.38], exemplified by subsequent presentations[4.22,4.27,4.39], which are summarized here. A chronological account of previous work, listed according to assumptions used, is presented by AGASSANT and AVENAS[4.37].

In addition to the set of assumptions usually specified for a shear flow analysis of polymers (Chapter 3), it is assumed further in the case of calendering that:

(1) $h \ll R$; the lubrication approximation[4.27] for channels whose dimensions change very gradually, is therefore adopted;
(2) the sheet width fully occupies the available width (W) of the calender rolls.

In Figure 4.28, zero pressure is assumed at the point where the material bank first makes contact with both rolls (half bank thickness $h = H_1$ at $x = x_1$), and also at the location downstream of the nip where contact is lost with one roll ($h = H_2$ at $x = x_2$); the roll-surface separation at the nip is $2H_0$.

Considering only the changes in velocity v_X across the nip (parallel plate flow), we obtain:

$$v_X = V + \frac{y^2 - h^2}{2\mu} \cdot \frac{dP}{dx} \tag{4.24}$$

where h is the distance from the mid-plane to the roll surface and V, the tangential speed of the rolls, is given by:

$$V = (2\pi R) \cdot N \qquad (N = \text{roll rotational speed})$$

Volumetric output (per unit width), Q', can then be expressed as:

$$Q' = 2 \int_0^h v_X \, dy = 2h \cdot \left[V - \frac{h^2}{3\mu} \cdot \frac{dP}{dx} \right] \tag{4.25}$$

It can be seen that where the material leaves the upper roll at $x = x_2$, the pressure gradient dP/dx is zero and therefore the overall output, Q, is given by:

$$Q = 2WH_2 \cdot V$$

To obtain an expression for *pressure gradient* dP/dx which can be subsequently integrated, it is convenient to eliminate h by use of an expression in terms of x.

If $h = H_0 + q$, then (from nip geometry, Figure 4.28), $x^2 = 2Rq - q^2$. Since $R \gg q$, we can neglect the square term for q, such that $x^2 = 2Rq$.
Rearranging 4.25:

$$\frac{dP}{dx} = \left[V - \frac{Q}{2h}\right] \cdot \left[\frac{3\mu}{h^2}\right] \tag{4.26}$$

Parameter ϱ $(= q^{0.5})$ is introduced by McKelvey[4.39] to simplify the terms in the integrated form of expression 4.26; since $\frac{dP}{d\varrho} = \frac{dP}{dx} \cdot (2RH_0)^{0.5}$, the pressure gradient can be expressed in terms of ϱ:

$$\frac{dP}{d\varrho} = (2RH_0)^{0.5} \cdot \left[V - \frac{Q}{2h}\right] \cdot \frac{3\mu}{h^2} \tag{4.27}$$

Making the substitution $h = (H_0 + \varrho^2)$ into 4.27, we obtain:

$$\frac{dP}{d\varrho} = \frac{\mu V}{H_O} \cdot \left(\frac{18R}{H_O}\right)^{0.5} \cdot \frac{\varrho^2 - \lambda^2}{(1 + \varrho^2)^3} \tag{4.28}$$

where λ is the value of ϱ at the position $x = x_2$. λ may be evaluated thus:

$$\lambda = \left[\frac{Q'}{2VH_0} - 1\right]^{0.5} = \left[\frac{x^2}{2RH_0}\right]^{0.5}$$

Equation 4.28 can be integrated to yield the desired expression relating *inter-roll pressure* to melt viscosity, roll speed, nip size and the overall geometry of the calender:

$$P = \frac{3\mu V}{4H_0} \cdot \left[\frac{R}{2H_0}\right]^{0.5} \cdot [f(\varrho, \lambda)] \tag{4.29}$$

The geometric function $f(\varrho, \lambda)$, which includes the integration constant, is approximated to:

$$f(\varrho, \lambda) = \left[\frac{\varrho^2 - 1 - 5\lambda^2 - 3\lambda^2\varrho^2}{(1 + \varrho^2)^2}\right] \cdot \varrho + (1 - 3\lambda^2)\tan^{-1}\varrho + 5\lambda^3$$

The position of interest is at $\varrho = -\lambda$, where the pressure is a maximum; by substitution into Equation 4.29:

$$P_{Max} = \frac{15\mu V\lambda^3}{2H_O} \cdot \left(\frac{R}{2H_0}\right)^{0.5} \tag{4.30}$$

Although P_{Max} is more sensitive to a change in ϱ (an increase in ϱ increases both the maximum pressure and the overall width of the pressure profile through the nip), the effect of material *viscosity* on maximum pressure is also evident: the two parameters are in direct proportion, according to the solution of the Newtonian analysis presented in Equation 4.30. Clearly, for a given calender design, both material viscosity and roll speed contribute directly to the magnitude of maximum inter-roll pressure.

The *location* at which the maximum pressure is realized lies upstream of the nip; a summary of pressure-position relationships has been presented by McKelvey[4.39]:

(1) At $x = x_1$, $P = 0$ (upstream position where contact with second roll is made);
(2) At $x = -x_2$, $P = P_{Max}$ (see Equation 4.30);
(3) At $x = 0$, $P = \frac{1}{2}P_{Max}$ (at the nip);
(4) At $x = x_2$, $P = 0$ (downstream, where contact with one roll is lost).

4.3.3.2 Influence of Pseudoplasticity

The analysis presented above is based upon a series of simplifications, not least being the assumption concerning shear rate-independent (Newtonian) viscosity. It is not surprising that in consequence, many studies show only partial agreement between theory and practice. Previous work in this area has revealed the following[4.22,4.40,4.41]:

(1) Maximum pressure. Disagreement between practical observation of maximum roll-gap pressure and theoretical prediction from the Newtonian analysis (above) was initially found to be relatively small[4.40]. However, a more recent analysis[4.41] has revealed a proportionately smaller value of P_{Max} when the power-law index of plastics (having equivalent zero-shear viscosity) is decreased from unity. For example, n-values of 0.75 and 0.50 yielded P_{Max} values of only 40 % and 10 % of those which would be expected from an "equivalent" Newtonian fluid. Furthermore, for asymmetric calendering processes (such as *frictioning)* with unequal roll speeds, maximum roll pressures were seen to be less than those predicted by averaging the speeds of the adjacent rolls.

(2) Pressure gradient. Another serious disagreement between theory and practice is associated with the rate of pressure generation, in advance of position $x = x_1$ (where the "$n = 1$" locus contacts the ϱ-axis). Since the viscosity of any pseudoplastic polymer melt is higher in the "region of disagreement" than it is at the nip (shear rate is a maximum where roll-gap is smallest), pseudoplastic effects must be partly responsible for the observed experimental deviation from Newtonian theory, and must therefore be accounted for, when attempting to calculate roll-separation forces on this basis.

If flow curves are available, it is worth noting the expression for apparent shear rate at the nip[4.22]:

$$\dot{\gamma} = \frac{3V\lambda^2}{H_O}.$$

4.3.3.3 Roll-Separation Forces and Methods of Compensation

The pressure acting on the melt between the rolls induces a normal force across the roll surfaces, which has a tendency to increase *roll-deflection* during processing. Accurate estimation of the *roll separation force* is necessary if deflection is to be estimated. Force is evaluated by integrating Equation 4.29 over the surface area which is in direct contact with the pressurized melt:

$$F = W(2RH_0)^{0.5} \int_{-\varrho_0}^{\lambda} P \, d\varrho \tag{4.31}$$

This gives a solution:

$$F = \frac{3\mu VRW}{4H_0} \cdot f'(\varrho, \lambda) \tag{4.32}$$

Expressions for the geometric function f' are given elsewhere[4.39]. For purposes of simplicity, f' may be approximated by an exponential model of the form:

$$f'(\varrho, \lambda) = A \cdot e^{B\lambda}$$

where the constants A and B (for a typical range of practical values of λ between 0.2 and 0.4) are given by: $A = 1.64$; $B = 14.56$

Overcoming Roll-Deflection. It is clear that, due to the generation of high pressure and roll-separation forces, severe bending moments are created which induce roll-deflection and, if not adequately compensated, contribute to sheet thickness fluctuation. Some means of overcoming or complying with roll deflection have been developed; these are summarized below and are illustrated in Figure 4.30.

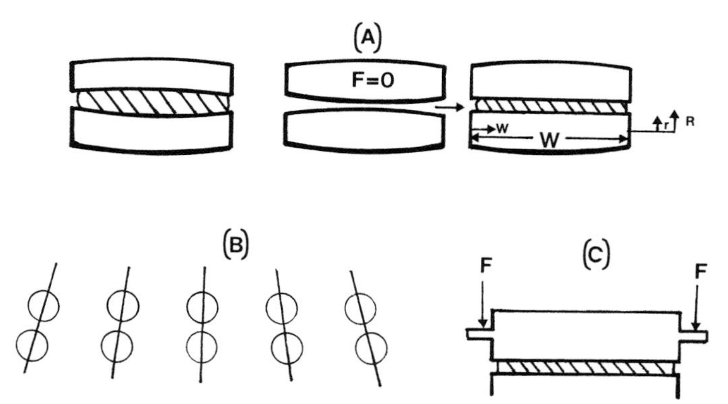

Figure 4.30 Calender roll-deflection, and some modes of correction; (A) roll grinding; (B) roll crossing; (C) roll bending.

(A) Roll Grinding – A reverse-profile crown is machined, to accommodate the deflection along the roll-axis. This method is obviously limited to specific combinations of compound formulation, machine settings, output and roll-gap.

(B) Roll-Crossing – Adjacent rolls are held slightly off-axis; this method has the advantage of accommodating large and potentially variable roll-deflections, and lends itself easily to closed-loop thickness control, since the cross-axis angle can be used as a response variable to correct a fluctuating thickness.
(C) Roll-Bending – This system requires the application of a compensatory bending load to the roll(s), to increase the effective crown and compensate for small deflections.

4.3.4 Machine and Formulation Variables

In common with most other plastics processing techniques, a calendering line can be represented by an individual *process viability window,* for any given combination of feedstock formulation and machine design. Before progressing to consider an example of a typical viability-format diagram, some important aspects of machine settings and formulation variables are summarized:

(A) Machine Settings

Temperature. The *heat content* of a calender roll-stack at typical processing temperatures for PVC (150 – 200 °C) is extremely high. In consequence, thermal equilibrium is approached very slowly from a cold start. Closed loop control of temperature, using a circulating heat exchange medium (usually oil) is therefore essential. Response time of the control mechanism must be short: the effect of a slight fluctuation on thermal history may result in differential viscosity and "swelling" effects, causing sheet thickness distribution problems.

Procedures used for optimizing temperature gradients must include some reference to viscous heat generation, which will become more significant as roll speed or polymer viscosity is increased. Special care is required for rigid vinyl compositions, which are noted for their heat-generating characteristics: in some situations, a net heat energy transfer may occur from the polymer to the steel roll, thus dissipating the viscous heat ultimately by radiation from the rolls to the environment.

To summarize, temperature increases are likely to:

(1) decrease viscosity, thereby diminishing the inter-roll pressure and bending force on the rolls;
(2) reduce the likelihood of stress-related defects (eg. local melt rupture due to fluctuation in take-off velocity;
(3) increase the acceleration of complete fusion of PVC, but also increase any risk of thermal degradation;
(4) increase the appeal and quality of sheet surface finish, especially for filled compositions.

Calender rolls. Optimum choice of rotation speed, frictioning speed ratios, roll crossing and appreciation of roll deflection are fundamental to the effectiveness of the production unit. Roll speed particularly can be utilized as an independent response variable for control of web thickness, such is the sensitivity of sheet gauge to a change in this parameter. An increase in roll speed, at a calender nip, is likely to induce:

(1) significant shear heat which, even if not excessive, may allow a reduction of the roll temperature setting ie. a less-severe temperature profile;
(2) an increase in output, coupled with corresponding increases in pressure (leading to process instability) and shear rate (giving the possibility of melt fracture or excessive swelling);
(3) shorter residence times, possibly detracting from heat transfer and the development of mixing/fusion characteristics.

Calendered sheet adheres to the faster rotating roll, so that the velocity distribution within an individual nip is modified if a positive *friction ratio* (unequal rotational speeds of adjacent rolls) is applied. Loss of drag flow symmetry leads to increases in shear rate; the associated consequences are greater fluidity and increases in microstructural homogeneity. This is beneficial when manufacturing composite products, since the degree of impregnation of fabric reinforcement plies increases as the friction ratio is increased from unity (for "surface coating"); compositions of higher viscosity may then be successfully used as *impregnation materials* if the friction ratio is increased further.

A generalized *process feasibility diagram* for plastics calendering, with reference to roll speed and temperature, is presented in Figure 4.31; the limitations imposed are illustrated by numerical identification, and may be summarized as follows:

(1) Unstable processing; excess pressure, roll deflection and thickness variation.
(2) Insufficient output; economically-unviable process.
(3) Surface defects; insufficient homogenization.
(4) Surface degradation; air-entrapment.
(5) Thermal degradation.

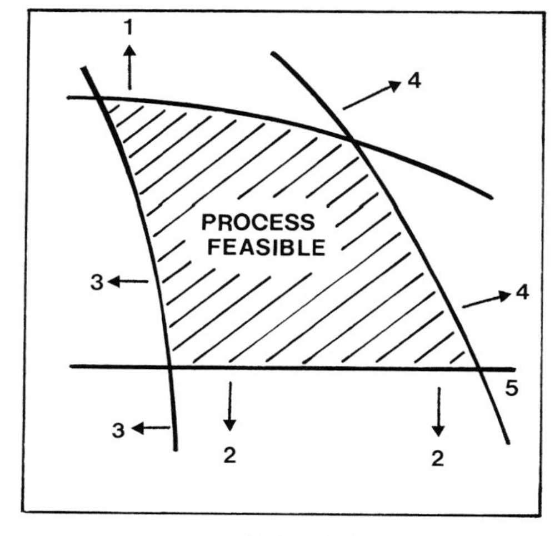

Figure 4.31 A generalized process feasibility window for the plastics calendering process (see text for definitions).

(B) Formulation Effects

The whole rationale of changing formulation constituents is based on the ability of plastics compounds to be modified according to properties demanded. For purposes of calendering, the reader is referred to the monograph by ELDEN and SWAN[4.36] for a thorough consideration of formulation variables. However, the summary presented below lists the more important formulation effects which can be specifically considered in terms of their direct relevance to processibility during calender operation.

Polymers. Whilst various co-monomers will react with vinyl chloride (eg. VDC, VAC.), vinyl copolymers are usually specified for end-use properties, rather than for any notable processing characteristics. The effects of copolymerization on processing (at least for copolymers of high VC composition) are often overshadowed by influences of molecular weight. In turn, these relationships are of less significance if plasticizers are added in increasing proportion. Typical K-values (a K-value is a measure of average molecular weight, derived by solution viscometry) for calendering compositions are, therefore[4.36]:

Rigid PVC – K 55 – 65
Flexible PVC – K 65

Plasticisers. The range of plasticised PVC compounds for calendered sheet and film can be split into two main groups: flexible compounds containing large amounts of plasticizer (up to 100 – 150 parts, per hundred parts resin, pphr) and semi-rigid compositions containing only minor proportions (less than 5 – 10 pphr) of plasticizer.

In the former group, high molecular weight polymeric plasticizers are often preferred, to avoid volatility during processing. For compounds in the latter category, plasticizers are often added to modify stability or to change physical properties, but may also assist the dispersion of fillers and pigments during shear working. Applicability of relatively low concentrations of plasticizer is sometimes limited by an embrittlement phenomenon called *antiplasticisation*[4.42].

Stabilizers. Protection against thermal or photo-degradation of PVC is afforded by incorporation of stabilizers (metal soaps, Ba, Cd, Zn, and Pb compounds and complexes), though the role of such compounds during the process is inevitably passive. Possible exceptions are compounds based on stearates, which often impart a further degree of lubricity at the roll-surfaces.

Lubricants. Some important advantages can be gained by incorporating lubricants (metal soaps or synthetic waxes) into rigid PVC compounds for calendering. These include an enhanced *friction release* characteristic from embossed or "matt-finish" rolls. Also, flow modification and modest increases in flow rates are feasible, since lubricants induce a "plug-like" character by increasing compound velocity at the roll-surface. This may also diminish local shear rates and viscous heat generation, but no less important, lubricants may retard the development of complete fusion.

Fillers. The extent to which viscosity increases with filler content is determined by the size, the size distribution and by the surface chemistry of the filler. Such effects are usually only important commercially in flexible PVC compounds, since only relatively low filler levels have hitherto been incorporated into calendering compounds from rigid formulations.

The effectiveness of fillers in calendered products is very sensitive to the adoption of correct and exacting procedures during the process: for example, china-clay or chalk fillers are sometimes used as compound scrubbers to maintain roll surfaces free of bloom or plate-out, yet if adequate levels of filler dispersion are not achieved during premixing, sheetstock quality and appearance can deteriorate dramatically. To assist dispersion, many classes of fillers are often surface treated: stearates, or chemically grafted silane or siloxane groups are often used. A further possible effect of using highly loaded or poorly dispersed fillers is to decrease the melt rupture strength, thus limiting the drawdown ratios which may be achieved during film manufacture.

Process Aids. Abrasive, friction-generating acrylic polymers, typically added at 1–5 pph, are effective process aids for rigid PVC compositions. Fluxing and gelation are accelerated and melt profile stability (eg. during drawdown, stretching processes, or where self-support is necessary) and melt elasticity are each enhanced[4.43,4.44].

4.3.5 Some Defects in Calendered Products

Although finished products are often only recognized as being defective during subsequent quality control or inspection procedures, many types of flaws can originate during the calendering process as a result of the sensitive relationships between formulation, machine settings and product properties. Whilst it is impossible to present a universal list of "defects and remedies" for calendered plastics, the generalities presented within this section should assist initial investigations of a fault-finding, or troubleshooting nature. Typical defects are listed and are succeeded by a technological interpretation, and other suggestions relating to the origin or correction of the stated imperfection.

4.3.5.1 Uneven Thickness Distribution

This is the single, most encountered problem in many calendering processes. Assuming that *crowning* (due to roll deflection in uncompensated roll stack designs – see Section 4.3.3) is not responsible, the effect is usually associated with differential thermal or shear history of feedstock, prior to and during processing. Reworking of scrap materials (whose rheological characteristics are inevitably modified, in comparison to virgin resin), is a notoriously frequent cause of exaggerating this problem.

If an uneven gauge distribution is observed along (rather than across) the web, localized viscosity fluctuations, or an imbalance between tangential roll speed and linear take-off velocity are likely causes. Also, a retarded roll-temperature control response is also possible, the effect being one of fluctuating melt viscosity.

4.3.5.2 Surface Blemishes and Defects

Discoloration. Yellowing of light shades of calendered sheet almost inevitably points to polymer decomposition. Excessive temperature or roll speed, or too small a nip leads to high shear rates during processing and excessive viscous heat generation; inadequate stabilization/lubrication may also be responsible.

Crows' Feet. Multiple "V"-marking of sheet is induced by material which is overturned and reprocessed in the calender bank, when the stock temperature is too low to allow effective re-welding of newly-folded sheet zones to occur. Inadequate dispersion of some types of coarse additives tends to worsen the severity of defects so formed.

"Fish-Eyes"/Lumpiness. Under-processing (inadequate mixing and/or gelation) may be apparent.

Bank Marks. By definition, an uneven surface rippling or pitting effect originates from the feedstock having had an uneven shear history in a calender bank.

Plate-Out. This term represents intermittent surface migration of additive agglomerates, particularly lubricants and fillers; further formulation optimization is usually required.

Surface Roughness. Micro-roughness arises as a consequence of local melt rupture and reformation, during the residence time spent under pressure in a calender nip. It is analogous to similar defects in extrudates and may be controlled by increasing temperature, or by decreasing roll speed and minimizing drawdown effects.

Air Bubbles. The presence of air inclusions in calendered sheet is exaggerated by higher roll temperatures, but usually originates from micropores introduced into the calender bank. Gaseous degradation products may also become visible in this way.

Some evidence has been established to link the presence of certain defects (air inclusions and surface roughness) in calendered sheet, to the rheological environment imposed upon typical grades of PVC[4.37,4.45]. For example, the likelihood of air inclusions in sheet is diminished if a critical pressure in the calender bank is exceeded:

$$P_{Max} = k \cdot \left(\frac{2R}{H_0}\right)^{0.5} \left(\frac{2n+1}{n} \cdot \frac{V_d}{H_O}\right)^n \cdot \varepsilon(n) \qquad (4.33)$$

(k and n are power law indices, R = roll radius, V_d = calender speed and $\varepsilon(n)$ is a dimensionless function of n. NB: $\varepsilon = 0.5$ if bank thickness, $H = 10 \cdot H_0$).

In general, P_{Max} must exceed approximately 120 bar (12 MN/m^2) to rid the stock of air inclusions, if typical figures for PVC are inserted into Equation 4.33.

Furthermore, the initial sheet thickness at which the appearance of a matt surface is apparent can be related to a maximum shear stress (τ_c; see Equation 4.34), or to a maximum pressure (P_{Max} in Equation 4.33), depending on formulation:

$$\tau_c = k\left(\frac{2n+1}{n} \cdot (r-1) \cdot \frac{V_d}{H_O}\right)^n \tag{4.34}$$

(r = overspread height $= \frac{h^*}{H_0}$)

4.3.5.3 Post-Fabrication Shrinkage

A longitudinal reversion effect, due to relaxation of elastic strain, may be induced when rapid cooling accompanies drawdown, and is accelerated at relatively high ageing (storage) temperatures.

A more rigorous analysis of processing defects in rigid PVC is presented in the paper by BOURGEOIS and AGASSANT[4.45].

4.4 Pultrusion

Pultrusion is the only *continuous process* for the manufacture of high performance fibre-reinforced composites which is of any industrial significance. It is used to produce constant cross-section products continuously, and is capable of producing a wide range of both solid and hollow profiles. A pultruded product consists of reinforcing materials which are bound together by a resin; basically, the process involves "pulling" the impregnated reinforcement through a die to produce the finished profile.

Fibre reinforcements used in pultrusion include glass, aramid and carbon in various forms; eg. rovings, fabric and mat. The matrices used are invariably thermosetting plastics, and over 90 % of all pultruded products are based on glass reinforced unsaturated polyesters.

4.4.1 Description of the Process

There are six basic steps in the process[4.46 – 4.47], as shown in Figure 4.32. These are:

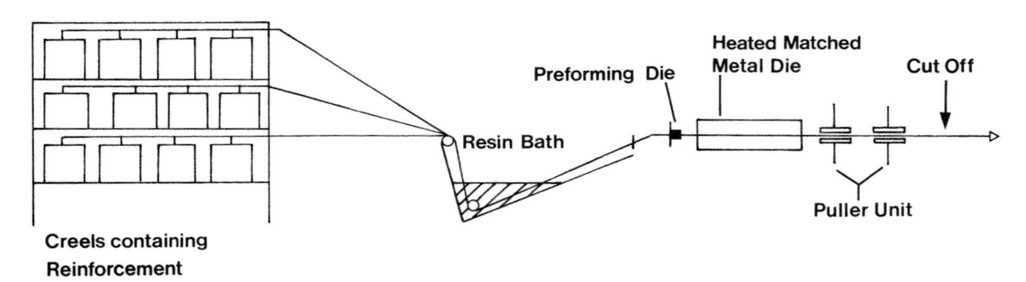

Figure 4.32 Schematic diagram of the pultrusion process.

(1) Reinforcement handling system (creel);
(2) Resin impregnation bath;
(3) Preforming die;
(4) Heated matched metal die;
(5) Continuous pulling mechanism;
(6) Cut-off.

The fibres are pulled from the *creels* through a resin bath where they must be thoroughly impregnated. Any excess material can be removed by a set of rollers at the end of the bath. The impregnated resin then passes into a *preforming die* which consolidates the wet laminate to the general shape of the required profile. It is usual to attach the preforming die to the *pultrusion die,* to ensure correct alignment. The laminate then progresses through the heated die to produce the final shape, and to cure the thermosetting resin. There must be a gap between the die and the pulling mechanism, to allow the profile to cool, and to develop sufficient strength to accommodate the clamping forces needed to grip the pultrusion. The pulling section typically consists of a pair of continuous caterpillar belts, similar to the type generally used in extrusion. Finally, the product is cut to the required length. For hollow profiles, a mandrel is positioned ahead of the resin tank, and extends into the die. The process for thermoplastic resins is very similar, but excludes the crosslinking step.

The reinforcement can be in the form of axially-oriented fibres, which produces a very stiff and strong product in the pultrusion direction. This type of construction tends to be used as a solid section for such applications as electrical insulators and fishing rods, where high tensile strength is required. However, for many structural profiles, a combination of axial fibre and woven cloth is used to improve transverse and shear properties of hollow sections, I-beams, tubes, channels etc.

4.4.2 Analysis of the Process

At present, pultrusion is essentially an "art-form", but a more scientific understanding of the process is being developed, in particular by METHVEN[4.48], and by SUMARAK[4.49]. The process is undoubtedly quite complex and any model of the system must consider the interaction between the reaction kinetics, heat transfer, flow properties of the resin, and the friction between the profile and the die wall[4.47].

A simple model of the process has been developed[4.48], in an attempt to quantify pulling speed and pulling force. The model is shown schematically in Figure 4.33. The total pulling force (F) is given by:

$$F = F_C + F_V + F_f \tag{4.35}$$

where
F_C is the *compaction force* at the die entry (preform region);
F_V represents the *viscous forces* developed up to the gelation stage; and
F_f is the *frictional force* between the die and the gelled profile.

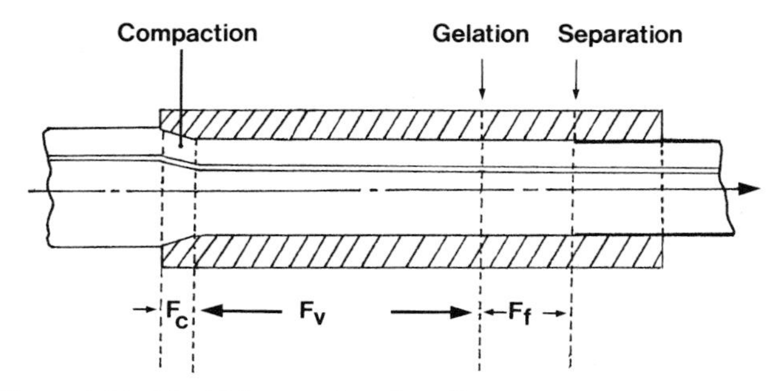

Figure 4.33 Model for a pultrusion die.

F_C is estimated from the pulling speed, the geometry of the die entry region, the fluid viscosity and the volume fraction of fibre. The viscous forces (F_V) arise as a result of *drag flow* induced by the fibres, and by back flow caused by thermal expansion. Each of these terms can be calculated from the process parameters and the fluid properties.

After the resin gels, the solid profile exerts a normal force on the die wall as a result of thermal expansion, and a frictional force (F_f) must be applied to overcome this effect. The thermal expansion arises as a result of conduction from the die wall and from the exothermic curing reaction. Eventually, the profile cools and shrinks away from the wall, and the position at which this occurs is determined from the shrinkage and thermal expansion characteristics of the resin. The positions where the various mechanisms take over are evaluated from heat transfer and resin cure kinetics.

This model represents an oversimplification of the process, but it does indicate that F_f is the major contributing factor to the total pulling force requirement; this appears to be consistent with experimental observations[4.47]. In addition, the separation point can be predicted, which in effect determines the appropriate length of die.

4.4.3 General Comments

The process is very versatile and can produce a wide range of sections which have excellent mechanical properties. The process is not as fast as conventional extrusion because of the high forces generated, but line speeds up to 3 m/min. (0.05 m/s) can be attained. It is possible with current commercial equipment to produce sections up to 600 × 200 mm, but this is not seen as a serious limitation and new machines are being developed to handle much larger sections.

A major feature and advantage of the pultrusion process is the range of resins and reinforcements which can be handled, providing a wide variety of properties in the finished products. This offers great flexibility to the design engineer. For smaller cross-sections, the profile can be wound continuously onto a spool. Tension members and small diameter fibre optic cable have been produced in this manner, in lengths of up to 2 km.

A variation of the basic pultrusion process is being developed for the manufacture of curved sections. A standard pultrusion line is used to produce a partially-cured profile, which can be moulded in a further, secondary process into a curved shape.

Overall, the market for pultruded products is still relatively small, but it appears to be the fastest growing area for high-performance composites.

References

4.1 FISHER, E.G.: *Extrusion of Plastics,* Illiffe, London, (1958).

4.2 LEVY, S.: *Plastics Extrusion Technology Handbook,* Industrial Press, New York, (1981).

4.3 RAUWENDAAL, C.: *Polymer Extrusion,* Hanser Publishers, Munich, (1986).

4.4 GIBSON, A.G., and WARD, I.M.: *The Manufacture of Ultra-High Modulus Polyethylenes by Drawing Through a Conical Die,* J. Mat. Sci., **15,** 979, (1980).

4.5 PAN, S.J., BROWN, H. R., HILTNER, A., and BAER, E.: *Biaxial Orientation of Polypropylene by Hydrostatic Solid-State Extrusion. Part 1: Orientation Mechanism and Structural Hierarchy* Polym. Eng. Sci., **26,** 997, (1986).

4.6 PAN, S.J., TANG, H.I., HILTNER, A., and BAER, E.: *Biaxial Orientation of Polypropylene by Hydrostatic Solid-State Extrusion. Part 2: Morphology and Properties,* Polym. Eng. Sci., **27,** 869, (1987).

4.7 JANSSEN, L.P.B.M.: *Twin Screw Extrusion,* Elsevier, London, (1978).

4.8 MARTELLI, F.G.: *Twin Screw Extruders: A Basic Understanding,* Van Nostrand, New York, (1983).

4.9 GAVIS, J., and MIDDLEMAN, S.: *Origins of Normal Stress in Capillary Jets of Newtonian and Viscoelastic Liquids,* J. Appl. Polym. Sci., **7,** 493, (1963).

4.10 HAN, C.D.: *Flow of Viscoelastic Fluids Through a Rectangular Duct,* Am. Inst. Chem. Eng. J., **17,** 1418, (1971); **18,** 1286, (1972).

4.11 HAN, C. D.: *Rheology in Polymer Processing,* Academic Press, New York, (1976).

4.12 ORBEY, N., and DEALY, J.M.: *Isothermal Swell of Extrudate from Annular Dies; Effects of Die Geometry, Flow Rate and Resin Characteristics,* Polym. Eng. Sci., **24,** 511, (1984).

4.13 MICHAELI, W.: *Extrusion Dies: Design and Engineering Computations,* Hanser Publishers, Munich, (1984).

4.14 LINDT, J.T.: *Mathematical Modelling of Melting of Polymers in a Single-Screw Extruder,* Polym. Eng. Sci., **25,** 585, (1985).

4.15 ALLSOPP, M.W.: in: *Manufacturing and Processing of PVC,* BURGESS, R.H.,Ed., Applied Science, London, (1982).

4.16 COVAS, J.A.: *Processing of UPVC in Single and Twin Screw Extruders,* Ph.D. Thesis, Loughborough University of Technology, (1985).

4.17 COVAS, J.A., GILBERT, M., and MARSHALL, D.E.: *Twin Screw Extrusion of a Rigid PVC Compound - Effect on Fusion and Properties,* Plast. Rub. Proc. Appl., **9,** 107, (1988).

4.18 MATTHEWS, G.: *Polymer Mixing Technology,* Applied Science, London, (1982).

4.19 TODD, D.B.: *Residence Time Distribution in Twin Screw Extruders,* Polym. Eng. Sci., **15,** 437, (1975).

4.20 TODD, D.B. and BAUMANN, D.K.: *Twin Screw Reinforced Plastics Compounding,* Polym. Eng. Sci., **18,** 321, (1978).

4.21 TADMOR, Z., and KLEIN, I.: *Engineering Principles of Plasticating Extrusion,* Van Nostrand, New York, (1970).

4.22 TADMOR, Z., and GOGOS, C. G.: *Principles of Polymer Processing,* John Wiley, New York, (1979).

4.23 CROWSON, R.J., FOLKES, M.J., and BRIGHT, P.F.: *Rheology of Short Glass Fibre-Reinforced Thermoplastics and its Application to Injection Moulding 1. Fibre Motion and Viscosity Measurement,* Polym. Eng. Sci., **20,** 925, (1980).

4.24 EDWARDS, M.F., ELLIS, D.I., GEORGHIADES, S., and SMITH, R.: *Cooling and Solidification of Extruded Pipe Sections,* Presented at PRI International Conference Plastics Pipes V, p.8/1, York, UK, (1982).

4.25 MIDDLEMAN, S.P.: *Fundamentals of Polymer Processing,* McGraw-Hill, New York, (1977).

4.26 KRIETH, F. and BOHN, M.S.: *Principles of Heat Transfer,* 4th Ed., Harper & Row, New York, (1986).

4.27 POWELL, P.C.: *Engineering with Polymers,* Chapman & Hall, London, (1983).

4.28 BECKMANN, G.: *Extrusion and Properties of Cellular and High-Impact UPVC Profiles for the Building Industry,* Presented at 3rd International PRI Conference PVC '87, p.13/1, Brighton, UK, (1987).

4.29 FISCHER, P., and WORTBERG, J.: *Extrusion Line for the Production of Multilayered Foam Pipes,* **ibid,** p.14/1.

4.30 DESPER, C.R.: *Structure and Properties of Extruded Polyethylene Film,* J. Appl. Polym. Sci., **13,** 169, (1969).

4.31 MADDAMS, W.F., and PREEDY, J.E.: *X-Ray Diffraction Orientation Studies on Blown Polyethylene Films: I. Preliminary Measurements; II. Measurements on Films From a Commercial Blowing Unit; III. High-Stress Crystallisation Orientation,* J. Appl. Polym. Sci., **22,** 2721; 2739; 2751, (1978).

4.32 GILBERT, M., HEMSLEY, D.A., and PATEL, S.R.: *Effect of Processing Conditions on the Orientation and Properties of Polyethylene Blown Film,* Brit. Polym. J., **19,** 9, (1987).

4.33 TEMPLEMAN, J.E., SWEETLAND, D.B., and LANGLEY, P. A.: *The Application of High Strength Polyolefin Grids in Civil Engineering,* Plast. Rubb. Proc. Appl., **4,** 99, (1984).

4.34 HEMSLEY, D.A., HIGGS, R.P., and MIADONYE, A.: *UV Fluorescence Microscopy in the Study of Poly(Vinyl Chloride) Powders and Extrudates,* Polym. Comm., **24,** 103, (1983).

4.35 MENGES, G., BERNDSTEN, N., and OPFERMANN, J.: *Extrusion of PVC: Consequences of the Particle Structure,* Plast. Rubb. Proc., **4,** 156, (1979).

4.36 ELDEN, R.A., and SWAN, A.D.: *Calendering of Plastics,* Illiffe, London, (1971).

4.37 AGASSANT, J.F., and AVENAS, P.: *Calendering of PVC: Prediction of Stress and Torque,* J. Macromol. Sci; Phys. Ed., **B14,** 345, (1977).

4.38 GASKELL, R.E.: *The Calendering of Plastic Materials,* J. Appl. Mech., **17,** 334, (1950).

4.39 MCKELVEY, J.: *Polymer Processing,* Wiley, New York, (1962).

4.40 BERGEN, J.T. and SCOTT, G.W.: *Pressure Distribution in the Calendering of Plastic Materials,* J. Appl. Mech., **18,** 101, (1951).

4.41 KIPARISSIDES, C., and VLACHOPOULUS, J.: *A Study of Viscous Dissipation in the Calendering of Power Law Fluids,* Polym. Eng. Sci., **18,** 210, (1978).

4.42 Mascia, L. and Margetts, G.: *Viscoelasticity and Plasticity Aspects of Antiplasticisation Phenomena: Strain Rate and Temperature Effects,* J. Macromol. Sci; Phys. Ed., **B26,** 237, (1987).
4.43 Cogswell, F.N., Player, J.M., and Young, R.C.: *The Influence of Acrylic Processing Aids on the Extensibility of PVC Melts,* Presented at PRI/BSR International Conference Practical Rheology in Polymer Processing, Loughborough, UK, (1980).
4.44 Gould, R.W., and Player, J.M.: *Methyl Methacrylate Copolymers as Processing Aids for Rigid PVC,* Kunststoffe, **69** (7), 10; 393, (1979).
4.45 Bourgeois, J.L., and Agassant, J.F.: *Calendering of PVC: Defects in Calendered PVC Films and Sheets,* J. Macromol. Sci; Phys. Ed., **B14,** 367, (1977).
4.46 Meyer, R.W.: *Handbook of Pultrusion Technology,* Chapman & Hall, London, (1985).
4.47 Martin, J.D., and Sumerak, J.E.: in: *Engineered Materials Handbook Vol.2: Engineering Plastics,* ASM International, USA, (1988).
4.48 Methven, J.M., and Atalay, O.: *A Simplified Model of Pultrusion,* Presented at 16th BPF Reinforced Plastics Congress, Blackpool, UK, (1988).
4.49 Sumerak, J.E.: *Understanding Pultrusion Variables for the First Time,* Presented at 40th Annual Conference of Reinforced Plastics Composites Institute, Atlanta, USA, (1987).

Chapter 5
Processing II – Moulding Techniques

5.1 Introduction

In the introductory section to the previous Chapter, the salient principles and different phases of plastics processing were described. In virtually all processes, a judicious selection of manufacturing conditions (temperature, pressure/force and rate effects) determines the analysis of plastics flow, deformation and cooling, according to the thermal and rheological properties of the material. We have split the processing content of the text into two individual Chapters, according to the *continuous* or *intermittent* nature of the manufacturing operations. Clearly, the dynamics of processing depends upon the sequencing of the equipment and the thermal environment which the material endures during shaping.

In this Chapter, the technology of processing methods is extended to consider *moulding techniques,* where the complete, 3-dimensional form of plastics products is determined by shaping within a closed mould. Whilst the general route to product manufacture (mixing, melting, shaping and shape stabilization) defined in Section 4.1 is equally applicable to moulded products, the analysis and interrelationships between processability, microstructure and final part properties are unique for any given moulding process. Furthermore, most moulding techniques, being cyclic or intermittent in nature, are associated with unsteady situations involving melt flow and heat transfer. In consequence, simple theory and practical observation are rarely compatible and it is found universally that the microstructure and properties of moulded parts are extremely sensitive to any fluctuation (whether deliberate or unavoidable) in material specification or processing conditions. Bearing this in mind, it is perhaps not surprising that the plastics moulding industries are amongst the first to have realized the benefits (in terms of quality, consistency and productivity) to be gained from automatic, closed-loop control of manufacturing lines.

In view of its size and importance, and due to the complex nature of its technology, the injection moulding process is considered first. Blow moulding and rotational moulding, two techniques used for the manufacture of hollow components, are described in turn and subsequently, discussions focus upon thermoforming, and on a family of fabrication processes for manufacturing with reinforced thermosetting plastics.

5.2 Injection Moulding

A vast array of modern plastics products are manufactured by injection moulding: this process is conceptually simple and is associated with relatively few constraints of product size, shape or complexity. Polymer feedstock is melted by external and internal (shear) heating and is forced under high pressure into a mould held tightly closed, where shape stability is achieved by *cooling* (thermoplastics) or by *chemical crosslinking* (thermosetting plastics). As long as a suitably-designed injection mould can be constructed to allow the part to be shaped, consolidated, then ejected at the end of the moulding cycle, the function of the process designer is to operate the process to yield parts of acceptable quality, whilst maintaining overall *cycle times* which comply with economic constraints. Although the working principle is straight-forward, there are many situations in which a compromise between quality and productivity must be achieved: in the context of manufacture, the factors which contribute towards an ideal solution of this type (material flow characteristics, moulding parameters) form the theme of this sub-chapter. They are highly *interactive* and are influential in most aspects of the process.

The first commercial injection machines were designed to operate with a plunger mechanism to activate injection. Whilst this system offered a "rapid injection" advantage, early experience demonstrated that the disadvantages associated with the following points led to inconsistent and unreliable processing:

- slow heating rates in the plunger assembly;
- lack of adequate shear mixing of feedstock;
- inconsistent feeding, since plastics granules were fed to the barrel on an intermittent basis;
- variation in pressure requirements, due to a combination of all these points.

Virtually all modern plastics injection moulding machines are now based upon a reciprocating *screw injection* unit: the screw acts as a ram in order to comply with the criterion of rapid injection, and the advantages of screw extruders described in the previous chapter (consistent feeding, effective compression, melting and laminar mixing) are also realized. The range of commercially-available screw injection machines offers a high degree of precision and repetitive accuracy; machines can be designed to meet the melting and shear flow requirements of a vast range of plastics materials.

Most of the discussion within this section focusses upon the *screw injection moulding* process for *thermoplastics,* to which the following objectives apply:

- to acknowledge the importance of thermal and melt flow properties to the stages of melting, mould filling and cooling;
- to develop mould design principles on the basis of unsteady-state heat transfer and non-isothermal melt flow;
- to feature the most important interrelationships between processing conditions, polymer microstructure and relevant physical properties of moulded parts.

5.2.1 Machine Description and Principles of Operation

In order to derive a full understanding of the important elements of injection moulding, reference is first given to a simplified schematic diagram of the screw injection unit (Figure 5.1), whose sequence of operation is described below.

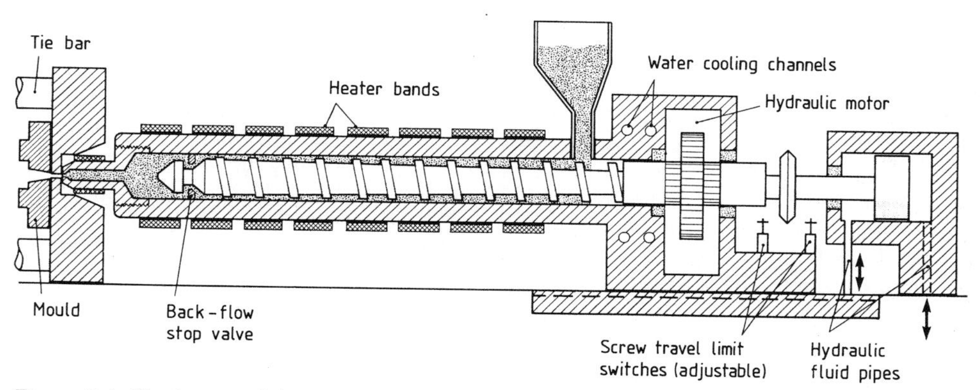

Figure 5.1 Single screw injection moulding machine: the injection unit. (Reproduced by permission from ICI Chemicals & Polymers Ltd., Welwyn, UK).

5.2.1.1 The Operating Cycle

- Consider the cycle from the point where the charge of molten plastic has been injected and is cooling: the screw is powered by a hydraulic motor and as it rotates, material is drawn from the hopper to be heated and compressed in the extruder screw channel.
- Since the *sprue* (the entry channel for melt flow into the mould) is already occupied by rapidly-cooling melt, the material being pumped forward by the rotary shear action of the screw generates a positive pressure in front of the screw tip, which (once it exceeds the *back pressure* set in the hydraulic cylinder at the rear of the injection unit) forces the screw back down the barrel (whilst still rotating) to a pre-set limit which determines the swept-volume available for the next shot, the *stroke.* The screw rotation phase is usually termed *screwback.*
- Soon after screwback is completed, the mould is opened about its *parting line* and the solidified injection moulded part is ejected from the machine. The floating (centre) mould platen is then moved to close the mould and a high clamping force is applied to maintain mould-face contact, about the plane of the parting-line, during subsequent injection.
- Injection of the next shot then occurs, usually at a constant volumetric displacement rate (*injection speed)* which is pre-set by the operator. In this way, the injection time can be controlled to comply with external constraints.
- Further consolidation of the injected melt is achieved during the *packing phase,* immediately following injection. Further material is packed into the mould under high pressure, to counterbalance the thermal shrinkage which would otherwise occur as the moulded part cools towards ambient temperature. The *packing time*

(alternatively termed the *dwell-phase,* or *hold-on time),* is effectively terminated when the flow path to the mould cavity becomes sealed due to material solidification in the gate region.
- Whilst the moulding is cooled, the screwback phase is reinitiated and the moulding cycle repeated.

The utilization of this basic machine cycle has allowed the following advantages of screw injection moulding to be realized:

(1) Accurate and consistent stroke, since the limit-switch positions for screw retraction can be set with high precision;
(2) Simple and effective feeding and melt plastication, using screw extrusion principles;
(3) Elimination of uneven temperature distributions and inconsistent pressure losses; these factors, through their combined tendency to induce shot weight inconsistency, were the prime disadvantages of plunger-type machines;
(4) Sufficient versatility to handle a wide range of plastics materials, including high-temperature thermoplastics, PVC (and other heat-sensitive polymers), filled/reinforced compounds and colour masterbatches (which require good dispersive mixing characteristics during processing), thermosetting polymers (where precise temperature control must be imposed to avoid pre-cure) and materials of a wide range of molecular weight.

The sequence of the injection moulding cycle is common to most processes. Divergence from the overall principle of "melting/injection/packing/cooling" is rare, but the fine detail of virtually every injection moulding technique (eg. choice of injection speed, melt temperature profile, clamping pressure requirement etc.) depends entirely upon the polymer and grade specified, the size and shape of the part and the production/quality specifications.

5.2.1.2 The Functions of some Machine Components

Before examining some of the individual phases of injection moulding in more detail, brief reference is given first to the constituent parts of a typical machine. There is an element of overlap with screw extruders, so that emphasis has been placed upon those elements of machine hardware which are specific to injection moulding units.

- **Hydraulic System.** Many moving parts of injection machines are actuated by hydraulic power. These include:

Injection. Linear displacement of the screw by high pressure in the rear cylinder.

Screwback. By hydraulic actuation of the screw motor. Although hydraulic motors are preferred, electrically-driven screws are still common.

Mould opening/closing/clamping. Tool movement is a "low pressure-large stroke" function, whereas mould clamping requires much higher pressures to suppress the tendency of injection/packing pressures to separate the mould faces and induce *flashing.*

Additional functions. Hydraulic power is also used to energize a number of other parts of the system, such as the ejector unit, moving mould cores and the ability of the entire injection carriage to reciprocate to and from its contact with the mould bushing.

Overall, the hydraulic system required to drive the machine functions listed consists of a pump (usually a rotary vane-type) acting upon a hydraulic oil, a series of valves to control the flow or pressure of the hydraulic fluid and the relevant actuation devices to convert the energy into various mechanical movements. Hydraulic actuation holds several important advantages for injection moulders, such as wear protection for mechanical parts, control of speed/force in stepless increments and easy control sequencing by analogue or digital methods. It serves as an easily controllable and efficient means of force transmission and multiplication.

Theoretical amplification of force is easily calculated, using the clause from *Pascal's Law* which states that "pressure is transmitted undiminished in all directions, and acts with equal force on equal areas of application". Force (F) is therefore evaluated by:

$$F = P \cdot A \tag{5.1}$$

(P = hydraulic pressure, A = area of cylinder).

Some prior knowledge of machine-part dimensions and line pressure allows the magnitude of force amplification to be estimated. Simple calculations of this type are made to estimate ejector pin diameter, cylinder size for required mould clamping and injection (melt) pressures prior to mould filling.

- **Injection System.** The injection unit consists of a feeding device, a means of fully heating and mixing the feedstock and an appropriately designed *nozzle* through which the melt is injected into the mould cavity. Much of the equipment has common design and working principles to those described earlier for screw extruders in Section 4.2.2.1; only the important additions or changes for injection moulding equipment are therefore described here. The following comments refer mainly to the in-line injection unit.

Feeding. Since the screw is only rotating for a fixed proportion of the total cycle time (unlike screw extrusion, where screw rotation and gravimetric feeding are continuous), the possibility of feedstock tending to "bridge" across the feed throat due to local surface melting of granules is increased. Continuous cooling of the block immediately above the feed position is therefore recommended for injection moulding (see Figure 5.1). Many of the *hygroscopic* thermoplastics (PETP, PBTP, PA6, PA66 etc.) are moulded on machine units which are integrated with a recirculating, hot-air drying loop (often with automated back-up hopper-loading facilities) to develop further guarantees against downstream inconsistencies.

Screw and barrel assembly. The technology of screw design and operation is very similar to the analogous situation in single-screw extrusion: typical length to diameter (L/D) ratios lie between 15:1 and 20:1, for injection moulding. If the injection machine has the facility to change to screws of higher/lower diameter, it must be recognized that the L/D ratio will itself alter, since the screws must be of equivalent length. Maximum screw stroke is typically between 2 to 4 times the diameter, whilst the maximum rotation speed is generally higher than for screw extruders of equivalent size (up to 200-300 RPM).

A one-way check-valve mechanism is designed on the tip of injection screws (Figure 5.1) to allow throughput (past the tip) to occur during screwback, but to prevent any possibility of reverse flow during injection, which would otherwise lead to inefficient and inconsistent production. Rigid PVC is an exception in this respect: the presence of a check-ring could provide a site for material hang-up and possible degradation; the relatively high melt viscosity of this polymer negates some of the risks associated with reverse flow.

Release of volatile reaction products using barrel evacuation by *venting* from a decompression zone is now practiced during injection moulding. This process is especially sensitive to this type of problem because melting of adjacent granules can occur whilst the screw is stationary, thereby trapping air within the compressed polymer melt. Vented barrels and double-compression screws act as an alternative (or back-up) to high-performance drying units, for processing of polymers which are hygroscopic, or which release gaseous decomposition products.

Injection nozzles. The nozzle is attached to the front end of the barrel and acts as the connection between injection unit and mould, which the polymer must negotiate first during the injection phase. There are many different designs of nozzles (according to land length, degree of taper and facilities to induce mixing, filtration or prevent material seeping from the tip); these comply with the required functions of nozzles, which are:

(1) to provide a leakproof connection between the injection cylinder and the mould by decreasing the bore to the dimension of the sprue-bushing;
(2) to provide an additional, but independently controlled means of heat application;
(3) to prevent the possibility of shot weight variation due to material drooling from the nozzle orifice.

A standard "open" nozzle design includes a tapered zone, a land length of reduced section (which may have a slight reverse taper) and an electrical heater band of sufficient power to maintain the correct melt temperature. Some more-detailed designs include shut-off nozzles (which are fitted with spring loaded needle-valves) to prevent drooling, extended nozzles for directly-fed cavities and nozzles to accommodate filter packs to avoid contamination of moulded parts.

• **The Injection Mould.** The interface between polymer melt properties and mould design is of fundamental importance to the feasibility and success of any injection moulding process. With this in mind, some sections of this Chapter (5.2.3 and 5.2.4) have been dedicated to the technology of injection mould design on the basis of the physical properties of flowing plastics melts.

Mould clamping system. Plastics injection moulding is synonymous with the use of extremely high injection and packing pressures, to form and consolidate parts produced by processing high-viscosity polymer melts. Therefore, in order to prevent mould-face separation and flashing, it is necessary for a powerful *clamp unit* to be specified with any injection machine; indeed, the sheer physical size of mechanical clamping units is associated with a major cost constituent of the total injection machine. The functions of the clamp unit are as follows:

(1) to support and carry the constituent parts of the moulding tool;
(2) to open and close the mould;
(3) to generate sufficient mould locking force to prevent the possibility of tool-face separation, for all moulded parts likely to be produced on a machine of given shot-weight specification.

Clamping force is generated by any of several possible designs: the mechanical (toggle) lock, the all-hydraulic lock and some combinations of the two basic types. Toggle locks are preferred for medium range, rapid cycling machines. Direct hydraulic locks are generally limited to small range machines (these are less-expensive than mechanical systems) whereas the combined "lock and block" pressure intensification systems are only significantly important on the very large machines. Figure 5.2 reveals the following constituent parts of a typical horizontal mould clamping frame (toggle type):

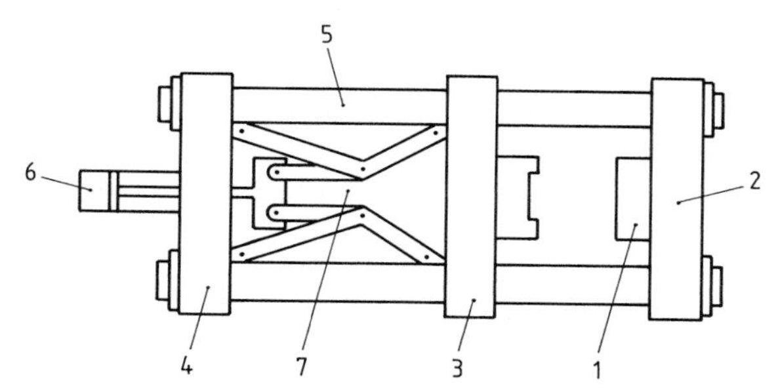

Figure 5.2 A typical mechanical (toggle-type) clamp unit for injection moulding. The moulding tool (1) is supported by the fixed platen (2) and by the moving platen (3); the position of the back plate (4) is adjustable. Other parts include the tie-bars (5), the actuating hydraulic cylinder (6) and the toggle mechanism (7).

Platens. The fixed platen is situated nearest to the injection unit and carries the tool cavity plate; the centre-axes of the platens, the tool and the machine nozzle are coincident. The moving or "floating" platen carries the core plate and is free to slide along the tie bars to open or close the mould. The third platen (the "back-plate") is held at a fixed location for specific tools and processes, but its position may be altered (*mould height adjustment)* to accommodate moulds of slightly differing size.

Tie-bars. Tie-bars absorb the clamping force applied to the mould during the machine cycle; they also support the platens and guide the floating plate precisely into position. Consequently, tie-bars are machined from high-tensile steel and are fully polished to minimize frictional resistance to the sliding platen. Usually, there are four tie-bars (one each towards the corners of the platens), at least one of which is removable to allow the insertion of large moulds.

An example calculation for a "165 cm^3" machine rated at 55 tonnes (540 kN) maximum clamping force, assuming 4 tie-bars of 50mm diameter shows:

Stress in each tie-bar = Force/Area $= 540 \times 10^3/4 \cdot (\pi R^2) = 69$ MN/m^2

Bearing in mind that this only represents the direct tensile stress due to clamping (effects of self-weight must also be considered), it is evident that the tie-bars must be extremely robust.

It is apparent that injection moulding machines are complex processing units. Considering the total number of functional parts and machine sequences, it must be accepted that optimum machine and mould design are fundamental to the viability of any new process. The role which polymer melt properties assume with respect to mould design is reviewed in a later section; for further information on general machine design and construction, the reader is referred to some detailed texts which are dedicated to this subject[5.1–5.5].

5.2.2 Thermal Requirements for Feedstock Melting

With respect to melt preparation and homogenization, modern injection moulding machines share some common principles with screw extrusion systems. Therefore, the interactions between material properties relevant to processing and the design and construction of various parts of the moulding unit have been developed initially along similar lines. However, for a more exact analysis of injection moulding, some additional factors have to be taken into account. These include:

(1) the intermittent (rather than continuous) nature of screw rotation;
(2) the facility to control or modify the back pressure, during screwback;
(3) the influence of the injection phase or residence time in hot runners or nozzles (in terms of overall heat and shear history) should be considered in addition to direct screw plastication.

In the previous Chapter (Section 4.2.4), typical steady-state melting mechanisms in single screw extruders were described, and the qualitative principles of screw compression (for various groups of plastics materials) were briefly categorized. In a general context, these principles also hold true for injection moulding, although it must always be appreciated that these statements can induce some additional discrepancies between theory and practice. This is especially true when shear rates during mould flow are excessively high, when the level of back pressure during the screwback phase is intense, or in instances where the screwback time is either very short, or where screwback represents a large proportion of the total machine cycle. Reviewed below are the material requirements for heating in the plastics injection moulding process[5.6 – 5.7].

Conductive Heating. Although polymer molecules, when subjected to high stress during processing, tend to dissipate mechanical energy as heat (*viscous heat* generation; see below) the primary heat energy source for most injection moulding processes is the array of barrel heaters. These are usually electrical resistance heaters for thermoplastics processes, but for thermosetting moulding powders, where barrel temperatures are modest (70 – 110 °C) and precise temperature control is required, circulating-oil heat-exchangers are often specified[5.8].

The heat flux (by conduction) across the interface between the barrel-surface and the polymer is given by[5.7]:

$$q_c = h \cdot A \cdot (T_M - T) \tag{5.2}$$

where h = surface heat transfer coefficient, A = total conducting area, T_M = set temperature of the barrel and T = temperature of the material absorbing heat energy from the machine.

The *actual* conducting area, $A' = (\pi DL) \cdot f$

(D, L = barrel diameter and length, and f is a factor which expresses the effective proportion of the total barrel length over which the polymer melt charge is receiving heat; ie. it accomodates the presence of a screw flight of finite axial width and ineffective heat transfer in the feed section due to throat cooling and uncompacted solids.)

Therefore:

$$q_c = h \cdot \pi DL \cdot f \cdot (T_M - T) \tag{5.3}$$

Heat flux can be expressed as the product of heat content (between the limits of desired melt temperature and feed temperature; ie. the enthalpy, $\triangle H$) and the overall mass output rate required to comply with the cycle, M':

$$q_c = \triangle H \cdot M' \tag{5.4}$$

Combining the expressions for heat flux yields the following equation, which predicts the *maximum processing rate* (on an injection machine of pre-specified size), for a polymer of given enthalpy at a particular processing temperature:

$$M' = \frac{(\pi D L f) \cdot h \cdot \triangle T}{\triangle H} \tag{5.5}$$

This expression is simply an initial approximation, for direct heat flow from the extruder barrel. It can also be manipulated in the opposite mode, ie. to estimate the minimum machine size requirement, for a demanded production rate. It is not surprising that an approach based upon an oversimplified argument produces accurate trends concerning the effects of a parametric change, but solutions involving the magnitude of dependent variables are subject to large errors due to the occurrence of shear heating during melt plastication.

Shear Heating. It is widely recognized that when viscoelastic fluids are stressed, a proportion of the energy requirement to achieve a given flow rate is dissipated as heat within the system. Due to the high friction resistance which exists when adjacent molecular chain segments deform, polymeric materials are capable of generating large amounts of heat in this way. Since most melt processes are based upon shear flows, the effect has become widely recognized by the term *shear heat*. Polymers are also poor conductors of thermal energy, so that the temperature increases observed due to the effects of shear heat are not easily equilibrated.

The rate of viscous heat generation per unit volume, for a molten plastic of viscosity μ, is given by:

$$q_v = \tau \cdot \dot{\gamma} = \mu \cdot (\dot{\gamma})^2 \tag{5.6}$$

where q_v is heat flux due to viscous dissipation, and τ, $\dot{\gamma}$ are local shear stress and shear rate (μ = apparent viscosity).

If the velocity profile, and therefore the shear rate distribution across the channel depth is known (this can be estimated from "true" shear rate data), the overall temperature rise can be evaluated by integration. Both τ and $\dot{\gamma}$ are functions of distance across the melt flow channel (eg. the screw channel, or a section of the mould feed system) but for pressure flows across simple sections, some approximations can be used to calculate the total effect.

Melt Plastication. The total heat flux required to melt a plastics feedstock material, at a demanded output rate, can be estimated using Equation 5.4. Once it is appreciated that the viscous heat effect (modelled by 5.6) can be superimposed upon the direct conduction of heat from the barrel, it follows that Equation 5.4 is likely to represent a *maximum* power requirement. Since the barrel heaters are closed-loop controlled, they can be envisaged as being responsible for "topping-up" the proportion of thermal energy which does not originate from the screw motor ie. the shear heat. Furthermore, if shear heating in the barrel is excessive (for example, if back pressure and screw rotation speed are increased to enhance the degree of dispersive mixing) the actual melt preparation temperature may sometimes exceed the "set" injection temperature, an occurrence which the temperature control system cannot usually compensate.

Melt Injection. The viscous heat phenomenon is also active during the injection phase, when particularly high rates of shear are apparent during rapid injection through constricted channels. For example, if a 50 cm^3 moulding is injected in 1 second (ie. at a volumetric output rate, Q of 50×10^{-6} m^3/s) through a gate of 1mm radius, the apparent shear rate ($\dot{\gamma} = \frac{4Q}{\pi R^3}$) is 63000 s^{-1}. Local temperature determines apparent viscosity and total pressure drop, so this effect must be balanced against two other occurrences:

(1) the increase in viscosity with increasing pressure;
(2) the simultaneous heat transfer to the mould cavity walls.

Injection moulding is therefore a process associated with extremely high shear rates, coupled with non-isothermal flow in the mould. These factors have historically detracted from successful modelling of the process to allow predictive analyses to be undertaken; it is only in recent times that computer software has been developed to account for these anomalous aspects of injection moulding[5.9–5.12]. Some of the theory upon which these packages are based will be referred to elsewhere in this section.

5.2.3 Injection Moulds; Factors Affecting Mould Design

Optimization of the design and manufacture of injection moulding tools is the key to successful part-manufacture. Injection moulds are physically quite large, but must be precision machined to extremely close tolerance levels, for optimum in-service use. Mould construction can be extremely capital-intensive; it is of strict importance that the tool manufacturers and designers work closely to ensure that the design specification is closely adhered to, and that the modifications necessary following proof-trials can be carefully controlled. These procedures have been common in the moulding industry for many years. It is only over a long time-span however, that the initial art of mould design based upon experience and rules of thumb, has been complemented by the requirement to design injection moulds on a more technological basis, defined by the flow characteristics of the specific polymers to be utilized in any given process.

Enhanced design procedures have led to an extension in the range of "mouldable" materials and products. For this reason, the focus upon injection moulds in this Chapter is firstly taken from a general standpoint, and then extends to examine closely how the physical characteristics of moulding resins contribute to what is a manufacturing process based upon very specific and unique principles.

5.2.3.1 General Functions of Injection Moulds

The moulded product takes its form by shaping molten plastics into a closed *mould cavity* formed by intimate contact of two or more mould faces. Melt flow through the *feed channels* must be optimized in such a way to diminish the total pressure requirement for injection and consolidation, thereby minimizing the clamping force necessary to prevent mould flashing and the overall running cost of the process. Overall, the primary functions of injection moulds can be summarized as follows:

- **Shaping and shape stability.** The flowing melt must be accommodated and distributed to all parts of the tool with minimum pressure drop in the system. Once the cavity is full, the design must allow *packing* to occur before freezing terminates this phase of moulding. A heat exchange system must be effectively designed and correctly operated, to promote melt flow during injection, yet induce rapid cooling of the part, once moulded. An effective means of component ejection, which minimizes the risk of damage to the part, is also an integral part of an injection tool.

- **Mechanical durability.** The clamping frame which supports the mould must accommodate the forces required to hold it closed against the internal cavity melt pressure. Furthermore, the flow path in the moulding tool should, wherever appropriate, have at least one axis of symmetry, in order to minimize mould deflection, when injection and packing pressures are operative.

These functions must comply with the overall objectives of the moulder:

(1) To mould products of acceptable dimensions, properties and quality, at an economically viable rate, minimizing the occurrence of scrap, or out-of-tolerance mouldings. Part properties may include surface and aesthetic appeal, mechanical demands for service and quality control, and freedom from a vast array of defects which can be induced if processing is not optimally controlled;

(2) To achieve consistency during production runs and maintain the desired component properties, in spite of material/tool changes and other forms of downtime. Enhanced machine control helps to achieve this objective, but it must be recognized that the origin of the feedback information upon which the controller acts, is best derived via in-cavity instrumentation.

5.2.3.2 Constituent Parts and Types of Mould

A simplified injection mould is shown in Figure 5.3; the mould cavities (the *impressions)* are defined by the contours of the adjacent *core* and *cavity* plates. The core (or male, positive) block is attached to the moving platen and the cavity (female, negative) plate is secured to the fixed platen and generally defines the external form of the moulding. At positions where the core/cavity plates mate, the mould *parting-line* is defined. Each of the mould impression blocks is usually machined internally to allow the passage of a heat-exchange medium to reach locations which require external temperature control; water, anti-freezing glycol mixtures and low viscosity oils are the preferred fluids, depending on whether the desired tool temperature is sub- or super-ambient. The reverse-tapered *sprue* is located in the cavity plate along an axis perpendicular to the parting line; it is usually machined from a separate component (the *sprue bushing)* with whose profiled face the injection nozzle makes contact. The *sprue puller* is designed as a self-gripping mechanism, so that the mouldings and sprue are always pulled towards the clamp side of the machine, when the tool is opened.

It is possible for the core and cavity plates to be machined in removable *inserts,* which are fixed into recesses in the mould bolster frame. This type of arrangement is proving very popular to assist moulders to comply with the demands of short runs, multi-product moulding and minimum down-time[5.13, 5.14].

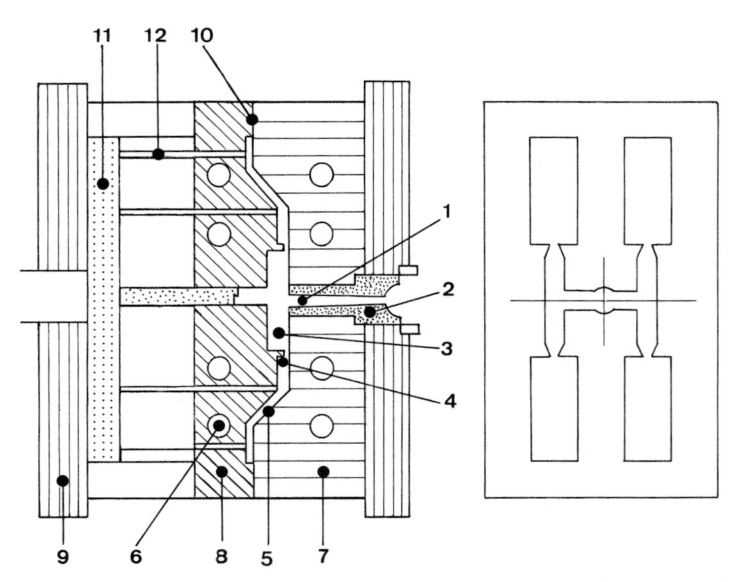

Figure 5.3 A simple two-plate injection moulding tool. The flow path consists of a reverse taper sprue, (1), machined through the sprue bush, (2), a system of runners, (3), feeding the mould cavities, (5), via the restricted gates, (4). The mould temperature is controlled by circulating coolant, (6). The cavity block, (7), core block, (8), and back-plate, (9), are also illustrated, together with the mould face parting line, (10). The ejector system consists of an ejector bar, (11), rigidly attached to a series of pins, (12).

The remainder of the overall mould frame consists of an *ejection mechanism:* hardened ejector pins are attached to an ejector plate, which is, in turn, actuated by a large-stroke hydraulic cylinder at the rear of the mould unit. When energized (at a pre-set rate), the pins move through the core plate and eject the solid mouldings into the *daylight* (about the parting-line) which is formed when the moving platen and core plate are retracted.

There are many other more complex variants of the basic injection tool described above; for purposes of reference, some of these are listed below in summary form. A more thorough treatment of mould design principles should be gained by consulting texts dedicated to mould design[5.3, 5.15 – 5.16], or by reference to machinery manufacturer's literature.

Three-plate moulds. Allowance is made for injection perpendicular to the parting line (a pin gate can be used for a multi-impression tool) and the scrap material in the sprue/runner is automatically separated from the moulding when the mould is opened.

Runnerless/sprueless moulds. If the material in the flow channel can be kept hot, the need to scrap or recycle a proportion of each cycle is generally alleviated, and the pressure losses during mould filling are reduced. These objectives are achieved by external heating of an insulated manifold (inside which the hot runner is machined), as part of the cavity block, or by use of a large-section "insulated" runner (25mm diameter, or greater) which maintains the charge in the molten state only because of the low thermal diffusivity of the plastic layer which immediately sets on the external surface of the runner.

Moulds containing cores and undercuts. Parts of complex shape (eg. components with screw threads, integral holes and moulded-in bosses) cannot always be ejected in the machine direction, so that the mould must be constructed to allow *retractable cores* (which may need to be rotated for threaded components) or wedges (activated by hydraulic or electromechanical circuits) to be withdrawn first.

Insert moulds. Plastics can be moulded around metallic inserts, thereby achieving a jointed, "composite" product in a single operation; robots are often used to place the inserts in position (usually in a vertically-clamped unit) before injection takes place.

Fusible metallic cores. One-piece sports raquet-heads with hollow shafts have been injection moulded around low melting-point metal-alloys, which are subsequently remelted to leave the moulded frame intact[5.17]. Short-glass fibre reinforced PA66 has been used in this application ($T_m = 265\,°\mathrm{C}$) together with a eutectic bismuth-tin (Bi-Sn) alloy ($T_m = 138\,°\mathrm{C}$).

Moulds for multi-colour parts. By judicious design of injection units and/or nozzles, correct machine sequencing (with tool design and layout to suit), allows components of more than a single material or colour to be moulded directly. Products include printed parts, marble effects, multi-colour (eg. car tail-light) clusters, and components made from more than a single polymer (e.g. single-piece pipe-fitting and rubber seal). The reader is referred to machine manufacturer's literature for more information on these interesting, but very specific developments[5.18–5.19].

5.2.3.3 The Mould Feed System (Flow Channels)

An injected shot of molten material must negotiate the channels which collectively make up the feed system of moulding tools; these were depicted in Figure 5.3 and consist of the following sections:

Sprue. The sprue is machined through the cavity plate, and since the material deposited there is likely to be recycled or scrapped, it should generally be kept as short as possible. To assist ejection, the sprue should have a reverse taper: the initial sprue bore should just exceed the diameter of the nozzle orifice (to ensure a good seal) and the final diameter should be not less than the smallest lateral dimension of the primary runner which it feeds.

Runners. Having reached a location at the centre-axis of the parting-plane, the melt flows down a series of channels (the *runners)* which radiate towards the appropriately-located entrances to the mould cavities (the *gates).* Since the sole function of runners is to divert the flow of polymer inside the mould, it follows that firstly, the total volume of material occupying the runners should be minimized; secondly, large, single-impression moulds can be fed directly from the sprue.

When designing a runner system for a new tool, the initially imposed constraints are as follows:

(1) total runner length should be controlled. This allows scrap material to be minimized and also controls heat and pressure losses which, if excessive, impose limitations upon the maximum part weight which can be successively moulded with an injection unit of given specification.

(2) the desired criterion for injecting multi-impression tools (regardless of whether all cavities are identical) is that the filling of all impressions should be just completed at precisely the same time. It is a desired situation that the runner system should therefore be correctly balanced (in terms of pressure drop), such that this occurs.
(3) lateral dimensions must not be too small: this would lead to high pressure losses and heat transfer-dominated melt flow. Neither should runner dimensions be too large: increased scrap levels and cooling times would inevitably result. A working compromise should be evaluated along the lines of the criteria discussed in the section on runner efficiency, presented later.
(4) the runner system should be designed to allow compatibility between the siting of impressions and cooling channels; it should never interfere with the choice of the overall mould-frame size.

It is apparent that the layout, shape and overall dimensions of runners are not only determined by the number and layout of impressions, but also by the thermal and flow characteristics of molten plastics. Toolmakers and moulders are becoming increasingly aware of the advantages which can be realized by designing moulds on the basis of prior knowledge of melt properties of injection moulding resins.

If we consider *cold runners* in more depth (those where the runner temperature, like the remainder of the tool, is kept significantly below the temperature of the flowing melt) it must be realized first that the actual temperature of the incoming polymer is subject to two competing effects: first, heat transfer to the mould walls occurs and second, shear heat is generated within the runner section. The first of these reduces average temperature, thereby increasing viscosity and pressure drop; the latter has the opposite effect, and also increases the risk of thermal degradation. Modern computer-aided mould design techniques are developed from flow and heat transfer models which predict the magnitude of these effects in quantitative terms[5.9–5.10].

Layout of Runners for Multi-impression Tools. Some principles of runner layout and design are illustrated in Figure 5.4:

In Figure 5.4 (A), design (1) would require the smallest mould frame, but assumes the disadvantage of parts being gated at the centre of one side. This leads to uneven or unsymmetrical flow, with the associated possibility of surface defects, anisotropic orientation and shrinkage, and associated distortion.

Design 2 overcomes the centre-feeding problem, but since the vertical axis of flow symmetry is lost, unbalanced mould face deflection (whilst injection/packing pressures are operative) is possible.

Design 3 restores flow symmetry and represents a desirable configuration, despite the possibility (for parts of high aspect ratio) of longer overall runner length.

As shown in Figure 5.4 (B), the use of branched runners at 120 °C forms the most convenient feeding system to multiple, triple-cavity moulds.

Design 1 in Figure 5.4 (C) illustrates the principle of using primary and secondary (branched) runners. This is always desirable in decreasing overall runner length, but unless the dimensions of the branches to the outermost cavities are different to those of the four inner impressions, the system will be unbalanced.

Design 2 is fully-balanced in theory; in practice however, the severe obtuse angles which the melt must negotiate to reach the inner cavities induce additional pressure drops (due to local, recirculating flows) which may detract from the attainment of a perfectly-balanced flow path.

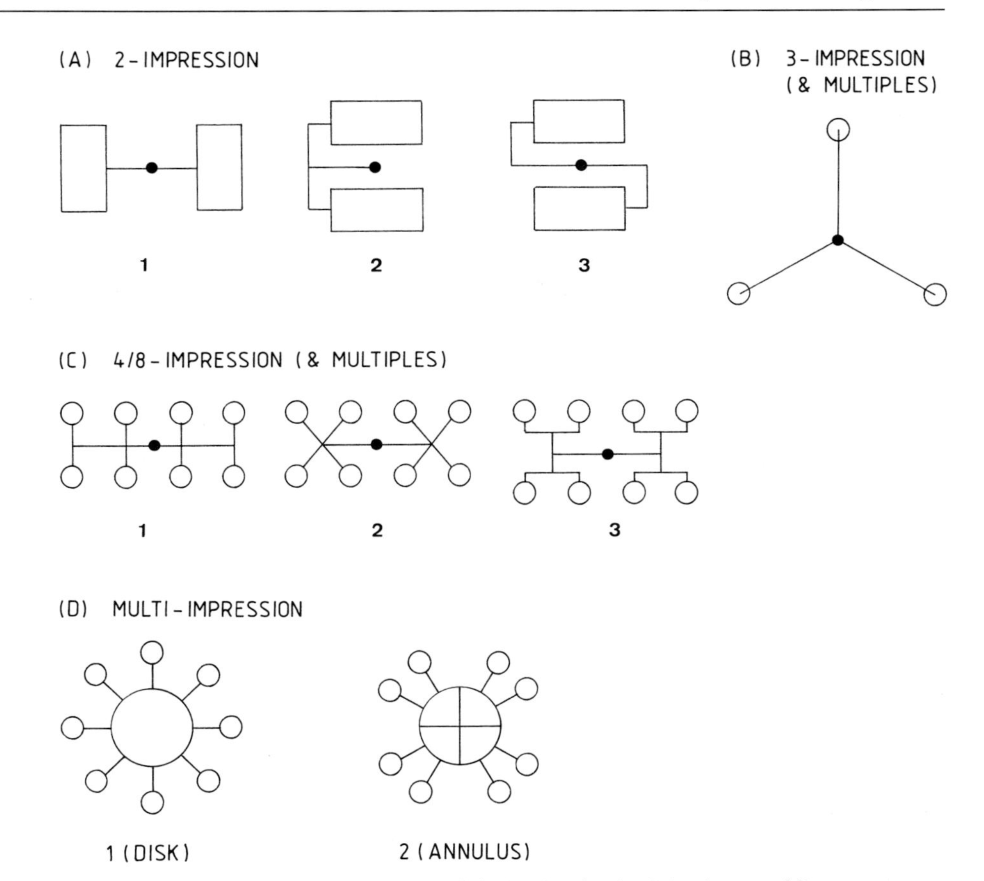

Figure 5.4 Some examples of runner layout and design in plastics injection moulding.

Design 3 is balanced, and is further optimized by the introduction of tertiary branched-runners. It is useful to radius all 90° bends, and to design a "run-off" well at the end of each main branch.

All designs (1 – 3) have the cavities lying in a (4×2) matrix, which eases the layout constraints associated with the coolant channels.

Disc-like, or annular runners are illustrated in Figure 5.4 (D), and are useful if the total number of impressions does not lend itself easily to balancing by branching, so long as material savings are feasible, or if the overall platen area can be diminished by utilizing such a design.

Family Moulds. Family moulds contain a number of cavities, differing in shape, thickness or volume. When designing runner systems to feed this type of tool, it is necessary for the entire flow paths (to the extremities of each of the cavities) to be associated with identical pressure drops during mould filling. In other words, runner balancing alone is insufficient; the pressure losses in the respective gates and cavities must also be quantified. In this way, it can be seen that the rheology of the moulding resin is absolutely fundamental to design procedures for family moulds.

Lateral Dimensions of Runners. Two immediate design constraints are:

- to minimize pressure loss. The Poiseuille equation for Newtonian flow through a circular duct of length L and radius R, predicts a pressure drop, P, given by:

$$\triangle P = \frac{8\mu LQ}{\pi R^4}$$

It follows that pressure drop is minimized when runner radii are generous.

- to minimize simultaneous heat loss, during injection. Fourier's equation for steady-state conduction equates heat flux across a path length x thus:

$$q = K \cdot A\left(\frac{T_M - T_C}{x}\right)$$

The surface (conducting) area of runners must therefore be minimized to reduce heat loss.

PYE[5.15] has proposed that the ratio of cross-sectional area and surface area (per unit length) of runner can be used to describe the *efficiency* of a given design. Using this methodology, data have been compiled in Table 5.1 which list some efficiency factors for some common examples of runner shapes.

Table 5.1 Efficiency of Some Simple Runner Cross-Sections

Shape	Dimensions	Efficiency factor
Circular	Diameter D	0.25 D
Square	Dimension D	0.25 D
Semi-Circular	Diameter D	0.15 D
Rectangular	D = d/2 (D,d = Long, Short Dim.)	0.166 D
Rectangular	D = d/4	0.125 D
Rectangular	D = d/n (general case)	D/(2n+2)
Trapezoidal	Depth, D; (Taper Angle, $\alpha = 10°$)	0.224 D
Trapezoidal	Depth, D; (Taper Angle, $\alpha = 20°$)	0.187 D
Opposite Trapezia	Depth, D; (Taper Angle, $\alpha = 10°$)	0.307 D
Opposite Trapezia	Depth, D; (Taper Angle, $\alpha = 20°$)	0.265 D

Although this definition of runner "efficiency" is somewhat simplistic, and cannot be used quantitatively (eg. polymer melts are non-Newtonian; heat transfer is unsteady and occurs by more than a single mechanism), the data in Table 5.1 allow an initial appreciation of the importance of runner section to be made. Final choice must be made with a degree of caution, however, since there are additional toolmaking problems associated with both square or rectangular sections (these must include a significant draft angle to assist part ejection) and also with circular or double-trapezoidal runners, which must be machined to form perfect alignment across the parting line.

Having decided upon runner shape and layout, the final decision concerns the lateral dimensions of the section. Many procedures are based solely upon experience, whilst others are rather arbitrary in nature. For example, Pye suggests that runner diameter should be a function of both part weight and length[5.15]. MENGES and MOHREN[5.16] suggest that the runner diameter should be at least 1.5 mm greater than the maximum thickness of the part. They also refer to a technique first developed by STANK[5.20], which allows runner dimensions to be estimated from a series of plots relating part weight and thickness to runner length, for various groups of plastics.

Gates. Polymer melts flow into individual mould cavities through one or more restricted *gates,* machined strategically at specific location(s). Overall, the functions of gates are as follows:

(1) to allow the polymer melt to pass into the mould at relatively high flow rates without inducing any flow-related defects to the moulding.
(2) to provide a site to separate the moulding from the material deposited in the feed system (automatically, if possible), and to minimize the effect of surface blemishes when this is done.
(3) to promote rapid freezing, once filling and packing phases are completed.

A shallow gate section complies with functions (2) and (3), but not necessarily with the first. Clearly, the effect of shape is fundamental to the successful use of gates; this is reflected by the points discussed below on the subject of melt rheology. As far as the complete range of possibilities is concerned, the number of different types and/or shapes of gate is quite large; these are described in detail elsewhere[5.3, 5.5, 5.15, 5.16].

Rheological Aspects. Because gate dimensions are very small in comparison to both the runner sections which feed them and the cavities which are fed by them, the flowing melt is very sensitive to changes in processing conditions as it passes through this part of the feed system. For a given volumetric flow rate, shear rate is inversely proportional to the fourth power of radius (circular section), or to the third power of depth (slit section). Also, flow in restricted sections is extremely prone to local temperature fluctuation, either by cooling effects during injection at low speed or more likely, by shear-heat dissipation in rapid-injection processes.

Shear rate. A combination of rapid injection and a restricted gate section leads to extremely high shear rates, typically in the range $10^4 - 10^6 s^{-1}$ for sections of 1-2 mm depth. The pseudoplastic nature of polymer melts dictates that viscosity decreases by this effect and also, since shear heat generation is also accelerated at high shear strain rates, further viscosity reduction is likely due to the indirect effect of increasing temperature.

Elasticity and defective flow. During flow through the gate, shear stress/rates are very high, but at the moment the flow-front first enters the cavity, an effect akin to "die swell" occurs as the shear stress is drastically reduced and sudden relaxation occurs. This effect is often of benefit to the practical moulder, since swelling promotes

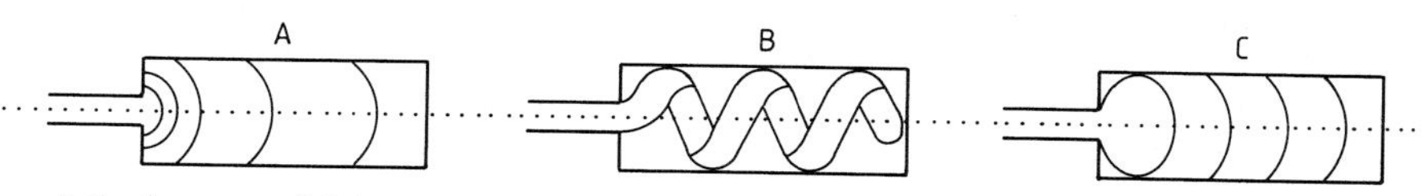

Figure 5.5 Aspects of injection mould filling: (A) Laminar shear flow – slow and controlled injection velocity. (B) Rapid injection leading to a buckled melt stream, ("jetting"), on contact with the opposite face. (C) Enhanced 'swell' effect, leading to laminar flow via melt contact at the cavity walls.

contact between melt and the cavity walls, leading to a controlled, laminar flow. However, another facet of rapid injection, the *jetting* effect (Figure 5.5), which leads to surface blemishes and to a multiplicity of internal defects induced by small-scale "welds" as contact points are formed between the folds of melt in the jet stream.

Gate freezing. The time taken for the polymer freeze-off in the gate effectively defines the available packing time. It is imperative, therefore, that the hold-time setting should not be less than the minimum freeze-time.

Location of Gates. Gating position is determined by the following factors:

(1) Part Geometry - packing is least effective away from the gate, so that it is sensible to design gating systems close to thicker sections.
(2) Mechanical Strength of Parts - molecular orientation distributions tend to be both severe and complex (due to packing effects) close to the gate. Gate location must be chosen so that impact loads (especially) are not likely within the immediate vicinity.
(3) Weld Lines - The formation of internal weld lines should be avoided wherever possible; multiple-gating and unsymmetrical flow promote the occurrence of weld lines.
(4) Aesthetics, Product Dimensions and Shrinkage - gate witness marks should be concealed, if consumer appeal is not to suffer. Also, since shrinkage of mouldings (due to ageing) varies according to the severity and direction of molecular orientation, a further constraint is that flow should be balanced and should occur about a single axis of symmetry at least (see Figure 5.6). This is especially true for crystallizable plastics and for fibre-reinforced resins, whose properties are particularly prone to the effects of anisotropy.

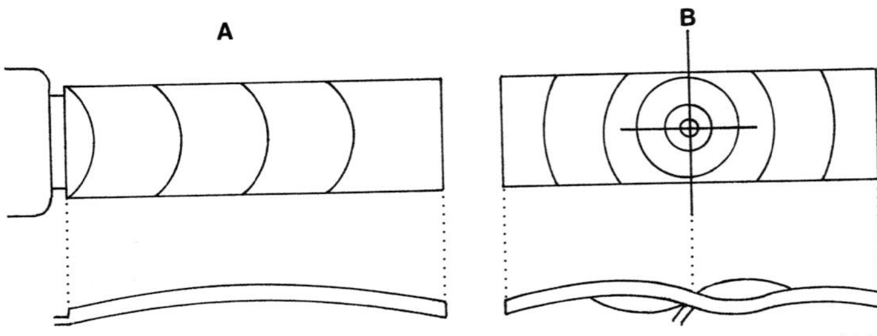

Figure 5.6 Effect of gate location on orientation, shrinkage and distortion: (A) End (film) gated – bowing may occur due to longitudinal orientation, and uneven cooling about the central-axis. (B) Central (pin) gated – bowing is evident towards the cavity extremities, whilst radial distortion occurs near to the gate.

Number of Gates. Since a single-gated mould is usually filled without the formation of weld-lines, it carries clear advantages. Multiple gating should therefore only be utilized when obvious necessities arise (eg. maximum flow ratio is excessive for a single feed) or when the filling of complex parts requires optimally-balanced filling to all extremities of the cavity.

Qualitative aspects of mould design can, to some extent, be predicted with limited accuracy from previous experience with a given product or polymer range. However, only with some knowledge of melt properties for specific plastics (and their dependence on temperature, rate, molecular weight etc.) can a full quantitative analysis of tool design be optimized. It is on this basis that Section 5.2.4 is constructed.

5.2.3.4 Mould Temperature for Thermoplastics and Thermosets

Shape stabilization is achieved by cooling thermoplastics, or by further heating thermosetting plastics to induce crosslinking. The correct thermal environment of the surrounds of the flowing melt is an important processing factor: thermoplastics must cool quickly, but not to the extent where excessive pressure losses lead to short mouldings. High temperature *pre-cure* of thermosetting plastics (which also causes short mouldings), in this case due to the rapid increase in viscosity which accompanies crosslinking should also be avoided. An overall range of typical processing temperatures, for each class of material, is summarized in Table 5.2.

Table 5.2 Typical Moulding Temperatures for Plastics

	Temperature (°C)	
	Shaping	Mould
Thermoplastics	150 – 350	10 – 150
Thermosets	60 – 110	150 – 210

As well as suppressing premature freezing and excessive pressure drops, a higher mould temperature for thermoplastics assists crystallinity development (eg. in PA, PETP, or in other plastics exhibiting relatively slow nucleation rates), promotes homogeneity of through-thickness structure and also contributes to a better quality of surface finish.

5.2.4 Melt Properties and Injection Mould Filling

Fluid flow during injection mould filling occurs predominantly under *shear* forces; the shear stress is related to the pressure acting upon any element of flowing plastic, and the shear rate is determined by the injection speed ("output rate") in a channel of given geometry. Apparent viscosity is the quotient of shear stress and shear rate; it expresses a material's resistance to viscous flow in a given flow field.

In Chapter 3, the assumptions upon which the classical expressions for shear stress/rate are derived, were introduced. Although these equations are used to model shear flow in many processes, there are some aspects of injection moulding which limit the validity and applicability of the assumptions quoted earlier. For example:

Isothermal flow. Injection mould flow is significantly non-isothermal, rendering this assumption particularly questionable.

Newtonian flow. Molten plastics are highly non-Newtonian, especially at high shear rates characteristic of restricted flow channels.

Incompressible melt. In many analyses, the fact that shear viscosity increases with increasing pressure (hindered molecular mobility) is counterbalanced by the corresponding reduction of viscosity due to viscous heat generation. Since the "isothermal flow" assumption is invalid for injection moulding, the effect of melt compressibility should also be considered, especially since the working pressures in this process are very high.

Steady flow. Injection moulding is a cyclic process; since significant heat transfer accompanies melt flow, causing time-dependent effects such as surface solidification, steady flows are rarely established.

Clearly, some modification of classical shear flow theory is necessary to accommodate the characteristics of injection moulding. In particular, the non-isothermal nature of the process can create significant errors when attempting to model from conventional principles. There have been many attempts to model the plastication, mould filling, packing and cooling phases of injection moulding; whilst the complexities associated with this process have defined the need for such simulative studies, they have also been responsible for the discrepancies observed between theory and practice. Our approach in the following section is to describe the assumptions and technology which has been used to simulate the injection moulding process; some notable works in this area are also acknowledged[5.21–5.24].

5.2.4.1 Pressure Drop Calculations for Isothermal Flow

It has been ascertained that polymers suffer from simultaneous heat transfer effects, during injection moulding into a tool whose temperature is different to that of the melt. However, it is convenient to analyse non-isothermal melt flow by first considering conventional flow theory, then discussing how heat flow detracts from this basis.

Shear and Elongational Flows. Although *shear flows* dominate injection moulding, *elongational flows* are also observed whenever the channel geometry is non-uniform; for example, elongational (or "tensile") flows accompany shear flows in the convergent section of injection nozzles, and when abrupt changes in section are encountered eg. at gates, or in cavities (see Figure 5.7). Thus, the total pressure requirement to fill the feed system and cavities depicted in Figure 5.7 is given by:

$$\sum \Delta P = \sum (\Delta P)_{\text{SHEAR}} + \sum (\Delta P)_{\text{TENSILE}} \tag{5.7}$$

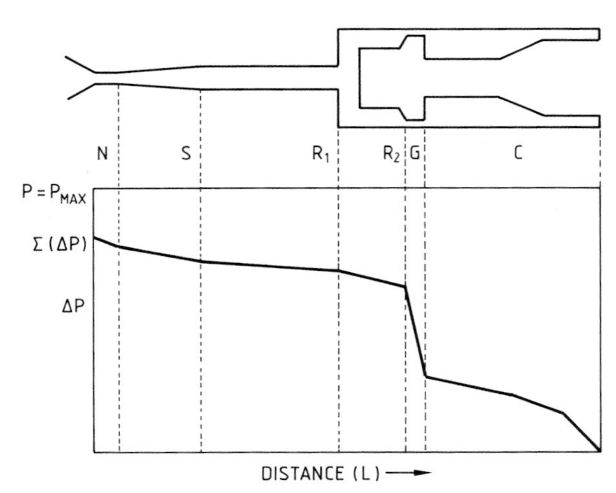

Figure 5.7 The flow channel in a plastics injection moulding tool: nozzle (N), sprue (S), primary and secondary runners (R_1 and R_2), gates (G) and cavities (C). A typical pressure distribution, at the point of mould filling, is also shown.

The total pressure drop due to shear flow, $\sum(\triangle P)$, is estimated by first calculating the individual shear flow pressure losses for each part of the flow system, then adding them together. For example, for a flow rate Q in a circular channel of length L and radius R, shear stress and apparent shear rate are expressed as:

$$\tau = \frac{R\triangle P}{2L}; \qquad \dot{\gamma} = \frac{4Q}{\pi R^3} \tag{5.8}$$

If the melt is assumed to obey a power law of the type $\tau = k\cdot(\dot{\gamma})^n$ (further refinement is usually considered unnecessary in injection moulding), then by substituting for τ and $\dot{\gamma}$ into the power law, and rearranging, we obtain the shear flow pressure drop:

$$\triangle P = \frac{2Lk}{(R)^{3n+1}} \cdot \left(\frac{4Q}{\pi}\right)^n \tag{5.9}$$

Similarly, for a slit section (eg. film gate of width W and depth H), where:

$$\tau = \frac{H\triangle P}{2L}; \qquad \dot{\gamma} = \frac{6Q}{WH^2} \tag{5.10}$$

The corresponding pressure loss is:

$$\triangle P = \frac{2Lk}{(H)^{2n+1}} \cdot \left(\frac{6Q}{W}\right)^n \tag{5.11}$$

Most flow channels can be represented, or at least approximated by circular or slit ducts. For example, a ring gate (thin annulus) is represented by a slit whose "width" is equivalent to its mean circumference.

The total pressure drop in the mould due to shear flows is given by:

$$\sum(\triangle P)_{\text{SHEAR}} = \triangle P_S + \triangle P_R + \triangle P_G + \triangle P_C \tag{5.12}$$

Subscripts S, R, G and C represent total pressure drops in the sprue, runner(s), gate(s) and cavity(ies); see Figure 5.7.

These individual elements may, of course, be summations of pressure losses in different parts of the cavity or feed system. For example, each of the cavities shown in Figure 5.7 has three sections contributing to the total *shear flow* pressure drop: one each in the thick and thin sections, and an additional component in the converging zone.

However, there is also an *elongational flow* active in the convergent section. In this instance, the pressure loss due to elongational flow must be added to shear flow pressure losses. POWELL[5.25] has presented a concise summary of equations for pressure losses (for both shear and elongational flows) for power law fluids in channels of both constant and changing section. Once more, taking Figure 5.7 as an example, the pressure losses in the converging (wedge) section are[5.25]:

$$(\triangle P)_{\text{SHEAR}} = \frac{\tau}{2n \cdot \tan\theta} \cdot \left[1 - \left(\frac{H_1}{H_0}\right)^{2n}\right] \tag{5.13}$$

$$(\triangle P)_{\text{TENSILE}} = \frac{\sigma}{2} \cdot \left[1 - \frac{H_1^2}{H_0^2}\right] \tag{5.14}$$

H_0, H_1, represent channel depths at the converging section entrance and exit, n is the power law index, α is the (half) taper angle and τ, σ represent the shear and tensile stresses at the exit position.

The overall criterion for mould filling is that the total pressure drop $\sum(\triangle P)$ should not exceed P_{MAX}, the maximum pressure which the injection unit is capable of generating ahead of the screw tip. Figure 5.7 also depicts the pressure profile which might be predicted for the feed system shown; since $\sum \triangle P$ is less than P_{MAX}, the mould filling condition is fulfilled. The severity of the pressure gradient in any section $\frac{dP}{dL}$, is related directly to the constrictive effect of shallow sections or, in the case of extensional flow, it is related to the taper angle (α). For many commercial moulding machines, P_{MAX} is of the order of 100 MN/m^2 (14,500 psi); it can be estimated for any machine if the dimensions of the hydraulic and screw cylinders are known:

$$P_{\text{MAX}} = f \cdot (P_H) \cdot \frac{D_H^2}{D^2} \tag{5.15}$$

P_H is the maximum hydraulic pressure and D_H, D are the diameters of the hydraulic and screw cylinders (respectively); f is a factor less than unity and indicates efficiency loss, for example due to friction of the screw. In all probability, f will lie in the range 0.75 to 1.0, with lowest values being observed at relatively low melt temperatures.

It can be seen from Equation 5.15 that for a given hydraulic cylinder specification, the maximum melt pressure available for injection is inversely proportional to the square of the barrel diameter.

5.2.4.2 Pressure Drop Calculations for Non-Isothermal Flow

So far in Section 5.2.4, isothermal conditions have been assumed throughout, so that any predictions of pressure drop made by Equations 5.9 and 5.11 (for example) are only strictly applicable to situations where the temperatures of the mould and flowing melt are comparable. Polymer melt flow in the injection *nozzle,* or in *hot runner* sections can therefore be modelled in this way with the expectation of high levels of accuracy.

For all other situations where a net heat transfer accompanies mould filling, some modification of the theory in Section 5.2.4.1 is required. Clearly, the effect of simultaneous heat losses depends upon the temperature differential across the mould wall surface, the speed of injection and the thermal diffusivity of the flowing melt.

GLANVILLE[5.26] first examined this situation and proposed that the temperature loss in the polymer be accounted for by "an effective increase in channel length". Experience and practical trials showed that the section length L could be raised to the power $L^{1.25}$ (and substituted into the relevant expression for pressure drop), to describe the behaviour of a range of materials in injection moulding. Although this approach helped diminish some of the errors between conventional theory and practical moulding, it was not applicable universally and did not distinguish between changes in processing conditions, or materials differing in thermal properties.

A different approach has been used by BARRIE[5.6,5.9,5.27] to form a quantitative basis to understand the interaction between melt flow and simultaneous cooling effects in injection moulding. The principle of the approach is as follows:

(1) The molten plastic is injected down a relatively cold flow channel; heat transfer from the melt is extremely rapid at the mould wall, such that the local temperature drops dramatically and the polymer solidifies.
(2) As we move away from the mould surface, the cooling rate diminishes, due to the thermal insulating character of the developing solid layer of polymer.
(3) More shear heat is being generated as fresh melt flows through the channel, which itself decreases in thickness as the frozen layer grows.
(4) Eventually, a pseudo thermal equilibrium is set up in the frozen layer, where the heat losses to the mould are counterbalanced by prolonged contact with the flowing melt; further growth of the frozen layer does not occur.
(5) Since these events take place relatively quickly, the thickness of the frozen layer is assumed to be constant as a function of distance along the cavity (thickness H, see Figure 5.8).

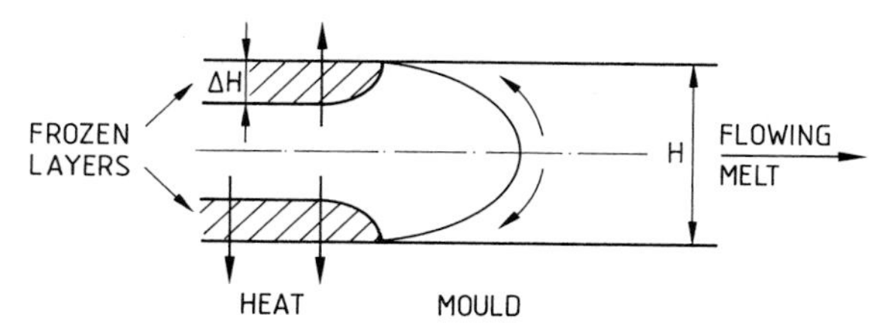

Figure 5.8 The 'frozen-layer' analogy for injection mould filling: the main melt stream remains fluid between the solidifying outer-layers, adjacent to the relatively cool mould faces.

The effective flow channel depth is given by:

$$H_{\mathrm{EFF}} = H - 2\triangle H \tag{5.16}$$

The *frozen layer* analogy is utilized by modern computer-aided mould design packages, such as those developed by BARRIE[5.9] and by AUSTIN[5.10]. Since molecular orientation is prevalent in layers close to the mould cavity surface (see Section 5.2.6), the thickness of the oriented layer can sometimes be an indication of the theoretical frozen layer referred to above.

Development of the Frozen Layer[5.6, 5.7, 5.9]**.** Heat flow is unsteady; reference is made to Fourier's uniaxial, unsteady-state heat flow equation, expressing temperature as a function of time and distance:

$$\frac{\delta^2 T}{\delta x^2} = \frac{1}{\alpha} \cdot \frac{\delta T}{\delta t}.$$

As detailed earlier in Chapter 2 (Section 2.4.1), thermal diffusivity (α) is expressed as:

$$\alpha = \frac{K}{\varrho \cdot C_P}$$

where K = thermal conductivity, ϱ = density and C_P = specific heat; all thermal properties should be specified appropriate to the temperature range under consideration.

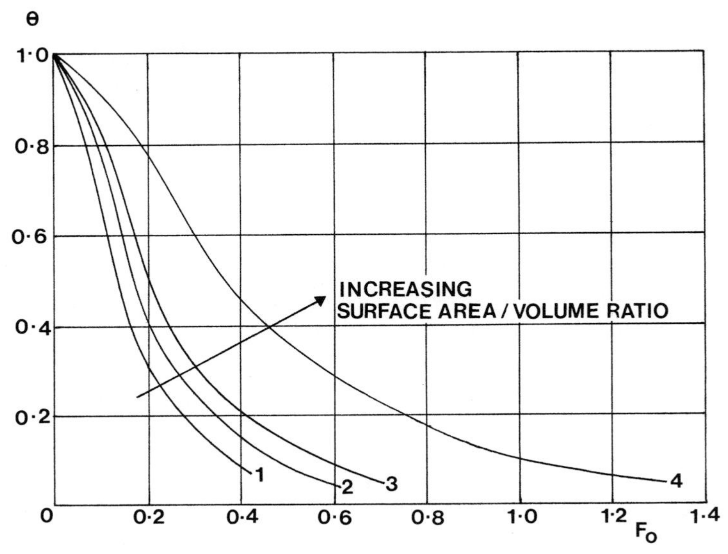

Figure 5.9 Relationship between Temperature Difference Ratio, (θ), and Fourier Number, (F_0), for unsteady heat transfer: (1) sphere; (2) cylinder (length L = twice radius, (R)); (3) cylinder (infinite); (4) flat slab.

A convenient solution to Fourier's equation can be expressed by graphical means, showing mathematical models relating dimensionless time (the *Fourier Number, Fo)* and dimensionless temperature (the *temperature difference ratio, θ).* Examples of cooling curves of this type, for different geometries, are shown in Figure 5.9. The dimensionless groups are expressed in terms of some other variables as:

$$Fo = \frac{\alpha t}{x^2} \tag{5.17}$$

$$\theta = \frac{T - T_C}{T_M - T_C} \tag{5.18}$$

x is the path length in the heat flow direction; T, T_C, T_M represent actual temperature, and the initial temperatures of the cavity and melt, respectively.

For values of temperature difference ratio, $\theta > 0.8$, a convenient relationship between the dimensionless groups is:

$$\theta = \frac{1}{2(Fo)^{0.5}} = \frac{x}{2(\alpha t)^{0.5}} \tag{5.19}$$

$$\therefore \quad x = 2(\alpha t)^{0.5}\theta$$

By definition, when the local temperature decreases to reach a level at which the melt cannot flow (the *freeze-off temperature, T_F*), x is equal to the frozen layer thickness (δh). Therefore:

$$\delta h(t) = 2\left(\frac{T_F - T_C}{T_M - T_C}\right) \cdot (\alpha t)^{0.5} \tag{5.20}$$

Equation 5.20 implies, therefore, that frozen layer growth is proportional to the square root of (injection) time. Experience has shown that due to the superimposition of shear heat generation, $\triangle$ H is more accurately modelled by:

$$\triangle H = Ct^{0.33}$$

where $C = 2(\alpha)^{0.5} \cdot \theta$. Substituting for $\triangle H$ into Equation 5.16, the *effective* channel thickness is expressed by[5.6, 5.7]:

$$H_{\mathrm{EFF}} = H - 2 \cdot (2\theta \cdot \alpha^{0.5} \cdot t^{0.33})$$

$$\therefore \quad H_{\mathrm{EFF}} = H - 4\theta \cdot [\alpha^{0.5} \cdot t^{0.33}] \tag{5.21}$$

Pressure drop is then calculated by substituting H_{EFF} for H, in the relevant expression for channel geometry. The freeze-off temperature, T_F, is the temperature encountered by the polymer (during the cooling mode) when the viscosity rises to a level at which the pressure drop is sufficiently high to induce a "no flow" situation. For amorphous plastics, T_F is usually some 20 – 50 °C above the glass transition temperature, whereas for semi-crystalline plastics, the freeze-off phenomenon is determined by the recrystallization behaviour of the plastic (ie. supercooling plastics such as PP will have T_F significantly (up to 60 – 80 °C) below the original crystalline melting point, T_m). For most groups of commodity plastics, we can approximate towards general T_F values as follows:

- Amorphous plastics $T_F = T_g + (20\ to\ 70)$°C
- Semi-crystalline plastics $T_F = T_m - (10\ to\ 80)$°C

5.2.4.3 Computer-Aided Mould Design

Aspects of fluid flow during mould filling which are peculiar to injection moulding (simultaneous heat loss to the mould, generation of significant heat by viscous dissipation at high shear rate) make accurate melt flow analysis extremely difficult to achieve and moreover, the pronounced sensitivity of moulding to polymer properties and running conditions induce more profound complexities. It is not surprising that the industry has had cause to benefit from the utilization of computer-assisted design and simulation methods in recent years.

For example, the SIMPOL system[5.9] is programmed to transform raw data on the polymer, mould geometry and machine conditions into quantitative, theoretical data predicting moulding feasibility (total pressure requirement) and the effects of changing processing conditions. From this information, machine and clamping frame requirements, cycle time constituents and an introductory cost breakdown can be estimated. The rheological analysis is based upon the theoretical treatment of non-isothermal melt flow presented in the previous section.

Some data are presented in Figures 5.10 and 5.11 which illustrate the technical points made previously; this analysis was carried out on end-gated, 2-impression, rectangular-spacer components, injection moulded using a grade of crystal polystyrene (MFI-11 g/10m.; Hoechst N3000). The main runners were trapezoidal, with a long dimension of 6.5mm. Increasing melt or mould cavity temperature each diminishes the total pressure requirement for mould filling (Figure 5.10); the former effect is strong, although the latter is not insignificant. Heat transfer effects start to dominate the flow response at low injection speeds, causing the flow characteristics to deviate from the linear plots which conventional flow theory predicts. In this way, an optimum processing rate (minimum pressure drop) is predicted, for each combination of moulding conditions.

The pressure drop - injection speed relationship for a cold runner system re-emphasizes the importance of simultaneous cooling effects: not only is the total pressure drop higher (at all injection rates), but the extreme sensitivity of pressure to rate (for slow injection) is immediately apparent. Figure 5.11 shows the individual contributory pressure drops from the runner system and the cavity, and emphasizes the sensitivity of mould filling feasibility to the lateral size of runners (see also Section 5.2.3.3). Choosing generous runners for this reason does of course, carry other penalties, ie. increased runner volume leads to more scrap (cold runner) or to increased residence time (hot runner). An additional effect here is the reduction of the filling pressure requirement for the cavity (alone), when runners are more restricted; this is directly attributed to the additional viscous heat dissipation in the feed system tending to increase the effective moulding temperature in the cavity.

Many other aspects of injection moulding can be predicted by computer methods, to alleviate the need for expensive and time-consuming practical proof trials. The flow analysis of the Moldflow package[5.10] is based upon a pre-defined, 3-dimensional finite element model of the part(s) to be moulded. Once the gates have been positioned and injection conditions set, the initial flow analysis is sufficiently rigorous and accurate to allow predictions of:

- mould filling dynamics;
- heat generation and mould cooling requirements;
- part distortion and mould face deflection.

A notable capability is the use of an enhanced computer-graphics system to study distributions of pressure and other rheological variables during injection. This allows a direct visual appreciation of weld-line position(s), thereby allowing the optimum feed system to be specified (ie. the number and position of gates; balancing the flow in the feed system and sections of cavities) without the costs and risk elements associated with injection mould manufacture.

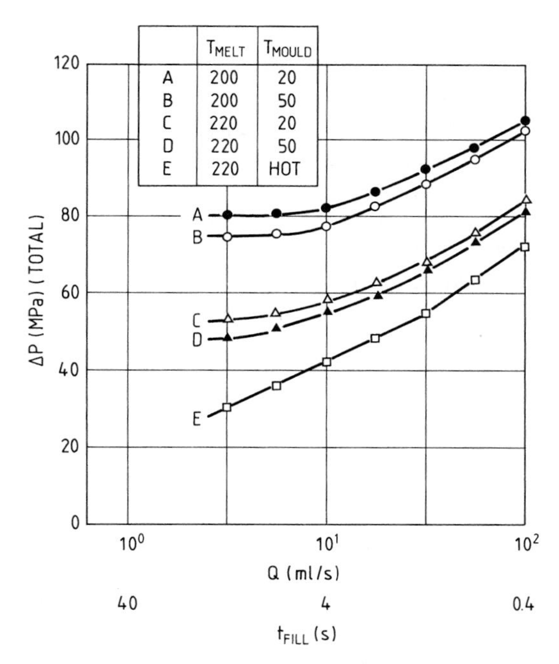

Figure 5.10 Predictions from SIMPOL © CAD system – relationships between pressure drop, ($\triangle P$), and injection rate (output Q, or mould filling time, t_{FILL}), showing the effects of changing melt and mould temperatures.

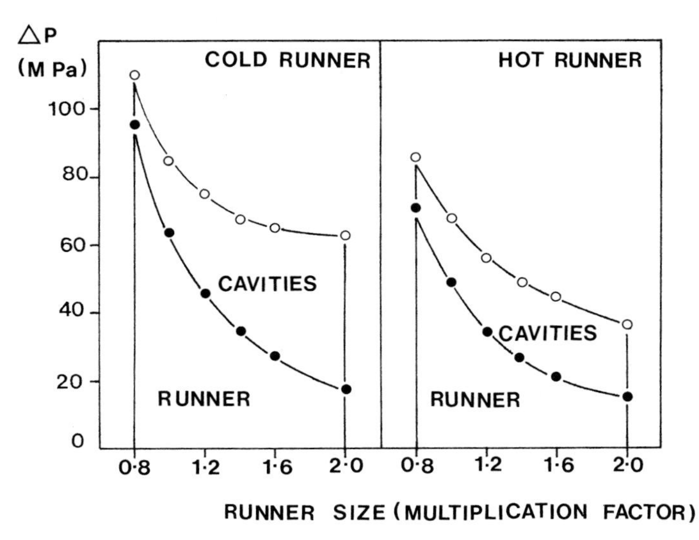

Figure 5.11 Predictions from SIMPOL © CAD system – relationships between pressure drop, ($\triangle P$), and the relative lateral size of runners, for both cold- and hot-runner systems. (A multiplication factor of unity, in this case, represents a trapezium 'long' dimension of 6.5 mm.)

5.2.5 Moulding Cycle and Machine Control

A typical injection machine cycle consists of a number of constituent time proportions; the magnitudes of these depend upon (for a given machine specification) part weight/thickness, the thermal properties of the polymer and the machine settings specific to a given process. A direct change, or a secondary adjustment of a parameter in any of these categories will have a strong influence upon one or more phases of the cycle and therefore, will determine the overall machine *cycle time* itself. Figure 5.12 may be used as a reference to study the make-up of what is considered to be a typical machine-cycle. Each of the respective time phases is considered in more detail, as follows:

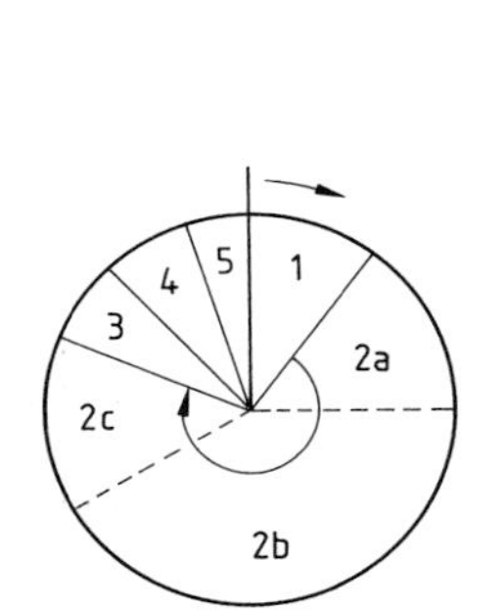

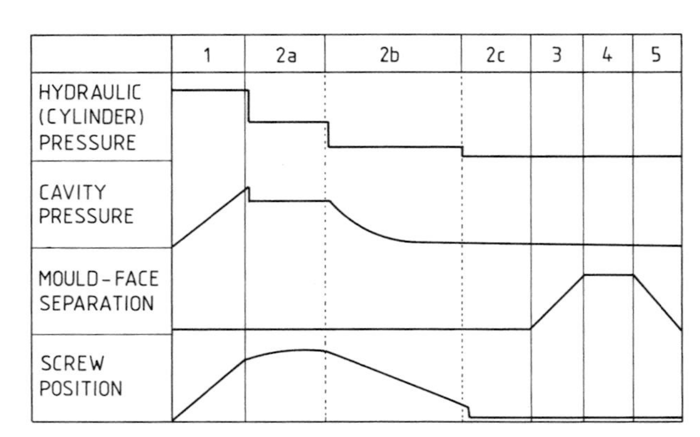

Figure 5.12 The injection moulding machine cycle in terms of time (left), pressure and stroke (right). Individual time-phases are: (1) injection; (2) consolidation and cooling (including (2a) packing/hold-on, (2b) screwback and (2c) carriage retraction); (3) mould opening; (4) ejection; and (5) mould closing.

(1) Injection. Molten plastic is injected into the mould at a predetermined rate (the *injection speed),* as a result of the axial motion of the screw, which is driven at a controlled velocity according to the oil flow rate into the hydraulic cylinder. Although direct reference is often given to an injected *shot weight,* a more accurate term is probably *shot volume,* since the mass delivered is dependent upon specific volume, which for plastics is temperature and pressure-dependent[5.16].

Most injection machines have the capability to maintain a constant (preset) injection speed; in general, fast injection is preferred for technical and economic reasons (it is essential for moulding packaging, and other thin-wall mouldings which are prone to excess pressure drop and premature freeze-off), but slower injection may be necessary to avoid turbulent flow in restricted zones, or to enhance the reformation of flow-induced welds.

Injection time is often used to describe filling speed; assuming that the screw speed setting is nominally constant, injection time is simply obtained by dividing total injected volume by the volumetric displacement rate. However, a control facility which is now offered on many modern injection machines is *injection speed profiling:* the stroke is divided into a number (usually 5 or 10) of equal substrokes, through which the injection speed can be individually controlled. Reasons for slowing injection speed at specific positions include avoiding turbulence and jetting at gates, to enhance the formation of homogeneous weld-lines after a divided flow and avoiding flashing at the end of the stroke.

(2) Cooling. The cooling phase, which (for the majority of thermoplastics moulding processes of part thicknesses greater than, say 2-3 mm) is often the longest part of the machine cycle, cannot be considered in isolation since a number of other functions occur simultaneously (see Figure 5.12). Before considering some of these, a brief review of cooling requirements is presented.

The total heat content of a moulding, q, is given by:

$$q = \triangle H \cdot m_P \tag{5.22}$$

($\triangle H$ = enthalpy between reference temperatures, m_P = mass of plastics part, including runners)

Most of this heat is extracted by oil or water in the adjacent cooling system:

$$-q = n \cdot (m_C \cdot C_C \cdot \triangle T_C) \tag{5.23}$$

($\triangle T_c$ = temperature rise in coolant; m_C and C_c represent the mass and specific heat of the coolant fluid, and n is a factor which expresses the proportion of the total heat content which is extracted by the coolant)

By a simple thermal energy balance, and rearranging, the total volume of coolant (of density ϱ_C) required is:

$$V_C = \frac{\triangle H \cdot m_P}{n \cdot \varrho_C . C_C \cdot \triangle T} \tag{5.24}$$

This approach gives a first approximation of the cooling requirements, for a given process.

Of greater significance to manufacturing feasibility is the *cooling time* for a moulded component. An approach equivalent to that described earlier (Section 5.2.4.2) for frozen-layer development can also be used to estimate cooling time, if the correct demould (or ejection) temperatures, T_{EJ}, for the centre-plane of the thickest part-section, can be specified. These represent the maximum temperatures at which mouldings can be ejected and remain form-stable. Ejection temperatures are related to the thermal transition temperatures referred to in Chapter 2 and can be summarized (for amorphous and semi-crystalline plastics, respectively) as follows:

- Amorphous $T_g \pm 20\,°C$
- Semi-crystalline $T_m - (20 \textit{ to } 40)°C$

The cooling time is defined by the time taken for the centre-plane to reach T_{EJ}: ie. $t = t_c$ when $T = T_{EJ}$ at x. By appropriate rearrangement of Equation 5.19, an expression for cooling time is developed:

$$t_c = \frac{x^2}{4\alpha} \cdot \left[\frac{T_M - T_C}{T_{EJ} - T_C}\right]^2 \tag{5.25}$$

(where x = heat flow path length, or half the section thickness).

Rather than use the relationship between Fourier number and temperature difference ratio modelled by Equation 5.19, other graphical solutions for different sample geometries (of the form given in Figure 5.9) should be used. In general, the lower the surface area/volume ratio of the moulding section, the nearer the cooling characteristic will be to that of the sphere.

A simple analysis to estimate cooling times in composite mouldings of more complex shape, has been presented by MENGES and MOHREN[5.16].

(2a) Packing/hold-on. When a moulding cools, significant *thermal contraction* is apparent. To counteract this form of shrinkage and ensure that part and cavity dimensions are more consistent, the screw imposes an additional "follow-up" pressure to pack sufficient material into the mould to compensate. The hold time is effectively terminated when the melt in the restricted gate region solidifies. Hold pressure and time can each be pre-set by the machine operator. Over-packing is not desirable; this can lead to excess orientation and distortion, or to flashed mouldings.

Packing-pressure profiling (on similar principles to injection speed profiling described above) is available on many modern machines, to optimize packing, shrinkage distribution and to comply with cycle time constraints. A further complication is the *mode* by which the machine senses the end of the injection phase, before switching over to packing pressure. This has traditionally been done on the basis of *stroke* (linear displacement measurements), *time* or *hydraulic cylinder pressure* but a method which should enhance the degree of machine control and part-consistency is switching by measured *cavity pressure*. In this way, closed loop adaptive control can be promoted, and a disturbance from ideal moulding conditions downstream of the screw (eg. polymer of fluctuating viscosity, due to varying MFI of the feedstock, or as a result of a heater malfunction) can be sensed and corrected.

More recently, control systems based upon prior knowledge of the pressure/volume/ temperature relationships of cooling thermoplastics (*P-V-T control*), have been developed to optimize the follow-up pressure profile in such a way to achieve moulded parts with desired properties (shrinkage, weight and part dimensions). There is only a single, optimum pressure profile for a given material/tool/machine setting combination (see Figure 5.13). A P-V-T diagram can be used to evaluate the hold-time programme necessary to obtain atmospheric pressure at the desired specific volume level, at the melt temperature specific to a given process[5.16, 5.28].

Successful packing has always been problematic in thick sections of injection moulded parts. However, a recent development (using *Oscillating Packing Pressure*[5.29, 5.30]) has allowed parts of up to 40mm thickness to be moulded successfully. High-frequency oscillations are applied to the solidifying melt in the mould cavity: the imposed energy is transformed into sufficient viscous heat to maintain

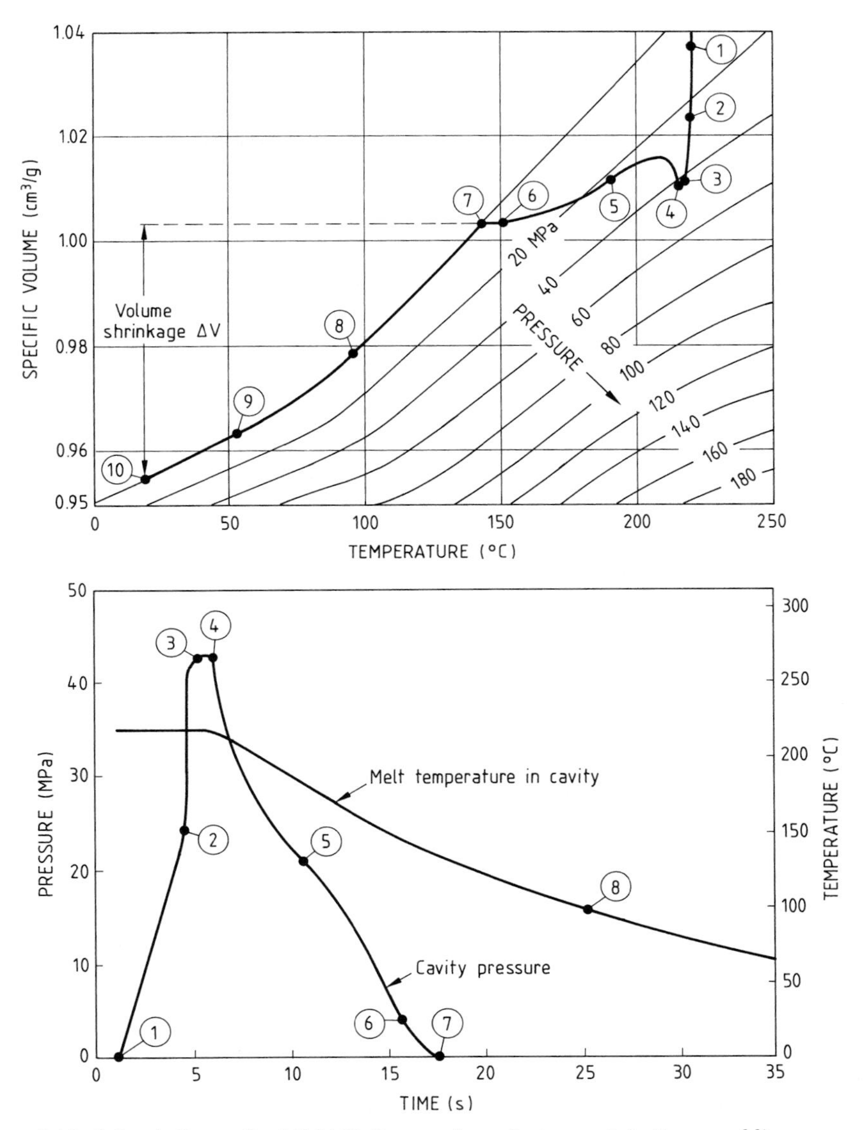

Figure 5.13 (*above*) Generalized P-V-T diagram for polystyrene injection moulding. (*below*) Pressure and temperature cycles for the same process (Figure reproduced from Reference 5.16, by permission of the publisher).

Legend:
(1) melt reaches pressure sensor close to gate;
(2) cavity is filled;
(3) pressure maximum attained;
(4) switchover from injection to packing phase;
(5) level of packing pressure attained;
(6) specific volume decreases until gate freezes;
(7) atmospheric pressure is attained – shrinkage initiates;
(8) solidification temperature reached at pressure sensor;
(9) ejection; and
(10) final state attained.

the necessary degree of fluidity to allow continued packing of thick sections to occur. This method is also effective in controlling the morphology of injection moulded plastics (notably heterogeneous, fibre-reinforced grades), particularly in the vicinity of weld lines.

(2b) Screwback. Screwback is the screw rotation and melt plastication phase which operates under a preset screw speed whilst the moulded part is cooling; this phase was described fully in Section 5.2.1.

(3,4,5) Mould opening, ejection and mould closing. Phases 3, 4 and 5 (Figure 5.12) are similar, in that the time constituent for each of them is determined primarily by the type and specification of mould clamp unit. Each of these parts of the cycle should be completed quickly to reduce the overall cycle, but there are some limiting factors which must be borne in mind:

- maximum mould closing speed should be attained after a gradual (rather than sudden) acceleration, to reduce the demands made on the hydraulic power unit. Also, a descreasing closing speed should be apparent immediately before mould face contact;
- ejection speed should be set at the maximum level which does not impart any sign of surface damage to the moulding.

Overall, injection moulding cycles consist of several constituent phases. The time factors associated with them are dependent not only upon the machine specification, but also by the physical properties of the moulding material and the machine settings for any specific product range. The effectiveness of a control system (and therefore the overall reliability and consistency of the process) is enhanced if the feedback source for closed-loop action originates from a physical measurement of a polymer melt characteristic: a cavity pressure-determined switchover from injection to packing phase is a good example.

5.2.6 Some Structural Features of Injection Moulded Plastics

Moulded plastics are utilized in a vast array of markets, many of which impose strict demands upon the component to fulfil all design criteria, often at extreme temperatures and under acute loading conditions. At an early stage, properties data relevant to a number of competitive materials are sought; it is an unfortunate characteristic of polymers that the material properties generated under controlled conditions on carefully-prepared 'standard' (and often homogeneous) test-pieces, often bear little relation to component behaviour in-service. This is because the precise detail of the manufacturing process (eg. barrel temperature, injection speed, shear history) has a profound influence upon the microstructure which is developed in moulded plastics, and also therefore, upon the physical properties of the part.

Injection moulded products are particularly sensitive to changes in processing, and may be susceptible to brittleness (for example) if the process is not optimized to comply with the characteristics of the polymer.

Mechanical properties especially are determined not only by the molecular weight and grade specification, but also by the effect that the rheological/thermal environment has on the developing microstructure. Some of the most important facets of microstructure, together with their dependence on processing conditions, are reviewed in this section.

5.2.6.1 General Microstructure of Semi-Crystalline Plastics

All microstructural characteristics of mouldings originate from the type and rate of flow during injection mould filling, and from the thermal history during plastication, or in the cooling or consolidation phases; only with some knowledge of the combined influences of thermal and flow-induced effects can defects be correctly diagnosed and remedial action successfully applied. In order to illustrate some of the structural characteristics of injection moulded plastics, particular reference is given to polymers which crystallize, since the microstructure of such materials is often quite complex, and is particularly prone to change when moulding conditions are altered (whether by desire or by default).

Figure 5.14 Polarized light micrograph of a typical injection moulded, semi-crystalline thermoplastic (section from a polypropylene tray).

Figure 5.14 illustrates a typical microstructure of an injection moulded, semi-crystalline polymer. By the usually adopted co-ordinate convention, x represents the flow direction (specimen "length"), y the "width" and z the thickness of a part section. Immediately, some characteristics are apparent:

- The presence of a birefringent (highly-oriented) layer, parallel to the flow direction, immediately below the sample surface. Promotion of orientation occurs in the melt (shear stress, which induces molecular alignment, is highest at the mould wall) and the orientation becomes immobile ("frozen-in") due to the extremely high cooling rate at this location. Orientation is discussed more thoroughly in Section 5.2.6.2.

- A discrete "textural interface" usually exists between the oriented region, and successive areas of different crystalline morphology further from the surface. Towards the interior (core) of the moulding, local cooling rates are lower and in consequence, nucleated crystallites are able to grow radially to form the aggregated *spherulites,* characteristic of slowly-cooled crystalline polymers. Spherulite size is inversely proportional to local cooling rate, so that crystalline texture becomes coarser in the moulding interior.

For some specific polymers, various forms of crystallization are feasible, some of which may be metastable. For example, a texture known as *β-form* can be promoted in polypropylene, and is believed to induce embrittlement. Isotactic polybutylene is associated with *polymorphism,* since crystallinity can exist in a number of polymorphic forms. Form II (tetragonal) is obtained by rapid cooling of the melt, but gradually transforms into form I (hexagonal) on room temperature ageing, with a corresponding increase in density (of about 4%) and change in physical properties. *Row nucleation* is also possible (see also Section 5.4.4) and the phenomenon of *trans-crystallinity* has become synonymous with crystal nucleation around fibres in glass-reinforced semi-crystalline thermoplastics such as PA66 and polypropylene[5.31].

Additional structural facets (which are particularly significant when injection moulding colour-masterbatch compounds, and which can easily be detected by simple microscopic techniques) are *pigment dispersion* and *pigment distribution.* The common-light micrographs in Figure 5.15 illustrate the effectiveness of dispersing a (polyethylene-based) titanium-dioxide (white-pigment) masterbatch in an injection moulded polypropylene component, using back pressure settings towards the lower and upper limits of the machine specification (Figure 5.15 (A) and (B), respectively).

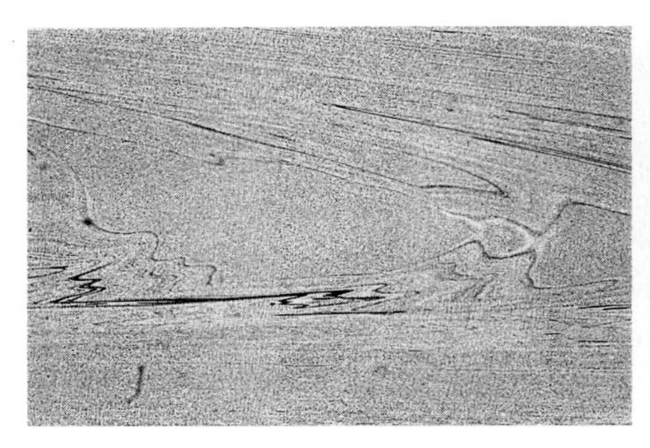
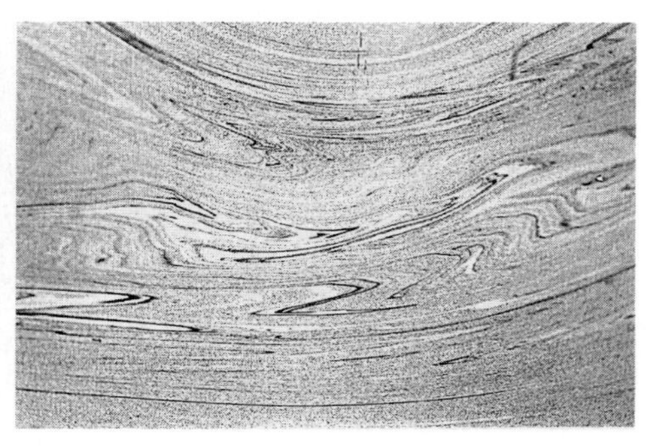

Figure 5.15 Micrographs showing poor (right) and improved (left) dispersion of a white titanium dioxide pigment masterbatch in injection moulded PP.

For further study, some reference citations are made with specific reference to the microstructure of injection moulded semi-crystalline plastics[5.32–5.34].

5.2.6.2 Orientation

On the basis of thermodynamic principles alone, long-chain molecules wish to conform to the *random-coil* state. In an unstressed melt, molecular chain segments have sufficient mobility to relax into this isotropic distribution. However, if the shear or tensile stresses active during moulding promote significant alignment (*molecular orientation)* in the melt, the oriented, anisotropic state can remain in the product if the cooling rate is sufficiently rapid to prevent significant relaxation taking place. It follows that the residual levels of orientation in moulded plastics components are determined first by processing conditions, but vary according to local cooling rates within the component.

Most models used to simulate the development of molecular orientation during mould filling have been developed from the pioneering work of BALLMAN and TOOR[5.35] (see Figure 5.8):

(1) The hot melt flows along a mould cavity, which has relatively cold surfaces. The pressure/shear stress distribution varies from zero (at the flow-front) to its maximum value at the cavity entrance. Since shear stress induces molecular orientation in the melt, radial deposition of unoriented material (from the melt-front) occurs at the mould surface.
(2) However, as injection continues, the material adjacent to the surface monolayer becomes highly sheared (shear stress is a maximum at the boundary of a flow channel) and highly oriented; cooling rate is rapid and the molecular orientation is locked into the structure of the solid part.
(3) Away from the cavity face, successive elements of material are subjected to lower shear stresses, and local cooling rates are also diminished due to the insulating characteristic of the frozen skin layer. Residual orientation decreases accordingly, as the centre-axis of moulded products is approached.
(4) Orientation, like shear stress at the instant of mould filling, is inversely related to the distance along the cavity flow path, from the gate-end. This type of orientation distribution is modified by the events during the packing phase, when relatively cold, high viscosity material is forced into the cavity under high pressure. A complex distribution of highly oriented material is therefore often seen adjacent to the gate.

This model suggests that the final degree of orientation in injection mouldings will be primarily determined by the processing conditions. Combining terms for shear stress (τ) and apparent shear rate ($\dot{\gamma}$) from Equation 5.10 (using a Power Law Model of the type $\tau = k \cdot (\dot{\gamma})^n$), we obtain:

$$\tau = \frac{H \triangle P}{2L} = k \cdot \left(\frac{6Q}{WH^2}\right)^n$$

where k (the consistency index) is related to shear viscosity (at a given shear rate), and is a strong, inverse function of melt temperature. On this basis, the following effects on orientation in the injection moulding process are usually seen:

• **Melt temperature:** orientation decreases with increasing melt temperature. Melt viscosity is diminished, and under rate-imposed conditions, so too is shear stress.

• **Mould temperature:** orientation decreases with increasing mould temperature. This is primarily a relaxation effect, related to a lower rate of heat transfer from the flowing melt.

• **Injection speed:** residual orientation in the moulded part decreases with increasing injection rate. The rheological effect which might be anticipated on the basis of the expression given above (increase in shear stress with Q) is often overshadowed by the fact that flow and significant cooling occur simultaneously in a slow-injection process, suppressing relaxation and freezing-in high orientation.

• **Packing pressure:** orientation generally increases with increasing packing pressure/time; the effects are often localized near to the gate(s).

It is commonplace in injection moulding research to attempt to quantify molecular orientation by *birefringence* measurements (double refraction caused by molecular anisotropy; see also Section 1.3.9); a polarizing microscope is a very useful tool in carrying out such studies. Examples of some detailed research work, on the subject of processing-orientation relationships in injection moulding, have been published elsewhere[5.36 – 5.38].

5.2.6.3 Residual Stress

The term "residual stress" is used to describe the presence, in a moulding, of a self-stressed state induced during the (inhomogeneous) cooling stage of processing. The local state and direction of energy-elastic stress varies from point to point within the moulding; the occurrence and magnitude of component deflection will depend upon the geometry of the part, in response to the magnitude of the stress distribution. Internal stresses of this type should not be confused with molecular orientation, nor should frozen-in orientation be classified as a type of stress. At best, orientation could be termed a "passive stress"[5.39], since a shrinkage stress would develop if, for example, an oriented (amorphous) sample were to be heated above T_g whilst held in a fixed position. Relaxation of orientation is an entropic effect, and is not feasible at temperatures below T_g.

STRUIK[5.40] has developed a simple model to account for the development of an energy-elastic residual stress distribution (see Figure 5.16). Cooling is assumed to be inhomogeneous, and occurs in two discrete stages. At $t = t_1$, the surface layers are quench-cooled to the temperature of the surrounding medium (akin to the true situation in injection moulding) and the polymer is free to contract as this occurs. When the internal core cools (assumed instantaneously at $t = t_2$), local thermal contraction is suppressed by the already-rigid (solidified) external layers; the core goes into a state of *hydrostatic tension,* balanced by *compression* in the skin.

Whilst the second cooling stage in this model is perhaps over-simplified, an adequate, qualitative interpretation of residual stress development can be derived from the approach. In practice, a part of simple geometry, cooled homogeneously from each face, would assume a near-parabolic stress profile, with maximum tensile stress (at the mid-plane) equal to half the maximum compressive stress (at either surface). Stress levels can be of the order of 5 MN/m^2, and must be considered superimposed upon any external stresses applied in service. Component deflection (distortion) due to residual stresses is likely if cooling rate from the two surfaces is uneven ie. bending will occur until the net bending moment about the long-axis is zero.

Measurement of sample curvature, as a function of the depth of material machined from one surface, is recognized as the most convenient method to generate

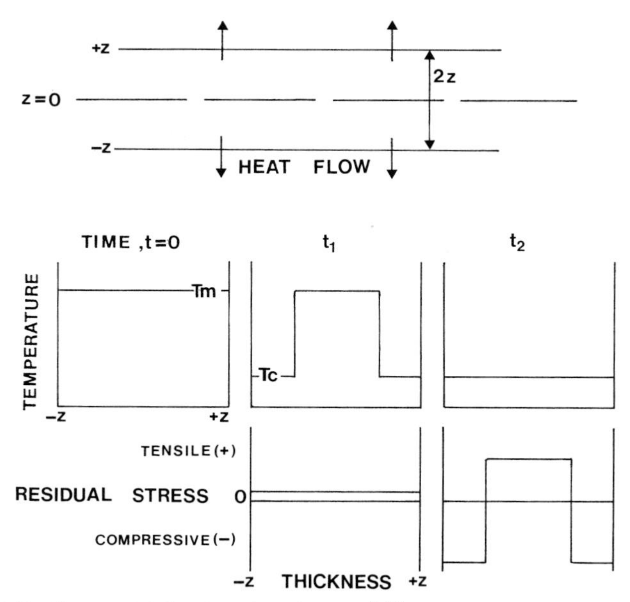

Figure 5.16 Cooling model developed by STRUIK[5.40] to simulate the development of residual stress in a plastics moulding cooled from both surfaces. Note the high compressive stress at the surface, balanced by hydrostatic tension in the core.

quantitative residual stress profiles in injection mouldings[5.39,5.41]. Typical through-thickness residual stress profiles derived by this method are illustrated in Figure 5.17, for injection moulded samples of PC produced by different moulding conditions.

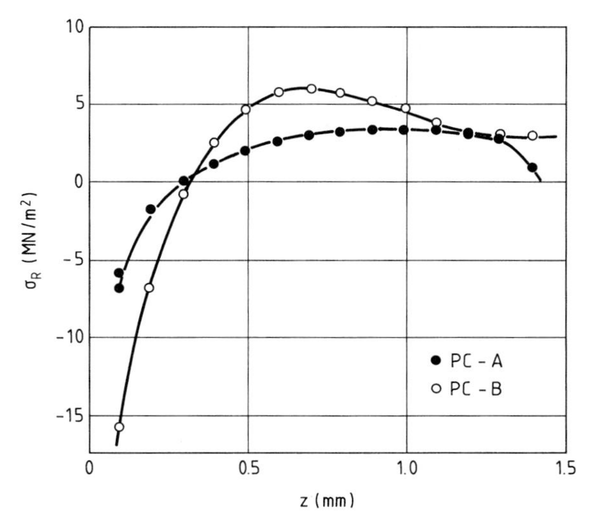

Figure 5.17 Residual stress (σ_R) as a function of sample depth (z) for injection moulded Polycarbonate:

PC-A Melt temperature 300 °C; mould temperature 115 °C;
PC-B Melt temperature 340 °C; mould temperature 75 °C.

The magnitude of residual stress is related to cooling rate, and is generally likely to increase when the difference between melt and mould temperature is greatest. Other moulding conditions are no less important in this respect (notably injection pressure[5.42, 5.43]), and residual stress distributions in moulded parts are also affected by physical ageing[5.40, 5.44].

5.2.6.4 Weld-Line Formation

Internal weld lines are formed in injection moulded parts whenever two flow-fronts meet. This can occur by several different routes:

(1) in multi-gated moulds;
(2) in single-gated annular, "frame-like" or cylindrical parts;
(3) recombination flows, after the melt stream has been divided around an insert, or if a particularly restricted zone has been negotiated.

In the first two examples, the flow fronts meet head-on, but some recombination flows (parallel melt streams) induce a weld seam whose axis is parallel to the flow direction; this latter type of defect has been specifically referred to as a *meld line*[5.10].

The formation of a typical internal weld-line is illustrated in Figure 5.18. Convex flow fronts are propagated from the extremes of the cavity; the development of orientation along the flow path is generally in accordance with the model described earlier (Section 5.2.6.2). At the position where the melt streams meet, the weld plane initiates near the centre-axis of the moulding and grows outwards towards the cavity walls. Locally, the melt is extended perpendicular to the overall flow direction; as a result, molecular diffusion across the interface is suppressed and *transverse orientation* is induced. For these reasons, together with the possibility of *air entrapment* at the interface, the weld plane is an inherent source of weakness for many moulded plastics.

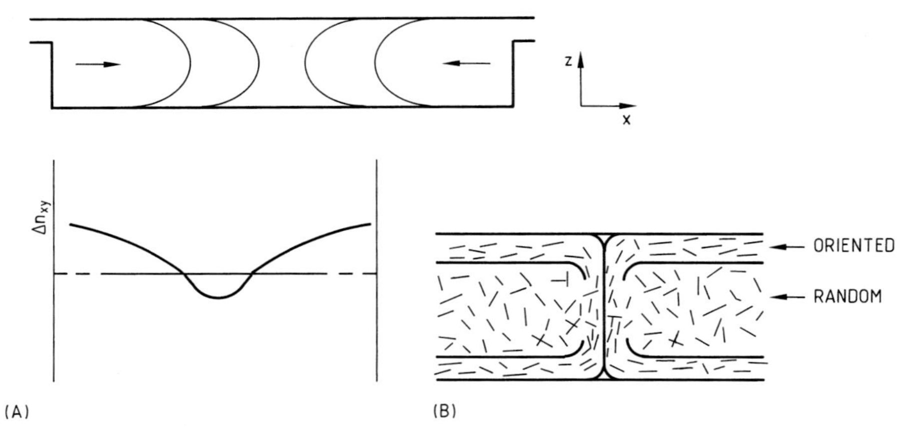

Figure 5.18 Formation and direction of orientation at internal weld lines in injection moulded plastics:
(A) Transverse molecular orientation (unreinforced resin);
(B) Orientation of short fibres (fibre reinforced grade) – transverse orientation is also evident at flow-front contact-plane.

Processing conditions can be optimized to control any embrittlement related to internal welds. It is usually beneficial to maintain a moderately high melt temperature, in order to assist interfacial molecular diffusion. Moulding at high melt/mould temperatures, and high injection speed (assuming that the cavity is well vented, and that the packing profile is specified accordingly) are all likely to enhance weld strength.

5.2.6.5 Shrinkage and Distortion

Thermoplastics are associated with relatively high coefficients of thermal expansion, since the strength of secondary bonds between adjacent molecular chain segments is very temperature sensitive. Semi-crystalline plastics are particularly prone to thermal shrinkage, due to the greater specific volume difference between melt and solid (crystalline) phases. It follows that components moulded from plastics can suffer from significant shrinkage effects during the cooling phase, which (for the context of injection moulding) can be defined as the ratio of specific volume of the part (at ambient temperature and pressure) to the total cavity volume.

If the degree of shrinkage in mouldings could be predicted with accuracy, it would be a relatively simple task to design precision components whose geometric shape could accommodate its occurrence. However, there are some factors which detract from the simplicity of such an approach:

- **Orientation.** If mouldings contain high levels of molecular orientation, shrinkage is anisotropic, since aligned chains shrink to a greater extent in the direction of orientation. Anisotropic shrinkage leads to *distortion.* A commonly-observed form of distortion associated with in-the-plane shrinkage is *warping;* this is common in large-area, thin-wall components, gated such that orientation is unsymmetrical or is predominantly uniaxial.

- **Part thickness.** The total deflection from a datum position, associated with a given level of shrinkage, is dependent upon section thickness. *Sink marks* on moulding surfaces are often seen in thicker sections, or at locations above ribs and internal fillets.

- **Other factors.** The absolute level of shrinkage, even for a given polymer, is not easy to predict. It is formulation-dependent (mineral-filled thermoplastics shrink much less than unfilled grades), and is very much determined by processing conditions. For example, a greater temperature differential between melt and cavity tends to increase the total volumetric shrinkage and locally, slow cooling rates (in thicker sections) enhance the degree of crystallinity, which itself has a similar, indirect effect.

Measures are taken to restrict the degree of shrinkage, of course, by suitable design of the packing phase (see Section 5.2.5). High cavity pressures are operative during this stage of processing, so that the effects of both *temperature* and *pressure* must be considered. These arguments form the basis of the P-V-T diagram, and are fully explained elsewhere[5.16].

5.3 Blow Moulding

Blow moulding is a generic term used to describe a range of processes for the production of hollow-form products. Although the packaging industries were amongst the first to benefit from the ease with which polymers could be processed by blow moulding, further advances in processing have expanded the technology into automotive, chemical, pharmaceutical, agricultural and general industrial applications. In parallel, innovative materials and process development have contributed not only to new applications, but also to wholly novel types of process: stretch blow moulding, and co-extrusion blow moulding are recent examples of this.

Most blow moulding processes are based upon a common theme:

- Conventional melt processes are used to make a preformed cylindrical part: the *parison* (extrusion), or *preform* (injection moulding).
- This is transferred at a controlled temperature to a split blow mould, where it is sealed and inflated (sometimes with mechanical assistance) to assume the internal contours of the mould.
- The part is cooled (whilst still under internal pressure) to achieve shape-stabilization, and is subsequently ejected when the mould opens.
- Processing diversity is evident since various methods are used in the following phases of production[5.45,5.46]:

(1) production of the parison/preform;
(2) transfer of the parison/preform to and from the mould;
(3) controlling the thermal cycle of the material;
(4) achieving inflation.

Some different processes will be reviewed for common blow moulding materials in subsequent sections. Also, since some sub-processes are associated only with specific groups of plastics, the important processing characteristics are discussed in the context of the demands imposed by the technique and the product range.

5.3.1 Techniques for Blow Moulding Plastics Products

Sustained growth of the blow moulding industry has occurred by materials development, interactive innovation in processing and by a growing acceptance of the advantages of plastics, on behalf of the consumer. Rather than categorize blow moulding according to products, markets or specific plastics, it is easier to subdivide the industry according to the types of process which have evolved.

Process requirements for blown containers are determined by the polymer grade specified, the shape/complexity of the component and the financial criteria associated with capital cost and potential output rate. Most of the important commercial processes are reviewed in this section.

5.3.1.1 Extrusion Blow Moulding (EBM)

EBM is based upon conventional screw extrusion principles, as described thoroughly in Chapter 4. Components are produced by first closing the mould around the extruded *parison,* then by inflating the parison by compressed air from a blow pin, whilst the ends of the parison remain sealed at the mould parting line (*pinch-off*). The parison deforms to assume the general shape of the mould and shape consolidation is achieved as the melt loses heat, primarily by conduction to the walls of the tool. Following cooling, the mould is opened, the component is ejected and the moulding cycle is repeated; often, the extruder operates continuously (see below), in parallel to the forming and cooling phases.

An extruder screw L/D ratio of 20:1 has become a typical standard size for blow moulding, but for dry colour masterbatch compounds, mixing can be enhanced by increasing L/D ratio (eg. to 25:1) and/or by incorporating a screw mixing head.

Optimum economic performance in EBM is achieved only if the process can be balanced to allow the extruder to run continuously. Since mould cooling is predominantly *unidirectional* (heat flow to the mould exceeds by far the non-conductive heat losses to the internal gas medium), for all but the very thin components the cooling phase dominates the complete cycle of any process, in systems where a single extruder feeds a single mould. Consequently, there are a number of techniques which allow *continuous extrusion* to take place, whilst reducing the effects of slow part-cooling: these include multiple die-head extruders, and the use of processes with a multiple-mould arrangement fed by a common extrusion die (eg "shuttle-type" reciprocating moulds, or a rotating "carousel"

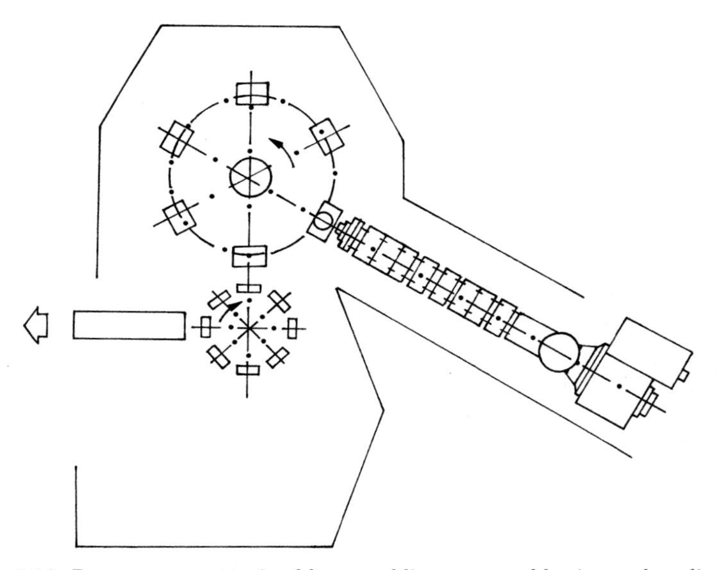

Figure 5.19 Rotary-type extrusion blow moulding process: blowing and cooling take place on the first six-station wheel, followed by deflashing and leak-testing on the eight-station finishing wheel. (Figure reproduced, with permission, from an original by Battenfeld Fischer Blasformtechnik, GmbH).

mould frame Figure 5.19[5.47]). The parison can be severed from the die-head by means of parison cutters such as stabbing (cold-cutting) knives, hot-wire cutters or shearing blades.

Intermittent extrusion is not popular, not only because of economic reasons, but also because of the increased risk of a stagnation zone, leading to melt decomposition. It is only chosen in 2-stage processes where an extruder feeds an accumulation system, capable of much more rapid parison production rates, for large (thick wall) products which would otherwise have severe parison sagging problems[5.46].

EBM technology has also been extended to *multi-layer co-extruded blow mouldings,* which has opened new markets particularly in "high-barrier", permeation-resistant packaging of foodstuffs, pharmaceuticals and speciality chemicals.

Many of these processes are described fully elsewhere, in dedicated texts[5.45, 5.46], or in machine manufacturers' literature[5.47, 5.48].

5.3.1.2 Injection Blow Moulding (IBM)

For equivalent productivity in the context of blow moulding, it is generally true that an injection moulding unit is more expensive than a single screw extruder. However, IBM is used extensively in situations which demand special product characteristics, which include:

(1) precision forming of screw-thread dimensions;
(2) avoiding flash, weld-lines and material waste at the container base;
(3) close control of surface quality and product dimensional accuracy.

Most modern injection-blow systems are designed to operate on a rotary table principle (horizontal parting-plane), with (usually) three stations, each associated with a different manufacturing phase (Figure 5.20):

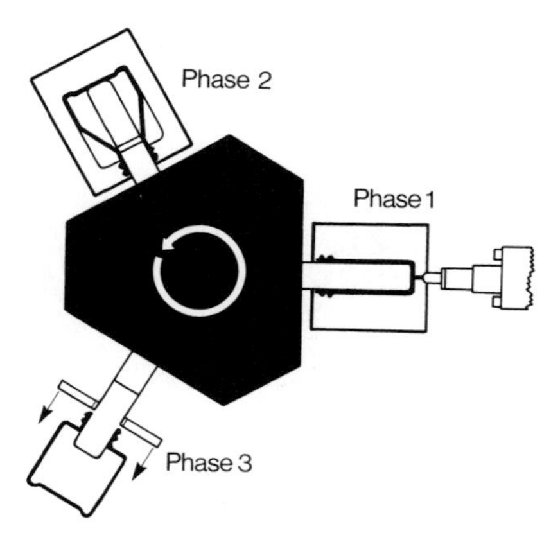

Figure 5.20 Operating scheme of a typical injection blow moulding system (Bekum SBM 340): Phase 1 – Preform injection moulding; Phase 2 – Rotation through 120°, followed by preform inflation into a container mould; Phase 3 – Following another 120° rotation, the container is stripped from the core and ejected. (Reproduced, with permission, from Bekum Maschinenfabriken GmbH).

- Phase 1 – conventional injection moulding; the core piece is attached to the rotary table. The mould may be fed by a *hot manifold runner* system, an especially-useful system for multi-cavity tools. When the temperature of the moulded *preform* is correct for blowing, the moulds are opened, the table is raised and the core-pins are rotated 120 °C to the next position.
- Phase 2 – preform inflation and cooling occur at station 2.
- Phase 3 – further cooling and part ejection.

Since the overall production cycle is split between three identical sub-cycles, its duration is restricted by the longest of these phases. It is claimed by the Bekum company[5.48] that preform production determines the overall cycle, in many processes.

5.3.1.3 Stretch Blow Moulding (SBM)

SBM processes have been developed in recognition of the enhanced properties which are obtained when molecular orientation is promoted, and its distribution in the product is optimized. This can be effected in the most controlled way by manufacture of a preform, controlled cooling or re-heating (see below), then *biaxial* (pneumatic and mechanical) *stretching,* followed by rapid part-cooling to freeze-in the orientation so-developed. Gas barrier resistance, optical clarity, impact strength and rigidity are all enhanced by the development of *biaxial orientation.*

Like conventional blow moulding, stretch-blow techniques are based upon extrusion or injection moulding, as applicable to specific plastics materials. The most important and extensive of these methods include:

- **Extrusion Stretch Blow Moulding (ESBM).** A parison is extruded and as the *pre-blow* mould closes, a mandrel descends to form the bottle neck. The parison is subsequently pinched and inflated (Figure 5.21), then conditioned at the optimum temperature for orientation in the next stage. Transfer of the blown and conditioned preform to the stretch-blow mould takes place; fabrication of the finished part is achieved by stretching in the circumferential direction (pneumatic inflation) and in the longitudinal direction (mechanical push-rod), followed by rapid cooling to

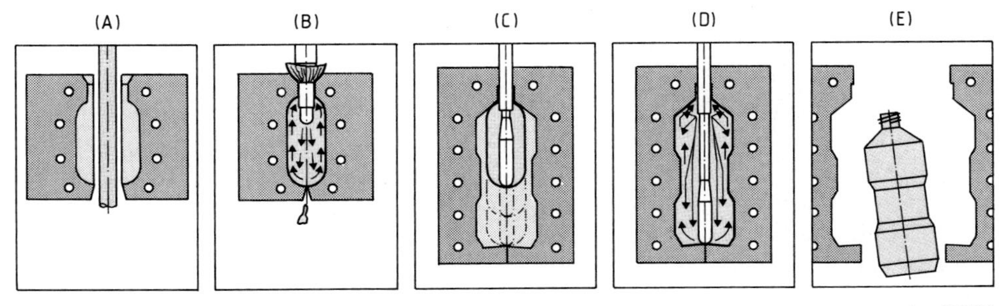

Figure 5.21 Operating scheme of a typical extrusion stretch blow moulding system for PVC, (Battenfeld Fischer VKI-3 ESB): (A) Parison extrusion; (B) Blowing and conditioning; (C) Axial stretching; (D) Blowing (hoop stretch); (E) Mould opening and ejection. (Reproduced, with permission, from Battenfeld Fischer Blasformtechnik, GmbH).

prevent relaxation of the (biaxial) molecular orientation. This process is centred on oriented PVC (OPVC) bottle manufacture, for transparent containers with enhanced clarity and greater resistance to gas permeation, higher impact strength and burst pressure.

• **Injection Stretch Blow Moulding (ISBM).** This process is based upon similar principles to ESBM, with the exceptions that firstly, the preform is injection moulded, and secondly, two-stage stretch-blow moulding allows the preforms and blown containers to be processed in autonomous stages, thus allowing greater overall production flexibility. Although OPVC and PAN bottles are manufactured by ISBM, the most successful exploitation of its processing technology has been in PETP; the manufacturing routes for stretch blow moulded PETP containers are described below, and are illustrated in Figure 5.22:

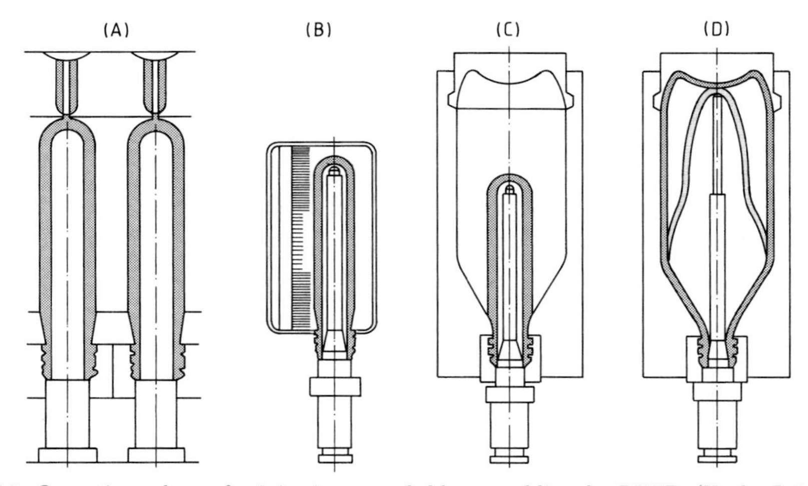

Figure 5.22 Operating scheme for injection stretch blow moulding for PETP, (Husky Injection Moulding Systems Ltd): (A) Preform injection moulding; (B) Preform heating by infra-red or radio-frequency system; (C) and (D) Biaxial stretching by mechanical push-rod (axial), and pneumatic inflation (hoop). The overall process can be run using a 'one-step' or 'two-step' sequence, (the latter involves preform storage, unscrambling and reheating), in order to balance production rates. (Reproduced by Husky Injection Moulding Systems, Ontario, Canada).

2-Stage process

Preforms are injection moulded to close tolerance, using multi-cavity, hot-runner tools. Cooling is rapid, thus preventing PETP from crystallizing. The amorphous preforms can be collected and stored indefinitely, before reheating and blowing. In the second stage, preforms are reheated by infra-red, radio-frequency or electrical resistance heaters[5.49], to a temperature above T_g which should not be so high as to induce spherulitic-type, *thermally-induced crystallization* before blowing. The temperature profile can be adjusted to obtain an optimum thickness distribution.

Biaxial stretching is then feasible in much the same way as for ESBM: mechanical (axial) stretching is achieved by a telescoping mandrel, and circumferential deformation by pneumatic inflation. During this phase, PETP crystallizes as a result of the deformation induced; the molecular ordering is bidirectional, with the c-axis of crystallites lying parallel to the stretch directions[5.50–5.52].

Single-stage process. In this technique, a convenient number of preforms (typically four) is moulded, conditioned to the required temperature, stretch-blown and cooled/ejected on the same unit: a four-station rotary table system is utilized.

Multi-layer, stretch-blow moulded containers are also produced on this basis by *co-injection moulding* preforms with different plastics in discrete layers.

5.3.2 Extrusion Blow Moulding – Materials and Melt Processing

In attempting to specify the optimum grade specification to achieve the desired range of product properties (mechanical properties of commodity blow mouldings include column stiffness, impact/puncture and environmental stress cracking resistance), further complexity is introduced by the demands made upon the feedstock during the various phases of processing. This section focusses upon some melt processing aspects of EBM, and highlights the rheological characteristics of plastics required at each stage of the process.

(A) Parison production and die swell. For any given blow ratio, the thickness of the moulding is directly proportional to the thickness of the parison (taking due account of the *die swell* which inevitably occurs). In consequence, the ability to control the elastic properties of the melt and predict the variation of die swell when the material, die design or extrusion conditions are altered, is necessary to ensure optimum thickness distribution in the finished part. The dependence of die swell on extrusion conditions has been summarized in Section 4.2.3.3. In the case of EBM it should be noted that an increase in die swell results in a lower *linear* output rate: since extrusion is usually continuous, further adjustments to the process dynamics are sometimes inevitable. Parison length sensors are available, to ease the effects of this problem[5.48].

(B) Extensional flow and parison sagging. There is a finite time lapse between melt extrusion and the moment at which pinch-off occurs. For any process with downward-vertical extrusion, an element of gravimetric self-weight loading occurs; the tensile stresses created tend to deform the parison longitudinally (*parison sagging*), especially towards the uppermost positions on the parison where the stress is highest. The effect of die swell is therefore opposed by sagging, but to an extent which is location dependent.

By relating the components of strain to the self-loading nature of parison sagging, the important melt properties which determine the extent of deformation can be specified as follows (see also Figure 5.23):

The total tensile strain, $\Sigma\varepsilon$, is the sum of the elastic (recoverable) and viscous (irrecoverable) components of deformation:

$$\Sigma\varepsilon = \varepsilon_E + \epsilon_V$$

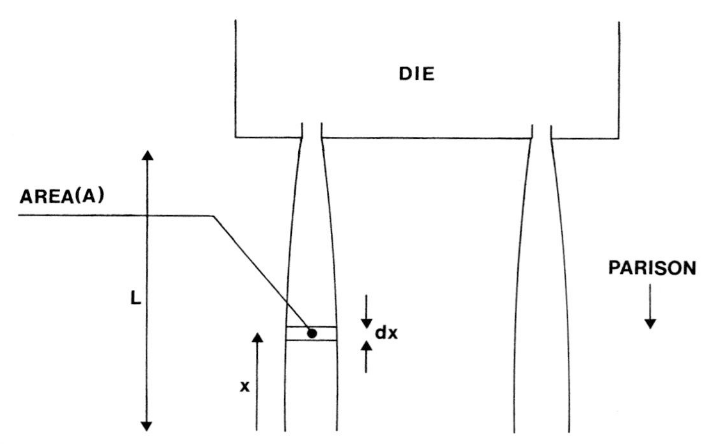

Figure 5.23 Geometry definitions for analysis of parison sagging in extrusion blow moulding.

Therefore:

$$\sum \varepsilon = \frac{\sigma(x)}{E} + \sigma(x) \cdot \frac{t}{\lambda} \tag{5.26}$$

($\sigma(x)$ is the tensile stress acting on an element of melt at a distance x from the free end; t = time after extrusion and E, λ are the elastic and viscous constants of the polymer, appropriate to the stress and processing temperature.)

In practice, $\sigma(x)$ is related to melt density (ϱ), the length of parison below the element (x) and gravimetric acceleration (g), by:

$$\sigma(x) = \frac{\text{Force}}{\text{Area}} = m(x) \cdot \frac{g}{A} = \frac{\varrho A x g}{A} = \varrho x g \tag{5.27}$$

Blow moulding polymers are chosen such that the elastic component of strain (ε_E) is dominant ie. the time-stability of the parison depends upon the elastic (modulus) constant. If the relaxation time $\frac{\lambda}{E}$ is significantly greater than the processing timescale for which elements of melt are self-stressed (typically below 10s.), then the response is, indeed, primarily elastic (and vice versa). It is of note at this point that unlike temperature dependence (where E and λ each decreases with increasing temperature), the relationship with stress is more complex, since modulus increases but viscosity often decreases with increasing stress. Clearly, sagging is more problematic at high stress, where significant viscous flow is much more likely.

If the parison deformation is primarily viscous (irrecoverable, time-dependent strain, ε_V), the total change in parison length can be evaluated[5.53]. Consider an element of fluid of length dx (Figure 5.23); since the elongational strain rate, $\varepsilon(x) = \frac{\sigma(x)}{\lambda}$, then the change in length of the parison (with respect to distance) is given by:

$$\frac{dL(x)}{dx} = \text{elongational strain,} \quad \varepsilon(x) = \sigma(x) \cdot \frac{t}{\lambda} \tag{5.28}$$

Substituting for stress from Equation 5.27 and integrating, we obtain the total degree of sagging ($\triangle L$):

$$\triangle L = \frac{\varrho g t}{\lambda} \int_0^L x\,dx = \frac{\varrho g t L^2}{2\lambda} \tag{5.29}$$

A more rigorous analysis of both elastic and viscous strains (taking local thinning of cross-sections into account) has been presented by COGSWELL et al[5.54].

Some data are presented in Figure 5.24 to show the time-dependent increase in parison length, for extruded PETP homopolymer and PETG copolymer, as a function of screw speed and extrusion temperature. Excessive viscous sagging in the homopolymer gradually occurs under elongational stress; the copolymer is clearly much more resistant to parison sagging under these conditions.

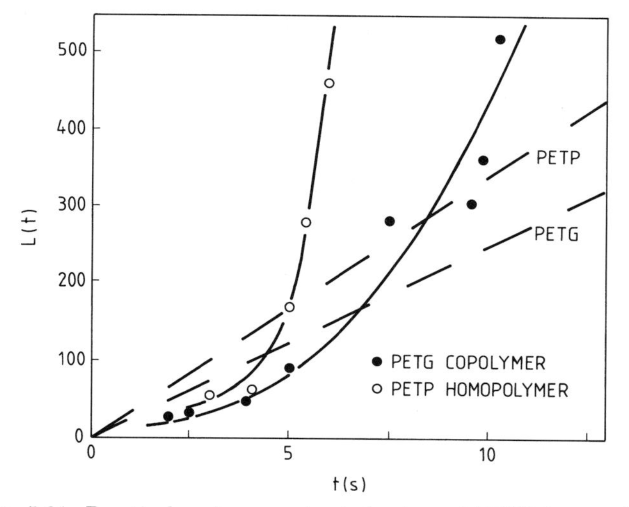

Figure 5.24 Practical parison sagging behaviour of PETP homopolymer (ICI Grade B90S) and PETG copolymer (Eastman Kodak grade 6763). The broken lines represent 'ideal' extrusion conditions, where parison length $L(t)$ is in direct proportion to time, (t), ignoring the effects of both die swell and parison sagging.

(C) Inflation. The ultimate criterion for the inflation stage is that the thickness of the moulded part be consistent and controllable. Progressive thinning must therefore be avoided, which implies that the material's resistance to deformation should increase in "thin-wall" areas, to compensate for the increase in local tensile stress which arises as deformation proceeds. Since elastic modulus increases with stress (and tensile viscosity usually does not), predominantly-elastic deformation is therefore preferable, necessitating a relaxation time $\frac{\lambda}{E}$ at a given applied stress which exceeds the time-span of inflation. Under high blow pressure (and therefore high tensile stress), a high tensile viscosity is therefore an additional requirement.

For a given blow pressure P_B, the tensile hoop stress in the parison (assuming thin shell theory) is evaluated thus:

$$\sigma = \frac{P_B D}{2t} \tag{5.30}$$

D, t represent parison diameter and thickness.

Although high blow pressure and rapid inflation are generally beneficial, the combination of pressure and parison geometry should not induce a tensile hoop stress which exceeds the rupture stress of melts (0.1 - 1.0 MN/m^2). Obviously, inflation becomes ineffective once melt rupture occurs.

An additional demand made upon the blowing phase is to suppress circumferential thermal shrinkage effects in the container by ensuring that an effective blow pressure remains operative whilst cooling is completed.

(D) Part thickness control. Estimation and control of the lateral dimensions of blown parts, in the simple case of circular-section mouldings, can be approached by considering the effects of parison swelling and simple circumferential deformation. It can be shown[5.55] that the elastic recovery which occurs following the extrusion of a rectangular section induces swelling in each lateral direction; the magnitude of swell in the narrow dimension is approximately equal to the square of that perpendicular to it. For the case of an annular section (using subscripts D, P and M to denote the dimensions of the die-gap, the swollen parison and the moulding), we have:

$$B(t) = B(D)^2$$

where B represents swelling ratio, in terms of thickness (B(t)) and diameter (B(D)). Therefore:

$$\frac{t_P}{t_D} = \left(\frac{D_P}{D_D}\right)^2 \tag{5.31}$$

If the possibility of sagging is discounted, and assuming constant volume deformation during inflation:

$$\pi D_P t_P = \pi D_M t_M$$

Therefore:

$$t_M = \frac{D_P}{D_M} \cdot t_P$$

Substituting for t_P from 5.31 and rearranging, the thickness of the moulding can be estimated by[5.55]:

$$t_M = [B(D)]^3 \cdot \frac{D_D}{D_M} \cdot t_D \tag{5.32}$$

Equation 5.32 is approximate, first since it ignores the possible effects of parison sag and pinch-off, and second, t_M represents the thickness of the moulded part assuming it were still molten. The degree of thermal contraction over the appropriate temperature span must be considered to obtain the correct estimation of the solid part wall thickness. The importance of swelling ratio on product dimensions is highlighted; as will be discussed below, if the relationship between swelling and processing variables (for a given resin) can be pre-specified, this relationship can be used as a principle for programmed control of part dimensions.

Uniform thickness distribution is more difficult to achieve in non-circular mouldings, due to the effects of progressive thinning. Melt properties can be designed to partly overcome this effect (as above), but a more effective solution is approached by re-designing the die mandrel (thereby modifying the lateral shape of the parison) to comply with the geometric constraints of the product to be blown.

In many processes, the parison dimensions are nominally held constant as a function of length. If an irregularly-shaped product is required, or if potential material savings are feasible by processing a non-uniform parison to achieve an optimally-shaped part, wall thickness control is achieved by *parison programming* whereby the die gap is adjusted by (electronically-controlled) vertical displacement of the mandrel assembly (Figure 5.25). Wall thickness programmers are able to provide the means to specify die-gap separation (and hence parison thickness profile) at up to (typically) 30 positions along the parison length. An additional, but less favoured route to thickness programming is by screw speed adjustment and its effect upon die swell. Although this principle is sound for a given material under specified extrusion conditions, applicability is limited due to excessive response time.

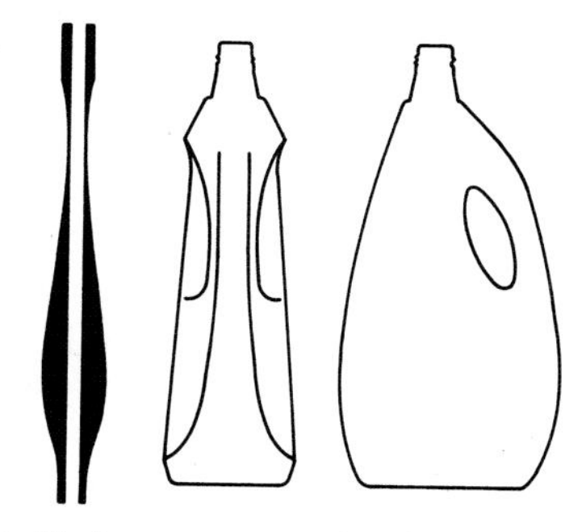

Figure 5.25 Parison programming in extrusion blow moulding – parison thickness profile is varied according to the container shape and to the wall thickness requirements. (Reproduced with permission from Bekum Maschinenfabriken, GmbH).

(E) Pinch-off and welding. When the mould is closed around the fully-developed parison, the base of a container will contain a horizontal weld-seam running parallel and adjacent to the split-line. Like other types of "internal" welds, pinch-off sites are potentially detrimental to container properties if they are not optimally processed.

(F) Summary of material requirements for EBM. There are varying, and occasionally conflicting melt property requirements for the different phases of the extrusion blow moulding process. Whilst the most important of these may vary according to the type/shape of container, and with the machine specification available, the summary below categorizes the general rheological properties which comply with the process environments described above.

Table 5.3 Rheological Aspects of Extrusion Blow Moulding

Processing phase	Melt property requirement
Single-screw extrusion	A relatively low shear viscosity at medium-high shear rate assists output capacity, per unit energy.
Die swell	Characteristics which are relatively insensitive to changes/fluctuations in temperature, screw speed and die geometry.
Parison sag	Elastic response required; high modulus and high elongational viscosity required at low stress.
Parison inflation	High relaxation time (low elastic modulus and relatively high tensile viscosity); elastic response to promote stability.
Pinch-off welds	Good weldability (low shear viscosity).

5.3.3 Some Technological Aspects of Stretch Blow Moulding

Optimization of processing to achieve the best or most balanced range of properties involves different principles to those described for conventional EBM, since the stretching phase is carried out in the "rubber-elastic" deformation regime ie. at temperatures just exceeding T_g. PETP is an interesting choice of material with which to examine processing technology, not only because of its dominance in the ISBM market, but also because, firstly, like other long-chain thermoplastic polyesters formed by polycondensation, PETP is hygroscopic and will absorb moisture extremely rapidly. In turn, at melt processing temperatures, thermo-hydrolytic degradation occurs, with a corresponding loss in molecular weight, rendering processing inconsistent and bottle properties inferior. Second, PETP is a semi-crystalline polymer but crystallizes only slowly into a spherulitic texture, and only in a restricted temperature span. However, PETP also crystallizes into chain-extended form as a function of strain: the biaxially-oriented, *deformation-induced crystallization* is primarily responsible for the excellent gas-barrier and impact properties which stretch-blow moulded PETP posesses.

If complete chemical reaction of any residual moisture in the PETP melt (during preform injection moulding) is assumed, it has been estimated[5.56] that only 16 ppm retained moisture in PETP granulate results in a molecular weight loss corresponding to a drop in intrinsic viscosity (IV) of 0.01 dl/g (M_n decrease of about 500 g/mol); see Figure 5.26. The need for complete and controlled drying before injection moulding is therefore the first step towards successful manufacture: dehumidifying (circulating air) dryers, operating at 170 – 180 °C for up to 8 hours, are usually specified. A further potential problem is the formation of trace quantities of acetaldehyde (AA) (which detracts from the flavour of packaged foodstuffs), formed by decomposition of hydroxyethyl end-groups or main-chain ester-linkages. In order to control the formation of AA, melt residence time, processing temperature and the developed shear rate should each be minimized and controlled.

Figure 5.27 shows the dependence of *isothermal rate of crystallization* upon temperature and molecular weight, in the cooling mode. Crystallization of this (spherulitic) type must be avoided, to ensure the production of wholly amorphous,

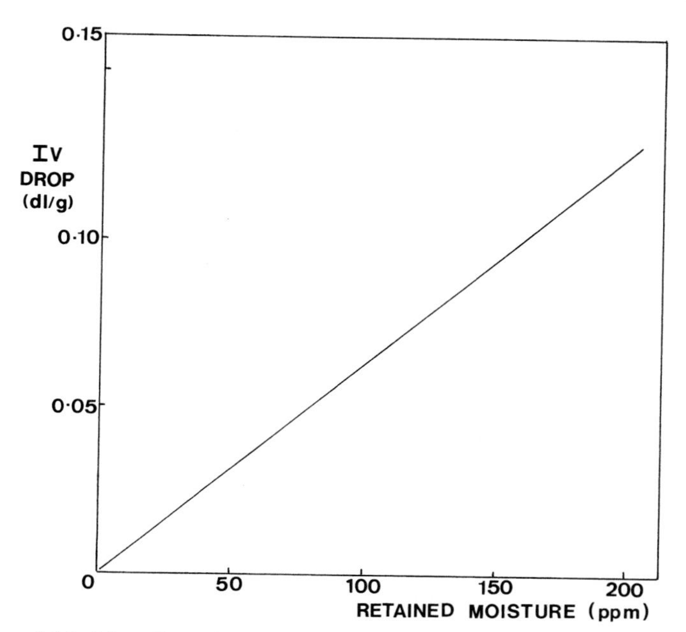

Figure 5.26 The effect of retained moisture on the theoretical drop in PETP intrinsic viscosity (IV), as a result of melt-phase hydrolysis. (Reproduced by permission of ICI Chemicals & Polymers Ltd.).

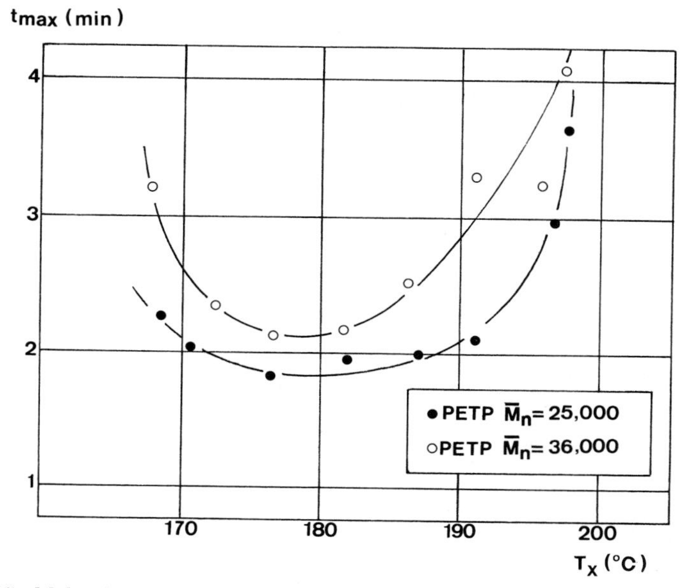

Figure 5.27 Molecular weight-dependent isothermal crystallization data for PETP at various temperatures. (The ordinate represents the time taken to reach a maximum rate of crystallization, at the temperature shown).

haze-free preforms which will subsequently draw in an even, predictable manner. Clearly, since a higher molecular weight resin tends to retard crystallization rate in this regime, the effect contributes to the compromise underlying the optimum choice of PETP intrinsic viscosity. *Copolymerization* is thought to have a similar effect to an increase in molecular weight, in the context of retarding crystallization[5.57].

Despite the aforementioned phases of production, each of which must be precisely controlled, the *stretching phases* remain the most important from the aspect of structural development in the container wall. Figure 5.28 illustrates how PETP draws in the thermoelastic region, as a function of (re-heat) temperature and tensile strain rate (ie. simulating stretch-rod speed, or inflation rate). To induce significant crystallization under any specific conditions, the *natural draw ratio* must be exceeded; this represents the position where the stress/strain gradient rises significantly, and is temperature and rate dependent, as shown. The strain level at which crystallization is initiated has also been shown to be inversely related to PETP molecular weight[5.58]. Once an element of material is drawn beyond the onset of strain-induced crystallization, the enhanced degree of stiffness is more than adequate to withstand the local increase in true stress due to gauge reduction; in consequence, further deformation becomes stable, and remaining material in the preform wall is gradually drawn beyond the natural stretch region. Uniform thickness distributions in container walls are therefore promoted by the onset of *strain-induced crystallization.* Typical preform stretch ratios are 2.4-2.6 (axial) and 4.0 – 4.5 (circumferential)[5.56].

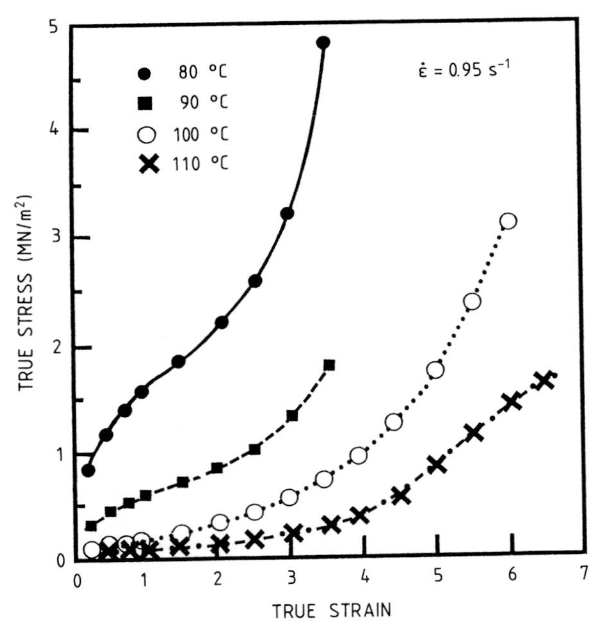

Figure 5.28 Elongational deformation of PETP at a nominal strain rate of $0.95 s^{-1}$; effect of temperature, (80 – 110 °C).

Thermoelastic stretching of PVC in the extrusion stretch-blow moulding process is depicted in Figure 5.29; a maximum degree of extensibility is apparent around 90 °C, so that molecular orientation is likely to be optimized if the preforms are accurately tempered within this region, then subsequently drawn at high strain rate to an overall *biaxial stretch ratio* (ie. ratio of bottle to preform surface area) of about 400 %. Mould cooling must be rapid, to preserve molecular orientation by suppressing relaxation effects.

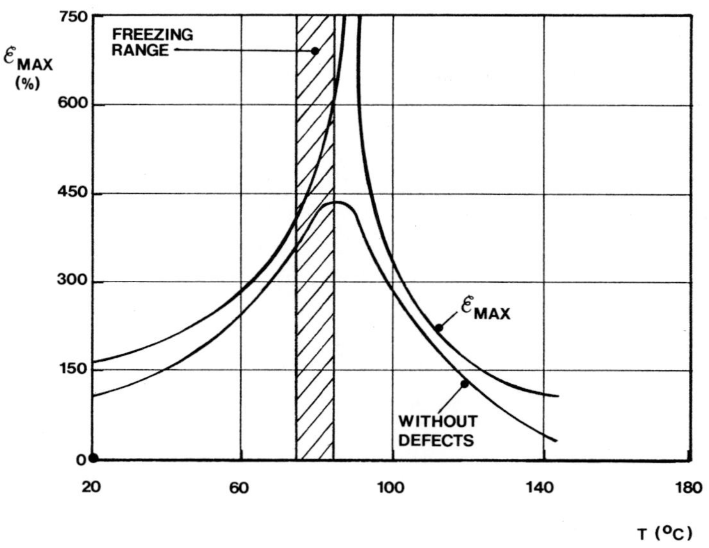

Figure 5.29 Effect of deformation temperature (T) on the ultimate elongation, (ϵ_{MAX}), for rigid PVC. (Reproduced with permission from Battenfeld Fischer Blasformtechnik, GmbH).

5.3.4 Blow Moulding and Microstructure

Having assessed the equipment and technological aspects of both extrusion-blow, and stretch blow moulding, and having highlighted the most important resin characteristics appropriate to each sub-stage of these techniques, it is important to specify some examples of microstructural development in blow moulded products. As will become evident in a later section (Chapter 7), premature failures can often be examined and the underlying causes related directly to *microstructural defects* arising from non-optimum processing conditions. Many features of moulded parts are very specific to given generic families and grades of plastics; nevertheless, the examples described should help to define principles which, to some extent, are applicable universally.

An extremely important element of commercial extrusion blow moulding processes is the working compromise between quality and productivity. This is especially relevant to relatively thick-sectioned parts made from semi-crystalline plastics, where any attempts to diminish or control cooling time tend to induce microstructural modifications to blow moulded components. A typical feature is the *crystalline textural interface* (for example, see the micrograph of a section through the wall of a HDPE petrol tank; Figure 5.30) situated some 200 μm inside the external surface of the component. Rapid cooling of the surface layers produces a quench-zone of *fine texture,* whilst the material in the core region is composed of a much more *coarse spherulitic texture* and is probably also more highly crystalline (and therefore has a higher density) overall. Increased cooling rate (higher parison temperature or lower tool temperature) is likely to exaggerate this effect, but in addition, the effect of blow pressure should not be underestimated, since higher blow pressure enhances thermal contact between the mould and the cooling melt, thus improving the thermal energy transfer rate across the interface.

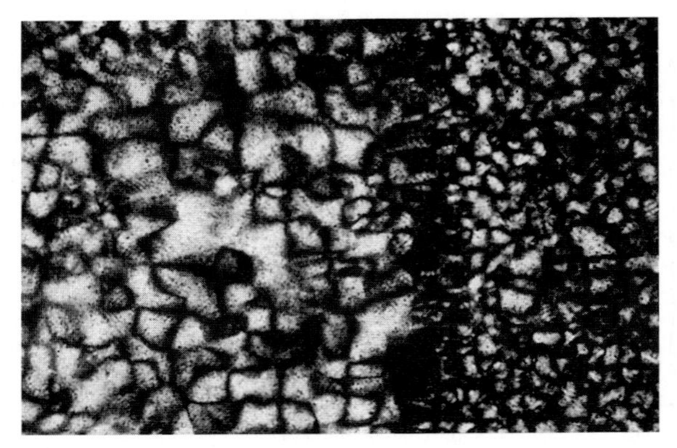

Figure 5.30 Optical micrograph of a blow moulded tank in HDPE, showing the development of different crystalline textures, according to local cooling rate changes during processing.

One means of diminishing cycle time without necessarily promoting different morphologies (see also Sections 2.4.7, 2.4.8) is to cool the container internally by forced air circulation, or by using liquid CO_2, whilst increasing the external mould temperature, in order to enhance the balance of bi-directional heat flow. Figure 5.31 shows the inner surface-texture of PP blow mouldings produced using different internal cooling methods. Static-air cooling (lowest heat transfer coefficient) produces an inner-surface upon which some relatively coarse spherulitic aggregates are visible. The crystallites become much smaller when forced air cooling is used, an effect which is enhanced much further if the inner surface is quenched by liquid CO_2. As is predicted by theory, spherulite size is therefore, inversely related to cooling rate.

The dominant feature in stretch-blown containers, and the key to understanding the link between processing and physical properties, is the characterization of the direction, magnitude and type of *molecular orientation* which is apparent. Due to finite differences in thermal and mechanical history, orientation is likely to vary from point to point within mouldings. Also, since highly-oriented PETP (especially) is prone to form layered structural planes as this material crystallizes under high strain, the thickness dimension of bottles may consist of several individual laminae, each of which may be oriented to a different extent.

 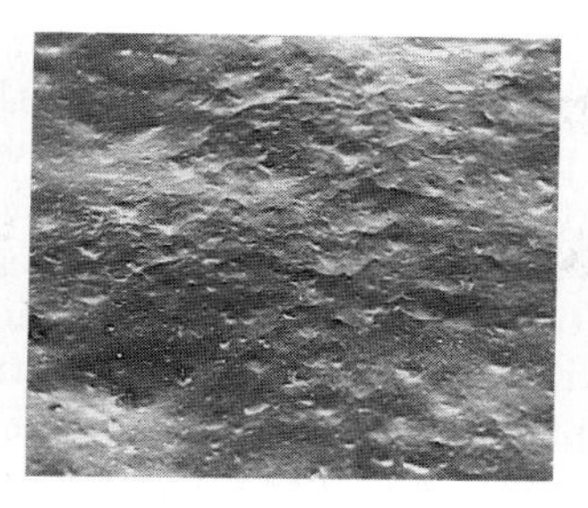 

Figure 5.31 Electron micrographs of the inside surfaces of extrusion blow moulded PP coolant tanks: (Left) Natural cooling (static air); (Centre) Circulating air; (Right) Liquid CO_2. Note the much finer crystalline texture (multiple nucleation, resulting in smaller crystallites), evident when cooling rate is increased.

For a given stretch ratio, orientation increases with increasing strain rate and molecular weight, and usually, orientation is also promoted when temperature is decreased (Figure 5.28). For biaxially-oriented containers such as the PVC bottles made by ESBM in Figure 5.29, a measurement of *birefringence* made through the plane of the wall indicates the degree of imbalance between the orientation in the axial and circumferential directions. The PVC bottles in Figure 5.32 illustrate how orientation and processing conditions can be characterized by polarized light: low birefringence represents relatively unoriented, optically isotropic material (Figure 5.32 A), whereas the much higher birefringence in Figure 5.32 (B) emphasizes the planar imbalance in orientation in the principle stretching directions, and a much greater level of planar orientation overall. Usually, the circumferential (hoop) stretch ratio and orientation is highest, to offer maximum resistance to the hoop stress component in pressurized containers.

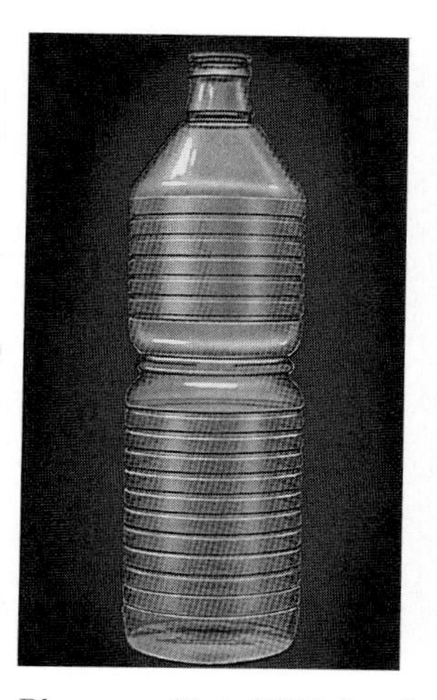 

Figure 5.32 Blow moulded PVC bottles viewed between crossed polars: (Left) Extrusion blow moulded, (melt phase deformation, resulting in low birefringence); (Right) Stretch blow moulded, (deformation above T_g, resulting in high orientation, characterized by high birefringence, (polarization colours) when the principal stretch ratios are different. (Reproduced with permission from Battenfeld Blasformtechnik, GmbH.)

Equivalent studies on PETP containers are made more complex because the material strain-crystallizes when highly oriented. In-the-plane birefringence data ($\triangle n_{XY}$) interpreted from polarized-light techniques are often negative: that is, the higher refractive index (and greater orientation) lies in the hoop direction (y-axis). This observation is consistent with expectation, since the process is designed such that the circumferential stretch ratio is the higher (see above). However, on reheating towards T_g, greater shrinkage is sometimes observed in the axial direction. Rather than implying that orientation is higher in this direction, the result emphasizes that the *crystallite orientation* is predominantly circumferential, and is more stable to subsequent increases in temperature. Thermal shrinkage is problematic in the bases and neck regions of PETP containers, where the local stretch ratio responsible for the (predominantly uniaxial) orientation is insufficient to develop significant crystallization.

It is clear that the interaction between material specification and processing conditions induces microstructural features which modify the physical properties of blow moulded products. These relationships are further explored, from the properties aspect, in subsequent chapters.

5.4 Rotational Moulding

As an alternative forming method to blow moulding for the fabrication of hollow (and often quite sizeable) components from thermoplastics, rotational moulding occupies a significant and increasingly-important part of the market-place due to a number of advantages which it holds over other methods. These include:

(1) Low tooling costs (rotational moulding is a low-pressure forming method) in comparison to moulds for equivalent-sized blow moulded or (especially) injection moulded products;
(2) Excellent and controllable thickness distribution in hollow components of complex external form, thus optimising the effectiveness of materials usage and contributing to an enhanced design-freedom;
(3) Moulding is carried out on a charge of material placed in-situ, and therefore no waste is produced.
(4) Roto-moulded products are relatively free of many of the microstructural defects (eg. orientation, residual stress, weld-lines) which are apparent in other competitive techniques.

Balanced against these points are the disadvantages that firstly, production cycles are inevitably quite long (the component is formed during consecutive heating and cooling cycles in the same mould) and secondly, much of the past experience of rotational moulding is confined to specific grades of a restricted group of plastics.

In the following sections, the process is first described, then the equipment design and operating principles are reviewed for some typical plastics materials. The specific properties of thermoplastics required for some physical aspects of processing are highlighted, and the structural characteristics of typical mouldings are discussed.

5.4.1 Production Techniques and Machine Cycle

The "low-pressure" character of rotational moulding arises because use is made of the ability of low-viscosity *plastisols,* or free-flowing *powder dispersions,* to be shaped easily under the force of gravity as the mould is rotated. However, like most other processes, the production cycle must be optimised, then carefully controlled in order to ensure product uniformity, optimum properties and economic manufacture. As will become evident in a later section, the rotational moulding industry has traditionally been based upon PVC plastisols *(Slush Moulding),* and latterly, on PE powders *(Rotocasting);* most aspects of this section are therefore dedicated to these areas.

Feeding. The charge of polymer is placed into the open split-mould, by manual or automatic means, using accurate dispensing units which are available for both powdered and liquid feedstocks of predetermined flow characteristics. There may be several individual moulds to be charged and processed simultaneously.

Heating. Having charged and subsequently closed the moulds, the entire mould assembly moves to a heating oven where the individual moulding tools absorb heat energy by convection from a heat source. Large scale production equipment uses propane or natural gas burners for this purpose, in conjunction with hot-air fan impellers. The heat is conducted through the wall of the mould, and is transferred to the polymer across the interface defined by the mould contours. Heating power is therefore an important machine specification: for example, the Orme company[5.59] quotes maximum heating capacities of between 1055 and 2637 MJ/Hr., for machines specified to carry between 360 and 800 Kg (per arm); usual power utilisation is between 60 and 70 %, and maximum internal air-temperature usually lies in the range 400 – 500 °C.

Mould rotation. Whilst being heated, the moulds are rotated simultaneously about perpendicular axes. Rotation speeds are modest (typically 25 RPM maximum)[5.59, 5.60], so that the feedstock is shaped by *gravitation,* rather than by centrifugal forces. The mould frame mechanism is geared so that the speed ratios about the primary and secondary axes (the arm speed and mould carrier speed, respectively - see below) can be adjusted to suit the shape of the product to be formed.

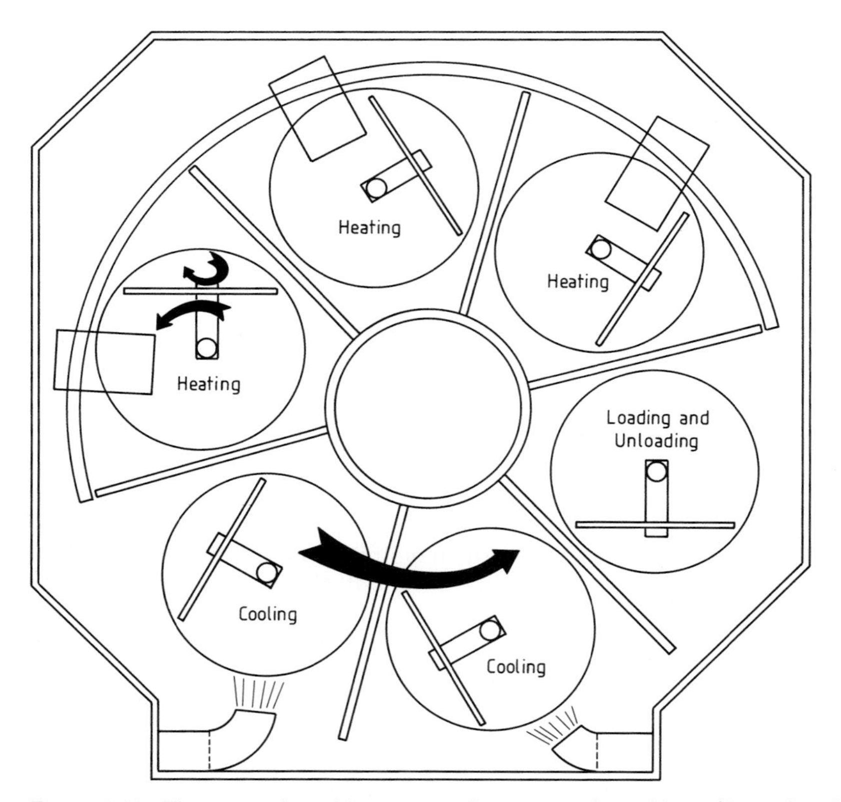

Figure 5.33 The carousel machine concept for rotational moulding. (Reproduced by permission from Orme Polymer Engineering, Northants, U.K.)

Cooling. Once the material has been homogenised into a continuous layer, the assembly is moved to a cooling chamber. Usual cooling media include cool-air jets and/or high-velocity water sprays. Significant convective air cooling should occur before the water-quench, in order to avoid inhomogeneous cooling and possible distortion in the product[5.61]. Cooling must continue until the internal surface temperature of the moulding is sufficiently low for it to be form-stable upon ejection. Mould rotation continues throughout the cooling cycle.

Ejection. At the end of the cycle, the moulds are opened and parts are removed manually or by vacuum-assisted automatic handling units.

Unlike blow moulding, where the material is heated in an extruder before being shaped and then cooled directly in the mould, rotational moulding is associated with more-substantial cycle times, due to the necessity of heating, shaping and hardening within the same mould unit. The problems related to heat input and extraction are exaggerated for thicker-section products. In consequence, most modern production units are based upon a *rotating carousel* system, with (typically) 3, 4 or 6 stations for heating/cooling, according to individual production time requirements. Charging and unloading are carried out at a single operator station. Figure 5.33 illustrates such an arrangement, and is a good example of rotational moulding systems used commercially.

5.4.2 Material Specification

The traditional, and still the most common plastics used in rotational moulding are PVC plastisols, and various types of PE powders (HD, MD, LD, LLD and crosslinkable grades). Other powder feedstocks include PP, PS, PA, PC and cellulosics; urethane plastics can also be moulded in solution.

PE powder properties have been reviewed by TOMO[5.62].Powder characterisation methods include reference to the following properties required by the rotational moulding process: particle size/distribution (300 – 500 μm is a typical range for ground PE), powder flow and bulk density. Polymer density is chosen by the desired levels of crystallinity and product property specification. Melt flow index is chosen to balance flow properties (low viscosity resins are a necessity, since heating time decreases as MFI is increased) with the physical properties of the product. Melt flow index of rotational moulding grades is generally similar to, or even higher than for injection moulding plastics. *Grinding* is now the popular route to powder manufacture: this additional preparative step is associated with a small cost supplement.

Products rotationally moulded from powders include automotive parts (arm rests, crash pads, fascias and tanks), road furniture (bollards, work-huts, trailers) and a multitude of industrial components such as storage tanks, drums, boats and flotation buoys. These are items of significant weight, requiring optimum thickness distributions, which are not produced in vast quantities.

Vinyl dispersions for rotational moulding are often custom-formulated; new markets or improved processing methodologies are easily explored on this basis. The plastisol should be sufficiently fluid (at low shear rates) to give an acceptable viscosity (2 – 3 Ns/m^2) to allow shaping to be completed before significant gelation occurs. PVC resins are therefore usually emulsion-grade fines of small particle-size (few microns) and resin-plasticiser ratios between 100 – 60 and 100 – 75 are acceptable. Typical formulations for applications examples are quoted elsewhere[5.63]. Vinyl goods with proven market success include flexible toys, footballs and water-resistant boots.

5.4.3 Heating and Melt Flow in Rotational Moulding

As is true in most other melt-shaping processes, the thermal properties and flow characteristics of polymers used in rotational moulding are fundamental to the production cycle. However, there are some aspects of the process which are significantly different to the engineering principles of competing fabrication techniques, requiring a different rationale for mould design and resin specification. Some of these points are reviewed briefly in this section.

5.4.3.1 Mould Heating and Melt Plasticisation

An analysis of the transient nature of the heating phase must be initiated with the assumption that, at zero time, the external face of the mould is at ambient temperature (T_0). THRONE[5.64] has developed a heating model for *forced-air convection,* which is derived by considering a heat energy balance: the change of internal heat energy (in the mould material) is equal to the convective heat flow from the gaseous environment (per unit time). Thus:

$$\varrho \cdot V \cdot C_P \, dT = h \cdot A \cdot (T_\infty - T) \, dt \tag{5.33}$$

(ϱ, V, C_P and A are density, volume, specific heat and surface area of the *mould,* h is the convective coefficient of heat transfer and T_∞ is the equilibrium oven temperature).

Using the following conditions for integration:

(1) $T = T_0$ at zero time ($t = 0$);
(2) $T = T$ at any time, t; we have:

$$\int_{T_0}^{T} \frac{1}{T_\infty - T} dT = \frac{h}{\varrho C_P L} \int_0^t t \, dt$$

where L, the mould thickness, represents the path length for conduction; ie. $L = \frac{V}{A}$. Integration, and rearrangement gives:

$$\ln \frac{T_\infty - T_0}{T_\infty - T} = \frac{ht}{\varrho C_P L}$$

Therefore the mould temperature, at any time t, can be estimated by:

$$T = T_\infty - (T_\infty - T_0) \cdot \exp\left(- \frac{h \alpha t}{LK} \right) \tag{5.34}$$

(having substituted thermal diffusivity, α, and conductivity K for ϱ and C_P, given that $\alpha = \frac{K}{\varrho \cdot C_P}$).

Equation 5.34 predicts an exponential rise in temperature, at a rate determined by the oven temperature, mould thickness and by the thermal properties of the metal from which the mould is constructed. Clearly, the rate at which heat can flow to the *moulding feedstock* is limited by the mechanical design of the mould. A similar approach can be used to calculate cooling times, if the appropriate heat transfer coefficients are utilised.

Further studies on heat transfer have been noted[5.65,5.66], in which the objectives were to update the existing processing models, and to provide experimental data to verify the theoretical basis of the earlier methodology. CRAWFORD and SCOTT[5.65] determined the optimum burner position, gas flow rate and type of flame for mould heating and showed that the rotational speed of the mould has little direct influence on heat transfer and subsequent cycle times.

Melting of powders has also been considered in some detail by RAO and THRONE[5.61]. Fine particle size promotes melting rate, but carries some associated preparative disadvantages during grinding. It may also be associated with the build-up of electrostatic charge and with particle agglomeration. The ideal particle shape, from the aspect of heat transfer and fluid flow, is rounded-corner cubic. Overall, the melting mechanism is complex and depends upon thermal diffusivity of the powder and the ability of particle surfaces to melt, adhere to the internal wall of the mould and form a molten film across which further heat transfer occurs.

5.4.3.2 Melt Flow in Rotational Moulds

Once a molten polymer film (of thickness t and density ϱ) has formed on the mould, the maximum (wall) shear stress for flow induced by gravity is given by:

$$\tau = \varrho g t \qquad (5.35)$$

It is soon evident that typical shear stress levels are very low, in comparison to many other commercial processes: for example, an ethylene plastic of melt density $800\,\text{Kg/m}^3$ will be sheared by just $23.5\ \text{N/m}^2$ if a film thickness of 3mm has developed. "Zero-shear" viscosity is therefore the relevant parameter with which to model flow behaviour in this regime, although it has been shown[5.61] that due to the absence of any significant flow over a large shear viscosity span, the velocity of melts is virtually independent of radial position, and therefore very little motion of molten particles occurs in this process.

In consequence, melting can be simulated by a *"motionless sintering"* mechanism, whereby particle coalescence occurs to produce an intermediate, 3-dimensional porous network, which densifies as processing time increases. Further analyses of the sintering mechanism and void/bubble promotion and elimination are given elsewhere[5.61,5.66]. Reference to the presence of voids in the structure of rotational mouldings is made in the following section.

5.4.4 Notes on Microstructure of Rotational Mouldings

Rotationally moulded parts are characterised by extremely good thickness distributions and a general absence of defects which are often induced in other processes by high shear forces or inhomogeneous cooling (orientation, residual stress). However, the demand to minimise cycle time imposes production constraints which can lead to imperfections in the products which are rarely apparent from an external inspection.

When processing PVC plastisols, the initial rise in temperature induces solvation of the PVC particles with plasticiser, and is associated with a drop in viscosity in the temperature range between ambient and 50 – 60 °C. It is within this phase that flow must be completed, to comply with the thickness distribution requirements prior to the onset of gelation. A steep rise in viscosity indicates the onset of gelation (between 50 and 100 °C), but optimum properties are only achieved when a continuous, fully-fused network has formed (150 – 180 °C [5.62]). If heating rates for thick-sectioned parts are too rapid, the possibilities of external decomposition, or incomplete gelation close to the internal surface, become possible.

A different type of microstructural facet exists in many products rotationally moulded from powdered feedstock, linked to the failure to eliminate the presence of *internal voids* which are formed during the shaping phase. The structure shown in Figure 5.34 is a chemically-etched section of a rotationally moulded tank in HDPE, and shows a multitude of internal voids, some of which are up to 500 μm in diameter. Crawford and Scott[5.66] have recently presented results from an experimental study of the mechanism of bubble formation/elimination in a rotational moulding grade MDPE powder. The following conclusions were made:

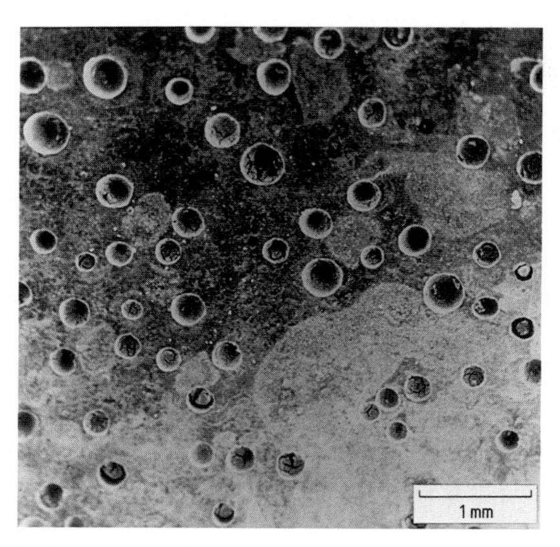

Figure 5.34 Etched section of a rotational moulding in HDPE, showing the presence of internal voids.

(1) Once melting and surface coalescence occurs, bubbles are formed due to air encapsulation, to an extent determined by the original particle size distribution, and resin molecular weight;
(2) Bubble size decreases with time due to a diffusion process; heating rate, temperature, bubble size and location are all parameters which influence the rate at which bubble size decays.

The polarised-light micrograph in Figure 5.35 shows the occurrence of *row nucleation* in rotationally moulded LDPE. Preferred crystallite nucleation of this type is inevitably heterogeneous in its formation, and is usually associated with imperfect distribution ("streaking") of compound additives, such as pigments, or fillers.

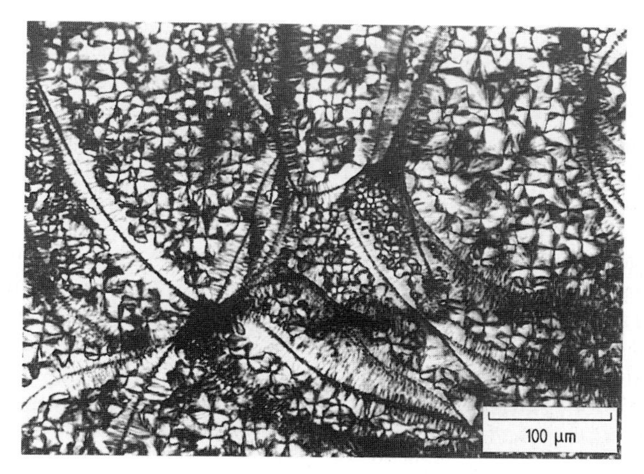

Figure 5.35 Polarised light micrograph of rotationally moulded LDPE, illustrating the phenomenon of row nucleation in semi-crystalline plastics.

5.5 Thermoforming Processes

The term "thermoforming" generally refers to a number of related processing techniques, whereby plastics sheets are heated to a temperature which is sufficiently high to induce a soft, compliant character, without introducing significant fluidity. Whilst in this *thermoelastic* state, the sheet is stretched by pneumatic (vacuum or compressed air) and/or mechanical forces to take the contours of a relatively cold, open mould. Heat is lost on contact with the moulding tool, so that the shape becomes form-stable as the thermoform temperature diminishes, and the material regains a significant degree of stiffness. Many of the most popular techniques use a vacuum as the driving force for sheet-stretching; the term *vacuum forming* has therefore become synonymous with products manufactured by a variety of methods of this basic type (see Section 5.5.1).

Product geometries of thermoformed goods are usually quite simple (profiled panels, trays, boxes and containers of relatively constant thickness without significant re-entrant sections), although deep-draws and low draft-angle containers may be produced in specific processes if forming conditions are optimised. Thermoforming is often considered as a competitive process to blow moulding and injection moulding, especially for relatively thin-wall, open-ended containers and vessels; the main advantages of this process are:

(1) The low capital cost of the thermoforming production unit, especially the moulding tool. This is entirely due to the low-pressure nature of shaping, relatively low forming temperatures and the use of an open mould;
(2) Large-area, thin-section parts can be shaped easily and quickly;
(3) Multi-impression and family moulds can be utilised to optimise the number of units produced per cycle.

Balanced against these attractions are some disadvantages:

(1) Product-shape ranges are limited;
(2) An even temperature distribution in the sheet is required but simultaneously, attempts must be made to avoid gravimetric (vertical) sheet deformation prior to forming;
(3) Stretching may induce progressive thinning, causing unacceptable thickness distributions (especially in deep-drawn containers) and excessive levels of molecular orientation;
(4) If the thermoforming unit is part of an integrated extrusion line (complete with sheet-die and chill-rolls), capital cost becomes very high. Alternatively, if the sheetstock is bought from the trade, a significant price premium over raw material cost is generally paid.
(5) The deformation on forming is predominantly elastic; although this is form-stable at relatively low temperature, subsequent heating close to a transition temperature (T_g) results in elastic strain recovery, exemplified by high levels of shrinkage. Thermoformed products are therefore associated with some strict limitations in service temperature.

5.5.1 Summary of Thermoforming Production Methods

Many different manufacturing routes are possible in thermoforming, according to the mode of sheet heating, the stretching mechanism (vacuum, compressed air, mechanical (plug-assisted) prestretching) and the sequencing of individual phases. These are used to obtain the best balances of productivity, thickness distribution and the desired physical properties in the component.

The most fundamental of the various processes are illustrated in Figures 5.36–5.39 and are summarised below.

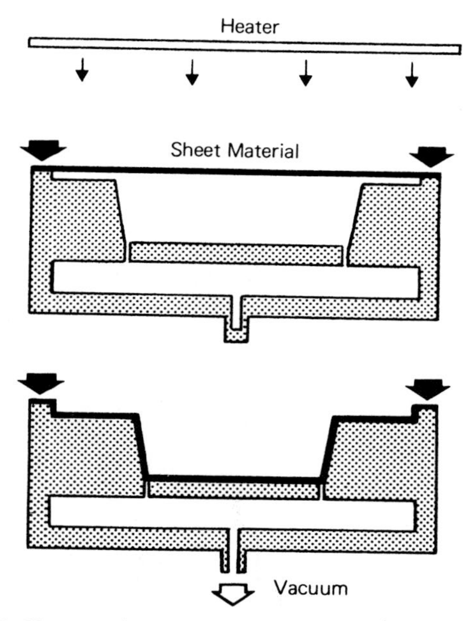

Figure 5.36 Negative (female) vacuum forming process.

(A) Negative forming (Figure 5.36). Sheet deformation is achieved by clamping to the mould top-surface, then evacuating the air-space between the sheet and the mould base; forming pressures are therefore limited to a single atmosphere (0.1 MN/m^2). This process is rapid and reliable, but is limited by excessive thinning and is restricted to simple shapes such as shallow-drawn trays.

(B) Plug-assisted negative forming (Figure 5.37). A matching, pneumatically-activated plug contacts the sheet and initiates the forming process, which is completed by the application of a vacuum during the descent of the prestretching device. Since heat is lost from the sheet to the lower surface of the plug, the degree of thinning, in various positions, can be controlled more exactly than by direct vacuum forming. By being able to adjust plug temperature, descent rate and surface friction, greater production flexibility can be achieved; this process is used for optimising thickness distributions in relatively deep-drawn containers (drinks cups, margarine tubs).

(C) Positive forming (Figure 5.38). A male mould ascends, makes contact with and prestretches the heated sheet, before the vacuum is applied in the usual mode.

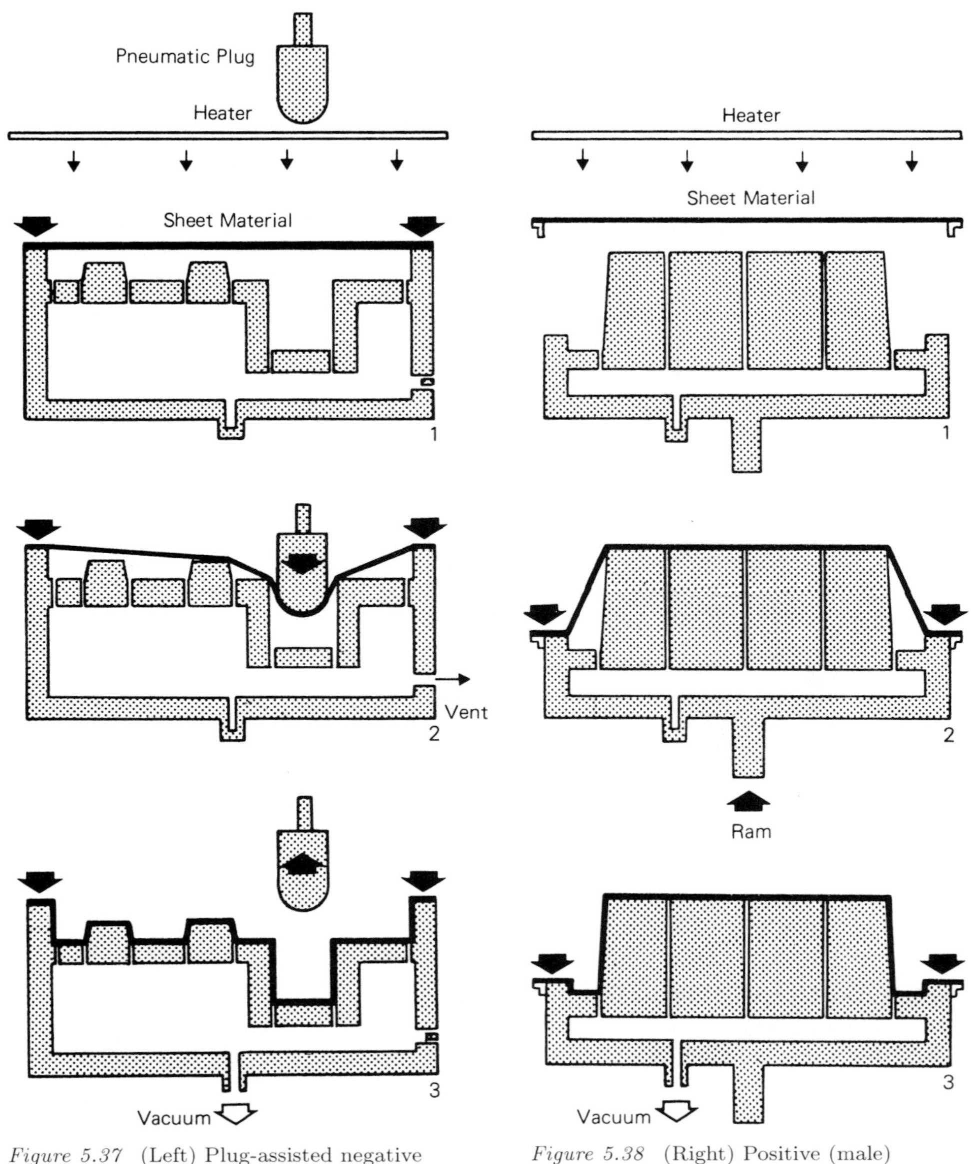

Figure 5.37 (Left) Plug-assisted negative (female) vacuum forming

Figure 5.38 (Right) Positive (male) vacuum forming

Although similar in concept to the technique described above, the typical thickness distribution is different to that obtained from negative forming, and more complex parts (severe surface-profile changes and gentle undercuts) can be produced (eg. chocolate-box trays).

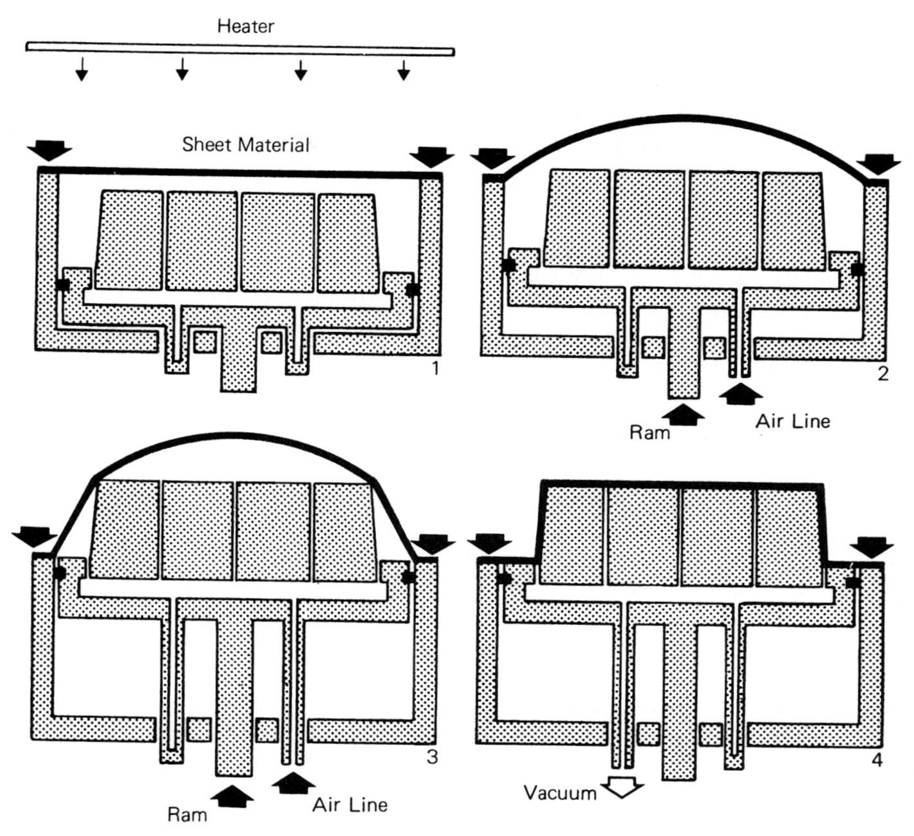

Figure 5.39 Air-slip forming. (Figures 5.36 – 5.39 have been reproduced by permission from ICI Chemicals & Polymers, Welwyn, UK.)

(D) Air-slip forming (Figure 5.39). In this process, the sheet is first expanded biaxially by positive air-pressure. Then, the male mould ascends and the vacuum is applied to complete the shaping operation. Since the end result is an excellent material distribution, this method is used to minimise material requirements for large components such as refrigerator liners.

More detailed discussion is provided on these, and some more advanced forming techniques, in other, more dedicated texts[5.46, 5.67–5.68]. Some of the important sequences during thermoforming (radiant heating, stretching and analysis of thickness distributions) are considered further, in this chapter.

5.5.2 Thermoforming Materials and Product Applications

Conflicting requirements are evident for successful plastics thermoforming: the heated sheet must be sufficiently *compliant* to be shaped easily and completely (usually under low stress), yet must retain an inherent degree of *shape stability* to avoid excessive viscous flow and sheet sagging during the pre-heat period.

The choice of optimum thermoforming temperature is therefore critical to successful manufacture; it follows that the best plastics are those which exhibit a predominantly *elastic* response to applied stress, over a wide *softening* range. It transpires that amorphous plastics are especially suitable for thermoforming in the region just above T_g, where the processing window is generally wider than for crystalline materials. Forming temperatures should be at the "lowest temperature at which an acceptable shape can be obtained using the available shaping force"[5.68], and are usually between 30 and 60 °C above T_g for most amorphous plastics, and more immediately above T_m in the case of semi-crystalline materials. Commodity sheet materials include PS, ABS, PVC, PMMA, PETP and PC, which are all amorphous.

Semi-crystalline plastics such as polyethylenes and nucleated PETP may also be vacuum-formed. According to BRYDSON[5.69], high molecular weight polyolefins, or branched resins of somewhat lower molecular weight, offer the best thermoforming characteristics to comply with the criteria laid down in the previous paragraph. Sagging is avoided when forming into crystalline PETP at high temperature, by initiating and controlling the development of crystal nuclei in the amorphous sheet during the heating stage.

Typical applications for thermoformed products include:

(1) Packaging; frozen foods, sweets and confectionery trays, egg-boxes, yoghourt/margarine tubs, ovenable cookware trays, cosmetics boxes, beakers and base-cups for PETP bottles;
(2) Semi-structural items; baths and shower-base trays (foam-filled), refrigerator liners, fascia panels and boat/sailboard hull components.

5.5.3 Analysis of Sheet Heating

Although conductive and convective modes of heating plastics sheets for thermoforming are feasible (the former by low-pressure contact with metal heater-platens, and the latter by rack-heating in circulating-air ovens), the preferred heating method is to use the absorption characteristics of plastics to *radiant* heat energy. This requires some prior knowledge on the radiant absorption spectra of relevant polymers; there is evidence that pigmentation and molecular orientation can modify the absorption characteristics of plastics sheets[5.70]. The emission of radiative heat energy is generally in the wavelength range 1.5 – 7.5 μm, chosen according to the absorption characteristics of the sheets[5.71].

THRONE[5.70] has described a simple technique to estimate the radiant heat transfer characteristics of plastics sheets during thermoforming. This is based upon a graphical solution to the transient-state heat-conduction equation (see also Section 5.2.4) which can also be used for convection heat transfer (see below) or indeed, for radiation heating if the relevant "heat transfer coefficient" can be determined.

Consider the data in Figure 5.40, where dimensionless temperature (θ) is plotted against dimensionless time (Fourier number, Fo) for a sheet of thickness $2x$, at $x = 0$ (ie. the mid-plane of the sheet). For *convection heating,* the Biot Number (Bi) determines the relationship between θ and Fo (Figure 5.40), and is defined by THRONE[5.70] as:

$$Bi = \frac{K}{hx} \tag{5.36}$$

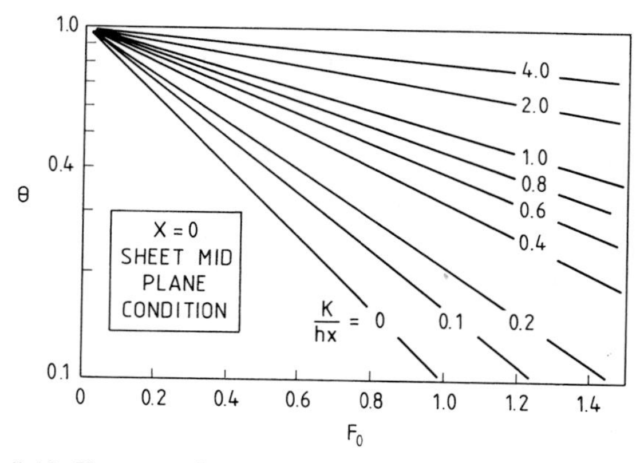

Figure 5.40 Heat transfer data for plastics sheets, showing temperature difference ratio (θ) versus Fourier Number (F_0) for various conditions (after THRONE[5.70]).

(K = thermal conductivity, h = convection heat transfer coefficient and x is half the sheet thickness)

In forced air convection, h usually lies between 10 and 50 W/m^{2}K. By knowing the heater and sheet temperatures and the thermal properties of the polymer, the time taken for the centre-plane to reach a desired forming temperature can be estimated by reading Fo from the appropriate Biot Number locus. (Note that the representation of Bi=0, where $h \to \infty$, represents *conduction* heat transfer). Similar heat transfer data can be presented for other positions in the sheet (eg. surface) and an identical technique used to calculate heating time[5.70].

For *radiant heating,* the analysis is complicated by surface reflection and the effects of air convection between the heater-bank and the sheet. However, a *"composite heat transfer coefficient"* (h') can be defined for radiation, to be used in a similar mode to that described above for convection[5.72]:

$$h' = h_c + h_r \tag{5.37}$$

h_c is the convection coefficient, and h_r is an artificial, but assumed-'equivalent' coefficient for radiant heat transfer, given by:

$$h_r = (1-r)e\sigma\frac{T_h^4 - T^4}{T_h - T} = (1-r)e \cdot F_T \tag{5.38}$$

(e = emissivity, which is unity for black bodies and 0.9 – 0.95 for most plastics; σ = Stefan-Boltzmann constant; r = reflectivity at surface (often assumed zero); T_h/T are the heater/sheet temperatures; F_T = "radiation correction factor", which is usually in the range 20 – 100 W/m^{2}K and has been presented graphically, as a function of T_h and T, by THRONE[5.70]).

Coefficient h' can then be used to obtain an effective Biot Number for radiation, and graphical heat transfer solutions can be used to generate heat transfer data at various positions within the sheet. As is often evident, h' values can be significantly higher than for convection, so that if the heater temperature is very high (300 – 700 °C, depending upon the element) the surface temperature of the sheet can itself become excessive, during the heating time required to raise the centre-plane temperature to the thermoforming range. THRONE[5.70] cites other workers who have suggested various ways of lessening the severity of thermal gradients in the sheet: these include surface shielding, rapid "on-off" cycling and forced-air surface cooling.

Heaters are usually arranged in matrix formation about each surface of the sheet; each element usually has individual power control settings. Using this approach, different temperatures can be promoted in various parts of the sheet (*"zone shading"*), in order to induce preferential stretching for optimum properties and thickness distribution in the thermoformed part.

5.5.4 Stretching and Product Thickness Distribution

Effective material usage is always an important prerequisite of product design procedures. This is especially true in the case of high-volume applications such as packaging, where products made by thermoforming techniques occupy an important sector of the industry. The desire to achieve the optimum *thickness distribution* in vacuum formed components is therefore crucial to the feasibility of successful manufacture. Some aspects of thickness distribution in thermoformed products were briefly considered in the introduction to Section 5.5, and are more comprehensively reviewed elsewhere[5.67 – 5.68]; clearly, the best manufacturing process for a given product depends upon the locations in the component where significant thickness is required for purposes of mechanical strength.

When a softened sheet is sucked into an open mould, or (in air-slip forming, for example) when a positive pressure is applied to one side of the sheet, the deformation process can be described as *free-surface biaxial extension.* Since this deformation (especially for most amorphous materials stretched in the temperature region just above T_g), is predominantly *elastic,* simple stress-strain relationships can often be used to simulate the drawing behaviour of plastics in this process. In practice, a 2-dimensional grid of predetermined spacing can be drawn on to the surface of an undrawn sheet, then remeasured to investigate the degree of deformation (as a function of position) during the forming process: see Figure 5.41.

If constant-volume deformation is assumed, the extension ratio in the thickness dimension (λ_z) is expressed as a function of the other principle extension ratios by:

$$\lambda_z = \frac{1}{\lambda_x \cdot \lambda_y} \tag{5.39}$$

SCHMIDT and CARLEY[5.73] have compared theoretical and practical data on inflated bubbles of various thermoplastics. Since principle extension ratios λ_x, λ_y are very location-sensitive, thickness was also shown to vary according to position on the part; the agreement between the practical and theoretical data confirms the predominance of elasticity during high-strain free-forming.

At temperatures just exceeding the glass transition, amorphous plastics exhibit non-linear drawing behaviour, which can be simulated by the statistical theory of rubber elasticity[5.74]. Following the proposal by SMITH[5.75], that the flow stress of rubber-like systems is separable into strain and time effects, LAI and HOLT[5.76 – 5.77] applied the deformation model to the free vacuum forming characteristics of

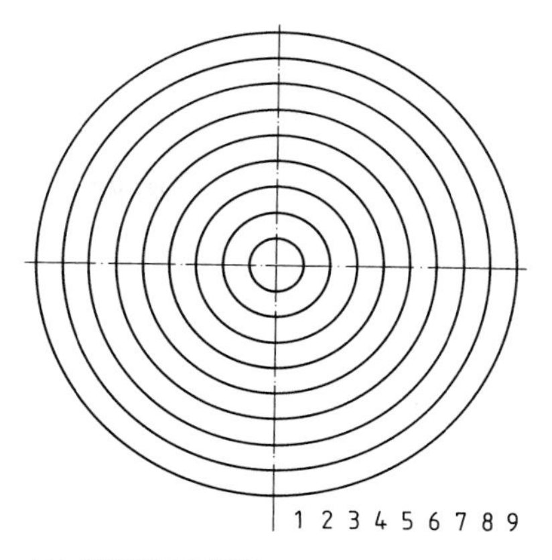

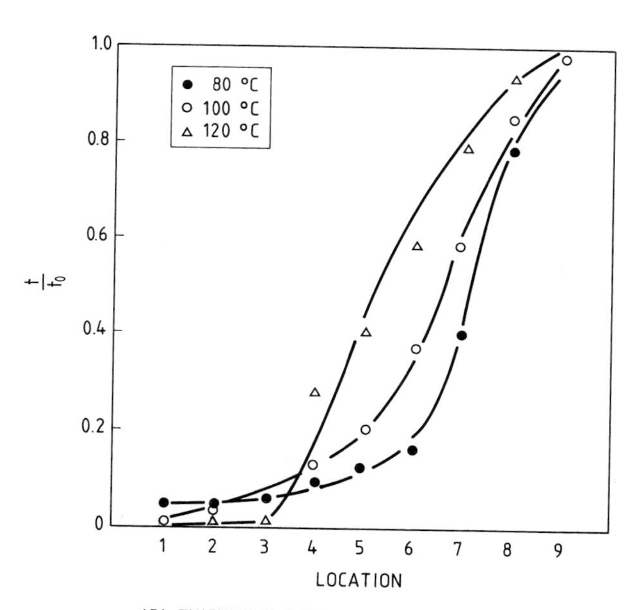

Figure 5.41 Experimental analysis of thickness distribution in vacuum formed plastics: (A) Grid definition; (B) Thickness reduction $\frac{t}{t_O}$ versus grid position for amorphous PETP, showing the effect of sheet temperature.

thermoplastics, to show the dependence of true stress (σ_T) on true strain (ε_T) and time (t) by:

$$\sigma_T = K \cdot t^m \cdot (\varepsilon_T)^n \tag{5.40}$$

(K is a constant; exponent m, the *stress relaxation index,* describes simultaneous relaxation effects during forming, whereas n, the *strain hardening index,* reflects the tendency towards chain-stiffening at high strain).

Some characteristic exponents for thermoforming materials are shown in Table 5.4. A large (negative) value of m, and/or a small strain hardening index detract from thickness uniformity in thermoformed parts; this assumption was validated, comparing PMMA and HIPS, by some practical trials reported by LAI and HOLT[5.77]. The large strain hardening indices for PETP reflect the tendency of this material to *strain-crystallise,* a process which is sensitive to temperature.

Table 5.4 Deformation Data for Thermoformable Plastics

Material	Temperature °C	*S.R.Index* (m)	S.H. Index (n)
PMMA	165	– 0.05	1.0
HIPS	122	– 0.33	1.1
PETP	80	– 0.13	7.7
PETP	90	– 0.44	7.0
PETP	100	– 0.23	3.3
PETP	110	– 0.29	3.1

For practical vacuum-forming processes however, there are some factors which require a more rigorous approach to determine thickness distribution. First, when any deforming element of a heated sheet makes contact with the mould surface, heat is immediately lost (local compliance diminishes) and further deformation is mainly restricted to that part of the sheet which has not contacted the mould. In conventional vacuum forming therefore, wall thickness decreases with distance from the clamping mechanism and assumes its minimum value at the intersection of the base and the centre-axis. A second, but similar heat-transfer effect also occurs if a plug-assisted technique is used: in this case, the container base freezes on contacting the plug surface, and the sides are progressively thinned as forming continues.

Extending the principles first proposed by SHERYSHEV et al[5.78 – 5.79], TADMOR and co-workers[5.80 – 5.81] have analysed the free-forming process to predict thickness distributions in vacuum-formed parts on the basis of the progressive thinning effect described above. Assumptions made include wholly-elastic and isothermal deformation, incompressibility (constant volume), uniform thickness in the free bubble and no slip on the mould surfaces. Theoretical predictions of wall thickness, as a function of position, are given in Reference 5.81 for various mould shapes, including conical, shallow or deep-truncated moulds. For example, in a conical mould, the thickness t at any position (see Figure 5.42) is given by:

$$t = t_0 \cdot \frac{1+\cos\beta}{2} \cdot \left(\frac{H - Z\sin\beta}{H}\right)^{\mathrm{Sec}\,\beta - 1} \tag{5.41}$$

(where t_0 is the initial sheet thickness, β is the cone angle to the vertical; see Figure 5.42 for other geometric definitions.)

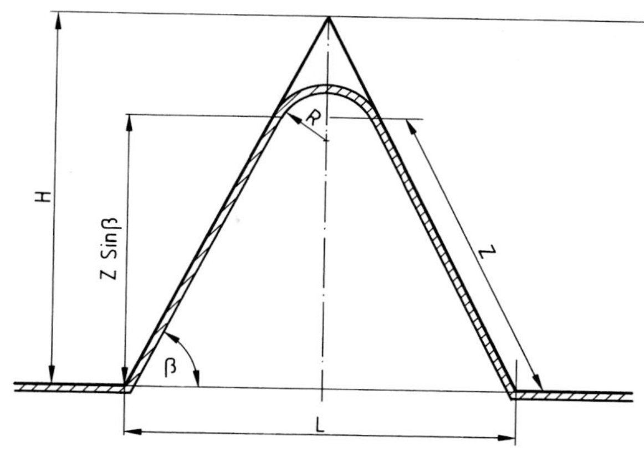

Figure 5.42 Definition of geometry for thickness distribution analysis in a conical vacuum formed component (after ROSENZWEIG et al.; Reference 5.81).

5.5.5 Orientation and Heat-reversion

Thermoforming techniques are restricted to fairly simple shapes. Consequently, the mechanical properties of vacuum-formed products cannot be improved by shape-refinement. For example, component stiffness cannot be improved by rib incorporation or by any other such design methods typical of injection moulded products; therefore, the stuctural properties such as molecular orientation introduced during biaxial stretching represent the key to understanding whether the desired levels of specified physical properties are achieved.

It is recognised that elasticity represents the predominant mode of deformation in thermoforming, over typical spans of strain rate and temperature. Assuming that local cooling rates are sufficiently rapid to suppress significant relaxation, the temperature descent through T_g effectively locks the molecular alignment into the morphology of the solid, thermoformed part. The molecular orientation in vacuum-formed components is therefore dependent upon the magnitude and direction of stretching: it is usually *biaxial* (although not necessarily equi-biaxial) and is highest in regions which are most highly-strained (and therefore thinnest) during the shaping stage. In terms of in-line strength and stiffness, high levels of molecular orientation induced by large strain compensate for the load-bearing shortcomings caused by gauge-reduction.

The dominance of elastic deformation may be ascertained by releasing the driving stress during shaping (by rupturing the developing thermoform), or by reheating the part to a temperature higher than T_g, which is usually sufficient to allow molecular recoiling to occur. It will be observed that this "memory" effect results in virtually complete reattainment of the original sheet gauge dimensions.

The tendency towards elastic recovery (which results in part shrinkage, and distortion) at elevated temperatures limits the practical use of many thermoformed containers to applications where significant in-service temperature rises are not envisaged. Figure 5.43 illustrates how the degree of shrinkage (in this case exemplified by a change in container volume, $\triangle V$) depends upon thermoforming conditions, for PETP. Dimensional stability appears to be enhanced at low sheet forming temperature (where strain induced crystallinity is enhanced), in combination with an elevated tool temperature, which allows simultaneous relaxation of orientation to occur in the amorphous phase[5.82].

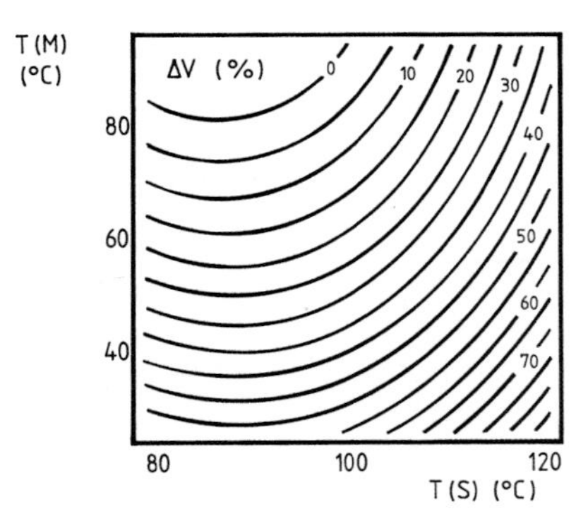

Figure 5.43 Two-variable contour plot showing the dependence of PETP dimensional stability on sheet temperature T(S) and mould temperature T(M) in vacuum forming to a draw ratio of 2 : 1. ΔV represents the degree of volumetric shrinkage following reheating to 80 °C.

5.6 Processing Methods for Fibre-Reinforced Plastics

There are numerous techniques available for the manufacture of thermoset plastics components; many of these have principles in common with thermoplastics processes (eg. injection moulding), but others have been developed to satisfy a particular market need. In this final section, we concentrate on the manufacturing techniques which are available for *fibre-reinforced thermosets,* where use is made of the low viscosity characteristic of *liquid resins* in order to *impregnate* and adhere to fibrous reinforcement materials by *lamination,* or by other moulding techniques.

(A) Manufacturing constraints

The process selected for the manufacture of fibre-reinforced plastics (FRP) depends upon a number of factors, which include:

- component numbers required
- item complexity
- number of moulded surfaces
- type and amount of reinforcement

These points are highlighted throughout the discussions on individual processes which follow.

(B) Materials

Resins – There is a wide range of polymer types available, but the most common matrix materials used in practice are thermosetting plastics such as unsaturated polyesters, epoxies, phenolics and polyurethanes. Composites based on polyurethane are considered elsewhere[5.83]. Thermoplastic materials (typically PP, PA and other "engineering" resins) have traditionally been reinforced with modest loadings of short fibres, to comply with the flow requirements in injection moulding; hence short-fibre reinforced thermoplastics, (SFRTP). More recently, however, thermoplastic materials such as PEEK have been utilised as impregnation materials for continuous reinforcement mat, to be processed subsequently by "heat forming", rather than by moulding.

Reinforcement – This is usually based on fibres such as glass, carbon or aromatic polyamide (aramid), which are available in various forms: continuous filament, continuous roving, chopped strands and in fabricated forms: chopped strand mat and woven or knitted goods[5.84]. Other materials may be added to modify certain properties of reinforced plastics composites, such as flame retardants, inert fillers and pigments.

5.6.1 Hand-lay, or Contact Moulding

In this process, low-viscosity liquid resin and reinforcement are placed in an open mould and are consolidated by brush or roller. The reinforcement is usually glass fibre, in the form of a *random mat,* which is classified by fibre length and weight per unit area[5.85]; the length of the fibre in the mat is typically 50 – 100mm.

The various steps in the process are:

(1) Coating. The mould is coated with a proprietary *release agent* to prevent the item sticking to the mould surface (the adhesion between metallic surfaces and crosslinkable resins is generally very effective);

(2) Gel Coat Application. Gel coat resin is usually applied by painting or spraying to a thickness of 0.5 – 0.6 mm. This is a pigmented resin (usually chemically-similar to the laminating material) which will give the required quality of finish to the component, provide a degree of environmental protection (ie. prevent the ingress of liquids or vapours which might contribute to fibre-matrix debonding) and prevent the fibre pattern showing through the external part surface;

(3) Resin Application. The laminating resin and reinforcement are worked into the surface by roller and brush, to the required thickness. At this stage, the objective is to ensure that all the reinforcing fibres are fully coated (*"wetted-out"*) by the resin, yet equally important, the overall resin-fibre ratio must be controlled, such that the mechanical properties of the composite material are optimised (see also Section 6.6.2). Expulsion of air is also an important requirement;

(4) Crosslinking. The item is allowed to cure before being removed from the mould. *Curing* (or *crosslinking)* is usually carried out at room temperature in most commercial operations and may continue for several hours, depending on the size and complexity of the item. The *cure rate* can be adjusted by controlling the resin chemical formulation ie. increases in the catalyst and accelerator levels will enhance the degree of crosslinking, and also the rate of cure. The component will continue to cure following removal from the mould, for a time interval of up to 7 days. This process can be accelerated by carrying out a *post-cure* for several hours (typically 8) at about 70 °C.

Inexpensive tooling is used in this process (for example, reinforced polyester or epoxy/sand), thereby making it suitable for short production runs, and the fabrication of large items. Contact moulding offers a great deal of design flexibility, since inserts, undercuts etc. can also be included. Capital costs are low, but it is necessary to provide extraction equipment in accordance with national legislation.

The main disadvantages of the process are that first, it is very operator-dependent, and second, there are difficulties associated with the desire to achieve high amounts of reinforcement. However, *local reinforcement* using continuous fibre or pultrusions is possible. Contact moulding is suitable only for short production runs, but has been used successfully in boat-building and in other transport industries, and also for architectural components.

5.6.2 Spray Moulding

This is a modest development of the hand-lay process, where the resin and glass are deposited simultaneously on the moulding tool[5.85]. Glass fibres in the form of a *continuous roving* are chopped to the required length by a rotating cutter and are mixed with the resin at the spray-head before being deposited on the mould surface. The fibre length is usually about 20 to 30mm. Subsequent consolidation of the laminate is achieved by rolling, in a similar manner to the hand-lay process.

Spray moulding is much faster than hand-lay, and utilises the same type of cheap tooling; this method is particularly suitable for large components. However, capital costs are higher, and there tends to be a larger variation in the mechanical properties compared with hand-lay techniques. Once again, it is very operator-sensitive, but does lend itself to a team approach where one operator sprays and others can consolidate the laminate.

5.6.3 Cold Press Moulding

The previous processes are associated with some major limitations:

- only a single moulded surface is produced;
- the techniques are relatively slow and are operator-dependent.

However, *cold press moulding* overcomes these limitations to some extent since it is a closed mould technique which is carried out at low pressure and at ambient temperature. Glass fibre (in the form of a *continuous filament mat*) is placed in a tool, and the thermosetting resin is poured in. The mould is closed and the resin spreads through the reinforcement, thereby *impregnating* and fully wetting-out the glass fibre strands.

Production rates are much higher than in the previous methods, and cycle times of 20 minutes can be easily achieved. Cold press moulding is particularly suitable for production runs in the range 1000 to 20,000 units. For relatively short runs, it is possible to use inexpensive forms of tooling (see Section 5.6.1). However, for longer runs, nickel faced tools produced by electroforming are preferred. The process is based on a low-cost hydraulic press, which can be modularised for large components. Pressure levels required for moulding are typically around 10^5 Pa. It is possible to mould large items, up to a flat area of about 25 m^2. Gel coat can be applied to the mould faces before the moulding operation commences, or alternatively, the item can be painted subsequently.

5.6.4 Hot Press Moulding

A premixed *sheet* or *dough* material containing resin, fillers and fibrous reinforcement is inserted between matched metal tools and the item is *compression moulded* at elevated temperature and high pressure[5.86].

5.6.4.1 Thermoset Moulding Compounds

There are several different types of moulding compounds used in hot-press moulding, as detailed below:

- **Dough Moulding Compound (DMC)** – This material is supplied to the moulder as a "dough", which has been precompounded in a Z-blade, or similar mixer[5.87]. A typical formulation might contain 15 to 20 % (by weight) of glass fibres of length 6 – 12 mm, polyester resin (25 %) and inert fillers (55 – 60 %). Great care must be exercised during the mixing stage, in order to ensure that whilst an effective mixture is obtained, the shear history is not excessively severe, which would otherwise induce *fibre attrition,* thereby reducing the average fibre length, which would diminish the reinforcing characteristics.

- **Sheet Moulding Compound (SMC)** – The moulding compound is produced in sheet form where the glass fibre is sandwiched between two resin layers, and the whole structure is then homogenised by "kneading" between specially contoured rolls. The sheet is produced continuously and is supplied to the moulder between two thermoplastic carrier films to prevent adhesion and contamination. SMC contains a much higher loading of glass fibre (30 % by weight) and longer fibre length; thus it usually offers better mechanical properties than DMC.

- **Bulk Moulding Compound (BMC)** – BMC is similar in concept and formulation to DMC; it generally consists of thermosetting polyester resin, mineral filler and chopped glass fibre strands of up to 12mm length. Other additives (pigments, catalyst etc.) are often included in such formulations, according to application and production requirements.

Hydraulic presses are generally used in hot press moulding: these must be capable of applying sufficient force to create pressures of between 3.5 and 15 MPa. Final closure of the press during the cycle must be slow to allow the compound to flow in a uniform front at a rate which will allow air to escape. If closure is too slow however, the compound may have a tendency to *premature curing.* In practice, a compromise is achieved and moulding conditions are optimised by a combination of experience and processing trials.

Hot press moulding can produce very fast cycle times but capital costs for both the press and the tools are relatively high; for this reason, it tends to be used for long production runs, in which the rapid cycling characteristic may be used to full advantage. SMC products tend to be large-area, rather planar components with good mechanical properties. DMC is a more versatile material which, because of the relatively short length of the fibres, can be injection moulded, thereby offering the designer the potential for quite complex shapes. However, shorter fibre length is also associated with inferior mechanical properties, in comparison to SMC.

It is clear that the optimum choice of process depends not only on materials and property requirements, but also upon direct running costs and the capital expenditure associated with the machine and tool investments. These issues are illustrated first in Figure 5.44, where the relatively cheap hand-lay technique is often economically preferable if the total production requirements are modest. The

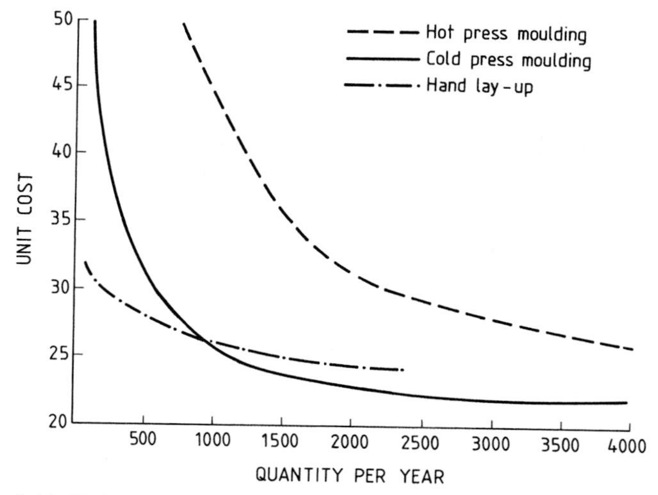

Figure 5.44 Unit cost comparisons for the manufacture of reinforced plastics composites by various routes.

greater capital outlay associated with hot press moulding (see also Table 5.5) can only be justified if the anticipated production rates are significantly higher.

Table 5.5 Economic Aspects of Thermoset Moulding Processes

Process	Mould cost	Daily output	Mould life
Hand Lay-Up	1	3	200
Cold Press Moulding	13	30	15 000
SMC Moulding	115	80	100 000

5.6.4.2 Microwave Heating of Sheet Moulding Compound

Preheating of SMC is an extension of a technique which has been used in the rubber industry to prepare thick charges for efficient moulding. *Microwave preheating* reduces thermal gradients in the mould and improves flow with no major disadvantages. Preheating provides a charge of uniform temperature which is capable of uniform, continuous flow, leading to reduced fibre orientation and more homogeneous cure. In turn, this inevitably produces improved mechanical properties; also, preheating reduces the overall cycle time and energy input, with obvious economic benefits.

Applications of *dielectric heating* in polymer processing were reviewed in 1980 by SIADAT[5.88]; many benefits were listed, most of which were referred to in the previous paragraph. In view of this knowledge and the related experience, it seems exceptional that so little work on SMC preheating has been reported.

Thermal gradients in SMC in a mould cause flow differentials, leading to different rates of cure and the setting up of thermal stresses. Thermal and velocity gradients at the mould surface were described by LEE et al[5.89], whilst NEWMAN et al[5.90] were the first to observe the benefit of more homogeneous flow, consequent on preheating. MALLICK et al[5.91] preheated the charges in a hot-air oven; however, this does not promote a uniform temperature gradient throughout the material, and thereby misses some of the main advantages of preheating.

COSTIGAN and BIRLEY[5.92] have reported the results of an experimental programme carried out on a microprocessor controlled servohydraulic press, the development of which is covered in papers by KANAGENDRA, FISHER and CHAPMAN[5.93 – 5.95]. The instrumentation associated with this press allows reliable measurements to be made of critical variables during controlled operation of the press; these include in-cavity measurements of temperature and pressure, ram velocity, mould platen displacement and separation, as a function of time.

A comparison of flow characteristics resulting from standard or preheated charges has been made by capturing the movement on a video recorder; the preheated charges flow more smoothly with a relatively flat flow front, compared with the disintegrating front (with "fibre tumbling") observed on compressing the standard charge[5.92].

Whilst obviously preheating is an additional process, requiring time (for heating), equipment and manpower, it offers some advantage in cycle time reductions on the expensive press, but above all, it gives mouldings with improved, and more consistent mechanical properties. This is summarised in Table 5.6, the data for which are taken from the research of COSTIGAN and BIRLEY[5.92]:

Table 5.6 SMC Moulding - Effects of Preheating

Charge Temperature (°C)	20	50	75
Preheat Time (s)	0	30	45
Time to Peak Exotherm (s)	67	59	55
Cure Time (s)	90	74	69
Ultimate Tensile Strength (MPa)	64	74	77

Additional advantages accrue from the lower viscosity of the preheated charge which can then be processed more easily on a small press; the lower forces can also lead to improved parallelism of the platens. Furthermore, products of greater thickness can be moulded satisfactorily.

5.6.5 Resin Transfer Moulding

In this process, resin is injected into a closed mould containing the reinforcement preform. The resin can either be injected under pressure[5.96], or introduced under vacuum[5.97]. It represents a relatively new technique for fibre reinforced materials but has been identified as one of the most suitable processes for mass production of large structural items. The potential advantages of this process are:

- low mould costs, in comparison to SMC and DMC;
- inserts can be incorporated;
- low pressure requirements;

- accurate fibre placement, and opportunity for selective reinforcement;
- automation possibilities;
- versatile technique.

The main disadvantage however, is the need for a separate, labour intensive operation to produce the preform.

5.6.6 Prepreg Moulding

Fibre is impregnated with resin, usually polyester or epoxy, to produce a flexible *prepreg* sheet containing a high volume fraction of fibre[5.85]. The reinforcement will be in the form of continuous fibre, or woven cloth. The prepreg sheets can be assembled in a mould and compression moulded at elevated temperature using matched metal tools. An alternative method uses a flexible bag to consolidate the prepreg preform. A pressure differential is created by vacuum, which can be augmented by positive pressure in a heated autoclave; the use of vacuum alone does not give such a well-consolidated item.

The *vacuum-bag* technique is particularly suited for prototype components and for limited production runs of large surface area items. Very high volume fractions (of about 60 %) of reinforcement can be achieved with good control of fibre orientation and excellent mechanical properties.

5.6.7 Filament Winding

This process is useful for the manufacture of axisymmetric hollow components which require a high volume fraction of fibre with good control of *fibre orientation*[5.98–5.99]. Continuous fibre is passed through a resin bath and then wound around a rotating mandrel. Precise placement of the fibre is achieved by controlling the speed of both the *winding* and *traversing* operations. For the production of pipe sections, the mandrel is slightly tapered to allow it to be removed easily at the end of the operation. For hollow items such as pressure vessels, a collapsible mandrel with segmented sections is used.

Filament wound composite components such as pressure vessels (pipes; storage tanks) and freight tanker bodies have excellent mechanical properties and can be produced in a wide range of sizes up to 2.5 metres in "diameter" and up to 17 metres long. Fibre placement is fast, accurate and reproducible and the process is very cost effective compared with hand-lay or prepreg moulding. However, the technique is limited to *axisymmetric components*, although novel methods are being developed to provide a wider range of shapes[5.100]. Process design must allow for mandrel removal and the external surface of the component is generally quite rough, which may be unacceptable in some applications.

The various techniques used for the manufacture of fibre-reinforced thermosetting plastics are extensive, and as we have seen, the technologies associated with them (and the properties derived from these techniques) are markedly varied; a great deal of flexibility and choice is therefore available to the designer. Mechanical properties of the products depend on fibre orientation, on the volume fraction of reinforcement and ultimately therefore, on the method of manufacture.

Some of these issues will be discussed later in the text; for example, the theory and anisotropic effects of fibre reinforcement (Chapter 6), and the occurrence of blistering in GRP laminates (Chapter 10).

References

5.1 RUBIN, I.I.: *Injection Moulding – Theory and Practice,* Wiley, New York, (1972).

5.2 BOWN, J.: *Injection Moulding of Plastics Components,* McGraw-Hill, London, (1979).

5.3 WHELAN, A.: *Injection Moulding Machines,* Applied Science, London, (1984).

5.4 JOHANNABER, F.: *Injection Moulding Machines, a User's Guide,* Hanser Publishers, Munich, (1983).

5.5 ROSATO, D.V., and ROSATO, D. V.: *Injection Moulding Handbook,* Van Nostrand, New York, (1986).

5.6 BARRIE, I.T.: *The Rheology of Injection Moulding,* in: *Polymer Rheology,* LENK, R. S. Ed., Applied Science, London, (1978).

5.7 *Injection Moulding,* Unit 7 of PT614 Polymer Engineering, Open University Press, Milton Keynes, UK, (1984).

5.8 MONK, J.F. Ed., *Thermosetting Plastics: Practical Moulding Technology,* Godwin, London, (1981).

5.9 BARRIE, I.T.: *SIMPOL – The Injection Moulding Simulation Program,* Ian Barrie Consultancy, Hertfordshire, UK, (1985).

5.10 AUSTIN, C.: *MOLDFLOW – Computer Aided Engineering for Plastics: The Moldflow Philosophy,* Moldflow Pty. Ltd., Victoria, Australia.

5.11 MENGES, G., and MOHREN, P.: *The IKV CIM-Concept,* Intern. Polym. Proc., **2,** (1988), 162.

5.12 BOWERS, S.: *Prediction of Elastomer Flow in Multi-Cavity Injection Moulding,* Plast. Rubb. Proc. Appl., **7,** (1987), 101.

5.13 LANGECKER, G.: *Mouldfix – Introduction to Battenfeld's Fast Mould Changing System,* Presented at Technical Symposium: Rationalisation by Automation: New Developments in Injection Moulding, Battenfeld Company Literature, Meinerzhagen, Germany, (1982).

5.14 DAVIES, S.J.: *Automation of the Injection Moulding Process into the 1990's,* Presented at PRI Processing Workshop, Stoneleigh, UK, (1988).

5.15 PYE, R.G.W.: *Injection Mould Design,* 2nd. Ed., Godwin, London, (1978).

5.16 MENGES, G., and MOHREN, P.: *How to Make Injection Molds,* Hanser Publishers, Munich, (1986).

5.17 HAINES, R.C.: *Volume Production with Carbon Fibre Reinforced Thermoplastics,* Plast. Rubb. Proc. Appl., **5,** (1985), 79.

5.18 Battenfeld Technical Literature: *Injection Moulding Machines, Application-Related Designs and Special Processes,* Battenfeld Maschinenfabriken GmbH, Meinerzhagen, Germany.

5.19 Metalmeccanica Technical Literature; *Injection Screws for Marbled Effect,* Metalmeccanica Plast Spa, Como, Italy.

5.20 Stank, H.D.: *Requirements of the Sprue, its Functions, Design of the Injection Moulded Part, Sprue and Gate Problems in Injection Moulding,* Series Ingenieurwissen VDI-Verlag, Dusseldorf, (1970).

5.21 Kamal, M.R., and Kenig, S.: *The Injection Moulding of Thermoplastics: Part I – Theoretical Model,* Polym. Eng. Sci., **12,** (1972), 294.

5.22 Kamal, M.R., and Kenig, S.: *The Injection Moulding of Thermoplastics: Part II – Experimental Test of the Model,* Polym. Eng. Sci., **12,** (1972), 302.

5.23 Richardson, S.M.: *Moulding,* in *Computational Analysis of Polymer Processing,* Pearson, J.R.A. and Richardson, S.M., Eds., Applied Science, London, (1983).

5.24 Mavridis, B., Hrymak, A.N., and Vlachopoulos, J.: *Deformation and Orientation of Fluid Elements Behind an Advancing Flow Front,* J. Rheol., **30,** (1986), 555.

5.25 Powell, P.C.: *Processing Methods and Properties of Thermoplastics Melts,* in *Thermoplastics Properties and Design,* Ogorkiewicz, R. M.: Ed., Wiley Interscience, London, (1974).

5.26 Glanvill, A.B., and Denton, E. N.: *Injection-Mould Design Fundamentals; Volume 2 – Tool Construction,* Machinery Publishing Ltd., London, (1963).

5.27 Barrie, I.T.: *Understanding how an Injection Mould Fills,* S.P.E. J., **27,** (1971), 64.

5.28 Elbe, W., Stitz, S., and Wübken, G.: *Injection Moulding of Plastomers II,* Internal Publication, IKV, Aachen, Germany.

5.29 Allan, P.S., and Bevis, M. J.: *The Production of Void-Free Thick-Section Injection-Flow Mouldings I – Shot Weight and Dimensional Reproducibility,* Plast. Rubb. Proc. Appl., **3,** (1983), 85.

5.30 Allan, P.S., and Bevis, M. J.: *Multiple Live-Feed Injection Moulding,* Plast. Rubb. Proc. Appl., **7,** (1987), 3.

5.31 Folkes, M.J.: *Short Fibre Reinforced Thermoplastics,* Research Studies Press, London, (1982).

5.32 Clark, E.S.: *Morphology and Properties of Injection Moulded Crystalline Polymers,* Appl. Polym. Symp., **24,** (1974), 45.

5.33 Bowman, J., Harris, N., and Bevis, M.J.: *An Investigation of the Relationships between Processing Conditions, Microstructure and Mechanical Properties of an Injection Moulded Semi-Crystalline Thermoplastic,* J. Mater. Sci., **10,** (1975), 63.

5.34 Tan, V., and Kamal, M.R.: *Morphological Zones and Orientation in Injection Moulded Polyethylene,* J. Appl. Polym. Sci., **22,** (1978), 2341.

5.35 Ballman, R.L., and Toor, H.L.: *Orientation in Injection Moulding,* Mod. Plast., **38,** (1960), 113; 205.

5.36 HAN, C.D., and VILLAMIZAR, C.A.: *Development of Stress Birefringence and Flow Patterns during Mould Filling and Cooling,* Polym. Eng. Sci., **18,** (1978), 173.

5.37 DIETZ, W., WHITE, J.L., and CLARK, E.S.: *Orientation Development and Relaxation in Injection Moulding of Amorphous Polymers,* Polym. Eng. Sci., **18,** (1978), 273.

5.38 KAMAL, M.R., and TAN, V.: *Orientation in Injection Moulded Polystyrene,* Polym. Eng. Sci., **19,** (1979), 558.

5.39 HAWORTH, B., HINDLE, C.S., SANDILANDS, G.J. and WHITE, J.R.: *Assessment of Internal Stresses in Injection Moulded Thermoplastics,* Plast. Rubb. Proc. Appl., **2,** (1982), 59.

5.40 STRUIK, L.C.E.: *Orientation Effects and Cooling Stresses in Amorphous Polymers,* Polym. Eng. Sci., **18,** (1978), 799.

5.41 TREUTING, R.G., and READ, W.T.: *A Mechanical Determination of Biaxial Residual Stress in Sheet Materials,* J. Appl. Phys., **22,** (1951), 130.

5.42 KUBAT, J., and RIGDAHL, M.: *Influence of High Injection Pressures on the Internal Stress Level in Injection Moulded Specimens,* Polymer, **16,** (1975), 925.

5.43 SANDILANDS, G.J., and WHITE, J.R.: *Effect of Injection Pressure and Crazing on Internal Stresses in Injection Moulded Polystyrene,* Polymer, **21,** (1980), 338.

5.44 COXON, L.D., and WHITE, J.R.: *Residual Stresses and Ageing in Injection Moulded Polypropylene,* Polym. Eng. Sci., **20,** (1980), 230.

5.45 FISHER, E.G.: *Blow Moulding of Plastics,* Illiffe, London, (1971).

5.46 FRADOS, J. Ed., *SPI Plastics Engineering Handbook 4th Ed.,* SPI, Van Nostrand, New York, (1976).

5.47 ONASCH, J., and RUSTON, M.: *Latest Developments from Battenfeld Fischer Blasformtechnik,* Battenfeld Technical Literature, Lohmar, Germany.

5.48 Bekum Technical Literature: *The New Generation of Single- and Twin-Station Blow Moulders,* Bekum Maschinenfabriken GmbH, Berlin, Germany.

5.49 CHUNG, T.: *Principles of Preform Design for the Stretch Blow Moulding Process,* Polym. Plast. Technol. Eng., **20,** (1983), 147.

5.50 YEH, G.S.Y., and GEIL, P.H.: *Strain-Induced Crystallisation of Poly(Ethylene Terephthalate),* J. Macromol. Sci. (Phys.), **B1,** (1967), 251.

5.51 CAKMAK, M., SPRUIELL, J.E., and WHITE, J.L.: *A Basic Study of Orientation in Poly(Ethylene Terephthalate) Stretch Blow Moulded Bottles,* Polym. Eng. Sci., **24,** (1984), 1390.

5.52 WHITE, J.L., and SPRUIELL, J.E.: *The Specification of Orientation and its Development in Polymer Processing,* Polym. Eng. Sci., **23,** (1983), 247.

5.53 POWELL, P.C.: *Engineering with Polymers,* Chapman and Hall, London, (1983).

5.54 COGSWELL, F.N., WEBB, P.C., WEEKS, J.C., MASKELL, S.G., and RICE, P.D.R.: *The Scientific Design of Fabrication Processes – Blow Moulding,* Plastics and Polymers, **39,** (1971), 340.

5.55 CRAWFORD, R.J.: *Plastics Engineering, 2nd Ed.,* Pergamon, Oxford, (1987).

5.56 ICI Technical Literature: *Melinar® PET – The Need for Controlled Drying,* ICI Chemicals and Polymers, Wilton, UK.

5.57 MEEHAN, T.P., and MCBRIDE, P.: *Advantages of Cyclohexanedimethanol (CHDM)-Based Copolyesters for use in Plastic Bottles,* Presented at 9th SPE Intern. Conference, High Performance Plastic Containers, London, (1988).

5.58 BONNEBAT, C., ROULLET, G., and DE VRIES, A.J.: *Biaxially Oriented Poly (Ethylene Terephthalate) Bottles: Effects of Resin Molecular Weight on Parison Stretching Behaviour,* Polym. Eng. Sci., **21,** (1981), 189.

5.59 Orme Technical Literature: *RotoSpeed Six,* Rotational Moulding Equipment, Orme Polymer Engineering Ltd., Northants, UK.

5.60 BRUINS, P.F.: *Basic Principles of Rotational Moulding,* Gordon and Breach, New York, (1971).

5.61 RAO, M.A., and THRONE, J.L.: *Principles of Rotational Moulding,* Polym. Eng. Sci., **12,** (1972), 237.

5.62 TOMO, D.: *Rotational Moulding of Polyethylene Powders,* in *Basic Principles of Rotational Moulding,* BRUINS, P.F., Ed., Gordon and Breach, New York, (1971).

5.63 KRAUS, T.J.: *Formulating Plastisols for Rotational Moulding;* **ibid**

5.64 THRONE, J.L.: *Rotational Moulding Heat Transfer – An Update,* Polym. Eng. Sci., **16,** (1976), 257.

5.65 CRAWFORD, R.J., ,and SCOTT, J.A.: *An Experimental Study of Heat Transfer During Rotational Moulding of Plastics,* Plast. Rubb. Proc. Appl., **5,** (1985), 239.

5.66 CRAWFORD, R.J., and SCOTT, J.A.: *The Formation and Removal of Gas Bubbles in a Rotational Moulding Grade of Polyethylene,* Plast. Rubb. Proc. Appl., **7,** (1987), 85.

5.67 THIEL, A.: *Principles of Vacuum Forming,* Illiffe, London, (1965).

5.68 BRUINS, P.F.: Ed., *Principles of Thermoforming,* Gordon and Breach, New York, (1973).

5.69 BRYDSON, J.A.: *Plastics Materials,* 4th Ed., Butterworth Scientific, London, (1982).

5.70 THRONE, J.L.: *Thermoforming: Polymer Sheet Fabrication Engineering Part 1. Solid Sheet Forming,* Plast. Rubb. Proc., **4,** (1979), 129.

5.71 HOGER, A.: *Warmformen von Kunststoffen,* Hanser Publishers, Munich, (1971).

5.72 KREITH, F., and BOHN, M.S.: *Principles of Heat Transfer,* 4th Ed., Harper and Row, New York, (1986).

5.73 SCHMIDT, L.R., and CARLEY, J.F.: *Biaxial Stretching of Heat-Softened Plastic Sheets using an Inflation Technique,* Int. J. Eng. Sci., **13,** (1975), 563.

5.74 TRELOAR, L.R.G.: *The Physics of Rubber Elasticity,* 3rd Ed., Clarendon Press, Oxford, (1975).

5.75 SMITH, T.L.: *Nonlinear Viscoelastic Response of Amorphous Elastomers to Constant Strain Rates,* Trans. Soc. Rheol., **6,** (1962), 61.

5.76 LAI, M.O., and HOLT, D.L.: *The Extensional Flow of Poly(Methyl Methacrylate) and High Impact Polystyrene at Thermoforming Temperatures,* J. Appl. Polym. Sci., **19,** (1975), 1209.

5.77 LAI, M.O., and HOLT, D.L.: *Thickness Variation in the Thermoforming of Poly(Methyl Methacrylate) and High-Impact Polystyrene Sheets,* J. Appl. Polym. Sci., **19,** (1975), 1805.

5.78 Sheryshev, M.A., Zhogolev, I.V., and Salazkin, K.A.: *Calculation of Wall Thickness of Articles Produced by Negative Vacuum Forming,* Soviet Plast., **11,** (1969), 30.
5.79 Sheryshev, M.A., and Salazkin, K.A.: *Calculating the Differences in Wall Thickness of Articles Produced by Pressure Forming,* Soviet Plast., **12,** (1970), 28.
5.80 Tadmor, Z., and Gogos, C.G.: *Principles of Polymer Processing,* Van Nostrand, New York, (1979).
5.81 Rosenzweig, N., Narkis, M., and Tadmor, Z.: *Wall Thickness Distribution in Thermoforming,* Polym. Eng. Sci., **19,** (1979), 946.
5.82 Birley, A.W., Haworth, B., and Ola, A.O.: *Thermoforming of Amorphous PET Sheet: Factorial Experimental Design Approach towards Process Optimisation,* Intern. Polym. Proc., **5,** (1990), **124**
5.83 Becker, W.E.: *Reaction Injection Moulding,* Van Nostrand, New York, (1979).
5.84 BS 3496, (1989), *E-glass Fibre Chopped Strand Mat for Reinforcement of Polyester and other Liquid Laminating Systems,* British Standards Institution, London.
5.85 Borstell, H.J.: *Hand Lay-up, Spray-up and Prepreg Moulding,* in : *Engineered Materials Handbook Vol.2 – Engineering Plastics* ASM International, USA, (1988).
5.86 Meyer, R.W.: *Polyester Moulding Compounds and Moulding Technology,* Chapman and Hall, New York, (1987).
5.87 Matthews, G.: *Polymer Mixing Technology,* Applied Science, London, (1982).
5.88 Siadat, B.: *Application of Dielectric Heating in Plastic Processing,* Presented at SPE National Technical Conference, (1980), p.37-8.
5.89 Lee, L.J., Marker, L.F., and Griffith, R.M.: *The Rheology and Mould Flow of Polyester Sheet Moulding Compound,* Polym. Comp., **2,** (1981), 209.
5.90 Newman, S., and Fesko, D.G.: *Recent Developments in Sheet Moulding Compound Technology,* Polym. Comp., **5,** (1984), 88
5.91 Mallick, P.K., and Ragupathi, N.: *Effect of Cure Cycle on Mechanical Properties of Thick Section Fibre Reinforced Polythermoset Mouldings,* Polym. Eng. Sci., **19,** (1979), 774.
5.92 Costigan, P.J., and Birley, A.W.: *Microwave Preheating of Sheet Moulding Compound,* Plast. Rubb. Proc. Appl., **9,** (1988), 233.
5.93 Chapman, G.M., Kanagendra, M., and Fisher, B.C.: *Servohydraulic Control Applied to SMC Compression Moulding,* Presented at 13th BPF Reinforced Plastics Congress, Brighton, UK, (1982).
5.94 Kanagendra, M., Fisher, B.C., and Chapman, G.M.: *Microcomputer Control Applied to SMC Compression,* Presented at 14th BPF Reinforced Plastics Congress, Brighton, UK, (1984).
5.95 Kanagendra, M., and Fisher, B.C.: *Process Interactions for SMC Compression Moulding under Microcomputer Control,* Presented at 40th Annual Conference SPI RP/C Inst, Paper 16-C, (1985).

5.96 JOHNSON, C.F.: *Resin Transfer Moulding and Structural Reaction Injection Moulding,* in *Engineered Materials Handbook Vol.2 – Engineering Plastics,* ASM International, USA, (1988).

5.97 GOTCH, T.M., and PLOWMAN, P.E.R.: *Improved Production Processes for the Manufacture of GRP on British Rail,* Presented at 10th BPF Reinforced Plastics Congress, Brighton, UK, (1978).

5.98 ROSATO, D.V., and GROVE, C.S.: *Filament Winding,* Interscience, New York, (1964).

5.99 HUMPHREY, W.D., and PETERS, S.T.: *Filament Winding,* in: *Engineered Materials Handbook Vol.2 – Engineering Plastics,* ASM International, USA, (1988).

5.100 MIDDLETON, V., OWEN, M.J., ELLIMAN, D.G., and SHEARING, M.R.: *CAD for Non-Axisymmetric Filament Winding,* Presented at 16th BPF Reinforced Plastics Congress, Blackpool, UK, (1988).

Chapter 6
Mechanical Properties I – Deformation

6.1 Introduction

The initial steps of materials selection for component specification and design involve a careful consideration of a range of mechanical properties, together with parallel appraisal of additional factors such as processability, aesthetics/customer appeal and cost-effectiveness of given designs. Ever-increasing use of plastics materials for parts requiring guaranteed mechanical performance and reliability is now evident, since different families of polymeric materials offer some extremely impressive and versatile properties, which can often be controlled according to demand. Mechanical properties range from the soft, relatively compliant and elastic thermoplastic elastomers (TPE's) through hard, viscoelasatic solids to some reinforced plastics, which compete, in terms of strength, stiffness and toughness, with the most prominent of modern engineering solids.

However, for the majority of applications, the case for choosing plastics materials does not reside with a fulfilment of a mechanical property specification *per se;* additional advantages include low specific gravity, good resistance to chemicals and "corrosive" effects, heat and electrical insulation and optical clarity. In addition, plastics are associated with relatively low material costs, and especially, with low-cost and flexible manufacturing methods. Not all of these properties can always be derived from individual groups of plastics; nevertheless, polymers are sufficiently versatile to offer the possibility to tailor some *specific properties* (eg. stiffness per unit weight; strength per unit density or per unit cost) which make them eminently attractive to engineers and designers. Many of the physical properties which contribute to the attraction of plastics materials are featured collectively in this and subsequent Chapters.

When a mechanical load is applied to a polymeric material, its response may be either:

- a change in shape and dimensions, or
- fracture.

The former response is often reversible but may induce time-dependent changes, whilst fracture is irreversible but may take place with little deformation *(brittle failure),* or alternatively, considerable strain could be imposed on the specimen and it may have experienced gross yielding *(ductility).* It is convenient, therefore, to differentiate between the deformation behaviour and fracture at low strain, and large-strain response to mechanical loads: the *yielding* phenomenon is a convenient limit for the former.

The present Chapter will cover, therefore, the stress versus strain relationship up to, and including yield, but excluding fracture. Stiffness, strength and yielding of plastics are described and those factors which determine or modify these properties are reviewed. Also described is the philosophy by which the basic mechanics of solid polymers can be transformed into design data to accommodate the sensitivity of plastics materials to time and temperature.

Complementary to the section on deformation, Chapter 7 considers the concept of *"failure"* in detail. In addition to the more traditional analyses of ductile and brittle failures in plastics materials, Chapter 7 leans heavily towards the analysis of failure in *plastics products,* such that the effects of processing which were highlighted in Chapters 4 and 5 can be considered within the context of mechanical performance.

6.2 Deformation of Plastics Materials – Introduction

If a body is subject to an applied force, the resultant *stress* induces a finite deformation (a *strain*), within the body. According to the geometry of the body and the type of loading, the magnitude of stress and strain may vary considerably within the sample.

The simplest way to relate stress and strain in structures is by classical elasticity theory[6.1–6.3]. As will become evident in this Chapter, elastic theories are relevant to stress analysis and mechanical design methods for plastics, but only if those factors relevant to the analysis have been measured rigorously, and if the dependence of polymer properties upon temperature, time and environmental effects have been recognised. There are many possible stress systems which may be set up when a mechanical load is applied to a structure of given geometric shape; included in this context are some practically-relevant situations involving:

- tension (uniaxially-loaded joints and supports; fibres and filaments; pressure in pipes or in cylinders)
- compression (seals, struts and bearings)
- shear (mountings; adhesive layers in composites, and in laminated films and containers)
- torsion (shafts, gears)
- flexure (bending stress in beams, housings and plates) which contains tensile, compressive and shear components
- hydrostatic pressure.

Summarised in Table 6.1 (see also Figure 6.1 for geometry and co-ordinate details) are some of the simple stress and strain systems which are regularly encountered when plastics components are loaded.

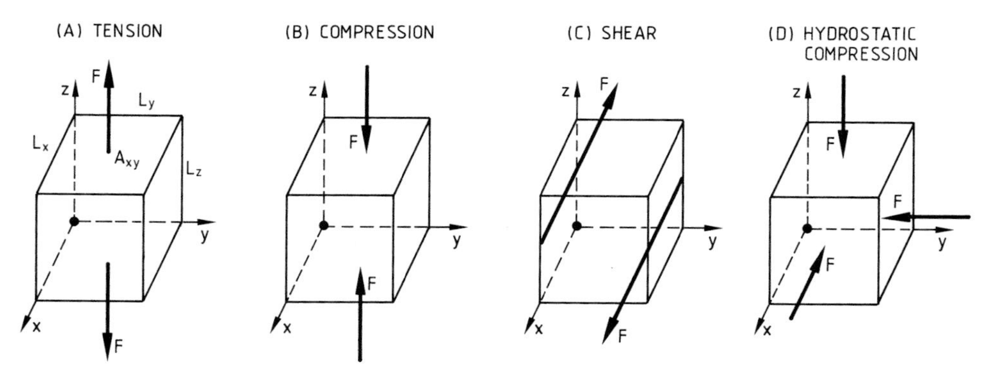

Figure 6.1 Modes of stress relevant to polymer engineers and designers:
(A) Tension, (B) Compression, (C) Shear, (D) Hydrostatic compression (see Table 6.1).

Table 6.1 Some Stress-Strain Systems in Loaded Components

	Stress	*Strain*	*Modulus*
Tension	$\sigma_X = \frac{F}{A_{YZ}}$	$\varepsilon_X = \frac{\triangle L_X}{L_X}$	$\text{E} = \frac{\sigma_X}{\varepsilon_X}$
		$\varepsilon_Y = \frac{\triangle L_Y}{L_Y}(-ve)$	
Compression	$\sigma_X = -\frac{F}{A_{YZ}}$	$\varepsilon_X = \frac{\triangle L_X}{L_X}(-ve)$	$E = \frac{\sigma_X}{\varepsilon_X}$
		$\varepsilon_Y = \frac{\triangle L_Y}{L_Y}$	
Shear	$\tau_{XY} = \frac{F}{A_{XY}}$	$\gamma_{XY} = \frac{\triangle L_X}{L_Y}$	$G = \frac{\sigma_{XY}}{\gamma_{XY}}$
Hydrostatic Compression	$\sigma_{X,Y,Z} = -P$	$\varepsilon_{X,Y,Z} = \frac{V-V_0}{3V_0}$	$K = \frac{PV_0}{V_0-V}$

(E, G, and K are the tensile, shear and bulk moduli; V_0 and V represent initial (unstressed), and stressed volume under a hydrostatic pressure P).

For any of these loading modes, stress and strain are related by a material parameter which describes the resistance to deformation under load. In the case of simple elastic solids, unique stiffness characteristics *(elastic moduli)* can be evaluated to characterise these relationships (as shown above) in any appropriate stressing mode. Many practical situations involve much more complex stress and strain distributions. For example, simple beam-bending induces a maximum tensile fibre stress at the outer surface, and a maximum compressive stress at the opposite face; similarly, torsion of solid cylinders represents a particular type of shear deformation, where the shear stress (and strain) vary from zero at the centre-axis, to attain a maximum level at the outer surface. In other situations, stresses can be resolved into tensile, compressive and shear components, some of which may be omitted by selective choice of co-ordinate axes.

However, when dealing with stress analysis and deformation in *plastics materials,* it must be recognised that stress and strain are not simply related, and therefore the use of a modulus or "stiffness" parameter must be restricted to specific application areas where directly-relevant data have been previously derived. In addition to the effects of molecular weight, crystallinity and the types and proportions of additives in the compound, the following factors influence the properties of solid plastics, and should be considered at an early stage:

(1) strain level; ie. stress is not *linearly* related to strain;
(2) loading time and rate effects (solid plastics are *visco*-elastic);
(3) incomplete and time-dependent recovery, when loads are removed;
(4) polymer properties are often *anisotropic;* and
(5) temperature, and other environmental factors.

Thermal and mechanical history during processing represent further variables which affect mechanical properties, as a result of the specific *microstructure* and *morphology* which develop according to shaping and cooling conditions. When assessing microstructure in plastics products however, extreme sensitivity to the local cooling regime is usually evident, such that the usual morphological state is heterogeneous; that is, it varies according to position within the component. These relationships will be discussed in greater depth throughout subsequent Chapters.

Simple stress-strain relationships in elastic solids are given by Hooke's Law:

Tension $$\frac{\sigma}{\varepsilon} = \text{Constant} = E \quad \text{(Tensile Modulus)} \tag{6.1}$$

Shear $$\frac{\tau}{\gamma} = \text{Constant} = G \quad \text{(Shear Modulus)} \tag{6.2}$$

Measurements of modulus and "failure" stresses and strains must be carried out not only in the mode appropriate to the component whose properties are being studied, but also in such a way to account for the likely dependence of properties on the factors listed.

Generalised expressions for the elastic behaviour of solids are available for study in more advanced texts[6.1–6.2]. The approach usually adopted is to define the resolved components of (normal and shear) stress and strain by tensor notation, so that these can be related by a stiffness or compliance matrix based on a series of *elastic constants* which include shear and tensile modulus, and Poisson's Ratio.

Before we advance to consider the factors which influence mechanical properties, it is useful to revise some basic terms which recur throughout the following discussions:

Stress. Force per unit area on a body; *engineering* or *nominal* stress is evaluated according to the *initial* cross-sectional area, whilst *true stress* takes account of the change in area which occurs as deformation progresses.

Strain. Sample extension, expressed as a ratio to its original length.

Stiffness. Resistance to deformation under applied loads, usually expressed as an empirically-derived *elastic modulus.*

Compliance. Opposite to stiffness, compliance reflects a material's propensity to deform under load. Compliance/stiffness parameters are therefore related directly to principal stresses and strains, under the appropriate modes of testing.

Poisson's ratio. Also known as the *lateral strain contraction ratio,* Poisson's Ratio relates the change in lateral strain to the longitudinal strain applied directly. Typical values are 0.30 – 0.33 for glassy polymers and 0.35 – 0.40 for semi-crystalline plastics; levels up to 0.50 are encountered for elastomers and rubber-like systems (eg. amorphous plastics at temperatures exceeding T_g).

It should be noted that linear elastic theory relates Shear Modulus (G) and Bulk Modulus (K) to Young's Modulus (E), using Poisson's Ratio (ϕ), by the expression:

$$E = 2G(1 + \phi) = 3K(1 - 2\phi) \tag{6.3}$$

Rather than dwell upon some detailed, and previously well-documented aspects of elasticity theory, our preferred approach is to concentrate upon the underlying physical factors which cause macromolecular networks to behave in a less than ideal manner, and to appraise the influential variables associated with component manufacture.

6.3 Temperature and Deformation

It has always been recognised that the temperature-sensitivity of physical properties of plastics materials is often quite severe. The mechanical stress-strain relationship is no exception; before examining the way in which modulus varies with temperature, it is worth recalling some concepts of polymer physics (see also Section 1.2) which can be exemplified in the present context of mechanical deformation.

6.3.1 Role of Primary and Secondary Bonding

With the exception of a small family of ionomers, most commercial plastic materials consist of *covalently* bonded polymer chains. Direct thermal breakdown of primary bonds is generally only feasible in the melt state, so that the temperature sensitivity of a solid polymer to a mechanical excitation would be very slight, in the theoretical situation where the stress were to be applied parallel to an array of perfectly-aligned polymer chains. (Although this principle is theoretically very significant from the point of view of realising exceptional axial strength, it is very difficult to utilise in commercial practice; however, the organic fibres industry has developed the means to mass-produce products with high uniaxial orientation eg. polyaramid "Kevlar"® fibres).

In reality, relatively few primary bonds in a macromolecular network are in perfect alignment with an axis of principal stress associated with a loading programme; consequently, the load-bearing properties are highly dependent upon secondary *inter-chain attractions* (dipoles, hydrogen bonds)[6.4]. Since the magnitude of these forces is very temperature sensitive (increasing molecular mobility leads to diminished bond strength by interchain-separation), we have an immediate explanation for the typical temperature-sensitive stiffness response observed in most commercial systems. The magnitude of this effect is somewhat dependent upon the specific type of inter-molecular bonding in a given class of material, by the main-chain chemical groups which determine the overall chain-flexibility, and also by the extent of molecular anisotropy present in finished components.

6.3.2 Transitions and Relaxation Temperatures

Important though the direct effects of temperature undoubtedly are, changes of greater magnitude are observed if deformation characteristics are studied through a temperature span sufficiently wide to include *primary thermal transitions* and other relaxation temperatures. Let us first consider some idealised relationships between temperature, crystallinity and "modulus" for a typical polymer (Figure 6.2): immediately, the dominating influences of the *glass transition temperature* T_g and the *crystalline melting point* T_m are apparent. Amorphous plastics change from hard, often brittle solids to highly extensible and predominantly elastic materials, on passing through the glass-rubber transition. A further temperature increase allows chain slippage to occur, thereby promoting viscous (fluid-like) deformation to a greater extent. The glass transition temperature is also highly influential in semi-crystalline plastics, where a "brittle-tough" transition in the solid phase is often observed, in response to an applied stress. Deformation resistance decreases dramatically on traversing T_m, as a melt is formed.

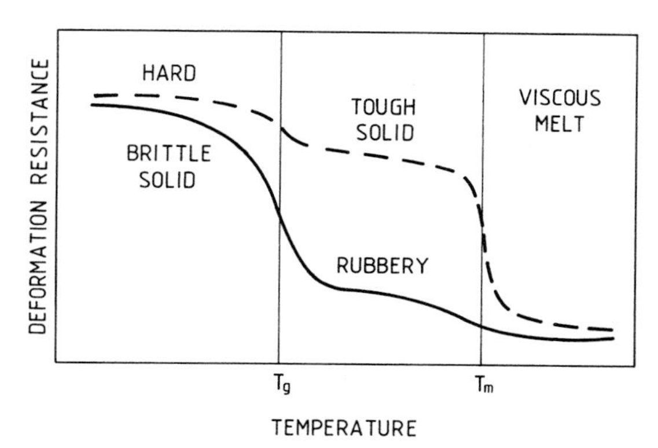

Figure 6.2 Generalised relationship between deformation resistance and temperature for amorphous (solid line) and semi-crystalline (broken line) high polymers.

Clearly, the extent of crystallisation determines the rate of change of such modulus-temperature characteristics, both within and across the primary thermal transitions. Note also that, within each temperature regime, the effects described in the previous paragraph are sufficient to induce less dramatic, but positive changes in compliance, as the temperature is raised.

It must be emphasised that the exact form of modulus-temperature relationships depends upon the mode of measurement, and upon rate effects (eg. loading frequency in dynamic tests, or strain rate in direct tension or flexure). As a result, some tests for "temperature resistance", based upon these general mechanical principles, have now assumed a standard format (see also Chapter 1). The *Deflection Temperature under Load (DTL)* is the temperature at which a given deflection is achieved (under standard 3-point loading conditions) in a beam-like sample; *Vicat Softening Point* is a test based upon similar principles, but the test-piece is a point-loaded circular disc, *with a concentrated centre-load.* It is essential to recognise that neither DTL nor Vicat measurements describe primary or secondary transition or relaxation temperatures; the data describe somewhat arbitary temperatures at which materials (under pre-defined loading conditions) exhibit a particular bending modulus.

The influence of subsidiary or *secondary relaxation temperatures* should also be appreciated. These represent the minimum temperatures at which it becomes feasible for small-scale motions of specific molecular segments to occur. In order of decreasing temperature, these transitions are labelled α (main-chain motion; T_g), β (side-group relaxation), γ, and so on. Mechanical properties may be modified by the possibility of these relaxations occurring over extended periods of time *(physical ageing)* during service life. Therefore, those factors which determine or shift the relaxation temperatures must be fully understood.

For example, flexible chain backbone structures (ether (–O–) linkages or extended ($-CH_2-$) sequences) will decrease the α-transition temperature, whereas rigid aromatic groups will have the opposite effect. Other influential factors for amorphous plastics which increase T_g include bulky, or highly-polar side-groups, whilst flexible side-chains, added plasticisers and additional free volume each contribute to a lowering of the glass transition. Relaxations in semi-crystalline plastics are made more complex by those induced by crystal defects and by crystallite size. In practice therefore, changes in morphology arising from non-uniform thermal history during processing are responsible for modification of mechanical behaviour, by their effects upon relaxation spectra. The reader is referred to the text by WARD[6.3], for a detailed appraisal of the effects of structure on molecular relaxations and properties: acrylic plastics, PETP, PTFE and polyethylenes are highlighted.

6.3.3 Effect on Stress-Strain Relationships

Temperature determines not only dynamic, short-term or low-strain moduli, but also the feasibility of plastics components undergoing yield, orientation hardening or rubber-like deformations. Consider the stress-strain relationships at gradually increasing temperatures in Figure 6.3: each of the loci A-E may be possible characteristics of the same polymer, according to the test temperature chosen:

A – at the lowest temperature, A represents a brittle but *viscoelastic* response, with failure occurring at maximum load and relatively low strain (typically 5 % or less), without gross yielding;

B – shows a distinct *yield point* (maximum load) with subsequent failure by neck-instability;

C – exhibits yielding (with a characteristic load drop), stable neck growth through *cold drawing* and *orientation hardening* (in reality, an *anisotropic stiffening* effect) and eventual failure in the highly-oriented neck at very high strain levels (which often exceed 300 %);

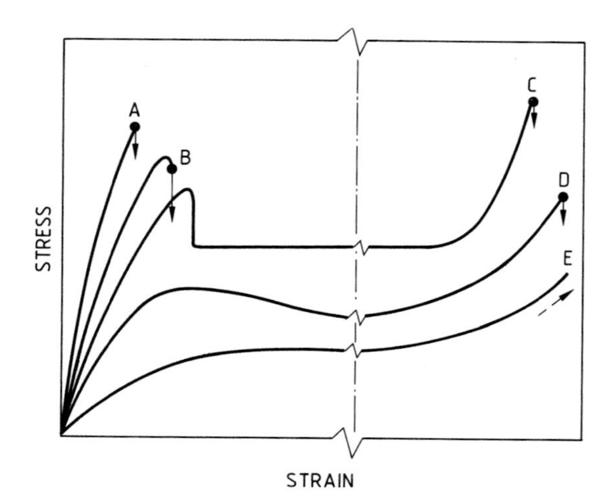

Figure 6.3 Generalised stress-strain characteristics of plastics: effect of temperature (see text).

D & E – are characteristics of a sigmoidal *rubber-elastic* response, typical of amorphous or low-crystallinity materials at temperatures just above T_g. These materials are characterised by a low drawing stress, and are highly extensible. Deformation may be predominantly elastic, unless significant stress-induced crystallisation occurs during the drawing process.

Similar families of stress-strain relationships could be derived by changing strain rate, rather than temperature; a degree of superimposition is often possible to relate these effects, since an increase in the former induces similar response changes to a decrease in the latter.

The extensibility data on PVC which were illustrated in the previous Chapter (Figure 5.29) represent an appropriate example of temperature-dependent mechanical properties; the high draw ratios which are attainable around the glass transition are used as a principle for commercial processing, to produce oriented products by thermoforming, or by stretch blow moulding.

Temperature changes undoubtedly have some profound influences upon the mechanical behaviour of plastics; it is therefore an important aspect to consider at the component design stage of any load-bearing part. However, as will be evident in subsequent sections, some interactive time-dependent effects should also be included in any design appraisal on parts manufactured from polymeric solids.

6.4 Time Dependence and Viscoelasticity in Solid Plastics

Time dependent properties are characteristic of long-chain structures and represent the main difference between the stress-strain responses of classical (elastic) solids such as metals, and the behaviour of plastics. Marked time dependence is found only in *thermoplastics,* since heavily crosslinked polymers show little, if any time dependence at ambient temperatures. The effect of time dependent mechanical properties is that high polymers exhibit characteristics of both elastic solids and viscous liquids. What is unusual is that elastic effects are often seen when *polymer melts* are deformed (eg. die swell in plastics extrusion; Chapter 3) and also, *solid plastics* suffer from viscous (time-dependent) effects, which limit the applicability of elastic theory in modelling the stress-strain behaviour of fabricated products.

Time dependent mechanical properties may become apparent in various ways:

- the *rate of loading* has a profound influence upon both low strain properties (eg. stiffness) and ultimate strength; for example, embrittlement under high-speed (impact) loading is a problem often encountered;
- solid plastics suffer from creep and stress relaxation effects, and also from incomplete and time-dependent recovery; the *timescale* of loading is therefore critical to mechanical design problems with plastics;
- in a high-frequency dynamic loading environment, viscoelasticity is evident by the existence of a phase difference between stress and strain.

These aspects of time dependent behaviour are considered in more detail in this section. It will be noted that a convenient way to simulate viscoelastic behaviour is by the use of mathematical models based upon an arrangement of idealised springs and dashpots, to represent a material's elastic and viscous "constants". Some molecular interpretations of this methodology are illustrated in Figure 6.4:

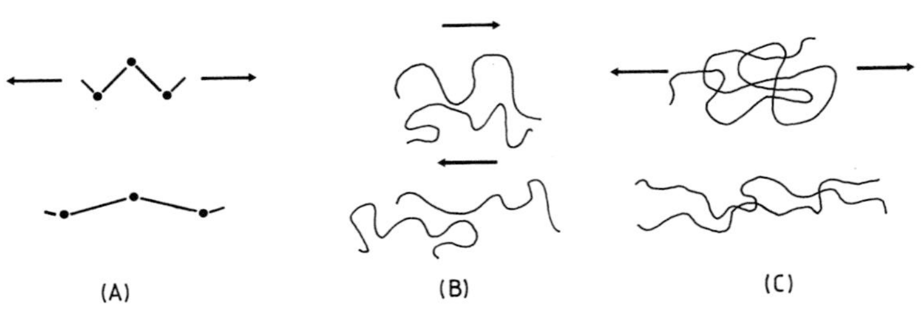

Figure 6.4 Molecular analogies of viscoelasticity in plastics materials: (A) Elasticity: (B) Viscous flow; (C) Chain uncoiling.

Elasticity (simulated by a spring) is a low-temperature, low-strain phenomenon in solid plastics; by definition, elastic deformation is wholly recoverable when the excitation is removed.

Viscous flow (dashpot simulation) is irrecoverable and contributes to creep, stress relaxation and other time-related responses (the energy of deformation is generally dissipated as heat); enhanced chain mobility at high temperature accelerates the progression and the extent of flow.

Chain uncoiling is favoured in the high strain (thermoelastic) regime close to the glass transition temperature of plastics. Bond rotation and segmental motion are exemplified by high-strain, low-modulus characteristics; deformation is largely recoverable, though not instantaneously so.

It would be desirable to use modelling techniques to predict the long-term (time-dependent) nature of some chosen mechanical properties of plastics components (eg. creep deformation in continuously-loaded, long service-life products). However, it is not possible to use a simple analytical tool to provide meaningful and accurate predictions; as the accuracy of any mathematical model increases, so too does its complexity. Comprehensive treatments of viscoelastic modelling techniques are given in the texts by ALFREY[6.5] and by FERRY[6.6], with more up-to-date comments (including a thorough analysis of some different types of excitation functions) by TURNER[6.7].

Subsequent sections feature some of the aspects of *viscoelasticity* in solid-phase plastics, and highlight the influential dynamic and material functions by discussing briefly how relevant test-data may be derived for the benefit of design engineers.

6.4.1 Deformation and Loading Rate

Constant deformation-rate loading tests are frequently used as comparative methods of distinguishing satisfactory quality from substandard products, and also to evaluate the effect of a change in formulation or manufacturing conditions upon mechanical strength; direct tension or flexural tests are most frequently employed in these circumstances. Test data derived from such techniques must be carefully standardised, however, since the viscoelastic nature of plastics is responsible for *strain rate sensitivity* of a number of deformation and failure parameters. In practice, the strain rate ($d\varepsilon/dt$) should be kept constant by controlling crosshead speed or drop velocity according to specimen dimensions.

The data in Figure 6.5 (tensile tests on samples of injection moulded high-impact polystyrene, as a function of elongational strain rate) illustrate typical sensitivity of thermoplastics to direct rate effects. (Also of note is the similarity between the temperature and strain rate effects in Figures 6.3 and 6.5; an increase in temperature is often assumed to be equivalent to a decrease in strain rate; a superposition principle, on this basis, has been proposed[6.7]). High loading rates are associated with a *stiffer* and an apparently *stronger* response, but also, an increased tendency towards *brittleness* is often evident. Strain rate sensitivity is particularly severe in the vicinity of a primary or secondary thermal transition (see also Chapter 1, and Section 6.3.2). Polymer relaxation is the key to greater understanding of rate-sensitive properties: if the relaxation time is low compared to the experimental (loading) timescale, energy is preferentially dissipated by viscous flow since segmental motion occurs more readily, and the macro-scale observation is a tendency towards lower strength and stiffness, and often, towards increasing ductility.

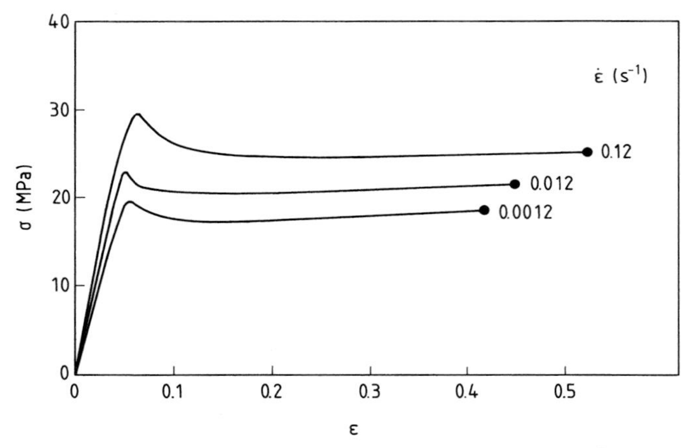

Figure 6.5 Effect of tensile strain rate ($\dot{\varepsilon}$) on the stress-strain characteristics of injection moulded HIPS.

It is strain-rate sensitivity which has necessitated that great attention be given to the resistance of plastics products to high-speed (impact) loading conditions, where the tendency towards embrittlement is most severe.

6.4.2 Deformation under Constant Load – Creep

An additional, and possibly the most important and relevant facet of viscoelasticity in commercial plastics products is creep: by definition, creep is a time-dependent strain increase to a constant applied stress. Many plastics materials creep readily at ambient or moderate temperatures; in consequence, product designers must be aware of not only the type, direction and magnitude of mechanical loads which are applied in-service, but also, the *time history* of such loading must be fully predetermined.

Creep deformation (and ultimately, *creep rupture* – see Chapter 7) occurs in a variety of practical situations: uniform or point-loading of beams and plates, internal pressurisation of pipes, tanks and bottles, compressive loading of sheet products, and direct tension in oriented fibres or grid structures represent just a few relevant examples.

6.4.2.1 Creep Deformation – The Voigt-Kelvin Model

Figure 6.6 shows simple loading-recovery diagrams for an elastic solid, and for a *linear viscoelastic* body. It is convenient to separate the deformation into components represented by elastic and delayed elastic strains, and by viscous flow. Whilst prior prediction of the magnitude of the latter strain component (for any given loading history) represents the major challenge faced by component designers, the existence of delayed elasticity imposes a further complexity, since the recovery of elastic strain is itself a time-dependent phenomenon.

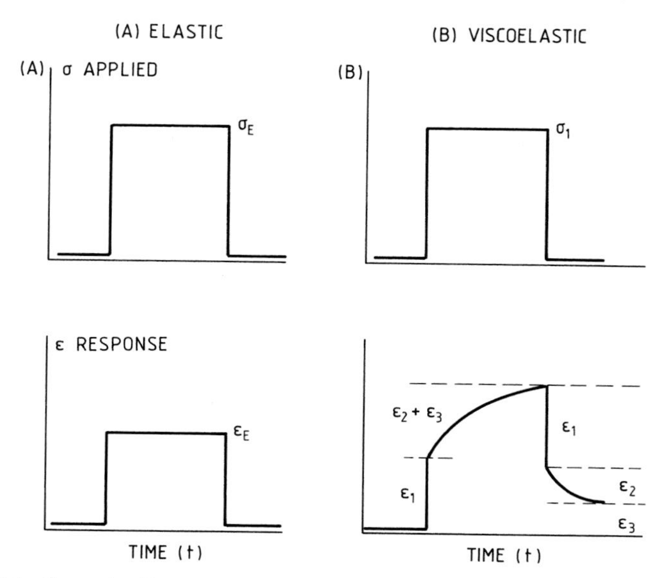

Figure 6.6 Creep loading programmes for (A) Elastic (subscript E); and (B) Viscoelastic bodies (subscripts 1, 2, and 3 refer to elastic, delayed elastic and viscous deformations for stress (σ) and strain (ε) respectively).

In order to enhance the accuracy of design procedures for new or prototype products, it is clear that the elements of time dependence during creep- loading and recovery periods be simulated by an appropriate mathematical model. Like many aspects of viscoelasticity, initial steps can be made by considering the response of an arrangement of "elastic" and "viscous" elements (springs and dashpots) to (in the case of creep), a constant applied force. The *Voigt-Kelvin* model consists of elastic and viscous elements in parallel (Figure 6.7); each element is characterised by a "constant" representing *elastic* (modulus, E) and *viscous* (coefficient of viscosity, μ) deformations, hence the subscripts E and V. Using the usual notation for stress (σ) and strain (ε) in tensile deformation, the stress-strain relationships are given by:

Elastic component $$\sigma = E \cdot \varepsilon \tag{6.4}$$

Viscous component $$\sigma = \mu \cdot \frac{d\varepsilon}{dt} \tag{6.5}$$

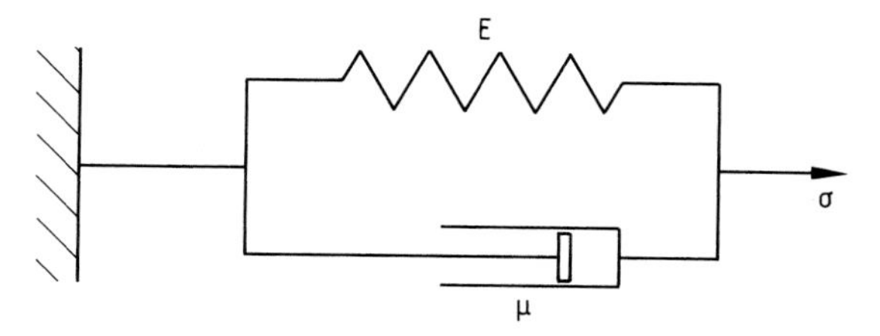

Figure 6.7 Voigt-Kelvin Model (simulation of creep deformation).

Creep loading. From the geometry of the model, the individual strain in each element is equal to the total strain; also, the applied stress is supported jointly by the spring and dashpot. Hence:

$$\varepsilon = \varepsilon_E = \varepsilon_V \tag{6.6}$$

$$\sigma = \sigma_E + \sigma_V \tag{6.7}$$

Substituting for stress (from Equations 6.4, 6.5) into 6.7, and rearranging, we obtain:

$$\sigma - (E \cdot \varepsilon) = \mu \cdot \frac{d\varepsilon}{dt}$$

Dividing by E and rearranging:

$$\int \frac{d\varepsilon}{(\sigma/E) - \varepsilon} = \frac{E}{\mu} \int dt$$

Integrating, and using the condition that at $t = 0$, $\varepsilon = 0$ (the constant of integration is therefore equal to $-\ln \sigma/E$):

$$-\ln\left(\frac{\sigma}{E} - \varepsilon\right) + \ln\frac{\sigma}{E} = \frac{Et}{\mu}$$

Therefore:

$$1 - \frac{\varepsilon E}{\sigma} = \exp\left(-\frac{Et}{\mu}\right)$$

ie.

$$\varepsilon(t) = 1 - \exp\left(-\frac{Et}{\mu}\right) \cdot \frac{\sigma}{E} \tag{6.8}$$

Time dependent strain, $\varepsilon(t)$, is therefore modelled by an increasing exponential relationship, and tends towards the ratio of stress and elastic stiffness "constant" $\frac{\sigma}{E}$ at infinite loading time.

Recovery. Using an identical model principle, the recovery characteristics of plastics can be simulated by considering a sudden removal of stress, at $t = 0$ (say). Using Equation 6.7, and since $\sigma = 0$ after load removal:

$$\varepsilon E = -\mu \frac{d\varepsilon}{dt} \qquad \therefore -\frac{E}{\mu}\int dt = \int \frac{d\varepsilon}{\varepsilon}$$

Integrating, and using the condition that at $t = 0$, $\varepsilon = \varepsilon_0$ (ε_0 is the creep strain at the instant of load removal), the constant of integration is equal to $-\ln(\varepsilon_0)$, which gives the solution:

$$-\frac{Et}{\mu} = \ln\frac{\varepsilon}{\varepsilon_0}$$

The magnitude of strain remaining in the sample during recovery is given by:

$$\varepsilon(t) = \varepsilon_0 \cdot \exp\left(\frac{-Et}{\mu}\right) \tag{6.9}$$

The creep strain and recovery characteristics predicted by the Voigt-Kelvin model are illustrated in Figure 6.8. In the absence of specific test data or more complex simulations, this analysis represents a reasonable first approximation to the response of typical linear viscoelastic systems. As will be evident in a later section, greater levels of accuracy are generally needed when attempts are made to extract information (eg. modulus data) to use for design purposes. At this stage, however, it must be emphasised that since creep strain increases with time, no single parameter can be used to characterise material stiffness completely; instead, we introduce the parameters below (which are *time dependent* and which may be used in *pseudoelastic design* techniques – see Section 6.7) to accommodate viscoelastic effects:

Creep modulus, $$E(t) = \frac{\sigma}{\varepsilon(t)}$$

Creep compliance, $$J(t) = \frac{\varepsilon(t)}{\sigma}$$

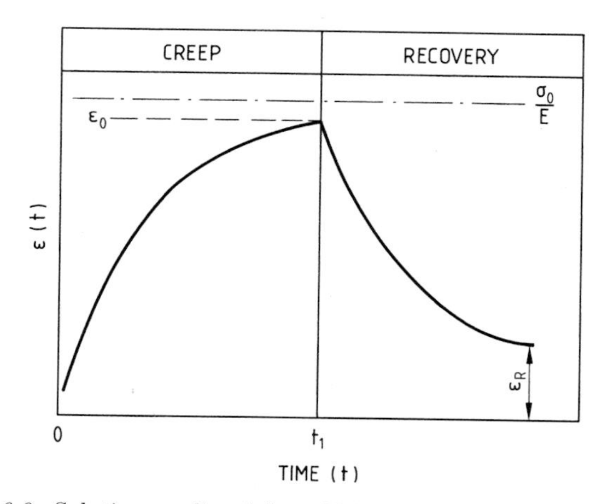

Figure 6.8 Solution predicted from Voigt-Kelvin model for creep and recovery (constant stress σ_O applied at zero time, and removed at $t = t_1$).

6.4.2.2 Creep Data – Measurement and Presentation

Creep data are easy to generate in principle, but in practice, a high degree of measurement expertise is usually necessary to ensure a reasonable degree of accuracy. The magnitude of strain, the thermal environment, the mode of stressing and the type/shape of sample (together, of course with the loading *timescale)* all contribute to the choice of test equipment.

Tension is the favoured testing mode due to the convenience of load application and the direct relevance to many situations characteristic of in-service conditions; stress is usually applied to a parallel-sided, high aspect-ratio test-specimen at a rapid but controlled rate, via a system of levers. The greatest equipment constraint is ensuring that strain measurement is accurate over long time-spans, by any technique which must not itself interfere with sample deformation; strain gauges, or *extensometers* based upon electrical, infra-red or optical measurement principles have proved particularly successful. Assurance must also be made that the specimen environment is strictly controlled: temperature and humidity are factors to which creep deformation is particularly sensitive. Further guidance on the practical aspects of creep testing is presented by TURNER[6.7], and more specific aspects are featured elsewhere[6.8, 6.9]; advice on test method accuracy is provided in relevant Standards Literature[6.10].

Creep tests in *compression* are carried out much less frequently, and generally only upon materials or products for specific end-use properties; plastics foams are materials which fall into this category.

Creep data are generally presented as *strain vs log (time),* for a series of tests, each carried out at constant stress. Additional variables considered include temperature (an increase in which promotes interchain motion and the rate of viscous flow) and humidity, which promotes creep rate by a chain-plasticisation effect, especially in hygroscopic plastics such as polyamides and thermoplastic polyesters. Figure 6.9 illustrates the typical non-linear creep behaviour exhibited by most thermoplastics.

Finally, it must be recognised that the creep data which are generated by standard test methods on idealised specimens (eg. isotropic, annealed, non-pigmented) may not accurately predict the likely service creep characteristics of inhomogeneous, fabricated parts from related, but dissimilar grades of material. Factors which may detract from the validity of making such correlations include *molecular orientation,* the form and extent of *crystallinity,* together with the presence of influential *additives* such as plasticising agents, fillers and fibrous reinforcements. Generally, an increase in most of these variables tends to increase creep resistance in the product, the only exception to the list given being the effect of plasticisers. However, the extent of any improvement may itself be temperature-dependent, and will certainly vary according to the nature of the polymer.

The relationships between creep and molecular orientation are quite complex, especially when anisotropy in semi-crystalline plastics is considered. For uniaxially-oriented plastics, creep resistance is greatest when the maximum principal stress is parallel to the chain-orientation direction, but often, the creep modulus assumes its lowest level (with respect to orientation) when stressed at around 45° to the orientation axis, where the chains in the oriented crystalline aggregates are thought to deform more easily by shear-dominated mechanisms along specific crystallographic planes[6.11, 6.12].

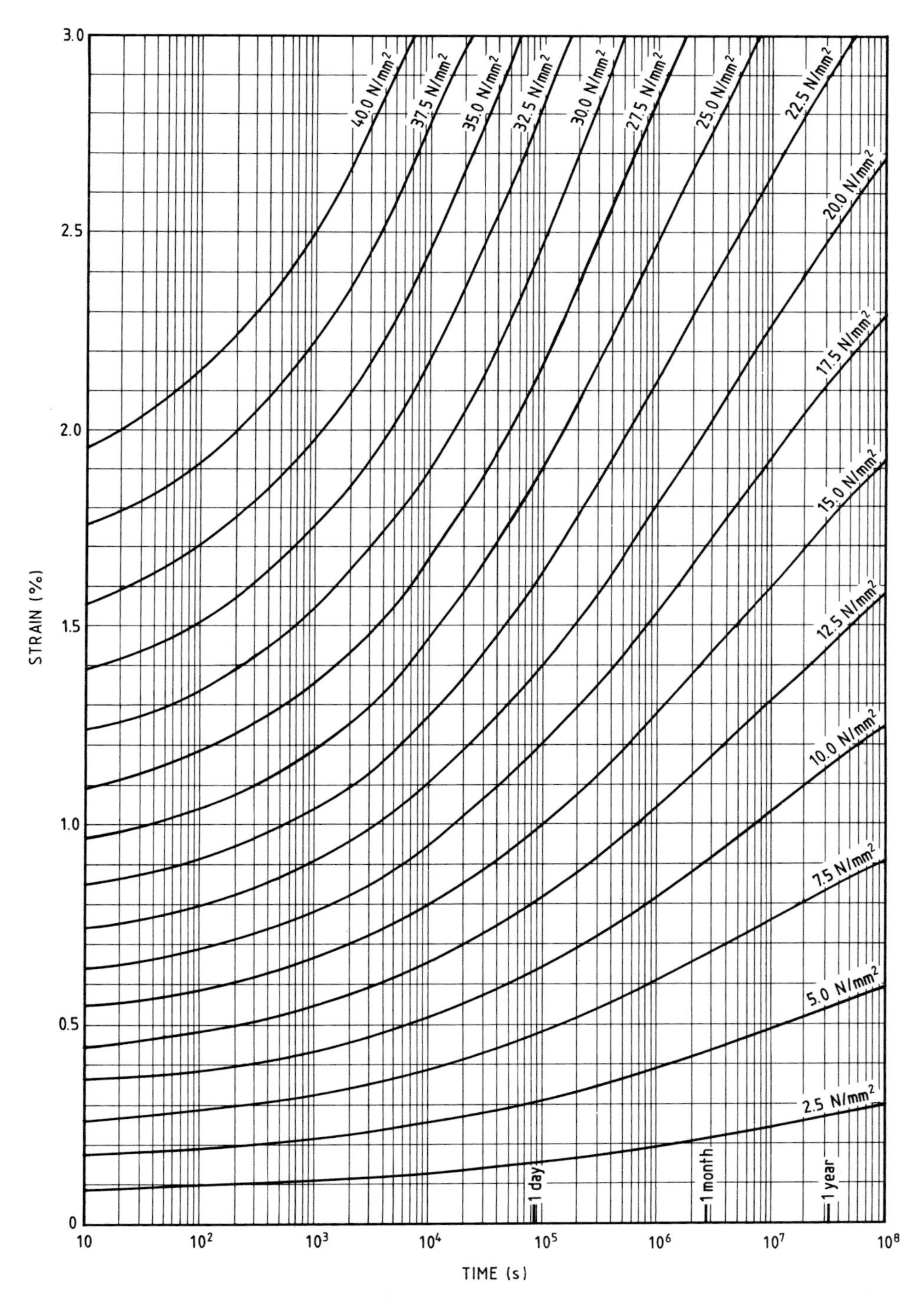

Figure 6.9 Tensile creep data for a grade of Kematal® Acetal Copolymer (at 20 °C and 65 % relative humidity). (Data reproduced by permission of Hoechst UK Limited, Polymers Division.)

6.4.3 Stress Relaxation (Constant Strain)

A distinct but related form of viscoelasticity in plastics materials is the phenomenon of *stress relaxation;* a time-dependent stress decay under constant strain conditions.

As is the case with creep deformation, relaxation effects demand that deformation time be incorporated into product design procedures. It is fair to suggest that stress relaxation has received less practical attention than creep (although theoretical modelling is probably as advanced); this is simply because there are less examples of practical applications for plastics for which stress relaxation is important. Many such "constant-strain" situations exist in joints between components in engineering applications; these include seals (eg. rubbery, compliant plastics deformed between more rigid products), bearings and threaded or "push-fit" connector parts.

6.4.3.1 Stress Relaxation – The Maxwell Model

Similar analogies of elastic deformation and viscous flow to those presented earlier can also be used in the stress relaxation mode (Figure 6.10). Since the active stress diminishes with time, the obvious practical problem associated with relaxation is that of mechanical joint integrity, as the sealing force drops to an unacceptably low level. By using prior knowledge of material characteristics, the *Maxwell* model (elastic and viscous constants in series – Figure 6.11) may be used to estimate the relaxation characteristics of linear viscoelastic polymers.

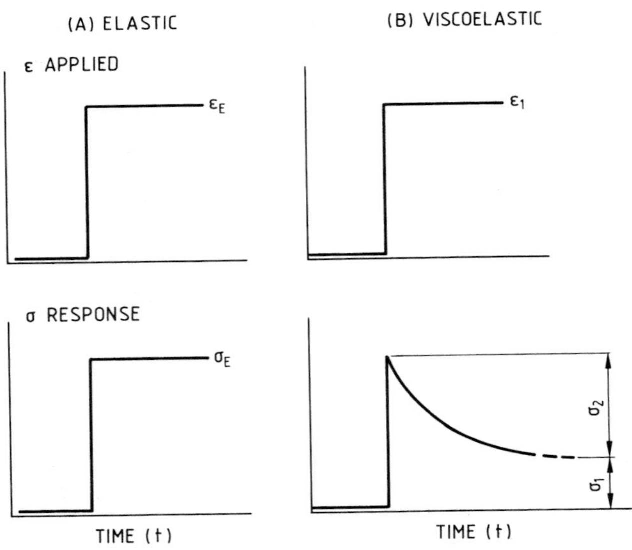

Figure 6.10 Stress relaxation programmes for (A) Elastic (subscript E); and (B) Viscoelastic bodies (subscripts 1 and 2 refer to elastic and viscous components of stress).

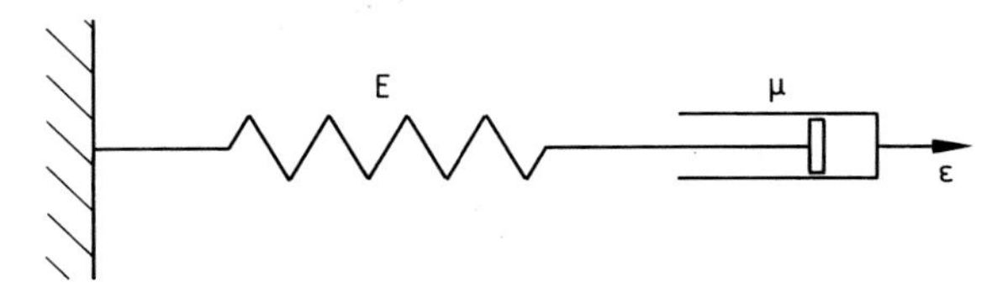

Figure 6.11 Maxwell Model (simulation of stress relaxation).

Stress relaxation. The component forces and deformations (respectively) are given as follows (using identical notation to that given previously):

$$\sigma = \sigma_E = \sigma_V \tag{6.10}$$

$$\varepsilon = \varepsilon_E + \varepsilon_V \tag{6.11}$$

Equation 6.11 may be differentiated and equated to zero (since $\frac{d\varepsilon}{dt} = 0$ in the stress relaxation mode):

$$\frac{d\varepsilon_E}{dt} + \frac{d\varepsilon_V}{dt} = 0 \tag{6.12}$$

From Equation 6.10 (substituting for stress from 6.4 and 6.5):

$$\varepsilon_E \cdot E = \mu \cdot \frac{d\varepsilon_V}{dt}$$

Since the derivatives of viscous and elastic strain components are equal and opposite (6.12), we have:

$$\int \frac{d\varepsilon_E}{\varepsilon_E} = -\frac{E}{\mu} \int dt$$

Subsequent integration (the condition that $\varepsilon = \varepsilon_0$ at zero time gives an integration constant $\ln(\varepsilon_0)$) yields the following solution:

$$\ln(\varepsilon_E) = -\frac{Et}{\mu} + \ln(\varepsilon_0)$$

Therefore:

$$\varepsilon_E = \varepsilon_0 \cdot \exp\left(\frac{-Et}{\mu}\right) \tag{6.13}$$

Stress relaxation is expressed as follows:

$$\sigma(t) = E \cdot \varepsilon_E = \sigma_0 \cdot \exp\left(\frac{-Et}{\mu}\right) \tag{6.14}$$

Once more, a simple exponential relationship is predicted by this type of model; it therefore serves as an approximation to in-service stress relaxation behaviour of plastics. An additional parameter which is often quoted for a test regime of this type is *relaxation time*, τ_R.

Since $\tau_R = \frac{\mu}{E}$, we can express stress relaxation in terms of a characteristic material relaxation time:

$$\sigma(t) = \sigma_0 \cdot \exp\left(\frac{-t}{\tau_R}\right) \tag{6.15}$$

(σ_0 is the initial stress (zero relaxation time) at constant strain.)

The Maxwell model predicts an exponential stress decay; the extent and rate of relaxation depend upon the characteristic relaxation time of the material, which is highly temperature dependent. Time-dependent stiffness, under stress relaxation conditions, is expressed thus:

Relaxation modulus, $$E(t) = \frac{\sigma(t)}{\varepsilon}$$

6.4.3.2 Stress Relaxation Data – Measurement and Presentation

Similar problems to those described in Section 6.4.2.2 are also apparent in stress relaxation tests. Achieving a pre-specified strain with accuracy is not always a simple practical exercise; this is especially true when it is considered that first, the initial strain rate should be quite high, in order to control the effects of *simultaneous relaxation* during the initial straining period; and second, low modulus materials (PPVC, thermoplastic elastomers) undergo large strains before reaching stress values of practical and commercial interest. Having achieved an appropriate "zero-time" deformation, stress decay is usually monitored via a co-axially mounted load cell. This arrangement is convenient when the deformation mode is *tensile* or *compressive;* the former mode is practically convenient, whilst the latter is more directly relevant to many commercial stress relaxation situations such as in pipe seals. Stress relaxation testing in compression is associated with some additional sample geometry constraints, the effects of which must be noted (if not completely overcome) during the experiment; these include "end-effects" such as lateral deformation restriction (due to sample-platen friction), and limitations of sample length due to column buckling effects.

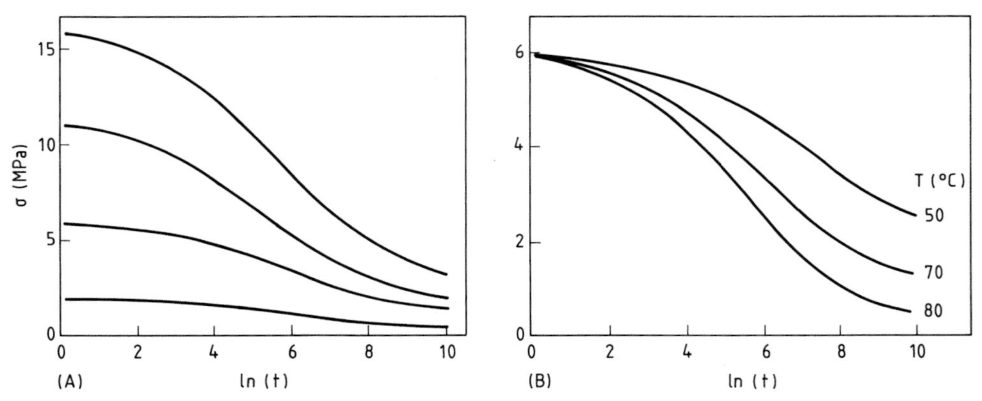

Figure 6.12 Stress relaxation data for injection moulded PMMA (ICI Diakon® grade MG102) expressed as stress (σ) versus natural logarithm of time (time in seconds): (A) effect of initial stress (σ_O) at 20 °C; (B) effect of temperature (T), at a constant initial stress (σ_O) of 6 MPa.

Whilst appreciating the different fundamental nature of creep and stress relaxation testing, but noting the similarities of test data sensitivity to temperature, humidity etc., it becomes apparent that a method of correlation between the tests would be desirable, wherever feasible, to avoid any duplication of effort; this point is considered further in Section 6.7.

Relaxation data are presented in a format (eg. *stress vs log (time)*) which reflects timescales of relevance to practical situations; a typical plot showing the relaxation characteristics of PMMA (as a function of time, temperature and straining mode) is presented in Figure 6.12.

6.4.4 More Complex Models – The Standard Linear Solid

Creep and stress relaxation can be simulated, albeit to a strictly limited level of accuracy, by the Voigt-Kelvin and Maxwell models (respectively). It is a simple matter to show[6.3,6.13] that the Maxwell element alone cannot describe creep and recovery, nor can the Voigt model be used to model relaxation effects. For enhanced accuracy, or in cases where more specific viscoelastic properties are to be studied, models of additional complexity can be proposed. Not surprisingly, as the demanded levels of accuracy are increased, so too is the degree of mathematical complexity associated with the enhanced simulation technique.

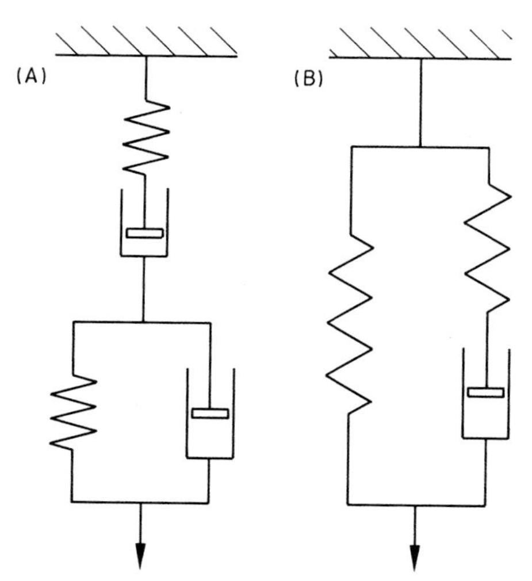

Figure 6.13 Multi-element viscoelastic models: (A) Series combination of Voigt-Maxwell bodies; (B) Standard Linear Solid.

Two examples of *Multi-Element Models* are shown in Figure 6.13; the first of these consists of a series arrangement linking the Maxwell and Voigt-Kelvin bodies, which has been shown[6.13] to describe *both* creep and stress relaxation with a higher degree of accuracy. Alternatively, predictions can be derived from the *Standard Linear Solid* arrangement in Figure 6.13(B) (Maxwell and elastic elements in parallel; first attributed to Zener[6.14]) based upon the following stress and deformation relationships:

$$\sigma_{E(1)} = \sigma_V \qquad \text{and} \qquad \sigma = \sigma_{E(1)} + \sigma_{E(2)}$$
$$\varepsilon = \varepsilon_{E(2)} = \varepsilon_{E(1)} + \varepsilon_V$$

By substitution and rearrangement, it can be shown[6.3] that the solution is:

$$\sigma + \frac{\mu}{E(1)} \cdot \frac{d\sigma}{dt} = E(2) \cdot \varepsilon + (E(1) + E(2)) \cdot \frac{\mu}{E(1)} \cdot \frac{d\varepsilon}{dt} \tag{6.16}$$

This has the general form:

$$a_0 \cdot \sigma + a_1 \frac{d\sigma}{dt} = b_0 \varepsilon + b_1 \frac{d\varepsilon}{dt} \tag{6.17}$$

where a_0, a_1, b_0 and b_1 are constants of the individual grade of material.

This model is therefore applicable to creep if $\frac{d\sigma}{dt} = 0$ (constant applied stress) and can be used for stress relaxation when $\frac{d\varepsilon}{dt} = 0$ (constant deformation). Further accuracy is only achieved by considering a greater number of constituent Voigt or Maxwell units in the overall model; this results in additional higher derivative terms in the general constitutive equation (equivalent to 6.17). Since each sub-unit in the overall model will be characterised by individual elastic and viscous constants, we therefore refer to *relaxation spectra* or *retardation spectra* (for stress relaxation and creep respectively), rather than to any unique relaxation/retardation times, in order to describe the deformation behaviour of plastics.

6.4.5 Superposition: Successive/Intermittent Loads and Recovery

In spite of the accuracy with which the prediction of creep and stress relaxation can often be made, the applicability of any theoretical model of these facets of viscoelasticity remains limited. This is because the majority of practical loading situations are more complex than those already described, involving successive or intermittent loads, together with periods of load removal and subsequent recovery. For linear viscoelastic materials, it must be noted that in more complex situations, each element of the *total loading history* continues to contribute to the time-dependent change in deformation; this statement forms the basis of the *Boltzmann Superposition Principle*[6.3], which is described (with some examples) below.

Boltzmann Superposition Principle – eg. Application to creep for linear viscoelastic materials:

"It is assumed that the overall magnitude of (time-dependent) strain consists of individual contributions from each independent input of stress; these are assumed to be active, even when additional stresses are applied, and superposition assumes further that the total strain can be calculated by simple addition of the individual strains."

This analysis is equally applicable to positive stress (loading) and to negative stress (unloading and partial recovery) and is therefore relevant to many practical loading systems.

Superimposed Stresses. Consider the applied stress programme in Figure 6.14: σ_0 is applied at zero time, and the stress is successively increased to σ_1 at time t_1, σ_2 at t_2 etc. The total strain response (bold line) is a function of each successive loading increment, and may be estimated (in terms of creep compliance) as follows:

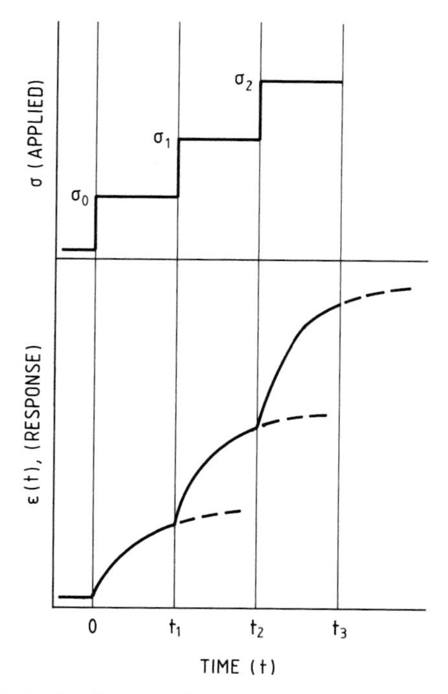

Figure 6.14 Application of successive stresses, and the strain response, to illustrate the Boltzmann Superposition Principle.

Creep strain (at any time, t) due to σ_0 (only):

$$\varepsilon(t) = \sigma_0 \cdot J(t)$$

Using the Superposition Theory, the *total* creep strain after the application of the second stress is equal to the strain component due to the additional stress, added to the anticipated strain for σ_0 alone (see above). That is:

Creep strain:

$$\varepsilon(t)(\sigma_0, \sigma_1) = \sigma_0 \cdot J(t) + (\sigma_1 - \sigma_0)J(t - t_1)$$

For a series of applied stresses, the total creep strain is expressed as:

$$\varepsilon(t)(\sigma_0, \sigma_1, \sigma_2 \ldots) = \sigma_0 \cdot J(t) + (\sigma_1 - \sigma_0) \cdot J(t - t_1) + (\sigma_2 - \sigma_1) \cdot J(t - t_2) + \cdots \tag{6.18}$$

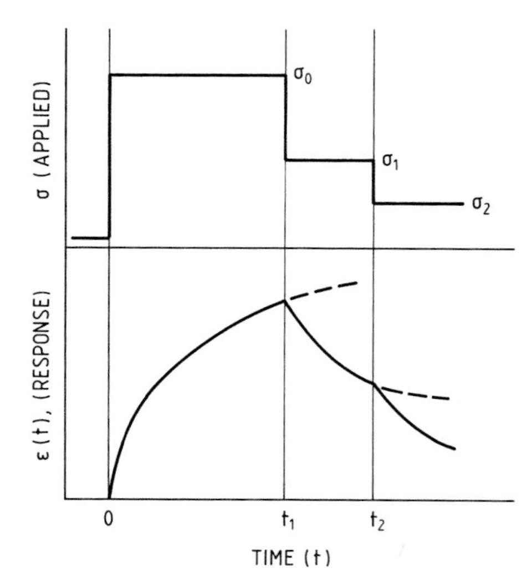

Figure 6.15 Superimposed stress-strain programme, involving creep loading and recovery phases.

Recovery. An equivalent analysis may also be used in situations where stresses are subsequently removed and partial strain recovery occurs; from Figure 6.15, the total creep strain after removing a proportion of the initial stress (σ_0) is given by:

$$\varepsilon(t)(\sigma_0, \sigma_1) = \sigma_0 \cdot J(t) - (\sigma_0 - \sigma_1) \cdot J(t - t_1)$$

For a series of recovery phases:

$$\begin{aligned}\varepsilon(t)(\sigma_0, \sigma_1, \sigma_2 \ldots) = \sigma_0 \cdot J(t) &- (\sigma_0 - \sigma_1) \cdot J(t - t_1) \\ &- (\sigma_1 - \sigma_2) \cdot J(t - t_2) - \cdots \end{aligned} \tag{6.19}$$

The magnitude of recovered strain is generally defined as the difference between the *anticipated strain* (due to loading) and the measured response. Using this definition, the recovered strain after the applied stress in Figure 6.15 has been diminished to σ_1 (for example) will be:

$$\varepsilon_R(t - t_1)(\sigma_0, \sigma_1) = \sigma_0 \cdot J(t) - [\sigma_0 \cdot J(t) - (\sigma_0 - \sigma_1) \cdot J(t - t_1)]$$

ie.

$$\varepsilon_R = (\sigma_0 - \sigma_1) \cdot J(t - t_1) \tag{6.20}$$

An alternative approach to characterise recovered strain, using an empirical *"Fractional Recovery"* parameter, is discussed in Section 6.7.

Intermittent Stresses. In practical situations where loading is periodic, successive strain recovery phases control the overall creep strain to a level which is often considerably less than would be the case for continuous loading under an equivalent stress[6.7]. Therefore, if we consider the loading "lifetime" in a periodic stress system as the total time under load, the approach leads to overestimation of total strain and is an over-rigorous design assumption. Instead, the *Superposition Principle* can be applied (as above) to such situations and the creep strain may be estimated for more complex stress histories. From Figure 6.16, it is appreciated that the maximum creep strain (at the end of each successive loading phase) is expressed as the strain due to the application of σ_0 for time t_1 $((t_1) = \sigma_0 \cdot J(t_1))$, added to the residual strain at the end of the previous recovery phase. Practical values of total creep strain for an intermittent loading programme usually fall between two limits: the upper limit is the theoretical creep strain plot for σ_0 (without unloading) and the lower limit is the creep strain observed for σ_0 applied for time t_1 with zero residual strain (ie. assuming complete recovery after each cycle). The latter situation is more desirable; such an ideal response is approximated under modest stresses, when the loading periods are short in comparison to each recovery phase.

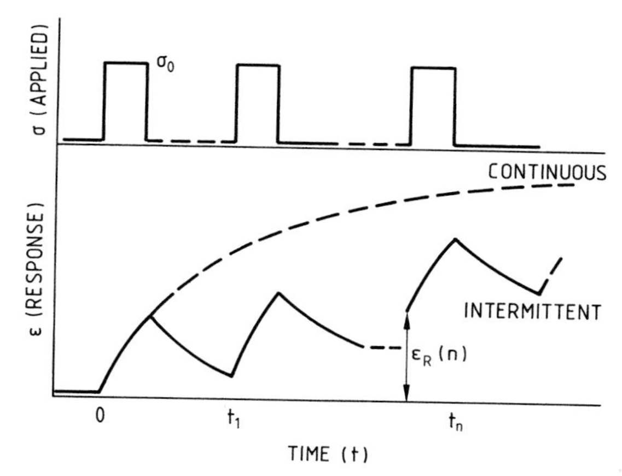

Figure 6.16 Creep loading programme, illustrating the effect of intermittent stresses.

Further treatment on the application of the Superposition Principle to periodic stress can be found elsewhere[6.15, 6.16].

6.4.6 Dynamic Mechanical Properties

Invaluable knowledge on polymer viscoelasticity can be gained by studying material response under conditions of fixed-frequency, oscillating stress. The effect of time dependence in linear viscoelastic solids is exemplified by the *phase lag* observed between stress and strain, which varies according to test conditions (frequency, temperature) and may be used to characterise structural motion and thermal transitions in plastics materials.

6.4.6.1 The Basis of Dynamic Mechanical Testing

Consider the deformation programme in Figure 6.17:

$$\varepsilon = \varepsilon_0 \cdot \sin(\psi t) \tag{6.21}$$

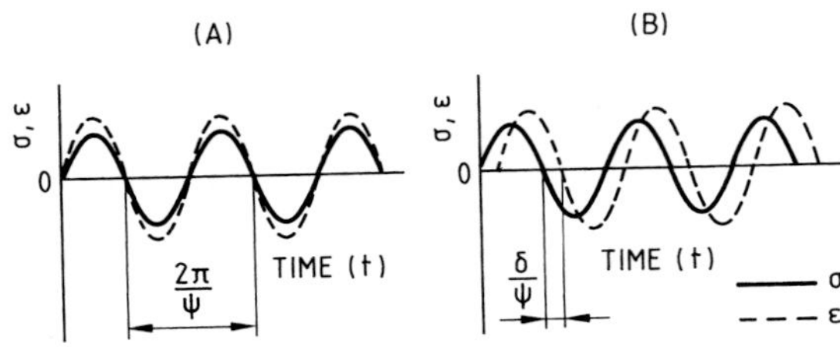

Figure 6.17 Dynamic stress-strain relationships for (A) an elastic body, and (B) a linear viscoelastic body. Note the phase difference between stress and strain, for the viscoelastic material.

Therefore, the stress in elastic and viscoelastic solids is:

Elastic body $$\sigma = \sigma_0 \cdot \sin(\psi t) \tag{6.22}$$

Viscoelastic body $$\sigma = \sigma_0 \cdot \sin(\psi t + \delta) \tag{6.23}$$

(ε_0, σ_0 are strain and stress amplitude, ψ is angular frequency)

For elastic solids (at any frequency), stress and strain are in-phase, and are related by a unique modulus value. On the contrary, no single parameter can be used to characterise the stress-strain relationship in viscoelastic materials, since strain lags stress by a *phase difference* (see Figure 6.17(B)). By making a trigonometric expansion in Equation 6.23, the stress may be resolved into two components:

$$\sigma = \sigma_0[\sin(\psi t)\cos\delta + \sin\delta \cdot \cos(\psi t)]$$

ie.

$$\sigma = (\sigma_0 \cos\delta) \cdot \sin(\psi t) + (\sigma_0 \sin\delta) \cdot \cos(\psi t)$$

Therefore:

$$\sigma = (E_1 \varepsilon_0) \cdot \sin(\psi t) + (E_2 \varepsilon_0) \cdot \cos(\psi t) \tag{6.24}$$

since the elastic moduli, assumed to be tensile in this case, are given by:

$$E_1 = \frac{\sigma_0}{\varepsilon_0} \cdot \cos\delta; \qquad E_2 = \frac{\sigma_0}{\varepsilon_0} \cdot \sin\delta$$

In Figure 6.18, the two components of stress are evident. Elastic modulus can be expressed in a similar manner (using complex notation) to relate stress and strain: E_1 describes stress-strain relationships which are in-phase, whilst the out-of-phase component (see Figure 6.18 B) is characterised by E_2. Furthermore:

$$\tan\delta = \frac{E_2}{E_1} \tag{6.25}$$

In complex notation:

$$\varepsilon = \varepsilon_0 \cdot \exp[i\psi t]; \qquad \sigma = \sigma_0 \cdot \exp[i(\psi t + \delta)]$$

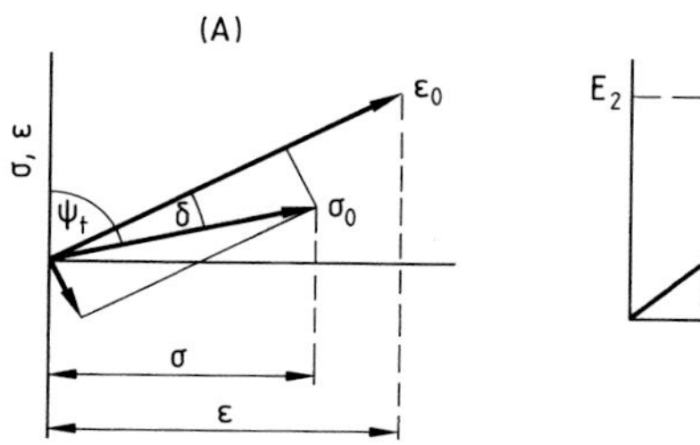

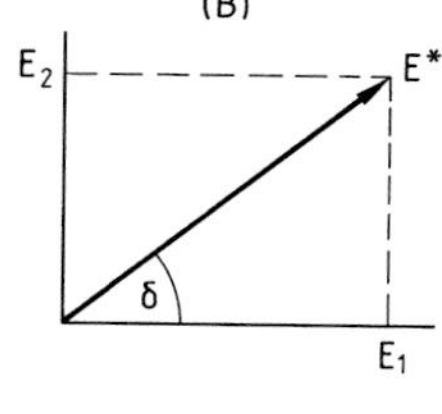

Figure 6.18 (A) Rotating vectors, whose projections represent stress (σ) and strain (ε) in sinusoidal oscillations, out of phase by a phase angle δ; (B) Vector notation for complex modulus (E^*), with real (E_1) and imaginary (E_2) parts.

Therefore the *Complex Dynamic Modulus* (E^*), consisting of a real and an imaginary component, is evaluated thus:

$$E^* = \frac{\sigma}{\varepsilon} = \frac{\sigma_0}{\varepsilon_0} \cdot \exp(i\delta) = \frac{\sigma_0}{\varepsilon_0} \cdot (\cos\delta + i\sin\delta)$$

Therefore:

$$E^* = E_1 + iE_2 \tag{6.26}$$

Clearly, under high-frequency dynamic loading conditions, plastics materials are characterised by two modulus components, and by the phase angle which, in reality, separates the sinusoidal stress and strain cycles.

E_1 is the real part of the complex modulus; it is often referred to as the *Storage Modulus,* since it represents the stored elastic energy in the sample. E_2, the imaginary component, arises due to viscous dissipation and is generally described by the term *Loss Modulus.*

Tan δ describes the significance of viscous dissipation in a material (at a given test temperature and frequency) and is often the parameter chosen to relate dynamic data to molecular or structural motion in plastics materials.

An additional parameter which can be estimated is the *cyclic energy dissipation* (per unit volume) as a result of inelastic deformation:

$$\text{Energy,} \quad U = \int_0^{\frac{2\pi}{\psi}} \sigma \cdot d\varepsilon = \int \sigma \cdot \frac{d\varepsilon}{dt} \cdot dt$$

Making the substitutions for σ (from 6.24) and $\frac{d\varepsilon}{dt}$ (by differentiating 6.21), we obtain:

$$U = \psi(\varepsilon_0)^2 \int [E_1 \cdot \sin\psi t \cdot \cos\psi t + E_2 \cos^2 \psi t]\, dt$$

This has the solution[6.3]:

$$U = \pi \cdot E_2 \cdot (\varepsilon_0)^2 \tag{6.27}$$

6.4.6.2 Test Techniques

Many dynamic testing methods have been developed; the general requirements are that the applied stress should be controlled to vary sinusoidally with time (preferably over a range of chosen frequencies), and that the resultant deformation be measured by a convenient method. Choice of technique depends upon the required frequency range, the available specimen geometry and the mode of applied stress/deformation. A convenient means of classifying dynamic test methods, adopted by the texts in References 6.3 and 6.7, is as follows:

Free-vibration methods	Frequency range 10^{-2}–10^{1} s^{-1}
Forced-vibration methods	Frequency range 10^{1} –10^{3} s^{-1}
Wave propagation	Frequency range 10^{3} –10^{6} s^{-1}

Two examples of dynamic test techniques are analysed in more detail below. The first (Torsion Pendulum) is a simple laboratory method which can be used to generate data at minimal cost, whilst the latter (DMTA) is a more modern and sophisticated technique which characterises properties over wide spans of frequency and temperature.

- **Torsion pendulum.** This is a free-vibration technique (see Figure 6.19) in which the strip or rod sample is suspended under a rigid frame; an inertia bar attached to its lower end is twisted initially about the centre-axis, then released and allowed to oscillate freely (at a frequency which depends upon the shear modulus of the polymer) whilst the amplitudes of successive oscillations are measured.

The viscoelastic character tends to damp the oscillations, resulting in a cyclic, time-dependent amplitude drop (see Figure 6.19).

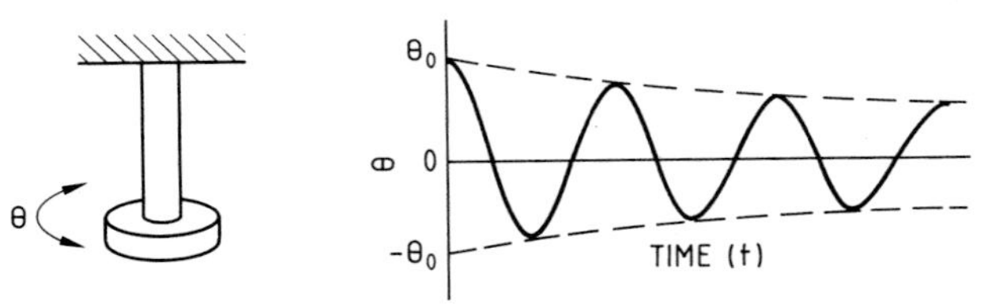

Figure 6.19 The Torsion Pendulum: loading mode and sinusoidal angular displacement (θ).

The equation of motion is a simple modification of that which represents an oscillating spring, since the dynamic modulus of viscoelastic polymers consists of both real and imaginary parts:

$$I \cdot \ddot{\theta} + A \cdot (G_1 + i \cdot G_2) \cdot \theta = 0 \tag{6.28}$$

(I = specimen moment of inertia, A is a geometric factor, θ represents the oscillation amplitude, and the real and imaginary components of shear modulus are G_1 and G_2 respectively)

Real data can usually be described by an exponential model of the form:

$$\theta = \theta_0 \cdot \exp(-pt) \qquad (\text{where} \quad p = i\psi - a)$$

(θ_0 is the initial amplitude of vibration, ψ is angular frequency and a is a constant)

Substituting for θ into 6.28, and separating real and imaginary parts (noting also that $p^2 = a^2 - \psi^2 - 2ai\psi$), we obtain:

$$G_1 = -I \cdot \frac{a^2 - \psi^2}{A}; \qquad G_2 = \frac{2a\psi i}{A} \tag{6.29}$$

The *Logarithmic Decrement* (λ) is given by:

$$\lambda = \frac{2\pi a}{\psi}$$

Substituting for a in Equation 6.29:

$$G_1 = \frac{I \cdot \psi^2}{A} \cdot \left[1 - \frac{\lambda^2}{4\pi^2}\right]; \qquad G_2 = \frac{i\lambda\psi^2}{\pi A} \tag{6.30}$$

Also:

$$\tan\delta = \frac{G_2}{G_1} = \frac{\lambda/\pi}{1 - (\lambda^2/4\pi)} \tag{6.31}$$

In practice, $\tan\delta$ can be estimated from logarithmic plots of θ versus time, whilst the components of the complex shear modulus, G_1 and G_2, can only be estimated when I and A have been determined by calibration with a material of known stiffness.

- **Dynamic mechanical thermal analysis (DMTA).** The measurement principle in DMTA involves the application of an oscillating strain (from a vibrating head) to a rectangular test-piece over a designated temperature span, at any of a range of available pre-selected frequencies. The mode of deformation is usually three-point bending, although some alternatives may be used for experiments of a more individual nature: eg. direct tension (for fibres and films), or a shear cell (to study adhesive joints). An example of a DMTA machine specification[6.17] is as follows:

Frequency range	$0.01 - 200$ Hz
Damping range ($\tan\delta$)	$0.0001 - 9.999$
Temperature range	$-150 - +300\,°C$
Young's modulus range	$10^5 - 10^{11}$ N/m^2
Displacement range	$0.01 - 0.25$ mm

Dynamic losses, associated with thermal transitions which accompany the onset of specific mechanisms of molecular motion, are usually evident at T_g (*α-relaxation peak*) and at a secondary process (*β-peak*) in the glassy state; to some extent, each mechanism is frequency dependent. The three-dimensional plot in Figure 6.20 illustrates the temperature and frequency dependence of $\tan\delta$, for a specimen of rigid PVC[6.17].

The DMTA principle represents an excellent example of how the viscoelastic nature of high polymers may be used to provide valuable information for purposes of materials development, process optimisation and component design[6.18, 6.19]; some practical applications are summarised below.

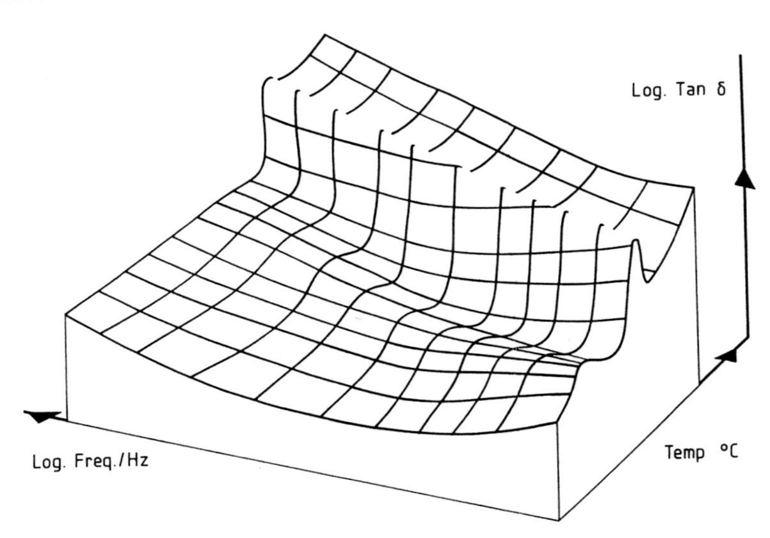

Figure 6.20 DMTA data: the frequency and temperature dependence of Tan δ for PVC. Note the different effects of frequency on the α and β transitions. (Reproduced by permission from Polymer Laboratories Ltd.)

(A) Characterisation: Relaxation peaks for many homopolymer systems are easily characterised; however, if the relaxation spectra of related blends or copolymers are also specified, the effect of compound modification is clearly evident. For example, if a two-component polymer alloy is *miscible,* there is a shift in the primary α-peak, to an extent determined by the blend composition, in relation to the glass transition temperature of each homopolymer.

(B) Analysis of Thermosetting Plastics: Dynamic measurements in the shear mode are useful in studying *crosslinking kinetics* and the *state of cure* in thermosetting resins. Cure analysis requires measurements to be taken (isothermally, or via a temperature scan) on specimens whose modulus changes dramatically as the physical form changes from a low viscosity liquid, to a hard, glassy solid, during crosslinking. The data in Figure 6.21 show T_g relaxation spectra in a carbon fibre-reinforced epoxy resin; the position and area of the loss peaks in composites indicate the degree and direction of reinforcement, in addition to the degree of cure in the matrix polymer.

(C) Orientation: Sample stretching can be accomodated under pre-specified conditions (ie. drawing above T_g to various strain levels, followed by controlled cooling) and subsequently, the relaxation spectra can be determined in the usual manner; this experiment is especially useful for analysing orientation in strain crystallisable plastics.

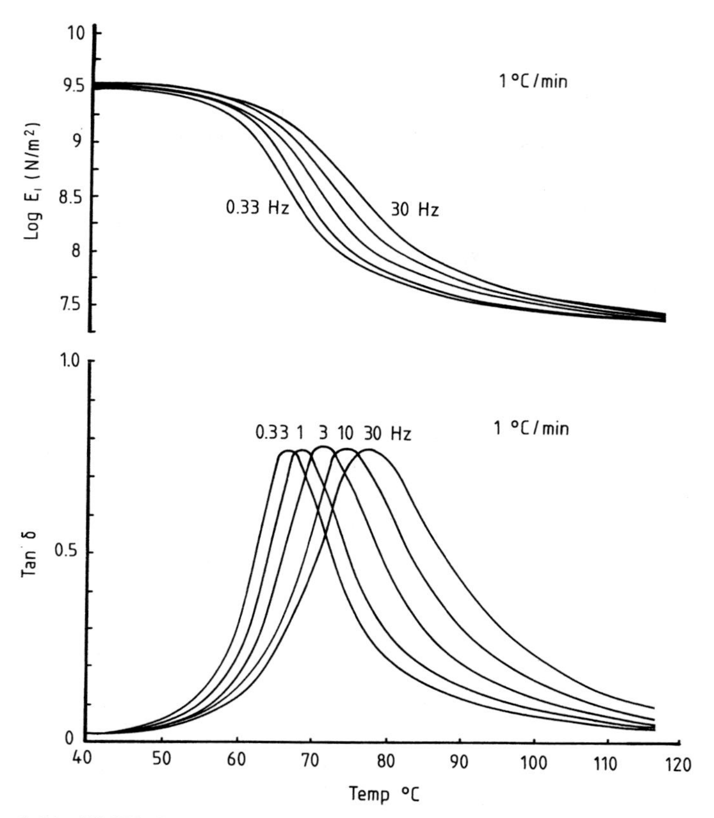

Figure 6.21 DMTA Data: the dependence of storage modulus (E_1) and $\tan\delta$ on temperature (across the α-transition), at various frequencies between 0.33 – 30 Hz, for a carbon-fibre reinforced epoxy resin. (Reproduced by permission from Polymer Laboratories Ltd.)

6.5 Yielding of Plastics

Yielding is synonymous with *plastic failure* in materials. There are some well-defined *yield criteria* (eg. Tresca, Von Mises, Coulomb) which are able to account for the onset of gross plastic collapse in traditional ductile materials such as steels. Furthermore, ductile failure can often be related to microstructural features (such as dislocations and other crystal lattice defects) and subsequently explained in terms of processing history and physical factors such as temperature, which influence the likelihood of yielding, and the observed yield stress.

In a more general context, plastics products differ from metal components in several ways:

(1) Plastics are very difficult to categorize as wholly "ductile" or "brittle" materials; this is because of the large number of factors which may contribute towards embrittlement of a predominantly ductile material (and vice versa). These include temperature and time/rate factors, and elements of microstructure, morphology, or material formulation such as molecular weight, crystallinity, orientation, chain branching, crosslinking, and the presence of fillers, plasticizers or fibrous reinforcements.
(2) The response of plastics up to the point of yielding is inelastic and is rate-dependent. Clearly, the onset of yielding will be influenced by the events which precede it.
(3) Microstuctures of fabricated articles are inevitably quite complex. Attempts to explain yielding in such terms have not yet reached a mature stage.

Whilst ductile yielding is a preferable and more predictable mode of failure for plastics products, the overall context of yield encompasses far more than a simple large-scale "failure" criterion; for example, *localized multiple yielding* is an effective means of toughening some families of commercial plastics[6.20, 6.21]. Also, it is possible to process thermoplastics by *forming* or by *drawing* in the uniaxial or biaxial mode; such processing methods utilize plastics materials' ability to undergo yield and large draws in a controlled, stable manner. Examples of products include monofilaments and oriented tapes from PP or PETP, and biaxially oriented films, sheets and bottles. It is for these reasons that yielding is considered as a facet of deformation in this chapter, rather than only as a type of failure in Chapter 7.

6.5.1 Aspects of the Yield Process under Tensile Stress

The most convenient practical means by which a material's resistance to an applied mechanical load can be studied is by using a constant deformation rate test in flexure, or in direct tension. We look at the phenomenon of yielding by considering the data from such tests, in two different ways.

6.5.1.1 Load-Extension Data

Most elements of ductile yielding can be seen clearly from load-extension (or, *nominal* stress-strain) data; it is a straightforward matter to relate the progression of the stress-strain plot and the changes in physical appearance to the various microstructural changes which occur. Consider the information in Figure 6.22: the numbers in parentheses indicate the link between the appearance of the test-piece and some points of interest on the load-extension trace during the experiment.

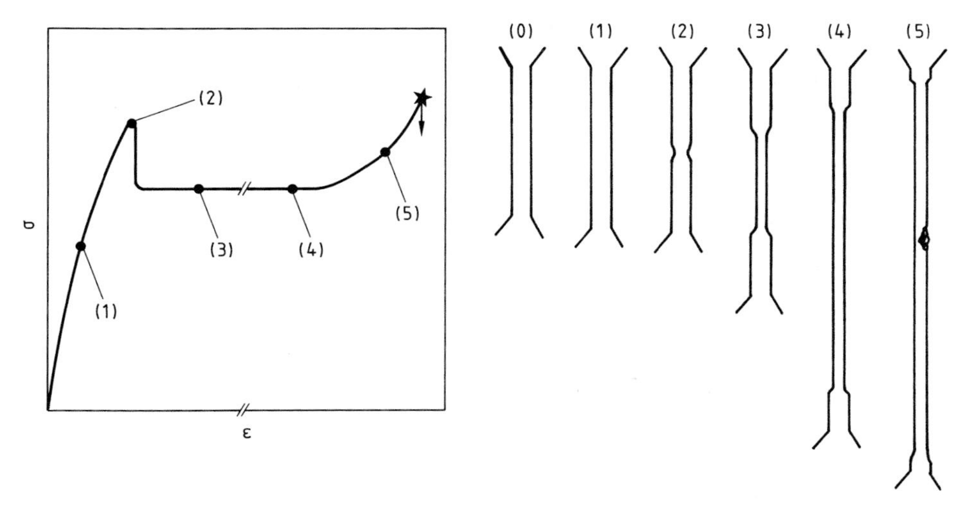

Figure 6.22 Nominal stress-strain relationship for a ductile plastic which undergoes yield and cold drawing. The physical appearance of the sample is also illustrated at various stages of the test (see text).

(1) Pre-yield behaviour. Materials which show distinct yield points inevitably exhibit time and temperature sensitive deformation characteristics (and non-linear stress-strain behaviour) under equivalent conditions. In contrast, stiff but brittle materials display a predominantly elastic (linear) response which ends suddenly (usually without any sign of gross yielding), as rapid crack propagation occurs at the moment of failure. The sensitivity of low-strain deformation to temperature and time has already been referred to in Sections 6.3.3 and 6.4.1.

(2) The yield point. The yield stress defines the maximum upper design stress limitation for any plastics material which is expected to fail in a ductile manner. At the yield point, molecular chain segments are able to slip past each other; such a deformation process, whether it occurs within crystalline or amorphous regions, is entirely irreversible and is exemplified by a sharp load-drop in the experiment. The lateral sample dimensions are immediately narrowed; this observation has become known as *necking* and is most often seen in crystalline plastics above T_g. (The criteria which determine the stability of necking are discussed in the following section).

(3),(4) Cold drawing. True stress rises in the vicinity of the necked region, so that further deformation may occur preferentially in the elements of material close to this point. However, the molecules become highly aligned in the direction of applied stress; local stiffness increases as a result, and a point is reached where the enhanced resistance to deformation in the plastically-deformed (anisotropic) region compensates the simultaneous increase in true stress. The post-yield necking process becomes stable in this way; further deformation results in neck propagation along the waist of the sample. This process is known as *cold drawing,* which occurs at a *drawing stress* often independent of strain level. The *orientation-stiffening,* or *strain-hardening* effect involves high local strains; consequently, cold drawing along the entire sample is reflected by an elongation of several hundred percent, or by a *natural draw ratio* of several times the original sample length.

(5) Fibrillation and fracture. At the point where the relatively compliant and isotropic material has all been highly strained (we often refer to a "chain-extended" crystalline morphology which forms during cold drawing), the load rises since the applied deformation is resisted by the uniaxially-oriented microstructure (of higher stiffness). Although the strength and stiffness properties parallel to the draw direction have been enhanced, adjacent segments of parallel chains are generally held only by secondary bond forces; longitudinal *fibrillation* is a usual precursor to eventual fracture, once the material in the remaining cross-section is unable to support the applied load.

6.5.1.2 True Stress-Strain Data – The Considère Construction

Clearly, necking stability is an important criterion which determines the propensity for cold drawing. Since the cross-section of material changes significantly as a result of yielding, it is appropriate to work in terms of *true stress,* rather than nominal. A procedure first suggested by Considère, based upon true stress-strain characteristics, may be used to assess both the likelihood of yielding and the feasibility of neck stability. The procedure is based on a constant-volume assumption; its validity therefore extends to yielding processes which do not involve large-scale crazing, which would otherwise result in significant dilatation. Using this assumption, we can state that the product of cross-section (A) and sample length (l) remains constant as straining takes place:

$$A_0 \cdot l_0 = A \cdot l$$

(subscript zero refers to the unstrained condition). The strain is given by:

$$\varepsilon = \frac{\Delta l}{l_0} = \frac{l}{l_0} - 1$$

Therefore:

$$\begin{aligned} A &= \frac{A_0 \cdot l_0}{l} \\ &= \frac{A_0}{(1+\varepsilon)} \end{aligned} \tag{6.32}$$

True stress (σ') is force per unit area, so that:

$$\begin{aligned} \sigma' &= \frac{F}{A} \\ &= \frac{F(1+\varepsilon)}{A_0} \\ &= \sigma(1+\varepsilon) \end{aligned}$$

There are two points of note here:

(1) At any finite strain, true stress always exceeds nominal stress;
(2) $(1+\varepsilon) = \frac{l}{l_0} = R$ (R is the *extension ratio*).

Since the drawing force, $F = A_0 \cdot \frac{\sigma}{1+\varepsilon}$, then:

$$\frac{dF}{d\varepsilon} = \frac{A_0}{1+\varepsilon} \cdot \frac{d\sigma}{d\varepsilon} + \frac{\sigma \cdot -A_0}{(1+\varepsilon)^2}$$

If we select the appropriate location for yielding on a nominal stress-strain plot (ie. where $\frac{dF}{d\varepsilon} = 0$), then:

$$\frac{d\sigma}{d\varepsilon} = \frac{\sigma}{1+\varepsilon} = \frac{\sigma}{R} \tag{6.33}$$

This condition is fulfilled if it is possible to construct a tangent to the true stress-strain curve from a position $R = 0$ ($\varepsilon = -1$) on the abscissa. Two examples of ductile yielding may be described (see Figure 6.23):

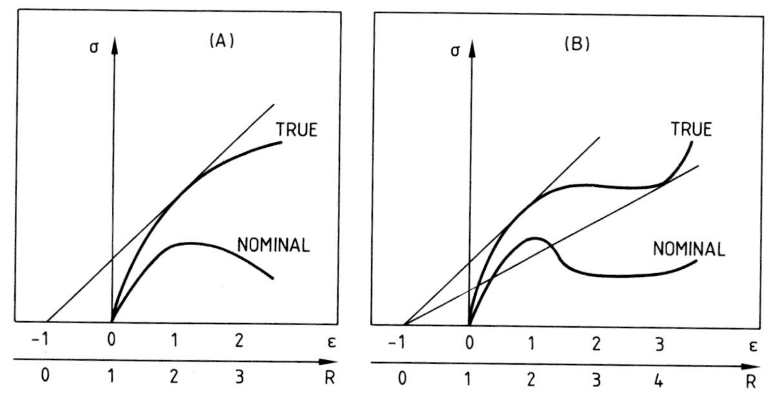

Figure 6.23 Use of the Considère construction (stress (σ) vs. strain (ε), or extension ratio (R)) to illustrate the conditions for unstable (A) and stable (B) yielding in ductile plastics.

Fig.6.23 (A). Equation 6.33 is satisfied; yielding is predicted at the point where the tangent from $R = 0$ meets the true strain locus. However, the neck is not stable; the sample cross-section becomes progressively narrower, until *necking rupture* occurs.

Fig.6.23 (B). Equation 6.33 is satisfied twice; this is the condition for stable neck propagation, leading to cold drawing.

This approach provides information on the likelihood of necking and cold drawing in plastics materials; in addition, it also helps define the true yield stress (true stress at maximum load) with enhanced accuracy. Further discussion on these points can be found elsewhere[6.3, 6.22–6.24].

6.5.1.3 Adiabatic Heating During Yield

It has long been recognized that a proportion of the absorbed deformation energy is transformed into heat during the yield process, causing significant local temperature increases with corresponding modification of properties[6.23–6.25]. However, attempts to explain necking and cold drawing purely on this basis are suspect, since the local temperature increases are rate-dependent, and are therefore quite modest at low strain rates. This effect can be appreciated quite readily by manually stretching a small sample of a low-modulus polymer (eg. LDPE rim-strapping from a cluster of aluminium drinks containers) at different rates.

WARD and others[6.3,6.26] have clarified an explanation of thermal effects during yield by showing that adiabatic heating influences the magnitude of drawing stress (especially at strain rates of approx. $0.1\ s^{-1}$) but does not modify the expected relationship between yield stress and strain rate. A drop in stiffness due to adiabatic heating is therefore not a single, adequate explanation for yielding and cold drawing; this mechanism does appear to influence the post-yield drawing stress, but only if the strain rate is such that the rate of heat generation exceeds the rate of heat-loss to the surrounding environment.

6.5.2 Crazing and Shear Yielding

Much of the early fundamental research on yielding in plastics materials was carried out by testing in direct tension, with the broad objective of increasing the level of understanding of those factors which contribute most to what was considered to be an incipient 'gross failure' process. However, since the recognition that any form of *micro-yielding* is synonymous with energy absorption, sometimes without apparent failure, much more attention has been given to studying preferential yield formation as a means of *toughness enhancement* in (predominantly) brittle plastics materials. This route towards improving impact and fracture properties (of major significance to amorphous plastics (styrenics, modified PVC) and to semi-crystalline 'engineering' thermoplastics (PA, POM, PBTP) is considered after an introductory section on the phenomenon of craze yielding.

6.5.2.1 Formation and Structure of Crazes

Crazing is a form of yielding usually, but not exclusively, observed in glassy thermoplastics when stressed at temperatures below T_g. Although craze yielding of products in service should always be considered undesirable (it occurs at a strain level of the order of 60 – 80 % of the "brittle failure strain", and therefore represents a precursor to eventual failure), its occurrence is not catastrophic since the microstructure of crazes is such that significant forces may still be carried.

Figure 6.24 depicts the edge-formation of surface crazes on the tension surface of a polymer (such as unmodified polystyrene), following the application of a flexural load; the growing craze propagates on a plane perpendicular to the direction of maximum principal (tensile) fibre stress. To the eye, crazes often appear to be similar to hairline cracks; although crazes may subsequently break down into crack-like defects under prolonged stress, they must not be considered in this context because the defects are bridged by *oriented microfibrils* – strands of yielded and drawn material which are oriented parallel to the stress direction – which

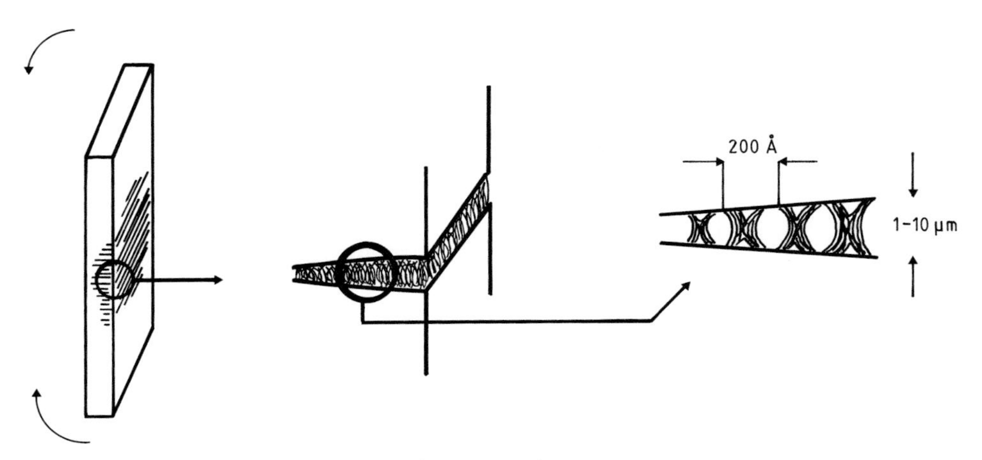

Figure 6.24 Formation and structure of crazes in plastics.

contribute significantly to further deformation resistance. The craze yielding process is *dilatational:* voids are formed between the oriented material which bridges the adjacent surfaces of the uncrazed bulk.

Crazing is commonly related to a *critical strain,* which increases if temperature is decreased, or if strain rate is increased. Stressing parallel to the orientation direction in a moulded part appears to have a similar effect, but the most striking observation is the extent to which crazing is accelerated when many plastics materials are loaded in the presence of an active environment. Surface vapour adsorption appears to assist the small-scale motion required for crazing by a pseudoplasticization effect. Amorphous plastics are particularly prone to *environmental stress crazing,* in combination with organic molecules such as ketones, chlorinated and other high-polarity solvents.

Crazing data for PES (at 150 °C) are superimposed upon conventional creep data in Figure 6.25; the time-dependent critical crazing strain (which represents a maximum design strain limit for this and similar materials, under given conditions) decreases from 1.4 to 0.8 %, if the loading timescale increases up to 10^6 s (12 days). The effect of molecular weight on crazing behaviour (for a given polymer) appears to hold only secondary significance.

The pioneering research on craze yielding is centred on work carried out by Kambour in the 1960's; much of this has been featured in an extensive review on the occurrence and effects of crazing[6.27]. The general mechanical criteria for craze yielding are beyond the scope of this text; an overview of some attempts to model craze formation is presented elsewhere[6.20].

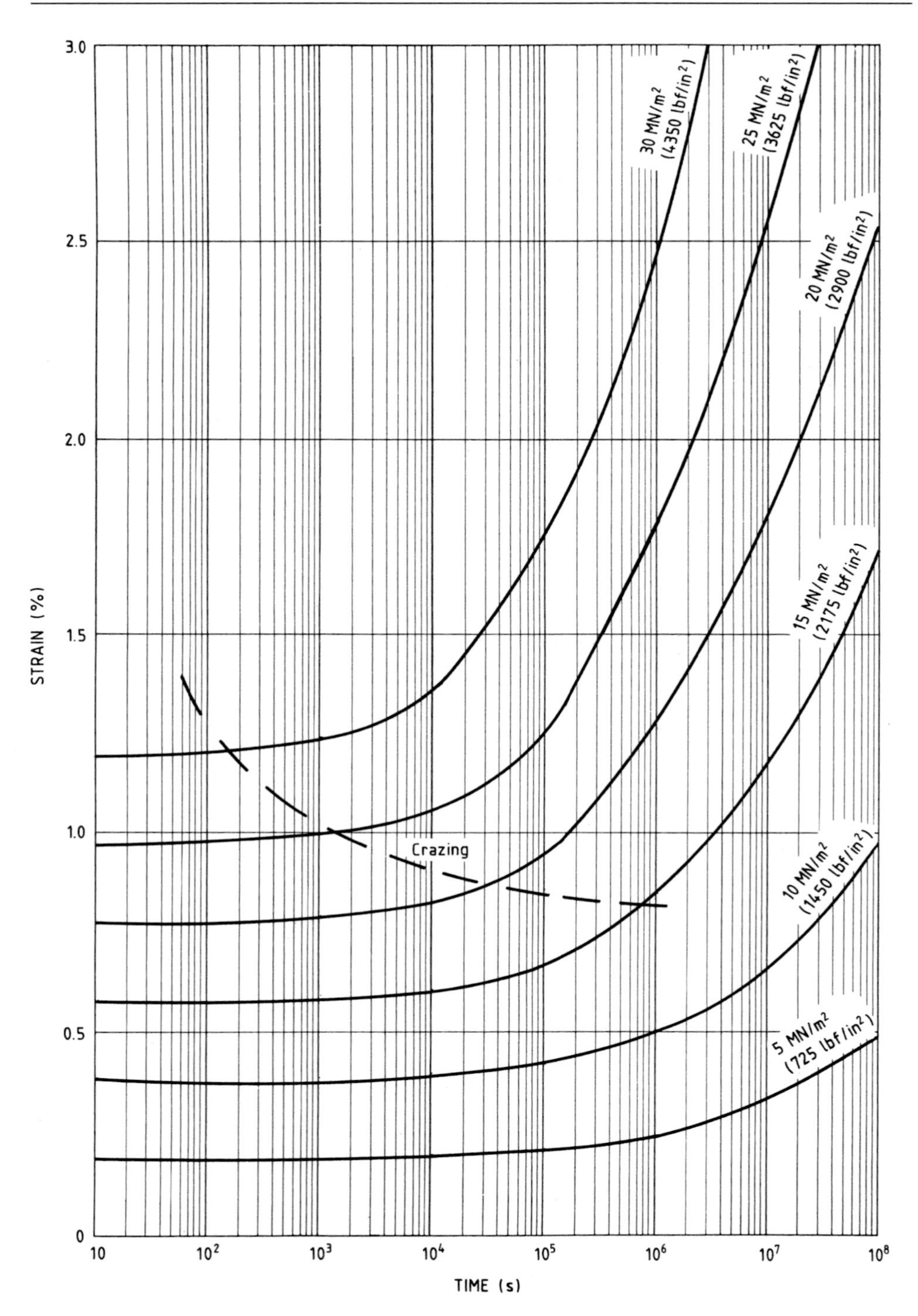

Figure 6.25 Superimposition of crazing locus on creep data, for 'Victrex'® PES (grade 300P) at 150 °C. (Data reproduced by permission of ICI).

6.5.2.2 Toughness Enhancement by Multiple Yielding

Yield criteria are often quoted in terms of *critical shear stress* parameters. Evidence of *shear yielding* is often seen in the form of shear bands, which occur on a plane of maximum resolved shear stress (see Figure 6.26). Unlike crazing, shear yielding is not dilatational and may be assumed to occur at constant volume; shear bands are deformation zones of high local strain[6.3,6.28]. In common with other forms of gross plastic deformation, the strain at which shear yielding occurs represents a mechanical design limitation for ductile polymers.

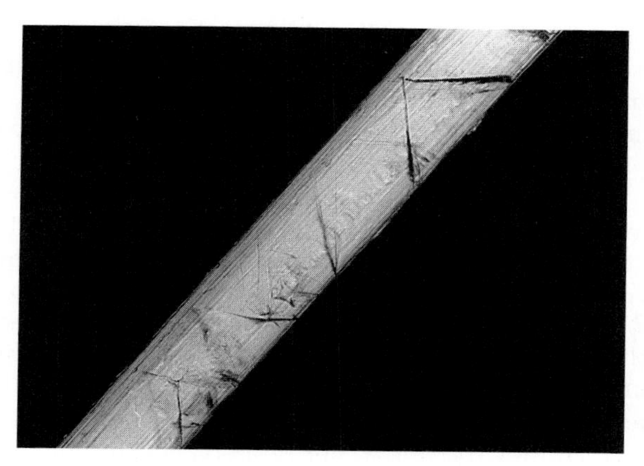

Figure 6.26 Light micrograph showing shear bands in a polyethylene monofilament, following excessive loading.

If yielding can be multiplied but localized, the proportion of stored energy which may otherwise contribute to a brittle (crack growth) failure mode can be greatly diminished. The desired condition to dissipate maximum energy in this way is to induce a multiplicity of *localized* yield zones, whether it be through crazing or by shear yielding. This is now a well-established *toughening* principle in thermoplastics, utilized by the addition of a rubbery phase by copolymerization (eg. butadiene in SAN, ethylene in PP) or by direct blending (eg. acrylic rubbers in PVC or PBTP). Some selection criteria for rubber-toughened plastics can be summarized as follows[6.20, 6.29]:

(1) The rubber should exist as a discrete second phase, usually as spheroids;
(2) Adhesion between the rubber and the matrix polymer must be optimized;
(3) The copolymerization or compounding processes must disperse the rubber phase effectively;
(4) The optimum particle size and distribution of the rubber may vary according to the yield characteristics of the matrix;
(5) For effective relaxation, the elastomer T_g should be well below the minimum expected service temperature of the toughened polymer.

There have been many attempts to predict the theoretical mechanism of rubber-phase toughening, some of which are inevitably specific to given combinations of polymers. As a more general indication, the following sequence may be adopted:

- Upon loading, the presence of the elastomeric phase concentrates and makes more complex the stress distribution in the surrounding matrix;
- Yield stress is reduced, and toughening occurs by the resultant plastic deformation, either by crazing *(Multiple Crazing Mechanism)*, or by the formation of shear bands *(Multiple Deformation Mechanism)*.

Multiple crazing is the underlying method of toughness enhancement in commercial HIPS compounds, and is characterized by severe *stress whitening* in the vicinity of maximum stress; this is caused by light scattering due to refractive index differences in crazed and unyielded material.

The effects of a shear-yielding toughening mechanism are less easily identified, since no such small-scale changes in specific volume occur. Moreover, separation of the two mechanisms is not always easy, since there is evidence to suggest that synergistic effects exist between them. Past experience does seem to show that accelerated shear yielding is an important element of impact and fracture-resistance in thermoplastics such as ABS and rubber-modified PVC.

It could, perhaps be anticipated that other more rigid fillers would induce yielding in brittle thermoplastics in a similar manner, assuming that all the above criteria (except (5), of course) were to be fulfilled. Whilst this is occasionally true, especially for ultrafine, chemically-treated fillers offering good adhesion to the parent polymer, it must also be recognized that the lack of an inherent energy-absorption characteristic within the rigid fillers limits the strain-accomodating properties of this type of composite material.

For a thorough and complete account of rubber-phase toughening, the reader is referred to the text by BUCKNALL[6.20], where property-enhancement in various plastics is referred to by individual citation. In addition, the specific nature of toughening in impact-modified PVC dry blends (a mechanism which is also extremely sensitive to the progressive elimination, during processing, of the particulate structure of PVC remaining from polymerization), has been studied by MENGES et al.[6.30]

6.5.3 Yielding in Semi-Crystalline Polymers

Much of the discussion on yielding within this section is primarily associated with glassy polymers whose structures show little, or a complete lack of small-scale molecular order. However, the level of understanding of yielding mechanisms in semi-crystalline plastics is not quite so advanced; this can undoubtedly be attributed to the fact that individual plastics materials crystallize in different forms (some of which are metastable), and to different extents. Consequently, they sometimes respond to an applied stress in a manner which is dominated by the relaxation characteristics of the amorphous phase.

Degree of Crystallinity. The importance of any crystal deformation mechanism increases with the overall degree of crystallinity. Also, if the amorphous phase is in its rubbery state ($T > T_g$), the crystalline lattice represents the main element of restraint in the microstructure.

Crystallite Orientation and Texture. Crystal lamellae consist of chain-folded structures whose (chain backbone) c-axis is perpendicular to the lamellar plane. For covalent bonds to remain unbroken under applied stress, a crystal deformation

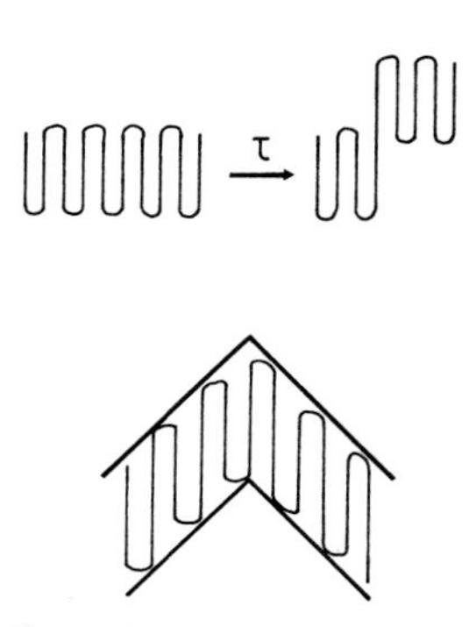

Figure 6.27 Crystal slip: an example of deformation in semi-crystalline plastics.

mechanism such as slip must occur with the slip-plane parallel to the chain axis; progressive, large strain deformations of this type result in the formation of the chain-extended crystalline form (Figure 6.27). The effect of crystalline texture on yield behaviour is probably overshadowed by the fact that large spherulites (produced by slow cooling from the melt) tend to induce a degree of brittleness, sometimes by a mechanism which is related to crack propagation through *spherulite boundaries.*

6.6 Property Enhancement by Orientation, Reinforcement and Foaming

The mechanical performance of plastics may vary considerably between groups of materials differing in structure and/or chain chemistry. Furthermore, most individual materials exhibit physical properties which can be enhanced or modified further, by several fundamental routes:

(1) Control of thermal and deformation history during processing allows the microstructure of parts to be changed to yield optimum performance; this is especially significant if molecular orientation can be induced to coincide with the likely direction of applied stress during service.

(2) The ability to enhance deformation resistance by compounding is a powerful selection criterion which plastics materials are able to satisfy. Improvements in specific stiffness characteristics can be generated by the addition of reinforcement fibres, particulate fillers, and by foaming.

Some introductory concepts of improved deformation resistance are reviewed in this section.

6.6.1 Molecular Orientation

Whilst deforming under applied stress during processing, molecular chains become preferentially aligned. The likelihood of a significant proportion of the *molecular orientation* remaining in the component microstructure depends upon the relative rates of relaxation and cooling; rapid cooling and/or delayed relaxation effectively locks the oriented molecules into the solid-phase structure.

Orientation developed in the melt-phase is stress-induced; therefore, a shear stress (τ) whose magnitude varies across the area of application (in injection moulding, for example, where τ is zero at the centre-axis and assumes its maximum value at the cavity surface) will apparently create a non-uniform orientation distribution. Also, since the cooling rate through T_g (or through the crystallization range) will be significantly different in many parts of a moulded product, the orientation and stiffness are likely to vary accordingly. The properties of fabricated plastics products are therefore not only *anisotropic,* they are also *heterogeneous.*

Molecular orientation in products can be classified loosely into several categories:

• **Low magnitude.** This type of orientation is predominantly uniaxial (according to flow) and is especially notable in injection moulded thermoplastics, but is also evident (though generally less critical) in some direct extrusion processes including extrusion blow moulding, where the orientation is induced by elongational stress, and lies in the inflation (hoop) direction. It is fair to suggest that this type of orientation occurs unavoidably, but can be controlled within limits during processing.

• **High magnitude.** In processes such as fibre-spinning and tape drawing (uniaxial), and in flat/tubular film manufacture, thermoforming and stretch blow moulding (biaxial orientation), orientation of higher magnitude is induced deliberately under high strain-rate tensile stress. Products manufactured from liquid crystal polymers (LCP's) may be included in this category.

- **Ultra-high orientation.** Although "ultra-drawn", or highly-oriented products are yet to gain significant commercial importance, the principle of introducing high orientation to coincide with stiffness requirements remains conceptually sound and ripe for further development. Research in this category is concerned with solution-grown chain-extended morphologies (eg. 6.31,6.32) and practical die-drawing techniques[6.33, 6.34].

(A "modulus" scale which helps to quantify this classification, with additional reference to other materials, is presented in Figure 6.28).

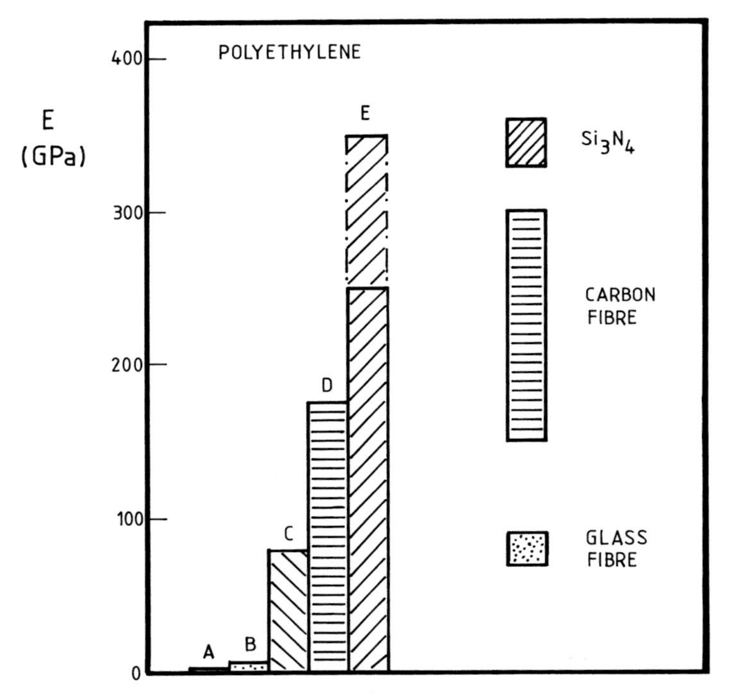

Figure 6.28 Tensile stiffness (E) of oriented polyethylene (A-E), in comparison to some other reinforcement materials: (A) – Isotropic PE (0.8 GPa); (B) – PE drawn fibre (3–5 GPa); (C) – Ultra-drawn PE; (D) – Solution-grown PE crystals; (E) – Maximum theoretical values.

Even in the first of these categories (which is the example most often encountered in practice), whilst the sensitivity of low-strain deformation resistance to orientation is probably less-severe than the equivalent relationships involving strength and toughness, it is always unwise to ignore the effects that changes in processing conditions have on the *extent* of orientation, and the subsequent magnitude of component stiffness.

A convenient method to study orientation effects is by measuring modulus in each principal deformation direction, relative to the stiffness of isotropic samples of the same polymer. Such *anisotropy ratios* reflect the enhanced properties which can be achieved by orientation; these are modest (around 1.5 – 2.0) in injection moulding, but can be considerably increased in the high-orientation categories described above (Figure 6.28).

6.6.2 Heterogeneous Reinforcement

An effective means to improve stiffness is by design and manufacture using *composites,* where a base (matrix) polymer is combined with a stiffer material, either before processing (by pre-compounding) or during manufacture. *Compounding* is preferred for volume-intensive processes using bulk moulding or extrusion-grade polymers (reinforced with particulate or short-fibre additives); continuously-reinforced products are often manufactured using *direct impregnation* techniques with low-viscosity liquid resins which are subsequently cured.

In general terms, a second phase is added to fulfil the following objectives:

- enhance physical properties
- reduce volume cost

6.6.2.1 Composite Materials

Commensurate with the overall objective of improving the mechanical properties of the base polymer by transferring applied loads to the stronger and stiffer reinforcement material, the functions of the matrix and reinforcement constituents are:

Matrix
- to support the fibres, preserving any preferred alignment;
- to transmit load to the fibres, by *interfacial shear;*
- to protect the reinforcement from environmental attack.

Reinforcement
- to optimize load-bearing properties, per unit weight and cost of material;
- to offer good adhesion fastness with the matrix, in order to promote and maintain interfacial load transmission.

Matrix polymers commonly used in composite systems include:

Thermoset resins. Unsaturated polyesters, epoxides and phenolic resins (wet-laid, continuous composites) together with *Dough/Sheet Moulding Compounds (DMC/SMC)* from polyesters and phenolics, and fibre-reinforced injection moulding powders from phenolic and amino resins. Thermosetting materials are particularly suitable as composite materials, due to their relative ease of fabrication and good adhesion characteristics.

Thermoplastics. Short-fibre reinforced, or mineral-filled grades of most families of thermoplastics are now available, predominantly for injection moulding of products for the high-performance "engineering" sectors.

Reinforcement materials in plastics composites are best divided into categories according to shape:

Particulate reinforcement (fillers). These include any particulate (sphere, cube, block), or platelet materials without an obvious "long-dimension". Stiffness enhancement is positive but modest, and is often offset by a reduction in toughness;

properties are improved by chemical-modification of particle surfaces to improve adhesion. Typical materials are inexpensive, bulk (inorganic) commodities such as calcium carbonate, talc, mica, kaolin, fly-ash etc[6.35].

Fibrous reinforcement. Much greater property enhancement is feasible by designing polymer composites reinforced by high aspect ratio fibres. Economic factors dictate that glass is most commonly used for this purpose; general properties of glass and other fibrous reinforcements (usually available at about 10 μm diameter) are presented in Table 6.2.

Table 6.2 Specific Modulus Data for some Commercial Fibres

	Elastic Modulus (E) $\frac{GN}{m^2}$	Density (ϱ) $\frac{g}{cm^3}$	Specific Modulus $\frac{E}{\varrho}$
Nylon 66	2 – 5	1.14	1.8 – 4.5
PETP	1 – 13	1.36	0.7 – 9.6
PP	1 – 6	0.91	1.1 – 6.6
E-Glass	70 – 80	2.5	28 – 32
Polyaramid	120 – 140	1.47	82 – 95
Si_3N_4	340 – 350	3.1	110 – 113
Carbon	200 – 400	1.9	105 – 210
Boron	380 – 400	2.5	152 – 160

6.6.2.2 Fibre Reinforcement and Deformation

Consider the mechanics of a simple fibre-reinforced composite system, where a load P is applied co-axially to an array of continuous, parallel fibres embedded in a plastics matrix. Using subscripts F, M, and C to denote the stress (σ), stiffness (E) or cross-section (A) of fibre, matrix and composite (respectively), we have:

$$P = \sigma_C A_C = \sigma_F A_F + \sigma_M A_M$$

Since the volume fraction of fibres (in a parallel array), $V_F = \frac{A_F}{A_C}$, composite stress is calculated by a *Rule of Mixtures:*

$$\sigma_C = \sigma_F V_F + \sigma_M (1 - V_F) \tag{6.34}$$

(If σ_F, σ_M are limiting values of failure *strength,* then so too is σ_C).

If we assume that a *constant strain* condition is applicable in each component, it can be shown that:

$$E_C = E_F V_F + E_M V_M \tag{6.35}$$

The rule of mixtures (sometimes known as the *homogeneous strain model)* gives an upper-bound limit for stiffness, which is related directly to volume fraction (V_F) of the reinforcement phase. Fibre content is calculated on a weight fraction basis, with loadings up to 40 % (by weight) being typical. In practice, longitudinal stiffness is generally less than that predicted by Equation 6.35; examples of imperfections which cause this include voids, resin-rich areas, misoriented or discontinuous fibres, and the possibility of incomplete wetting.

Wherever stresses are applied at an angle to the fibre-axis, the modulus diminishes to a value between limits defined by Equation 6.35, and an equivalent expression for *transverse modulus:*

$$E_T = \frac{1}{(V_F/E_F) + (V_M/E_M)} \tag{6.36}$$

Predictions made from Equation 6.36 are usually inaccurate and pessimistic, since the model represents an oversimplification. Whilst a full analysis of stress is outside the scope of this treatment, it can be shown[6.36] that a semi-empirical modification of Equation 6.36 yields:

$$\frac{1}{E_T} = \frac{1}{V_F + \eta_T V_M} \cdot \left(\frac{V_F}{E_F} + \frac{\eta_T \cdot V_M}{E_M} \right) \tag{6.37}$$

(η_T is the stress partition parameter, for which experimental data has shown 0.5 to be an accurate representation).

The deformation characteristics of fibre-reinforced composites are therefore highly *anisotropic;* the extent of mechanical anisotropy in continuous systems can, of course, be controlled by application of several laminae at successively different ply-angles. In the case of short-fibre reinforced moulding compounds, analysis of stiffness is potentially more problematic, since injection moulded products show evidence of preferential *Fibre Orientation Distributions (FOD)* (as a result of temperature and shear history during processing) in different through-thickness locations. It has been proposed that sufficient predictive accuracy for stiffness is usually obtained if a random, in-the-plane distribution is assumed[6.37].

6.6.2.3 Stiffness of Continuous Fibre Reinforced Composites

We have seen that fibre reinforced composites are anisotropic, and therefore mechanical properties will depend on fibre orientation. It is necessary for design purposes to have some method of representing the stress-strain behaviour of these materials. Initially, we will consider a unidirectional laminate, where all the fibres

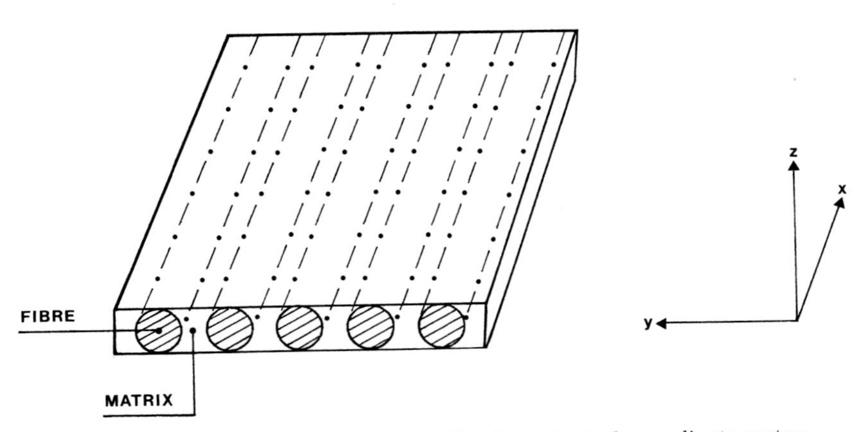

Figure 6.29 Basic unidirectional laminate showing principal co-ordinate system.

are oriented in the same direction (Figure 6.29). A suitable model for describing stress-strain relationships in a unidirectional laminate is given in matrix form[6.36], by:

$$\begin{bmatrix} \varepsilon_x \\ \varepsilon_y \\ \gamma_{xy} \end{bmatrix} = \begin{bmatrix} S_{11} & S_{12} & 0 \\ S_{12} & S_{22} & 0 \\ 0 & 0 & S_{66} \end{bmatrix} \begin{bmatrix} \sigma_x \\ \sigma_y \\ \tau_{xy} \end{bmatrix} \tag{6.38}$$

where:

ε_x, σ_x	– strain and stress in x (fibre) direction;
ε_y, σ_y	– strain and stress in y (transverse) direction;
γ_{xy}, τ_{xy}	– in-plane shear strain and stress;
S_{11}, S_{12} etc.	– components of the compliance matrix S_{ij}.

Thus the model is defined by four material constants: S_{11}, S_{12}, S_{22}, S_{66}.

The longitudinal modulus (E_L) and transverse modulus (E_T) discussed in the previous section are given by:

$$E_L = \frac{1}{S_{11}} \qquad E_T = \frac{1}{S_{22}} \tag{6.39}$$

We can represent the stress-strain relationship in terms of a stiffness matrix (E_{ij}):

$$\begin{bmatrix} \sigma_x \\ \sigma_y \\ \tau_{xy} \end{bmatrix} = \begin{bmatrix} E_{11} & E_{12} & 0 \\ E_{12} & E_{22} & 0 \\ 0 & 0 & E_{66} \end{bmatrix} \begin{bmatrix} \varepsilon_x \\ \varepsilon_y \\ \gamma_{xy} \end{bmatrix} \tag{6.40}$$

It should be noted that:

$$E_{11} \neq \frac{1}{S_{11}} \quad \text{(etc.)}$$

Typical values of the components of S_{ij} (and E_{ij}) for a range of composites are given by TSAI and HAHN[6.36].

(A) Off-axis loading.

The effective stiffness of a unidirectional laminate will depend on the angle of application of the stress. This can be determined as follows:

(1) Transform stresses and strains from fibre axes (x, y) to oriented axes (1,2) – Figure 6.30;

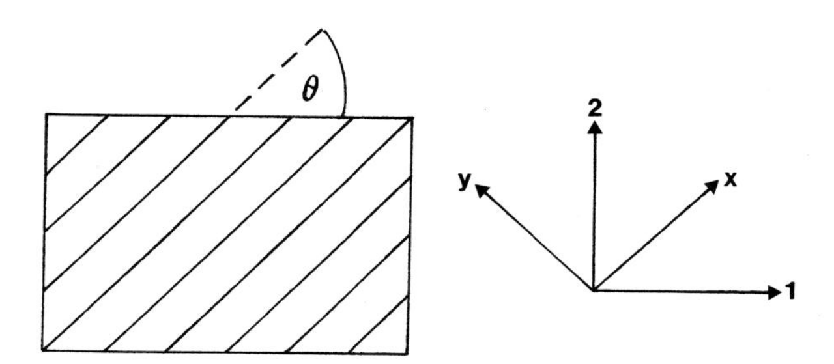

Figure 6.30 Off-axis co-ordinate system.

(2) Substitute for σ_x, σ_y, τ_{xy}, ε_x, ε_y and γ_{xy} in Equation 6.38, to obtain a transformed compliance matrix (Q_{ij}).

The transformed stress-strain relationship is:

$$\begin{bmatrix} \varepsilon_1 \\ \varepsilon_2 \\ \gamma_{12} \end{bmatrix} = \begin{bmatrix} Q_{11} & Q_{12} & Q_{16} \\ Q_{12} & Q_{22} & Q_{26} \\ Q_{16} & Q_{26} & Q_{66} \end{bmatrix} \begin{bmatrix} \sigma_1 \\ \sigma_2 \\ \tau_{12} \end{bmatrix} \tag{6.41}$$

where the components of Q_{ij} are functions of θ and S_{ij} only. A significant result from this expression is that off-axis loading will induce shear, in addition to direct strains. Let us consider the following stress pattern:

$$\sigma_1 \neq 0, \qquad \sigma_2 = \tau_{12} = 0$$

Substituting into 6.41, we find:

$$\begin{aligned} \varepsilon_1 &= Q_{11} \cdot \sigma_1 \\ \varepsilon_2 &= Q_{12} \cdot \sigma_1 \\ \gamma_{12} &= Q_{16} \cdot \sigma_1 \end{aligned} \tag{6.42}$$

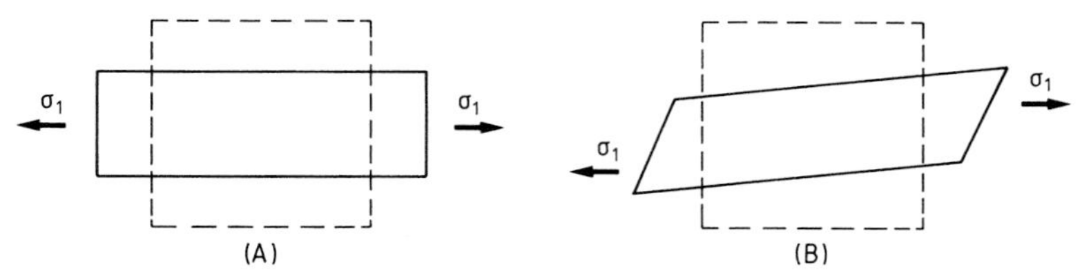

Figure 6.31 Deformed shape of composite element under uniaxial tension:
(A) Unidirectional in fibre direction – Equation 6.38 (ie. $Q_{16} = 0$);
(B) Off-axis experiment – Equations 6.41 and 6.42 (ie. $Q_{16} \neq 0$).
(----- undeformed; ______ deformed)

The deformed shape of the loaded composite is shown in Figure 6.31 (B), which illustrates the effect of the *shear coupling coefficient* (Q_{16}). The effect of orientation (θ) on the longitudinal modulus E_L ($= \frac{1}{Q_{11}}$) is shown in Figure 6.32.

A similar relationship exists for the transformed stiffness matrix:

$$\begin{bmatrix} \sigma_1 \\ \sigma_2 \\ \tau_{12} \end{bmatrix} = \begin{bmatrix} D_{11} & D_{12} & D_{16} \\ D_{12} & D_{22} & D_{26} \\ D_{16} & D_{26} & D_{66} \end{bmatrix} \begin{bmatrix} \varepsilon_1 \\ \varepsilon_2 \\ \gamma_{12} \end{bmatrix} \tag{6.43}$$

where:

$$D_{ij} = f(\theta, E_{ij})$$

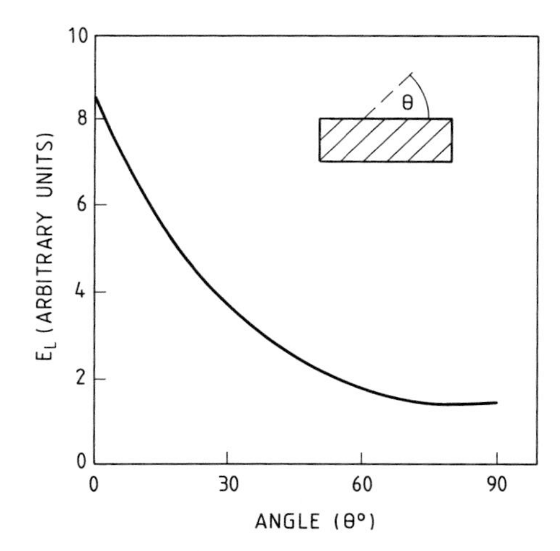

Figure 6.32 Effect of fibre orientation on longitudinal modulus (E_L) for GRP.

(B) In-plane stiffness of symmetrical laminates

In practice, composite structures are produced by stacking individual plies, each with a different orientation. There is an infinite number of possibilities, but we will only consider symmetrical laminates, which are the important systems commercially.

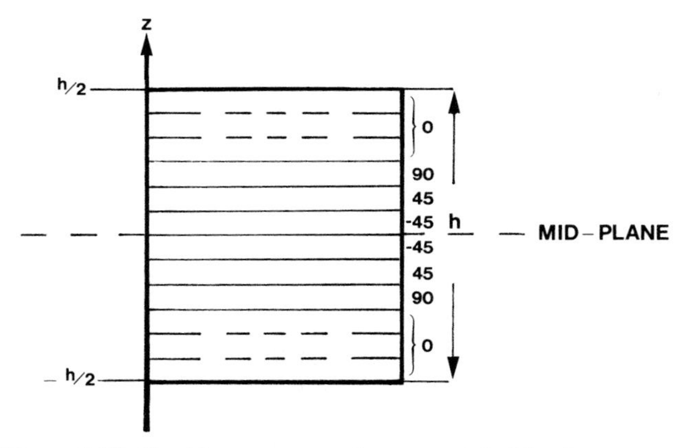

Figure 6.33 Stacking sequence of a symmetrical laminate.

A symmetrical laminate contains plies of the same type, orientation and thickness placed at equal distances either side of the central plane (Figure 6.33). These types of laminates simplify the stress analysis of a structure and eliminate the distortions which arise as a result of the shear coupling shown in Figure 6.31. The stacking sequence of a symmetrical laminate (as shown in Figure 6.33) can be represented by:

$$[0_3/90/45/-45/-45/45/90/0_3]$$

or:

$$[0_3/90/45/-45]_s$$

A laminate is symmetric when:

$$\theta(z) = \theta(-z)$$
$$Q_{ij}(z) = Q_{ij}(-z)$$
$$t(z) = t(-z)$$

(t = thickness of an individual layer)

It is assumed that the strains in the laminate are constant. An average stress is defined as follows:

$$\overline{\sigma}_1 = \frac{1}{h}\int_{\frac{-h}{2}}^{\frac{h}{2}} \sigma_1\,dz \qquad \text{etc.} \tag{6.44}$$

where h represents the total laminate thickness. Thus from Equations 6.43 and 6.44, we find:

$$\begin{bmatrix} \overline{\sigma}_1 \\ \overline{\sigma}_2 \\ \overline{\tau}_{12} \end{bmatrix} = 1/h \begin{bmatrix} A_{11} & A_{12} & A_{16} \\ A_{12} & A_{22} & A_{26} \\ A_{16} & A_{26} & A_{66} \end{bmatrix} \begin{bmatrix} \varepsilon_1 \\ \varepsilon_2 \\ \gamma_{12} \end{bmatrix} \tag{6.45}$$

where:

$$A_{11} = h\int_{\frac{-h}{2}}^{\frac{h}{2}} D_{11}(\theta)\,dz \qquad \text{etc.}$$

Equation 6.45 can be inverted to give the equivalent compliance matrix:

$$\begin{bmatrix} \varepsilon_1 \\ \varepsilon_2 \\ \gamma_{12} \end{bmatrix} = h \begin{bmatrix} a_{11} & a_{12} & a_{16} \\ a_{12} & a_{22} & a_{26} \\ a_{16} & a_{26} & a_{66} \end{bmatrix} \begin{bmatrix} \overline{\sigma}_1 \\ \overline{\sigma}_2 \\ \overline{\tau}_{12} \end{bmatrix} \tag{6.46}$$

The longitudinal (E_L), transverse (E_T) and shear (G) moduli are then given by:

$$E_L = \frac{1}{a_{11}h}$$
$$E_T = \frac{1}{a_{22}h}$$
$$G = \frac{1}{a_{66}h}$$

Full details of the analysis are given by TSAI and HAHN[6.36]. The technique will be illustrated with two common classes of symmetrical laminates: *angle ply* and *cross ply* (Figure 6.34). Angle ply laminates contain an equal number of plies oriented at $+\theta$ and $-\theta$ to the reference axis. Typical results for E_L and G are shown in Figure 6.35, in comparison to the off-axis data. The angle ply laminate is much stiffer because of the constraints imposed by adjacent layers which are not allowed to deform independently.

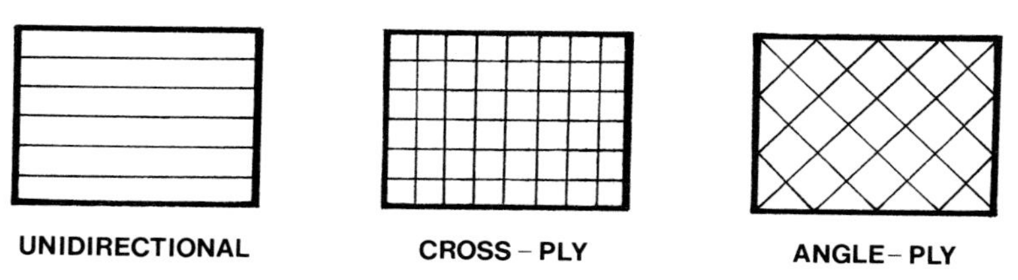

Figure 6.34 Examples of symmetrical lay-ups.

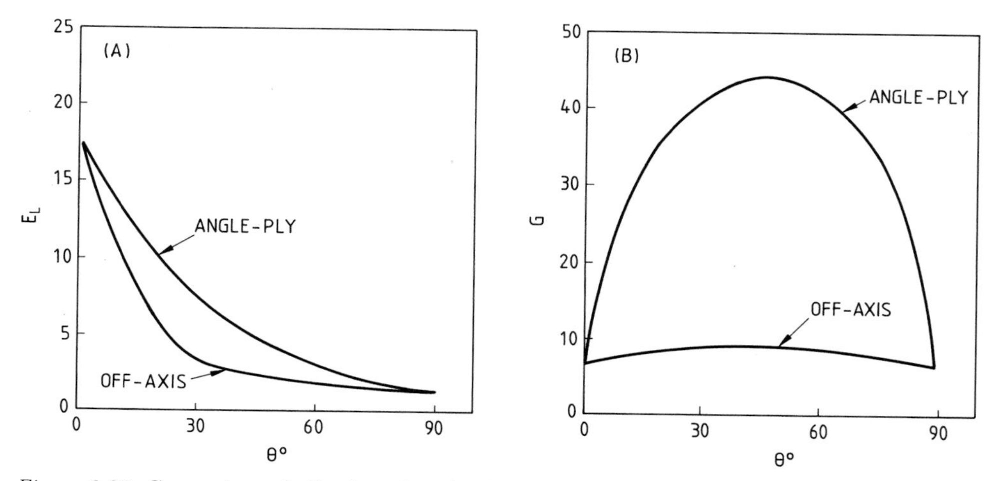

Figure 6.35 Comparison of off-axis and angle-ply laminates: (A) Longitudinal modulus (E_L), (B) Shear modulus (G).

In cross ply laminates, the ply orientations are either 0 ° or 90 °; typical results for E_L are shown in Figure 6.36. The longitudinal stiffness increases in proportion to the amount of unidirectional reinforcement.

Angle ply and cross ply composites are used extensively; they can be used in combination to produce a wide range of properties, as discussed by TETLOW[6.38]. Both angle ply and cross ply laminates can be represented by the model described by Equation 6.38, such that the in-plane stiffness of these systems can be represented by 4 constants (S_{11}, S_{12}, S_{22}, S_{66}).

We have only been concerned with the in-plane stiffness of symmetrical laminates. A similar approach can be developed to determine the flexural stiffness of laminates; this has been discussed in detail elsewhere[6.36, 6.39].

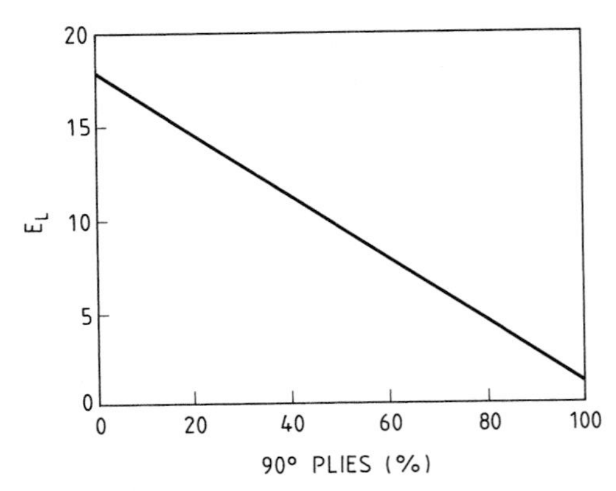

Figure 6.36 Longitudinal modulus (E_L) of cross-ply laminates.

The anisotropic nature of fibre reinforced plastics gives rise to an almost limitless range of stiffness characteristics. Whilst the analysis of such properties is often algebraically tedious, there are numerous commercial software packages available for use on desk top computers. Clearly, the composite designer has the opportunity to "tailor make" a structure, to meet any specific mechanical property requirements.

6.6.2.4 Mechanism of Fibre Reinforcement

Before advancing to consider the properties of any individual composite system, it is necessary to establish the basis of stress transfer by the interfacial shear mechanism. (We start by ignoring the load transfer effects at fibre ends, which are negligible in continuously reinforced composite materials.) Consider the forces on a single fibre element (length dx, radius r), embedded in a polymer matrix, when a stress σ is applied parallel to its long axis (Figure 6.37):

$$F_1 = \pi R^2 \cdot \sigma_F; \qquad F_2 = \pi R^2 \cdot (\sigma_F + d\sigma_F); \qquad F_3 = 2\pi R \cdot dx \cdot (\tau_I)$$

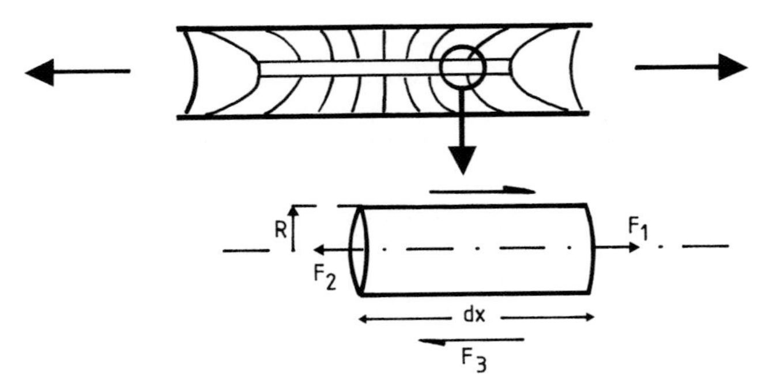

Figure 6.37 Stress-contour model of a single fibre in a composite material.

(F_1, F_2 are tensile forces relating to the tensile stress acting on the fibre (σ_F); τ_I is the shear stress at the polymer-fibre interface). In an equilibrium situation, we can equate the component forces to zero; rearranging, we obtain:

$$\frac{d\sigma_F}{dx} = \frac{2\tau_I}{R} \tag{6.47}$$

Therefore, the magnitude of tensile stress generated along the fibre depends upon the *interfacial shear stress* which is developed. Since the desired failure criterion is by fibre-fracture, the implication from Equation 6.47 is that we require a *critical fibre length,* in order that the total surface area for load transmission is sufficient to develop the tensile stress towards the theoretical tensile fracture strength of the fibres. Figure 6.38(A) illustrates the distribution of shear tractions and tensile stress, as a function of distance, as predicted by Equation 6.47. In reality, since τ_I is limited by the shear yield strength of the matrix (τ_M) or the interface, the simple stress distributions shown in Figure 6.38(B) are sufficient to describe the situation in the different examples where the actual fibre length is greater than, or less than (respectively) the critical fibre length L_C. Integration of 6.47 gives:

$$\sigma_F = \frac{2\tau_M}{R} \cdot x$$

Since σ_F is just equal to the fibre fracture strength (σ^*) at $x = \frac{L_C}{2}$, we have:

$$L_C = \frac{R\sigma^*}{\tau_M} \tag{6.48}$$

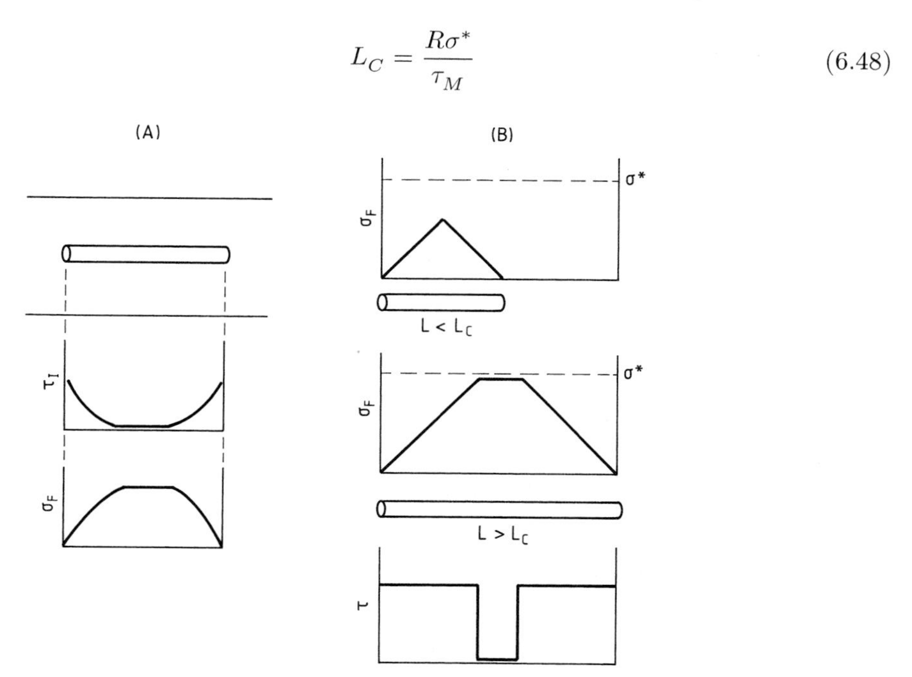

Figure 6.38 Development of stress in a fibre-reinforced composite:
(A) Interfacial shear (τ_I) and tensile fibre stress (σ_F);
(B) Theoretical tensile stress distributions for fibres of subcritical, and supercritical length.

For thermoplastics composites of interfacial strength in the range 20 – 80 MN/m^2 reinforced with glass fibres 10μm in diameter, typical L_C values lie in the range 0.125 – 0.5 mm. These values are close to the practically-attainable fibre lengths which are usually anticipated from typical melt processing methods such as screw extrusion or injection moulding; clearly, the performance of the composite is influenced not only by the volume fraction and orientation of reinforcement, but also by fibre length.

6.6.2.5 Particulate-Reinforced Composites

When particulate mineral fillers are added to plastics compounds, it is inevitable that physical property modification will occur. For a typical composite system with a particle content of up to 30 – 40 % (by weight), the generalized list below indicates the most likely effects of filler-addition:

(1) Modulus. Significant increases in stiffness (and in creep/deformation resistance and hardness) are usually apparent, including a greater level of stiffness-retention at elevated temperature; the effect on modulus of *high-aspect ratio fillers* (such as mica or talc) is particularly marked, but the effect of particle size is generally less important.

(2) Yield strength. Modest increases in yield or fracture strength (under tension, compression, or a flexural stress) in low strain rate tests, are often observed. The strength of filled compounds is a more difficult parameter to predict, since it depends upon particle size, shape, and on the polymer-filler interaction. Finer particle size materials have a much higher specific surface area, which reduces the degree of stress concentration in the matrix and results in a composite with higher mechanical strength. Adhesion between the filler and polymer has a significant effect which can be improved by the use of *coupling agents* (Figure 6.39).

(3) Ultimate properties. Impact resistance, fracture properties and toughness are related to the energy absorbing characteristics of filled composites, and are unlikely to replicate the enhancement generally seen in low-strain properties; indeed, filler addition may completely modify an otherwise-ductile mode of failure. However, more complex relationships with ultimate properties exist, since toughness enhancement is sometimes seen with low-additions of (especially) chemically-modified fine-fillers, but fracture resistance usually decreases sharply if coarser fillers are used, especially at higher addition levels.

Key factors which relate to the likelihood of successful property enhancement include both the *compound specification* (filler particle size/shape, volume loading), and some additional factors which are also dependent upon the processing conditions (*dispersion* characteristics and the degree of *adhesion* to the polymer matrix) which are used to mix, and subsequently mould finished products. Ductility of filled polymer-based systems is usually diminished by a reduction in crack propagation energy of highly-filled systems; this is directly related to the low interfacial strength between polymers and mineral fillers.

Bonding may be improved by *surface treatment* of the fillers, and effective *dispersion* (another important prerequisite for property enhancement) is achieved by destroying any tendency towards particle agglomeration using purpose-designed

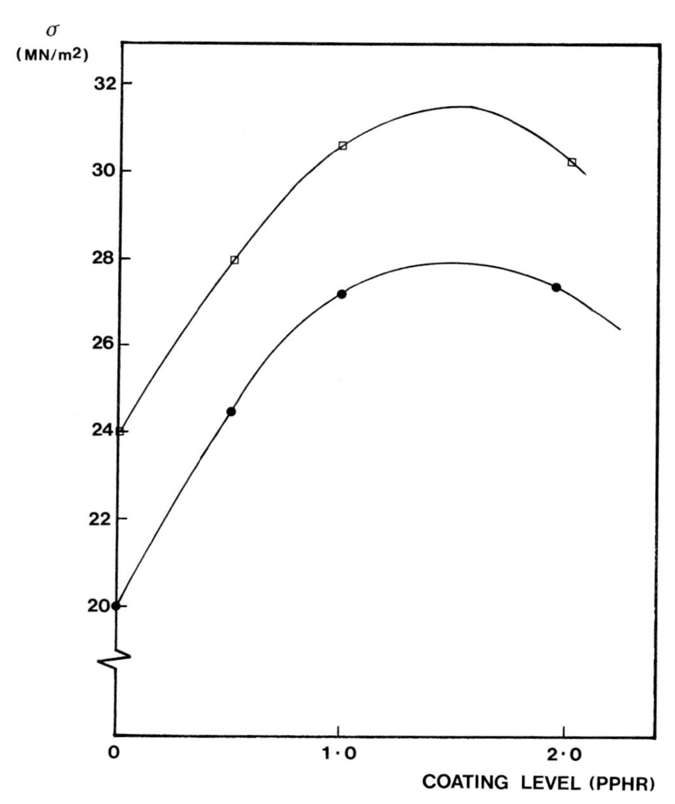

Figure 6.39 Effect of coating level on tensile strength (σ) of composites from PP homopolymer with 30 % wet ground mica. (Stress at yield (•) and break (□)).

screw configurations in compound extruders (Chapter 4). Two types of chemical treatment may be considered in order to achieve these objectives:

- Coupling agents – These react with the filler surface and have good compatibility with the polymer to provide a good interfacial bond; silanes are the principle materials used in industry for this purpose.
- Wetting agents – The function of these materials is to modify the surface energy of mineral fillers to improve dispersion and compatibility within the composite.

The precise mechanisms of surface treatment and modification are not well established, and currently form the basis of much active research.

6.6.3 Foamed Plastics

By incorporation of a suitable proportion of a chemical or physical *blowing agent,* many groups of plastics materials may be processed into finished or semi-finished products in the form of *foams* of controlled density. The *cellular structure* is produced by gas evolution when the blowing agent decomposes or boils; foaming is initiated during processing when the melt pressure drops below the partial pressure of the dissolved gas.

Mechanical properties of foamed materials and products may be modified by the controlled development of the microcellular structure. For example, polyurethane chemistry is sufficiently versatile to allow the production of foamed products which offer an impressive range of different properties:

Rigid foams – high compressive rigidity and resistance to collapse;
– relatively high density (up to 750 Kg/m^3).

Flexible foams – lower rigidity, but excellent elasticity and resilience;
– low density (10 – 100 Kg/m^3).

Polyurethane foams assume greater rigidity if the degree of crosslinking is increased, or if the chain segment length between urethane linkages is decreased.

Under free-expansion conditions, the properties of *thermoplastics foams* (eg. PE, PS) are density-dependent, and can be controlled by the amount of blowing agent used, and by melt processing conditions. Injection moulded *structural foams* or *sandwich mouldings* (eg. PP, HIPS, PS-modified PPO) are formed by injecting a short-shot of melt into a closed mould cavity; low-pressure expansion is then responsible for foamed-part formation. With this type of process, it follows that the *overall* density – which is usually controlled in the range 70 – 90 % of the solid density – is determined by the shot volume of melt; the *density distribution* between the rigid (solid) outer skins is less-easily controlled.

For typical semi-structural parts, the prime advantage offered by sandwich/ structural foam products is the increase in *flexural stiffness,* per unit mass of polymer. This route towards enhanced product stiffness is considered in more detail in later sections of this Chapter (6.6.4; 6.7.2.2).

6.6.4 Foam Sandwich Structures

It is possible to utilize the advantages afforded by heterogeneous reinforcement, and by foaming, in a single, composite structure: *sandwich structures* are well known in engineering, in which thin sheets of a high stiffness, high strength material are separated by a thicker core of material, usually of lower density, and with less strength and stiffness.

The outer faces make the largest contribution to flexural rigidity, by virtue of the very high second moment of area. The core, however, is not unimportant, and must fulfil various functions:

- It must have sufficient shear stiffness to prevent movement of the faces relative to each other in flexure; otherwise, the laminae behave as independent panels and the sandwich effect is lost.
- The core must also maintain the required separation between the stiff panels; otherwise, the benefit from the second moment of area may be lost.
- Finally, the core must be stiff enough to keep the panel faces approximately flat, in order to prevent local buckling in the compression regions.

Polymer foams, sandwiched between panels of *fibre reinforced plastics,* offer lightweight alternatives to metal panels, especially when stressed in the *flexural* mode.

However, the integrity of the sandwich structure is essential if optimum properties are to be secured; a theoretical treatment is given below.

The application of standard beam theory gives the relationship between bending moment (M), curvature ($\frac{1}{R}$) and flexural rigidity (D):

$$\frac{M}{D} = \frac{1}{R} \tag{6.49}$$

For an isotropic homogeneous beam, D is the product of the modulus of elasticity (E), and the second moment of area (I):

$$D = EI \tag{6.50}$$

The arguments used for a simple beam also apply to a sandwich beam, since the two materials constrain each other to deform in the same manner. Hence, for a symmetrical cross-section and isotropic face and core materials, the deflection curve will apply with the following flexural rigidity:

$$D = E_f I_f + E_c I_c \tag{6.51}$$

where f and c refer to face and core respectively. Substituting for the second moments of area in terms of the beam dimensions:

$$D = \frac{E_f wt^3}{6} + \frac{E_f wtb^2}{2} + \frac{E_c wc^3}{12} \tag{6.52}$$

where w is the width of the specimen, t is the thickness of one face, c is the thickness of the core, and $b = (c + t)$.

In practical sandwich constructions the first and third terms (associated with flexure of the faces individually about their own centroids, and flexure of the foam beam, respectively) may be neglected in favour of the second term. The assumptions made include:

- the faces on either side of the panel are identical;
- tensile modulus of the core is much lower than that of the faces; and
- faces and core are mechanically isotropic.

When a beam is subjected to flexure, *shear stress* is produced in any cross-section and, since the transverse shear moduli of these systems are usually much lower than the in-plane tensile moduli, the shear deformation, particularly of core material, makes a considerable contribution to the overall deflection. The maximum displacement of the beam due to shear deformation in the core has been obtained as[6.40, 6.41]:

$$\delta_S = \frac{WL}{4AG_c} \tag{6.53}$$

where $A = \frac{bd^2}{2}$, G_c is the shear modulus of the core, and the product AG_c is referred to as the *shear stiffness* (S) of the sandwich.

The total central deflection (δ) of the sandwich beam is thus:

$$\delta = \frac{WL^3}{48D} + \frac{WL}{4S} \tag{6.54}$$

The effectiveness of a sandwich structure can be tested by 3-point loading tests over a range of spans. (Note: It is important that the sample lengths should be varied in accordance with the span so that the "overhang" beyond the outer supports is both small and consistent; large overhangs contribute to the stiffness of the system). Equation 6.54 may be rearranged:

$$\frac{\delta}{WL} = \frac{L^2}{48D} + \frac{1}{4S} \tag{6.55}$$

A plot of $\frac{\delta}{WL}$ against L^2 should therefore be linear, with the slope being a function of the flexural rigidity, and the intercept a function of the shear stiffness. The sandwich system quoted below may be cited as a practical example of how the theory described in this section may be utilized.

6.6.4.1 GRP-PVC Foam Sandwich Panels

3.5mm GRP // 25mm PVC foam at 75 kg/m^3 // 3.5mm GRP

The testing of such a beam in flexure need not be an expensive exercise, can prove the validity of the analytical treatment outlined above, and may be used to characterize the contributions of foam core and rigid faces to the stiffness of the sandwich panel.

A beam of the above construction 500 mm long and breadth 25 mm is supported on parallel bars 460 mm apart; it is loaded by a 10 kg mass (98.1 N) at its centre, via a loading yoke. The deflection at the mid-point can be measured conveniently using a dial gauge with a sensitivity of 1 mm per revolution. The readings were taken 5 seconds after loading, and the specimens were unloaded immediately; within 30 seconds, recovery was complete and the samples had returned to their original positions. Starting at 460 mm span, readings were taken at successively lower spans, whilst machining the specimens to maintain a small and consistent overhang.

The standard plot is given in Figure 6.40 where, despite the relative simplicity of the equipment, a tolerable straight-line relationship is evident. Repeating the exercise with a 20 kg mass (196.2 N) gave the data in Figure 6.41; however, in this case the 500 mm specimen was not shortened for the lower spans. The derived values of D and S were:

Figure 6.40	D = 303 Nm^2	S = 19,230 N
Figure 6.41	D = 356 Nm^2	S = 18,797 N

These results could be interpreted so that the increased flexural rigidity in the second case is derived as a consequence of allowing the 500 mm specimen to overhang the outer supports by variable amounts. The shear stiffness is approximately constant, in each case.

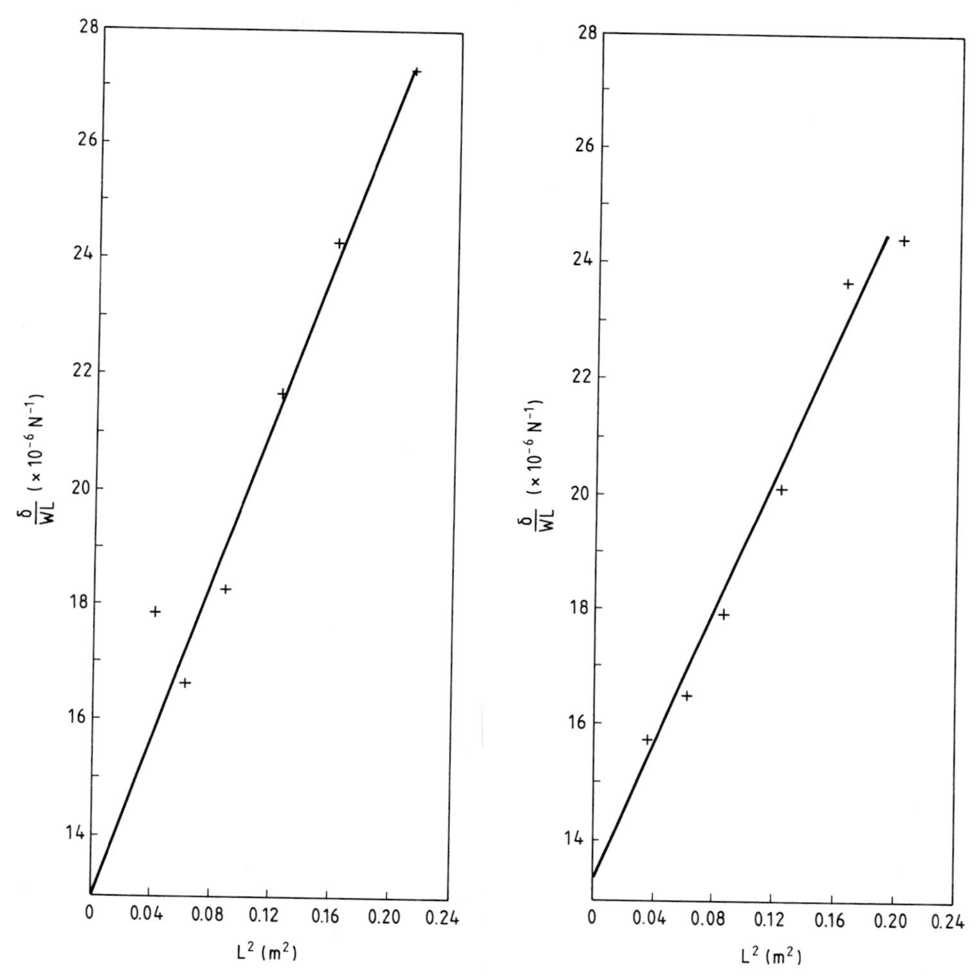

Figures 6.40 (left), 6.41(right) Flexural deformation data for GRP/PVC Foam/GRP sandwich laminates: Figure 6.40 – 10 kg mass, Figure 6.41 – 20 kg mass.

A further interesting point in the analysis is based upon the contributions made by D and S to the deflection, as the span is changed. These can readily be obtained by substituting the appropriate data in Equation 6.54, whence it is calculated that at the smallest span (200 mm), 85 % of the strain is due to shear deformation of the foam, whereas at 460mm span, the foam contribution has already been reduced to below 50 % of the total strain.

6.7 Deformation Resistance and Component Design

A design engineer who wishes to extend his expertise to practical design procedures with plastics-based materials must first be prepared to accept the wide range of factors which determine, or modify mechanical properties of products. Attempts have been made in this Chapter to account for some of these variables, and introduce – albeit qualitatively in certain cases – some areas of research in Polymer Physics which have provided a scientific basis on which to enhance design expertise and develop further applications areas.

When a component is stressed, it responds by deforming to an extent governed by its stiffness; or, if the load cannot be withstood, the component fails by yielding, or by brittle cracking. We consider mechanical design procedures in this section only in the context of small-strain deformation resistance; the point of fundamental importance is therefore how the *stiffness* of plastics materials can be used to design against an *"excess deformation"* criterion.

The objectives of this section are twofold: first, to show how conventional design methodology can be modified to include *viscoelastic* properties and second, to emphasize some practical design techniques by which deformation resistance in plastics products can be enhanced.

6.7.1 On the Measurement of Elastic Modulus

The extent of deformation in any material which can be accurately and uniquely characterized by a single stiffness parameter is easily predicted by suitable insertion of the elastic modulus into the relevant equation of elasticity. As has become evident for most plastics materials, however, the extreme sensitivity of stiffness to time and temperature effects, in addition to processing conditions and formulation variables, has posed an additional question relating to this type of design specification procedure: namely, how can we select a correct and accurate value of modulus appropriate to product end-use?

6.7.1.1 Review of Elementary Procedures

Despite the shortcomings of such data, most modulus measurements for plastics materials are derived from short-term tensile or flexural tests carried out under constant deformation rate conditions. Linear viscoelastic materials yield a nominal stress-strain characteristic which allows modulus determination by BS 2782 (302A): the modulus is the quotient of a change in stress and a change in strain, taken from a *linear* portion of a stress-strain curve. With most unreinforced plastics, time-dependent effects will obviously detract from the validity of this approach.

For a typical non-linear stress-strain characteristic (Figure 6.42), the following modulus parameters are of interest:

(1),(2),(3) Tangent Modulus measurements $\frac{d\sigma}{d\varepsilon}$ at 0 %, 0.2 %, 1.0 % strain (respectively);

(4),(5) Secant Modulus measurements $\frac{\sigma_i}{\varepsilon_i}$ at 0.2 %, 1.0 % strain.

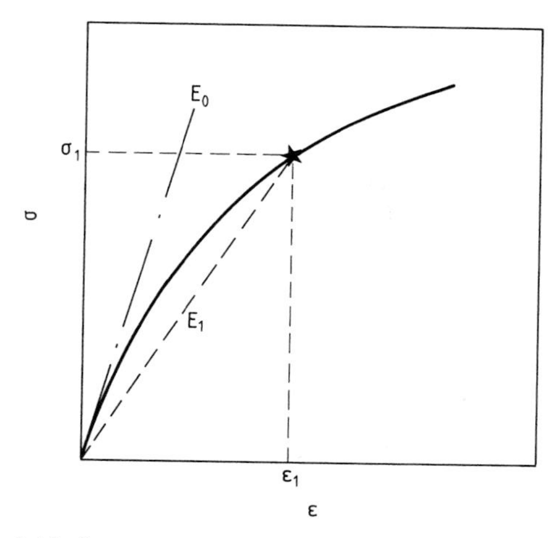

Figure 6.42 Some options for the determination of a short-term modulus in non-linear viscoelastic materials (see text for legend).

The most relevant, but rarely the most accurate measurement (due to practical constraints such as electrical noise and signal response time in recording equipment) is the initial *tangent modulus* (1); as alternatives, tangent moduli at other strains may be proposed. Definition of a *secant modulus* involves an approximation to a linear response over a narrow, but pre-specified and standardized level of strain. This type of test can, of course, be duplicated for any given combination of temperature, environment and strain rate. If measured accurately, such data carry useful comparative meaning (eg. to compare "equivalent" grades; to note the effect of formulation changes) and are utilized as part of many quality assurance programmes, but are of limited use with respect to long-term design computations.

6.7.1.2 Use of Viscoelasticity Data

Under constant stress conditions, the strain apparent in a non-linear viscoelastic material is given by:

$$\varepsilon = f' \cdot (\sigma, t)$$

(A) Tensile creep data. Relationships of state (between stress, strain and time) are complex, and are probably limited to individual groups of plastics; attempts to model these data are not especially well-advanced. In order to utilize time-dependent properties in situations where loading is applied throughout a significant component lifetime, the established method is to generate long-term *creep data* (the tension mode is favoured) and extract stress/strain/time information relevant to the anticipated loading and expected lifetime of the component material under surveillance.

(Similar principles may also be applied to *stress relaxation* data, and to creep data derived from other testing modes). This principle is illustrated, using PA66 as an example, in Figure 6.43:

(A) represents a basic family of long-term strain data, generated for a series of different stresses, plotted on a log-time axis.

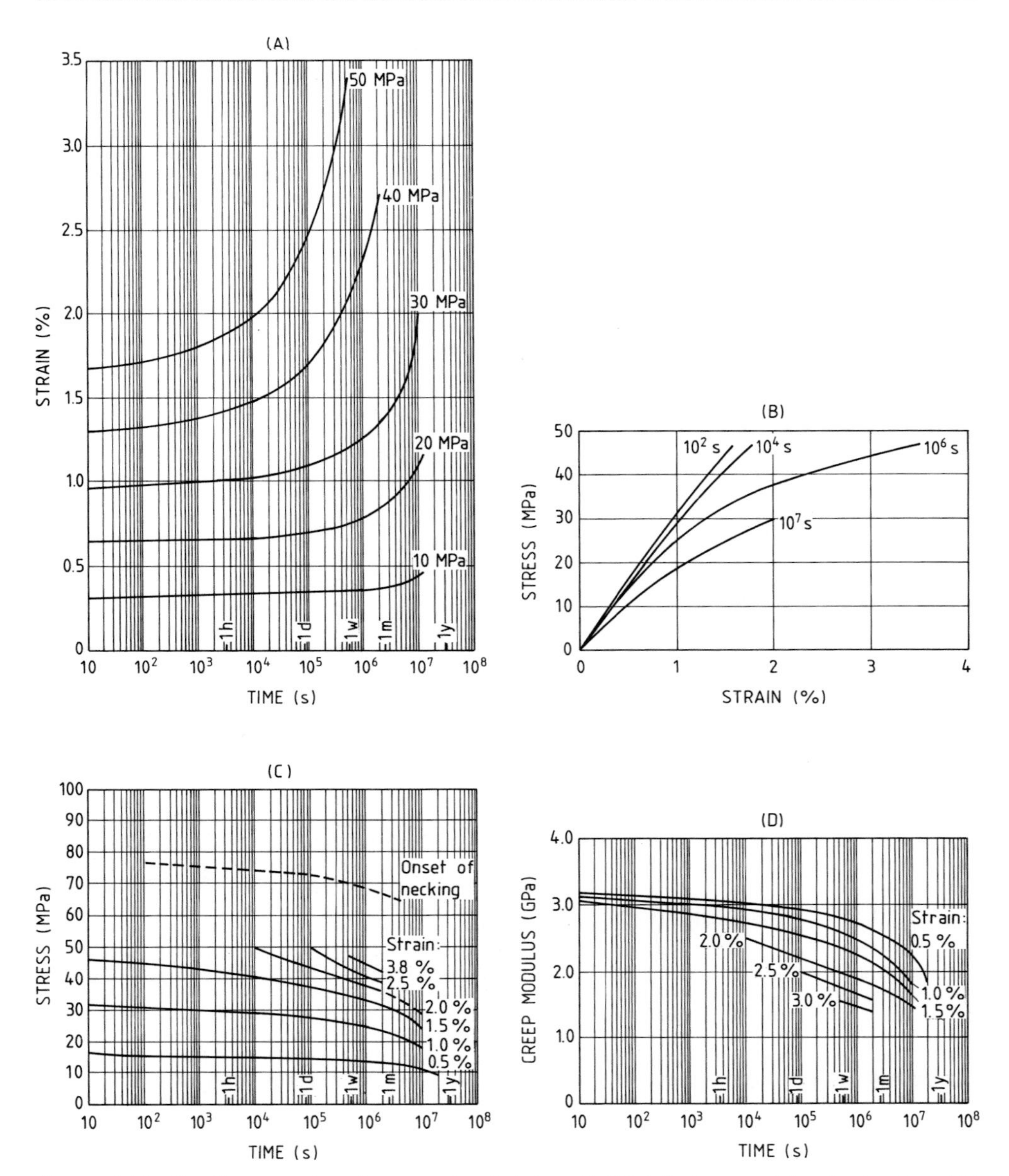

Figure 6.43 Tensile creep data for a grade of Nylon 66 (23 °C; dry conditions), showing: (A) Basic creep curves; (B) Isochronous stress-strain curves; (C) Isometric stress-(log)time plots; (D) Time-dependent creep modulus data. (Reproduced from the original data on 'Maranyl® A100', with permission from ICI)

The data in Figure 6.43 (B), (C) and (D) are taken from re-constructions of the basic creep curves in (A). These procedures are followed in order to increase the accuracy of interpolation, in situations where service stresses do not coincide with any of the stress levels used in the test.

(B) *Isochronous* (constant time) stress-strain plot; data can be read directly from plot (B) to relate to any pre-specified service life.
(C) *Isometric* (constant strain) stress-(log)time plots; once more, these can be constructed according to any strain level of interest.
(D) *Creep modulus* loci presented at various strains; the modulus is a "time-dependent secant" type, estimated from isochronous data at the relevant position.

Temperature is an additional, but by no means the only other variable which affects creep modulus; others include molecular weight, crystallinity, crosslinking and orientation of the resin, together with environmental effects such as relative humidity. Overall, this approach can be extended to include such variables and thereby generate time-dependent modulus data for the *pseudo-elastic design technique*[6.15] (see Section 6.7.2).

(B) Recovery. We have seen that the recovery characteristics of plastics are also time-dependent, and involve some degree of permanent (irrecoverable) strain. A technique first proposed by TURNER[6.16] and later adopted by the Standards' Authorities[6.10] has helped to normalize creep strain/recovery data to an extent where the effects of stress and loading time can be accounted for on a single master recovery plot. This method uses two additional parameters (refer to the stress-recovery diagram in Figure 6.8, for definition of terms), as follows:

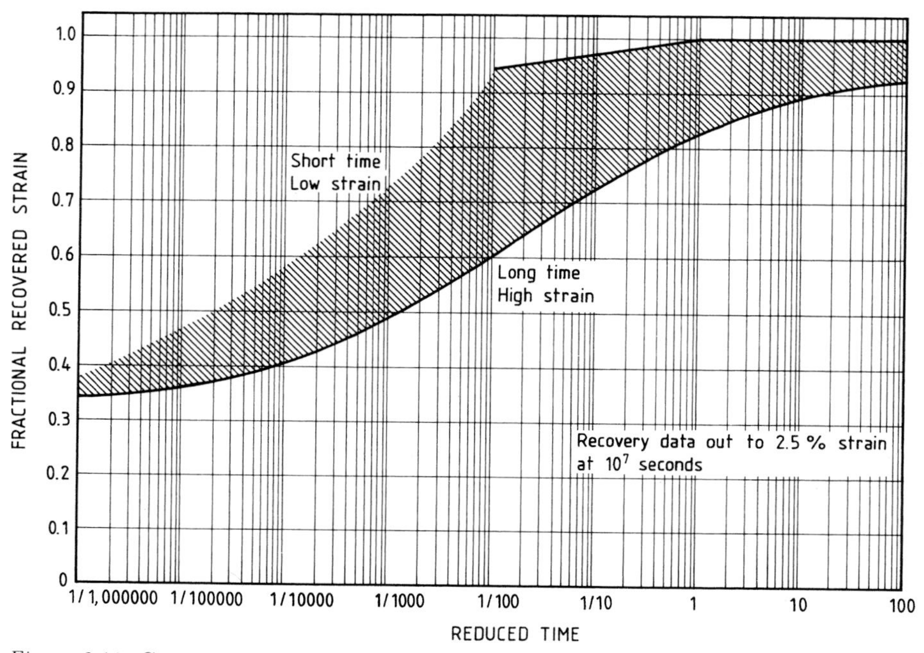

Figure 6.44 Creep-recovery data (Fractional recovered strain versus reduced time) for a grade of Kematal® POM at 20 °C. (Reproduced by permission of Hoechst UK Limited, Polymers Division.)

- **Fractional recovery,** (FR) = Recovered Strain / Max. Creep Strain

$$FR = \frac{\varepsilon_0 - \varepsilon_R}{\varepsilon_0} \tag{6.56}$$

- **Reduced time,** (t_R) = Recovery Time / Creep Time

$$t_R = \frac{t - t_1}{t_1} \tag{6.57}$$

Figure 6.44 shows how these parameters normalize creep-recovery data for any series of experiments carried out using a range of initial stress levels over different timescales of loading. Fractional recovery is positively related to reduced time: the data tend to reduce to a narrow band, allowing approximate calculations to be carried out, for any loading conditions. Design data of this type are available for many thermoplastics, especially those intended for engineering-type applications involving the application of intermittent stress.

6.7.2 Introduction to Plastics Component Design

In this section, we show how modulus data derived from long-term tests can be put to effective use by design engineers; also, some material properties and other factors which control *product stiffness* are highlighted.

6.7.2.1 The Pseudoelastic Design Principle

In the previous section, it was shown how creep data may be used to generate accurate, time-dependent stiffness data. A *pseudoelastic design* technique[6.7,6.10,6.15] has proved to be an invaluable means of accounting for the inelastic nature of plastics materials in the context of designing on a long-term basis; the principle states that conventional equations of elasticity may be used in all such computations, as long as the "modulus" insertion is appropriate to the envisaged conditions for which a design solution is sought. In other words, the pseudoelastic "constant" must be measured under creep (or relaxation) conditions, in the appropriate thermo-physical environment and preferably, under the mode of stressing appropriate to service life.

A design brief should consider the following points:

- **Loading history.** Given the degree of sensitivity which exists between creep modulus and time/temperature, the "worst-case" procedure should be adopted in the absence of more specific information; ie. assume loading is continuously applied, and that maximum temperature is apparent throughout.
- **Calculation of modulus.** Creep modulus (secant) can be calculated from:

$$E(t) = \frac{\sigma_i}{\varepsilon_i(t)} \tag{6.58}$$

(1) If σ_i can be estimated initially, ε_i is read from isochronous data (see Section 6.7.1.2), for the appropriate timescale;

(2) If σ_i is unknown, design strain limitations (for any grade of material) can be used to estimate σ_i (from isometric data) in the same way.

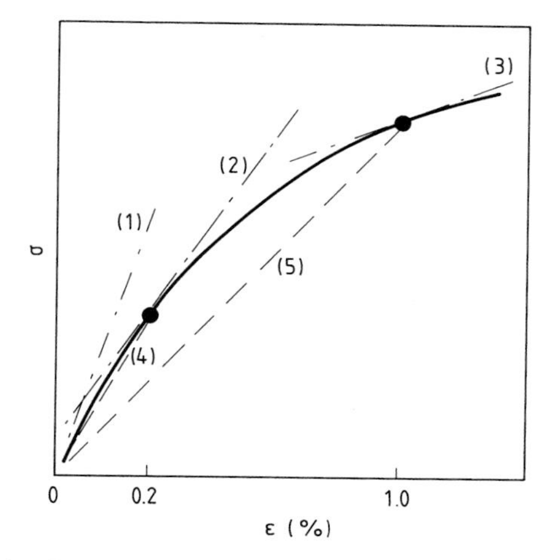

Figure 6.45 Isochronous stress-strain ($\sigma - \varepsilon$) plot, as used in an iterative technique to calculate creep modulus by the pseudoelastic design method.

Table 6.3 Some Elasticity Formulae for Deflection (δ) and Stress (σ)

Cantilever:		
End Load	$\delta = \frac{WL^3}{3EI}$	$\sigma = \frac{WLb}{2I}$
Uniform Load	$\delta = \frac{WL^3}{8EI}$	$\sigma = \frac{WLb}{4I}$
End-supported beam:		
Centre Load	$\delta = \frac{WL^3}{48EI}$	$\sigma = \frac{WLb}{8I}$
Uniform Load	$\delta = \frac{5WL^3}{384EI}$	$\sigma = \frac{WLb}{16I}$
Beam with fixed-ends:		
Centre Load	$\delta = \frac{WL^3}{192EI}$	$\sigma = \frac{WLb}{16I}$
Uniform Load	$\delta = \frac{WL^3}{384EI}$	$\sigma = \frac{WLb}{24I}$
Rim-supported circular disk:		
Uniform Load	$\delta = \frac{3W(1-\phi)(5-\phi)R^2}{16\pi Eb^3}$	$\sigma_R = \sigma_T = \frac{3W}{8\pi b^2} \cdot \frac{3+\phi}{\phi^2}$
Fixed circular disk:		
Uniform Load	$\delta = \frac{3WR^2(1-\phi^2)}{16\pi Eb^3}$	$\sigma_R = \sigma_T = \frac{3W(1+\phi)}{8\pi b^2}$
Annular tube:		
Uniform Pressure	$\delta_R = \frac{R}{E}(\sigma_H - \phi\sigma_A)$	$\sigma_H = 2\sigma_A = \frac{PR}{b}$
Sphere:		
Uniform Pressure	$\delta_R = \frac{R\sigma}{E(1-\phi)}$	$\sigma_R = \sigma_T = \frac{PR}{2b}$

(W is the applied load; E is elastic modulus, ϕ is Poisson's ratio and I is the second moment of area; geometric variables include length (L), thickness (b) and radius (R); subscripts R and T refer to radial and tangential stress, subscripts H and A to hoop and axial stresses in tubes)

When stress is not constant (eg. in flexure), an iterative approach can be used to estimate creep modulus, and ultimately, component deflection (see Figure 6.45). The iterative programme is as follows:

(1) Estimate initial tangent modulus (E_0);
(2) Calculate deflection (δ) using elasticity theory;
(3) Also using appropriate expression (Table 6.3), calculate maximum stress (σ_1);
(4) Read-off corresponding creep strain (ε_1) from isochronous data (Figure 6.45) and estimate new value of creep modulus (E_1) for the first iteration;
(5) Recalculate δ;
(6) Repeat (2) – (4) for i iterations, until adequate convergence is obtained.

Some examples of elasticity formulae are given in Table 6.3; for a more thorough review of such practically-relevant expressions, reference is given to the compendium by ROARK[6.42], and to more fundamental texts[eg. 6.41] on mechanics of materials.

6.7.2.2 Factors Affecting Deformation Resistance of Plastics Products

Before considering some of the many sources of design flexibility which are open to polymer engineers, it is worth returning briefly to the concept of *specific properties* to emphasize the importance of the lightweight, low-cost nature of most plastics-based products.

- **Stiffness per unit weight.** In automotive, aerospace and other applications where component weight must be strictly controlled, the magnitude of stiffness per unit mass is significantly enhanced by choosing engineering materials of relatively low density;
- **Stiffness per unit cost.** It is obvious that, assuming all else is equivalent, cheaper raw materials are more attractive to designers. What is not always quite so apparent to less-experienced engineers is that, since polymers (like any other materials) are bought by weight but are designed on a volumetric ("thickness") basis, the low-density character also carries a considerable economic advantage in this indirect context.

If we examine any elasticity formula, it is possible to consider individual ways by which product deformation resistance may be enhanced. For example, the maximum deflection (δ) in a simply-supported, rectangular-section beam (width w, depth b and length L) subjected to a constant applied load (W) at its centre, is given by:

$$\delta = \frac{WL^3}{48E(t)I} \tag{6.59}$$

I = 2nd moment of area = $\frac{wb^3}{12}$ for a rectangular section. Rearranging Equation 6.59 in terms of *flexural rigidity*, we have:

$$E(t) \cdot I = \frac{WL^3}{48\delta}$$

To minimize δ for a given load, beam length and component lifetime, we must increase either (or both) of the terms on the left side of this equation; individual consideration of these parameters provides a convenient means with which to consider design optimization.

(A) Material properties. Creep modulus is increased by the addition of:

(1) Reinforcing fibres: although stiffness of plastics composites exceeds the equivalent property in unreinforced resins, composite moduli inevitably become anisotropic. This can be used to advantage in hand-laid or moulded thermoset composites, but with injection moulded thermoplastics, the extent of fibre alignment is difficult to predict theoretically and is also a function of processing conditions[6.37,6.43].

(2) Fillers: modest increases in creep modulus (but more often, accompanied by significant increases in stiffness per unit cost) are derived by compounding plastics with low aspect ratio mineral fillers. A practical limitation tends to be the simultaneous tendency towards increased brittleness, if the filler is inappropriately chosen, dispersed imperfectly, or if an excessive addition level is used.

(3) Molecular Orientation: the contributions to creep appear to be retarded if the maximum principal stress is parallel to the main-chain orientation direction in oriented plastics products.

(4) Crystallinity: deformation resistance under long-term load will generally increase – albeit to a modest extent – if crystallinity is enhanced; for example, PP homopolymer crystallizes to a greater extent than commercial copolymer grades, and assumes improved creep properties as a result.

The extent of creep strain for a given stress, and the rate of increase of creep strain with time, are both increased by:

(1) Molecular Plasticization; for example, by ingress of moisture into hygroscopic resins or by plasticization of PVC compounds with appropriate additives.

(2) Copolymerization/Blending with a low modulus phase; creep rate increases with rubber content in toughened PS[6.20] and is also accelerated to a modest degree if PP is copolymerised with ethylene. These effects on modulus are, of course, offset by the dramatic improvement in toughness which also occurs.

(B) Use of shape and composite structures. There are significant advantages to be gained by assessing the optimum geometric section which contributes most (per unit section area) to the *second moment of area (I);* this is especially true in components stressed in flexure, but also carries relevance to situations involving more complex stress states overall. The "depth" dimension is particularly useful in this respect, a fact which promotes the appeal of introducing longitudinal *ribs* and *fillets* to commercial mouldings (Figure 6.46).

Composite or laminated "sandwich" structures offer an alternative means by which the *effective* second moment of area can be increased. Some typical material combinations (usually multi-layer products consisting of high-modulus "skins" bonded to low modulus core materials) are:

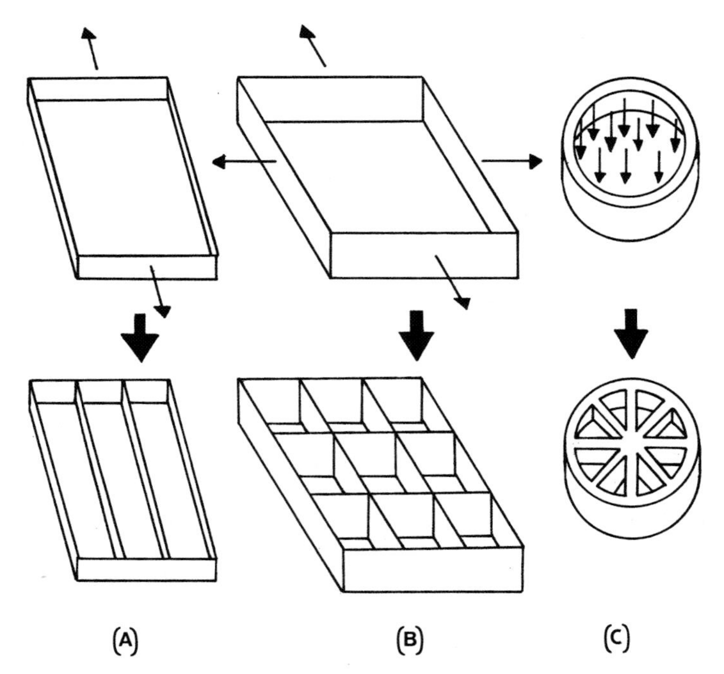

Figure 6.46 Use of ribs and fillets to improve stiffness of moulded plastics:
(A) – Uniaxial flexure of beams;
(B) – Biaxial stresses on panel-sections;
(C) – Uniform pressure acting on a circular disk.

Metal – Polymer – Metal	(eg. Aluminium/PE/Aluminium)
Solid Polymer – Foam – Solid Polymer	(eg. HIPS, PVC, PP or modified-PPO structural foams)
Reinforced – Foamed – Reinforced Polymers	(eg. GRP skins with PVC, PU or PF foam)
Oriented – Isotropic – Oriented	(eg. commercial injection mouldings)

The bending stiffness advantage gained by this route (assuming the discrete laminae to be perfectly bonded) has already been discussed (Section 6.6.4); it can be explained by second-moment effects using the concept of *equivalent sections* (see Figure 6.47): this assumes that the second moment of area of a two-ply laminate is *equivalent* to that of a "T" section of consistent elastic modulus, E_f. The waist dimension (w') of the equivalent, "constant-modulus" T-section is calculated on the basis of the modulus ratio between skin (or face, "f") and core ("c"):

$$\frac{w'}{W} = \frac{E_c}{E_f} \tag{6.60}$$

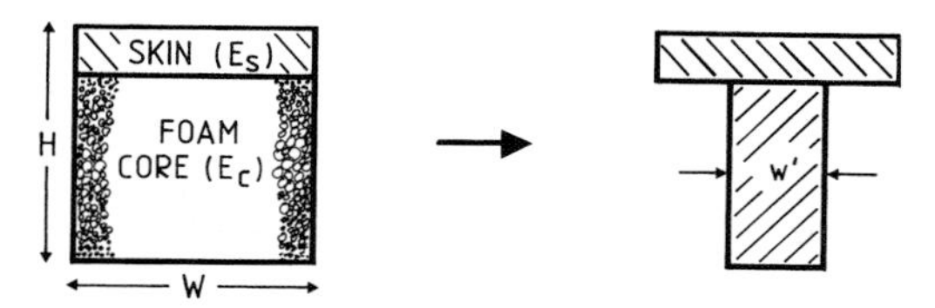

Figure 6.47 The concept of equivalent sections applied to a two-ply plastics laminate structure.

A general case solution for a "structural foam" second moment of area (I_{SF}) is:

$$I_{SF} = I_0 - I_1 \left[1 - \frac{E_c}{E_f} \right] \tag{6.61}$$

I_0, I_1 represent the second moments of area of: the actual (total) section (assuming it to be solid and homogeneous), (I_0); the section occupied by the core material alone, (I_1).

If the product consists of solid-foam-solid sections from the same material (*structural foam* moulding (eg. HIPS, PP, modified PPO), or *reaction injection moulding* (eg. PU, PF, "Nyrim") then the stiffness ratio E_c/E_f required for the design calculation can be estimated with accuracy from a square relationship with the equivalent densities[6.13,6.44]:

$$\frac{E_c}{E_f} = \left(\frac{\varrho_c}{\varrho_f} \right)^2 \tag{6.62}$$

The distribution of through-thickness density measurements is therefore of fundamental importance, to allow accurate predictions of flexural stiffness to be made in foamed plastics; since density distribution and the morphology of the cells are both dependent upon the type of process and foaming system used, the interrelationships between processing and component stiffness are usually quite complex. Recent work has examined these interactions in PP structural foams[6.45, 6.46].

These general principles may be extended to composite structures of any number of laminae (of various thicknesses), whose individual moduli may be chosen in order to attain the objective of increasing the specific stiffness of the component. Time-dependent deformation in polymer-based laminates can be accommodated in design computations by the pseudo-elastic method described above; the effect of viscoelasticity on stiffness, and the influence of shape and/or lamination on second moment of area, are therefore kept separate in these circumstances.

6.8 Hardness, Friction, Surface Abrasion and Wear

This chapter is concluded with reference to some tribological properties which, although encountered perhaps less-frequently than the more fundamental deformation characteristics (stiffness, viscoelasticity, yielding), carry enormous importance for components designed for applications involving surface contacts and relative motion at interfacial boundaries (eg. bearings, gears, sliding-friction parts, slurry-pipes).

Hardness testing will always assume some importance for plastics and rubber products, since its determination is relatively simple and data can be related to other bulk properties such as yield stress and modulus.

Surface lubricity adds to the appeal of many engineering thermoplastics for maintainence-free contact and bearing applications. However, friction properties of plastics are not easily characterized and are sensitive to temperature, rate effects, material formulation, surface roughness and to the contact-face pressure and zone area. In addition, since virtually all rigid plastics are softer than competitive metallic materials, any procedures to replace metallic components with thermoplastics must include direct reference to the *wear-resistance* requirements, including the possibility of surface micro-defects contributing to large-scale embrittlement.

6.8.1 Hardness

The attraction of hardness tests in the metals industry as a means of studying the effects of alloy constituents, fabrication and heat treatment on yield stress, is well established. Such tests are rapid and easy to execute, data are easily analyzed and test equipment is relatively cheap. However, if we consider the hardness properties of solid plastics, it must immediately be recognized that the viscoelastic nature introduces additional complexities. First, the temperature and timescale of loading/unloading and measurement must be completely standardized, a factor which restricts the use of certain types of hardness data to direct comparisons; second, a "yield stress" derived from *surface* measurement may not necessarily be indicative of the bulk properties required from a fabricated product; the development of molecular orientation and residual stress during processing, for example, detracts from correlations between surface and bulk properties.

6.8.1.1 Rubbery Plastics

Alternative views of hardness tests exist in the polymer industries, arising from the different hardness properties of rubbers and plastics; hardness data from the former group of materials (as relevant to plasticised PVC, and to Thermoplastic Elastomers) are directly related to the *modulus* of the material, albeit by relationships which are difficult to derive theoretically or prove in practice. It is, nevertheless, a relevant measurement made under load and inflicts no permanent damage on the material (in contrast to plastics hardness measurements, many of which record the amount of permanent strain sustained by the material).

Hardness tests for rubbers involve elastic deformations introduced by hemispherical or truncated-cone indenters. The latter are used in *Shore Hardness* measurements and the former often referred to as *International Rubber Hardness Degree (IRHD)* tests. For most of the time there exists a reasonable correlation between the two scales, but agreement should not be regarded as axiomatic; there are wide divergences for some thermoplastic elastomers, for example. Typical Shore "A" data (the suffix refers to the type of indentor which is used) for soft plastics are usually in excess of 95, expressed on a scale between 0 and 100; elastomeric materials can be formulated specifically for purposes of required hardness levels.

Rubber hardness tests may be used as indicators of compressive creep if time-dependent measurements are taken at appropriate intervals (10s, 100s etc.). Whilst standard hardness tests for rubbers are usually specified at 23 °C, quantitative information of greater use to practising designers can be gained easily by testing at different temperatures.

Plasticized PVC is unusual in that the *softness* is usually quoted, to help distinguish these grades from rigid PVC compounds. The *BS Softness Number* is the penetration (in mm $\times 10^{-2}$ ie. units of 10 μm) for a hemispherical indenter. The equipment is, indeed, identical with that for measuring IRHD, except for the dial gauge scale.

6.8.1.2 Rigid Plastics

Hardness tests for plastics are many and varied, ranging from the simple but effective pencil hardness test to a wide variety of tests developed from equivalent techniques for metals. The majority of these techniques rely upon an assessment of the permanent damage inflicted upon the test-piece (area, or depth of penetration) following indentation by a tool of specified shape (hemisphere, cone, pyramid) under defined loading conditions. For example, a simple pencil hardness test may correlate with some more rigorous testing procedures, since all rigid thermoplastics fall within the limits of 6B (which will not induce surface damage on LDPE) and 9H (which is sufficiently hard to scratch high molecular weight, cast acrylic sheet); see Table 6.4. Rockwell Hardness Number[6.47] represents the indentation depth (in arbitrary units on one of several reverse scales, according to the "major" load applied, and to the diameter – hence the stress concentration – of the spherical steel indentor) in a sample, following loading to a standard programme.

Table 6.4 Hardness Data for some Thermoplastic Materials

	Pencil	Rockwell 'R'
Cast PMMA	8H – 9H	120 – 125
PS	3H	115 – 120
PC	2H – 3H	115 – 120
PP	2H – 3H	100
HDPE	H – 2H	40
LDPE	2B – 3B	10

Crosslinked thermosetting plastics are harder than the range of pencils commonly available; typical Rockwell "R" hardness of filled or reinforced thermosets is in the range 120 – 130.

Although some test techniques and classification methods are rather superficial, they do have some empirical relevance to conditions experienced by moulded products in service. This is the key to assessing the potential use of hardness data or test methods: a degree of direct relevance to what is expected throughout the component lifetime is obviously desirable.

A more complete description of hardness tests for plastics is given in the text by CRAWFORD[6.13]; these include indentation tests (Brinell, Vickers, Knoop, Rockwell), techniques based on rebound characteristics, and scratch-hardness methods.

Figure 6.48 shows the decrease of hardness (in this case reflected by a *Hardness Number,* calculated as the reciprocal of a diamond-scratch width) with increasing moisture content in some PA66 resins; the moisture plasticization effect in polyamides is clearly demonstrated by these data. Similar trends are seen in grades A (unfilled), B (short glass-fibre reinforced PA66), and also in grade C (decrease in hardness due to lower crystallinity).

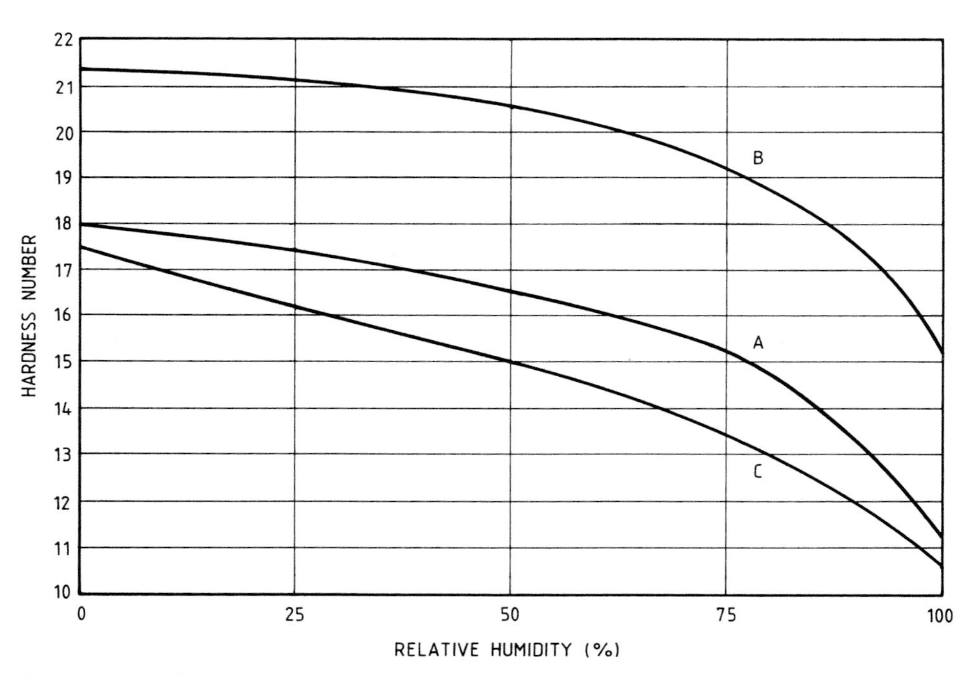

Figure 6.48 Effect of moisture absorption on the scratch hardness of some 'Maranyl' PA66 resins: (A) – Unfilled grade A100; (B) – Short glass-fibre reinforced grade A190; (C) – Controlled crystallinity grade F114. (Reproduced by permission of ICI).

An indentation hardness technique[6.48] has recently been reported to characterize the morphological changes which occur when rigid PVC powders are processed. The development of *gelation* and the type of crystallinity in the fabricated product depend upon shear history and processing temperature; these relationships are reflected by the maximum indentation and the amount of recovery following the hardness test. Figure 6.49 shows typical data derived from an instrumented test of this type; the decrease in maximum indentation with increasing extrusion temperature shows the increasing inter-particle resistance of the PVC network, exemplified further by the increasing degree of recovery apparent above 200 °C.

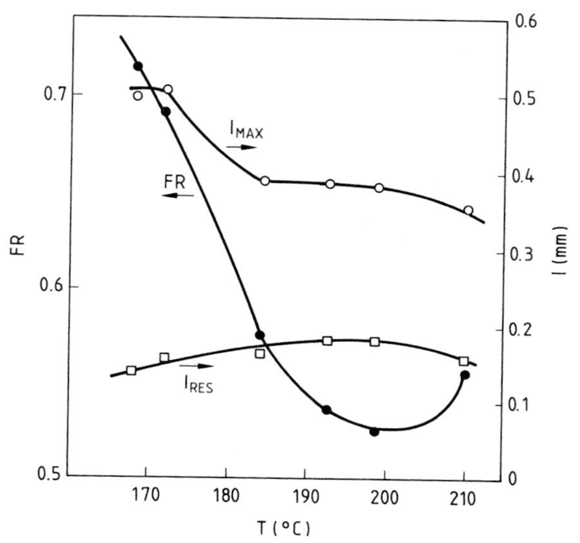

Figure 6.49 Graph showing hardness data for rigid PVC processed using a twin-screw extruder: maximum indentation (I_{MAX}), residual indentation following recovery (I_{RES}) and fractional recovered strain (FR), plotted against extrusion temperature (T). (After COVAS[6.48], reproduced with permission from Elsevier Applied Science Publishers.)

6.8.2 Other Tribological Properties – Friction, Abrasion and Wear

Friction characterizes the resistance to motion evident when two adjacent surfaces are moved with respect to one another; the *coefficient of friction* is therefore relevant to plastics melt-processing techniques (as described in Section 4.2.3.1 for single-screw extrusion) but is much more important to solid plastics parts which may be in contact with other components whilst under load or in motion. Frequently, considerable losses in power can be related to the need to overcome friction effects; additional problems associated with friction are *wear* damage and *heat* generation, which can cause siezing of surfaces. The ability to reduce or control friction is therefore important. This can be achieved by *heterogeneous lubrication* (as is the case for many other engineering solids, eg. ferrous-metal substrates lubricated by oil), by improved design, but also by *internal lubrication* using self-lubricating grades of plastics (eg. materials containing additives such as molybdenum disulphide, which are compounded into plastics for this specific purpose).

Conversely, there is sometimes a requirement to *increase* friction; for example, in shoe-soling compounds (PPVC, PU), bicycle tyres (urethane and other thermoplastic rubbers) and as a principle for processing, as in the friction-welding joining process.

The laws of friction are quite straight-forward and form the basis for appropriate measurements to be made. The frictional force (F) developed when one surface slides across another is proportional to the normal force (W) which opposes the movement:

$$F = \mu \cdot W \tag{6.63}$$

The coefficient of friction (μ, which is independent of contact area) is therefore equal to F/W, and represents an interfacial characteristic between the surfaces. Interpretation of friction data is partly made on the basis of adhesion theory; that is, the normal force induces a multitude of temporary contact points at the interface, which are ruptured by local plastic deformation induced by the frictional forces, as motion proceeds. The effective area of contact (A') is given by:

$$A' = \frac{W}{\sigma_Y} \tag{6.64}$$

(where σ_Y is compressive yield strength).

The frictional force is required to stem and thereby overcome the adhesion in the areas of contact. F is related to contact area and shear stress (τ) by:

$$F = A' \cdot \tau$$

Therefore we can state:

$$\mu = \frac{F}{W} = \frac{A' \cdot \tau}{A' \cdot \sigma} = \frac{\tau}{\sigma_Y} \tag{6.65}$$

Clearly, first principles demonstrate that friction behaviour is related to shear and compressive yield properties. However, due to the complexities introduced by viscous heat dissipation inducing local temperature increases, μ can rarely be considered to be constant and so the relevance of friction data to given practical situations is often confined to specific test conditions and configurations. In all examples, the requirement is to measure the frictional force (F) as a function of the normal force (W), whilst taking account of any conditions relevant to individual practical situations (temperature; motive, or static tests; presence of interfacial lubricants).

Whilst metals generally obey the basic laws of friction (ie. μ is independent of pressure) plastics generally show changes in friction coefficient with increasing load. Increases in temperature, sliding velocity and pressure each usually increase μ[6.49–6.51], although further increases in sliding speed (greater than (say) 0.5 – 1.0 m/s) may show a decrease in μ; rate effects and local temperature rises would appear to be interdependent, in this respect. Slip-stick characteristics are observed when the "static" coefficient falls below that at higher sliding speeds.

Friction properties are also critically dependent upon *surface roughness* or the *degree of lubricity;* Table 6.5 shows typical data for PA66 (by permission of ICI), illustrating the effects of sliding speed and lubrication.

Table 6.5 Friction Data for Nylon 66

Hemispherical Slider			Plate	Sliding Speed (mm/s)			
				0.03	0.4	3.0	16.0
(A)	PA66	(1)	PA66 (1)	0.63	0.69	0.70	0.65
(B)	PA66	(2)	PA66 (2)	0.42	0.44	0.46	0.47
(C)	PA66	(2)	Steel	0.39	0.41	0.40	0.40
(D)	PA66	(2)	PA66 (2)	0.27	0.24	0.21	0.19
(E)	PA66	(2)	Steel	0.20	0.23	0.22	0.18
(F)	PA66	(2)	PA66 (2)	0.22	0.15	0.11	0.08
(G)	PA66	(2)	Steel	0.26	0.15	0.07	0.04

Tests A – C are unlubricated;
Tests D and E are moisture-lubricated;
Tests F and G are liquid paraffin-lubricated

PA66 (1) – as-moulded
PA66 (2) – machined

Typical values of dynamic friction coefficient for most unlubricated plastics lie between 0.2 and 0.7. It is of interest to note the effect of molecular weight on the friction properties of PE: Ultra-High Molecular Weight Polyethylene (UHMWPE, characterized by weight-average molecular weight of around 4×10^6 g/mol) offers excellent surface mechanical properties for bearing and slideway applications. Although the friction coefficient of UHMWPE increases with surface pressure and rubbing velocity, it is possible to obtain friction coefficients between 0.1 and 0.2[6.51].

Having appreciated the influences of pressure and sliding speed on friction properties, a parameter which has been of significant use to bearing designers is the product of pressure (normal load per unit area of contact) and relative surface velocity, individual values of which ("PV" data, for a chosen grade of polymer) can be specified to represent an upper limitation for any such material in contact applications. It must be recognized that this parameter is critically dependent upon surface lubricity, so that in applications which may not be externally lubricated, the pressure-velocity specification will be a function of the surface characteristics of each individual grade of material, together with the additives which are present in the moulding compound.

Friction also determines the integrity of push-fit joints made from thermoplastics. If a circular (rigid) former of length L and radius R is connected into a smaller (annular) plastics tube, the axial separation force (F) is estimated by[6.52]:

$$F = (2\pi R) \cdot L \cdot \mu \cdot \sigma(t) \tag{6.66}$$

where the interfacial stress $\sigma(t)$, which is usually part of a stress relaxation situation, must take into account the time-dependent nature of the plastics bushing. Since $\sigma(t)$ will decrease with increasing contact time, so too will the interfacial pressure, and the integrity of the push-fit joint.

Wear is deemed to have occurred if solids particles are lost from the interfacial surface of a component in contact with another, during motion. Data have been derived[6.53] to show how dynamic wear rates in plastics vary with sliding speed and local temperature; it is concluded that wear rate is most accurately related to the product of tensile strength and failure strain ie. it is *strain energy* dependent. On this basis, we may speculate that high-strength (and relatively ductile) engineering plastics (PA, POM, PBTP, UHMWPE) offer very good wear performance in service. This can be further enhanced by the addition of carbon black or MoS_2 fillers, but compounding with mineral or glass fillers often limits wear resistance by changing the mode of damage from *adhesive* (unfilled resins) to *abrasive* wear.

Although it is often difficult to speculate or generalize upon wear characteristics of plastics, it is known that wear rate is accelerated by temperature, since the yield strength of most plastics is an inverse function of temperature. This effect is apparent in Table 6.5, since the friction coefficient for PA66 is generally lower when the nylon surface is in contact with steel; for polymer-polymer contact, the local temperature will rise to a greater extent when deformation occurs, and furthermore, the low thermal conductivity of the material will decelerate the dissipation of this heat.

Accelerated test data must therefore simulate not only the likely conditions under which surface damage occurs, but also the mechanism by which wear takes place. Some additional references are given for more detailed information on the friction and wear characteristics of plastics[6.54, 6.55].

References

6.1 TIMOSHENKO, S., and GOODIER, J.N.: *Theory of Elasticity,* McGraw-Hill, New York, (1951).

6.2 LANDAU, L.D., and LIFSHITZ, E.M.: *Theory of Elasticity,* Pergamon, Oxford, (1959).

6.3 WARD, I.M.: *Mechanical Properties of Solid Polymers,* Wiley, New York, (1971).

6.4 BILLMEYER, F.W.: *Textbook of Polymer Science,* 2nd Ed., Wiley, New York, (1971).

6.5 ALFREY, T.: *Mechanical Behaviour of High Polymers,* Interscience, New York, (1948).

6.6 FERRY, J.D.: *Viscoelastic Properties of Polymers,* 2nd Ed., Wiley, New York, (1970).

6.7 TURNER, S.: *Mechanical Testing of Plastics,* 2nd Ed., Godwin, London, (1984).

6.8 MILLS, W.H., and TURNER, S.: *Machines for Materials and Environmental Testing,* Presented at Joint I.Mech.E./Soc.Env.Eng. Symposium, Manchester, (1965).

6.9 SAUNDERS, D.W., and DARLINGTON, M.W.: *An Apparatus for the Measurement of Tensile Creep and Contraction Ratios of Small, Non-Rigid Specimens,* J. Phys. E: Scientific Instruments, **3,** (1970), 511.

6.10 BS 4618, (1970): *Recommendations for the Presentation of Plastics Design Data; Part 1,* British Standards Institution, London.

6.11 WARD, I.M., Ed.: *Structure and Properties of Oriented Polymers,* Applied Science, London, (1975).

6.12 BASSETT, D.C.: *Principles of Polymer Morphology,* Cambridge University Press, (1981).

6.13 CRAWFORD, R.J.: *Plastics Engineering,* 2nd Ed., Pergamon, Oxford, (1987).

6.14 ZENER, C.: *Elasticity and Anelasticity of Metals,* Chicago University Press, (1948).

6.15 POWELL, P.C.: *Engineering with Polymers,* Chapman and Hall, London, (1983).

6.16 TURNER, S.: *The Strain Response of Plastics to Complex Stress Histories,* Polym. Eng. Sci., **6,** (1966), 306.

6.17 Polymer Laboratories Technical Literature: *PL DMTA – MkII Dynamic Mechanical Thermal Analyser,* Polymer Laboratories Ltd., Loughborough, U.K.

6.18 WETTON, R.E.: *Dynamic Mechanical Method in the Characterisation of Solid Polymers,* Polym. Testing, **4,** (1984), 117.

6.19 GEARING, J.W.E., and STONE, M.R.: *The Dynamic Mechanical Method for the Characterisation of Solid Composite Polymers,* Polym. Comp., **5,** (1984), 312.

6.20 BUCKNALL, C.B.: *Toughened Plastics,* Applied Science, London, (1977).

6.21 BUCKNALL, C.B., COTE, F.F.P., and PARTRIDGE, I.K.: *Rubber Toughening of Plastics,* J. Mat. Sci., **21,** (1986), 301.

6.22 OROWAN, E.: *Fracture and Strength of Solids,* Rept. Prog. Phys., **12,** (1949), 185.

6.23 VINCENT, P.I.: *The Necking and Cold-Drawing of Rigid Plastics,* Polymer, **1,** (1960), 7.

6.24 VINCENT, P.I.: *Mechanical Properties of High Polymers: Deformation,* in *Physics of Plastics,* Ritchie P.D., Ed., Iliffe, London, (1965).

6.25 MARSHALL, I., and THOMPSON, A.B.: *The Cold Drawing of High Polymers,* Proc. Roy. Soc., **A221,** (1954), 541.

6.26 ALLISON, S.W., and WARD, I.M.: *The Cold Drawing of Polyethylene Terephthalate,* Brit. J. Appl. Phys., **18,** (1967), 1151.

6.27 KAMBOUR, R.P.: *A Review of Crazing and Fracture in Thermoplastics,* J. Polym. Sci: Macromol. Rev., **D7,** (1973), 1.

6.28 BOWDEN, P.B.: *The Yield Behaviour of Glassy Polymers,* in *The Physics of Glassy Polymers,* HAWARD, R.N., Ed., Applied Science, London, (1973).

6.29 ROZKUSZKA, K.P.: *An All-Acrylic Impact Modifier for Weatherable PVC Formulations,* Presented at International Symposium: Current Trends in PVC Technology, Loughborough University, U.K. (1986).

6.30 MENGES, G., BERNDSTEN, N., and OPFERMANN, J.: *Extrusion of PVC: Consequences of the Particle Structure,* Plast. Rubb. Proc. and Appl., **4,** (1979), 156.

6.31 KELLER, A.: *A Current Account of Chain Extension, Fibrous Crystallisation and Fibre Formation,* J. Polym. Sci: Polym. Symp., **58,** (1977), 395.

6.32 PENNINGS, A.J.: *Bundle-like Nucleation and Longitudinal Growth of Fibrillar Polymer Crystals from Flowing Solutions,* J. Polym. Sci: Polym. Symp., **59,** (1977), 55.

6.33 COATES, P.D., and WARD, I.M.: *Drawing of Polymers through a Conical Die,* Polymer, **20,** (1979), 1553.

6.34 GIBSON, A.G., and WARD, I.M.: *The Manufacture of Ultra-High Modulus Polyethylenes by Drawing through a Conical Die,* J. Mat. Sci., **15,** (1980), 979.

6.35 KATZ, H.S., and MILEWSKI, J.V.: *Handbook of Fillers and Reinforcements for Plastics,* Van Nostrand, New York, (1978).

6.36 TSAI, S.W., and HAHN, H.T.: *Introduction to Composite Materials,* Technomic Publishing, Westport, USA, (1980).

6.37 DARLINGTON, M.W., and UPPERTON, P.H.: *Designing with Short-Fibre-Reinforced Thermoplastics,* Plast. and Rubb. Intnl., **10,** (1985), 35.

6.38 TETLOW, R, in: *Carbon Fibres in Engineering,* Langley M, Ed., McGraw-Hill, New York, (1973).

6.39 ASHTON, J.E., HALPIN, J.C., and PETIT, P.H.: *Primer on Composite Materials: Analysis,* Technomic Publishing, Westport, USA, (1969).

6.40 ALLEN, H.G.: *Analysis and Design of Structural Sandwich Panels,* Pergamon, Oxford, (1969).

6.41 BENHAM, P.P., and WARNOCK, F.V.: *Mechanics of Solids and Structures,* Pitman, London, (1978).

6.42 ROARK, R.J.: *Formulas for Stress and Strain,* 5th Ed., McGraw-Hill, New York, (1976).

6.43 DARLINGTON, M.W., MCGINLEY, P.L., and SMITH, G.R.: *Structure and Anisotropy of Stiffness in Glass-Fibre Reinforced Thermoplastics,* J. Mat. Sci., **11,** (1976), 877.

6.44 ICI Technical Literature: *Propathene for Structural Foams,* Technical Service Note PP137, ICI Chemicals and Polymers Group, Wilton, U.K.

6.45 AHMADI, A.A., and HORNSBY, P.R.: *Moulding and Characterisation Studies with Polypropylene Structural Foam, Part 1: Structure-Property Interrelationships,* Plast. Rubb. Proc. and Appl., **5,** (1985), 35.

6.46 AHMADI, A.A., and HORNSBY, P.R.: *Moulding and Characterisation Studies with Polypropylene Structural Foam, Part 2: The Influence of Processing Conditions on Structure and Properties,* **ibid,** 51.

6.47 ASTM D785 (1970): *Rockwell Hardness of Plastics and Electrical Insulating Materials,* American Society for Testing Materials.

6.48 COVAS, J.A.: *Hardness Measurement as a Technique for Assessing Gelation of Rigid PVC Compounds,* Plast. Rubb. Proc. and Appl., **9,** (1988), 91.

6.49 OGORKIEWICZ, R.M. Ed.: *Engineering Properties of Thermoplastics,* Wiley-Interscience, London, (1970).

6.50 MASCIA, L.: *Thermoplastics: Materials Engineering,* Applied Science, London, (1982).

6.51 Hoechst Plastics Technical Literature:® Hostalen GUR, Hoechst Aktiengesellschaft, Frankfurt, Germany.

6.52 LOCKETT, F.J.: *Engineering Design Basis for Plastics Products,* HMSO, London, (1982).

6.53 LANCASTER, J.K.: *Estimation of the Limiting PV Relationships for Thermoplastic Bearing Materials,* Tribology, **4,** (1971), 82.

6.54 EVANS, D.C., and LANCASTER, J.K.: *The Wear of Polymers,* Treatise of Mat. Sci. and Tech., **13,** (1979), 86.

6.55 TANAKA, K., and NAGAI, T.: *The Effect of Counterface Roughness on the Friction and Wear of Polytetrafluoroethylene and Polyethylene, in Wear of Materials,* LUDEMA, K., Ed., ASME, (1985).

Chapter 7
Mechanical Properties II – Failure

7.1 Mechanical Failure in Plastics – An Introduction

Failure is a practical problem with products and implies that the component no longer fulfils its function. Frequently, the ability to withstand mechanical stress or strain (and thereby store or absorb mechanical energy) is the most important criterion in service; thus, it is with *mechanical failure* that we are usually concerned. Further, the occurrence of failure implies that at some time in its life the product was satisfactory. Failure by mechanisms other than mechanical will be included in later chapters, for example, electrical breakdown in Chapter 8.

There are several different types of mechanical failure which are applicable to plastics products. These include:

- **Excess deformation.** Many groups of plastics are low modulus materials and are capable of accommodating relatively high strains; for example, *neck stability* and *cold drawing* (following a yielding process) were illustrated in the previous Chapter. Moreover, the levels of strain associated with the yield point itself are often significantly higher than is the case in other, comparative engineering solids. In consequence, a designer may choose to attempt to fulfil a passive, strain-limiting *deformation criterion* (as opposed to an incipient "failure" criterion), in order to gain some assurance that the relatively low levels of stiffness in plastics do not lead to customer dissatisfaction.
- **Ductile failure.** Ductility reflects a material's propensity to undergo large-scale, irreversible, "plastic" deformation when under stress, and the *yield stress* describes the load intensity at which such effects occur. Since many families of plastics, under certain conditions, are able to exhibit yielding phenomena, the importance of defining yield stress as a practical upper limit for stress development in loaded components is apparent; it therefore represents a convenient means of describing the onset of ductile failure, though it must be appreciated that there are many factors which influence the magnitude of yield stress, and indeed whether this mode of failure is necessarily likely.
- **Brittle failure.** Brittleness in materials exemplifies a tendency towards failure induced by the creation and growth of crack-like defects in the system; analysis of such a failure mechanism (whether this involves fracture under a sudden impact load, or less-rapid crack propagation effects) is made more complex by the fact that both remote (average) stress and crack geometry determine the *stress intensity* in the sample, which is the dominating influence on eventual brittle failure. Many plastics materials can be included in the "brittle" category, so that a *fracture mechanics* approach is potentially applicable to failure analysis; however, as was the case with ductile failure and yield stress, there are many contributory factors

which can modify fracture resistance in fabricated plastics, and make the validity and applicability of conventional fracture theory somewhat limited.

On the basis of these arguments, it appears that conventional theory may be used to analyse failure in plastics components (whether ductile or brittle), as long as the appropriate *failure mode* can be anticipated. Obviously, plastics are associated with related property data which may be significantly different in comparison to other materials (for example, lower yield stress and modulus, but higher yield strain) but this may not detract from the general validity of a chosen approach. However, what is unusual about plastics materials, and of critical importance to a design specification, is the array of factors which are able to modify mechanical behaviour to the extent of being able to change the observed failure mode under stress.

Thus we are frequently able to see examples of *ductile-brittle transitions* in fabricated plastics, which are undesirable when associated with the tendency towards brittleness in normally ductile polymers. The likely effect of changing temperature, in this context, is illustrated in Figure 7.1, where it is evident that the observed failure mode (which will generally occur at lowest stress) is more likely to be brittle when temperature is diminished. The point of greatest importance is the possibility of a change in the *mode* of failure; since these effects are prevalent in many commercial plastics, and because there are many parameters other than temperature which influence a transition in mechanical failure data, it is imperative that all such parameters can be specified, and their effects understood.

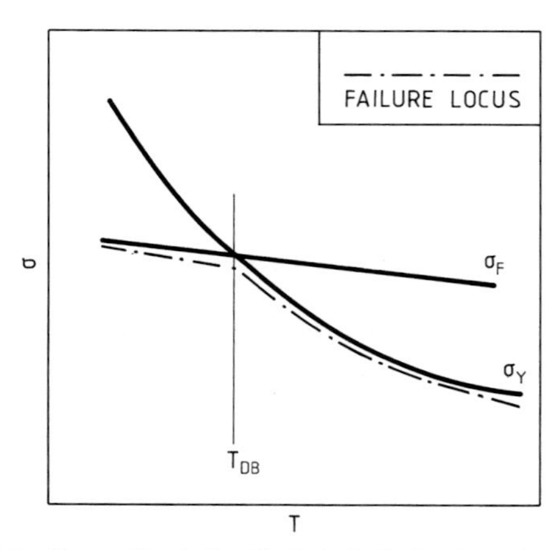

Figure 7.1 Generalized ductile-brittle behaviour in thermoplastics: dependence of yield stress (σ_y) and brittle-mode fracture stress (σ_F) on temperature. The failure mode will vary according to stress level, thereby defining an apparent ductile-brittle transition temperature (T_{DB}).

For example, any tendency towards brittleness may be accentuated under *high-velocity loading* conditions; Figure 7.2 illustrates the dependence of impact energy on temperature, for which a movement into the brittle mode is inevitably associated with a dramatic reduction in energy-absorbing capability. It is unfortunate, though important to note that such transitions in plastics' impact resistance often occur at temperatures close to intended service use conditions (typically -20 to $+50$ °C). Such transitions carry enormous significance towards the robustness of the component, but are often difficult to specify since the transition point (defined

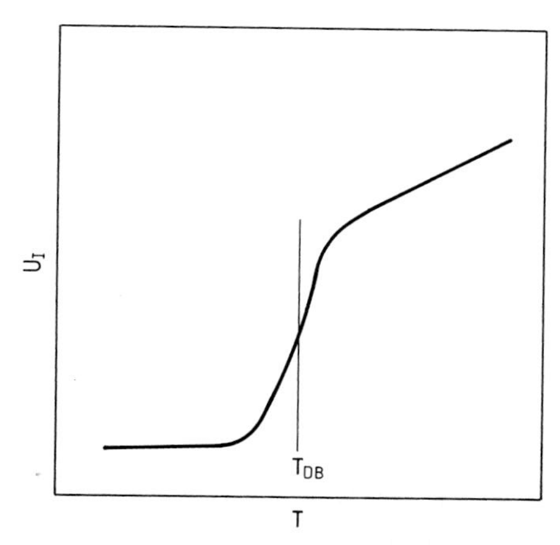

Figure 7.2 Ductile-brittle transition in plastics materials under high-velocity loading: impact energy (U_I) versus temperature (T).

as T_{DB} in the example given, but sometimes also associated with a primary or secondary relaxation effect) is shown to depend on other testing conditions such as the severity of stress concentration, and the loading velocity. Similar transitions are also apparent when a number of factors are subject to change:

- materials factors (microstructure; orientation, crystallinity, crosslinking, molecular weight, additives);
- component design and effects of processing conditions;
- environmental effects (presence of moisture, or photodegradation); and
- test conditions, such as strain rate and the stress system.

The respective influences of a number of these factors are discussed in greater detail in subsequent sections (7.6.1 – 7.6.3), with reference to some important microstructural features appropriate to a range of commercial products. First, it is appropriate to discuss the cases of ductile and brittle mechanical failure, which might be anticipated by extrapolation of the behaviour in laboratory tests.

For a more comprehensive introduction to the subject of polymer failure and fracture, the reader is first guided to the text by BROSTOW and CORNELIUSSEN[7.1], whilst the contribution by KAUSCH[7.2] offers a fundamental and extensive review of fracture mechanisms appropriate to high polymers.

7.2 Ductile Failure

7.2.1 Definition and Scale of Yielding

Ductile failure is, by definition, failure at high strain; it is by implication, beyond yield, which is the practical load-bearing limit for a ductile material, the point at which gross plastic deformation first occurs. In the case of plastics materials, it is perhaps more appropriate to describe the *ductile behaviour* of any polymer (rather than to refer to the material as "ductile"), since there are many variables which may modify, or change the yield characteristics.

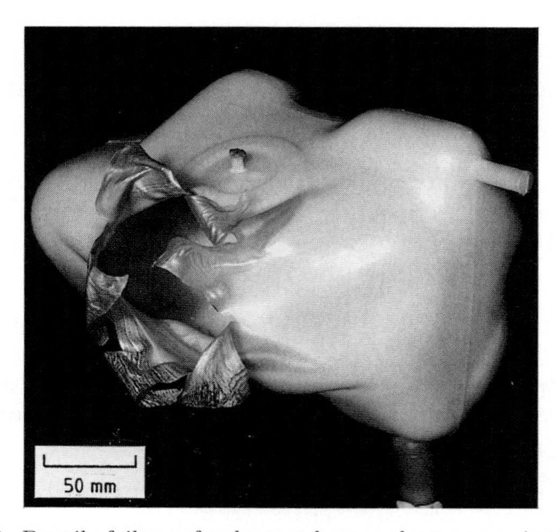

Figure 7.3 Ductile failure of polypropylene coolant reservoir tank.

On this basis, it is not surprising that mechanical properties data from a range of possible test methods are often subject to conflicting interpretations. In some cases, including failure of the PP pressure vessel illustrated in Figure 7.3, where the application of excess pressure at elevated temperature resulted in cold drawing prior to eventual plastic collapse, the occurrence of yielding is obvious; in other situations, evidence of yielding may only become apparent following additional, and more subjective investigations:

- **Macroscopic evidence.** Yielding may result in the promotion of a multitude of microscopic voids in the material, concentrated in the region of maximum stress; these may become visible to the human eye when sufficient light-scattering occurs, thus promoting the well-established effect of stress whitening;
- **Microscopic evidence.** Optical and electron microscopy techniques allow a more detailed examination of yield zones or fracture surfaces to be carried out. This type of information is valuable in showing exactly where yielding occurred on the sample, and to what extent; in other words, the precise nature and progression of failure may be investigated in this manner.
- **Mechanical test data.** For many polymers which behave as non-linear viscoelastic materials, it is not surprising that "yielding", or "ductile behaviour" is often subject

to rather arbitrary definitions. In general, a load-deformation plot which shows rupture occurring after passing through a maximum load loosely defines ductile behaviour (see section 6.5.1.1).

Some examples of load extension curves are presented in Figure 7.4; whilst the fracture data (σ_F, ε_F) in (A) and the nominal yield characteristics (σ_Y, ε_Y) in (B) are easily calculated, the specification of the appropriate yield parameters in Figure 7.4(C) requires additional interpretation. This might be typical of a ductile response from *oriented plastics* materials: as the original degree of orientation increases, the differences in stiffness of the bulk material before and after subsequent yielding become less noticeable, resulting in a more indiscrete yield point, as shown.

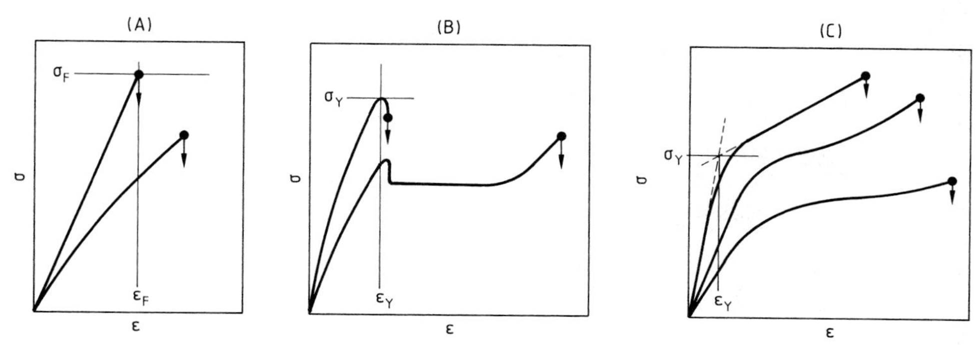

Figure 7.4 Nominal stress-strain relationships for plastics, showing: (A) Brittle fracture at maximum load (σ_F and ε_F represent 'brittle' stress and strain at failure); (B) and (C) Ductile yield failure (at stress σ_Y and strain ε_Y).

In these examples, the evidence for yielding as part of a ductile failure mechanism is relatively clear. However, small-scale plastic deformation is frequently present at the tips of growing defects during brittle fracture of plastics; whilst this observation is in no way unique to polymers, it is true to suggest that the relatively low yield stress of many commercial plastics leads to situations where the mode of failure may appear to be "brittle" on the macroscopic scale, yet the amount of deformational energy absorbed in microscopic yielding may be sufficiently high to limit the validity of a fracture mechanics approach to describe failure. Such validity criteria are dependent upon the size of the yield zone, relative to the overall specimen dimensions. This specific aspect of yielding will be addressed in more detail in a subsequent section (7.3.3.1).

7.2.2 Specification of Ductile Failure in Plastics

We have seen that many high polymers exhibit yielding, and that the local increases in stiffness which occur at the yield point are able to stabilize the necking deformation process, thereby allowing *cold drawing* to occur at a relatively steady drawing stress, until ultimate failure occurs in the drawn *microfibrils* created during, and following yield. In consequence, we can define more than a single failure characteristic for a ductile polymer: yield stress and ultimate rupture stress have been distinguished, and discussed in this context (using data from a uniaxial tensile test as an example) in Section 6.5.1.1 (Figure 6.22). In addition, it is useful to recall that small-scale *crazing* may also occur, at a stress level well below (typically 70 – 80 percent of) that at which macro-scale yielding or fracture becomes evident.

The yield stress characteristic is of most direct relevance to product design engineers, for the following reasons:

- it represents the point at which the external stress reaches a critical level, allowing sufficient molecular motion to promote large-scale, irreversible deformation, thus rendering the product unserviceable;
- it occurs before ultimate failure (ie. at lower stress/strain levels), although for many ductile plastics (especially in fabricated products associated with more complex or heterogeneous microstuctures, as a result of processing), yielding may progress to ultimate failure relatively quickly, without necessarily involving the development of high strain over a significant part of the sample.

In general, and in comparison to other strong solids, plastics materials can be considered to be effective strain-accommodating materials, with typical yield strain values of between 0.05 and 0.2 (5 – 20 %). Having established why the yield stress characteristic best defines the upper limitation upon which any programme of external mechanical loading (which is likely to lead to a ductile failure mode) should be based, it is useful to examine the criteria which determine yielding phenomena in plastics, and to obtain an appreciation of how some external factors are able to influence the ductile failure mode.

7.2.2.1 Yield Criteria

It has always been an important objective for engineers to establish accurate criteria to describe the general yielding behaviour of ductile solids. In this respect, polymers are no exception, and notable reviews of some attempts to modify existing yield criteria to accommodate some specific deformation characteristics of ductile plastics are acknowledged[7.3 – 7.5].

Whilst a detailed analysis and revision of established yield criteria (Tresca, Von Mises, Mohr-Coulomb) would fall outside the scope of this text, it is useful to make the following points concerning the general applicability of such criteria to yield processes in ductile plastics:

(1) Experiments to verify yield criteria in polymers have shown a marked dependence of yield stress on hydrostatic pressure, and on the stress system associated with various mechanisms of loading (for example, differences observed when testing in uniaxial tension, compression, or in pure shear);
(2) Specification of the most accurate and appropriate yield criterion for plastics is still subject to some debate.

Evidence has been derived to suggest that the Coulomb criterion describes the yielding behaviour of some glassy polymers[7.6], whilst a subsequent review suggests appropriate modification of Tresca and Von Mises criteria to account for the special characteristics of plastics[7.7]. In all these cases, the yield criteria refer to a critical component of *shear stress* (τ_Y), which is related to the principal stress components, and which can be used to define the condition for ductile failure in any of a range of test methods. The modified Tresca and Von Mises yield criteria for plastics may be represented as follows:

Tresca

$$\frac{\sigma_1 - \sigma_2}{2} = \tau_Y - \mu_S \cdot P \tag{7.1}$$

Von Mises

$$\frac{1}{\sqrt{6}} \cdot [(\sigma_1 - \sigma_2)^2 + (\sigma_2 - \sigma_3)^2 + (\sigma_3 - \sigma_1)^2]^{1/2} = \tau_Y - \mu_s \cdot P \tag{7.2}$$

σ_1, σ_2 and σ_3 are principal stresses, τ_Y is critical shear stress, μ_s is a constant which describes the sensitivity of yield stress to P, the hydrostatic component of stress, which is given by:

$$P = \frac{\sigma_1 + \sigma_2 + \sigma_3}{3} \tag{7.3}$$

BROWN[7.8] has presented yield stress data for several amorphous, and semi-crystalline plastics, as derived using the modified Tresca and Von Mises yield criteria presented above; typical values of μ_s were shown to lie between 0.10 and 0.25.

7.2.2.2 Some Influences on Yield Behaviour and Ductile Failure

Alongside other viscoelastic properties, yield and ultimate failure parameters in ductile plastics represent thermally-activated rate-process variables; in practice therefore, time and temperature are independent variables of some importance, which must be carefully standardized if data are to be used in a comparative manner. Of greater significance to long-term *design assurance* procedures however, is the decrease in yield/rupture stress evident when testing over long periods of time; clearly, it is desirable to be able to specify or model the long-term yield behaviour of plastics, for any form of loading (creep, stress relaxation, fatigue etc.). Some aspects of designing against long-term failure are featured in subsequent parts of this Chapter; at this point, it is useful to summarize the influence of some relevant test variables on ductile failure:

(1) Time/rate effects. The influences of time and rate effects may be manifested in different ways, according to the type of experiment. For example, yield stress generally increases with increasing strain rate, and to a first approximation for many plastics within the ductile mode, yield stress is linearly related to the logarithm of deformation rate. However, when samples are subject to constant stress in a creep experiment, we attempt to establish relationships between failure stress – *creep rupture* (see Section 7.5.1) – and time.
(2) Temperature. When temperature is increased, the additional molecular mobility facilitates the progression of plastic flow, resulting in a decrease in yield stress and generally, an increased sensitivity to strain rate effects.
(3) Pressure. Increases in hydrostatic pressure will result in a corresponding increase in yield stress, as described in the previous discussions on yield criteria.
(4) Microstructure and compound specification. There are many elements of microstructure which will contribute directly to observed changes in yield stress. Some of these have been referred to in Chapter 6, and since both yield stress and stiffness are often affected in a qualitatively similar manner when such changes occur, the comments made earlier (with respect to the effects of heterogeneous reinforcement, for example), are also thought to hold significance to yield stress data.

It should be stated however, that such generalizations do not necessarily assume universal relevance to all classes of polymers; the effect of individual changes in any specific experiment of interest (material, grade and method of manufacture; adopted test method and mode of stress) must be treated individually. For example, the yield behaviour of glassy polymers depends upon the extent of *free volume* in the amorphous phase, whereas the extent of crystallinity, and orientation of crystallites assumes far greater importance in the deformation characteristics of semi-crystalline plastics.

7.2.2.3 Choosing an Appropriate Test Method

Common test methods for the measurement of mechanical properties in the ductile mode include the following:

Long-term tests

- **Creep Rupture.** Constant stress experiments extended to include time-dependent rupture and/or crazing data.
- **Stress Relaxation.** Analogous to creep rupture, but carried out under constant strain conditions; this type of experiment is much less common than creep, firstly because it is not relevant to commercial loading conditions to the same extent, and secondly because failure times are extended (for equivalent "zero-time" stress) in the stress relaxation mode.

Short-term tests

- **Uniaxial Tension.** This technique is widely used to establish tensile yield data, using temperature and extension rate as independent variables.
- **Biaxial Tension.** In practice, this technique is more difficult to carry out, due to difficulties associated with instrumentation and gripping of continually-deforming surfaces.
- **Uniaxial Compression.** In terms of deformation mechanics, this test is similar to biaxial extension, though in practice its potential use is quite different. One of the most prominent advantages is the greater assurance that the ductile failure mode will be more likely to be preserved, under conditions which may otherwise lead to brittleness in tension.
- **Simple Shear/Torsion.** Problems associated with asymmetric specimens in simple shear may be overcome by the use of torsion tests.

All these methods are associated with individual advantages and shortcomings, according to the accuracy of stress and strain measurement, the likelihood of encountering external constraints (gripping in biaxial tensile tests, friction effects in uniaxial compression) and not least, ensuring that a change in failure mode does not become apparent. Clearly, the primary objectives associated with any chosen investigation must be carefully considered in order to specify the optimum test technique.

If we choose to initiate research centred upon the effect of a change in microstructure on yielding characteristics, it could be that one particular test method may be the most appropriate choice, perhaps for reasons of eliminating external constraint, thereby enhancing the sensitivity of a chosen yield parameter to the independent variable. However, if the objective is to obtain long-term data for design purposes, it is imperative that accuracy is optimized, and that the mechanism of testing is directly relevant to the envisaged loading conditions in service.

Design for ductile failure devolves therefore, into extrapolating the yield stress versus testing rate data to very low rates, that is, to very long times. Since yield stress measurements are generally accurate and reproducible, design can usually be undertaken with some confidence.

However, when failure in "ductile" plastics does occur, experience has generally shown that brittle fracture is often a much more dominant failure mode. The next sections of this Chapter are therefore dedicated to an appraisal of brittleness in plastics products, concentrating first on the principles of brittle fracture, and also giving detailed reference to some distinctive modes of brittle failure.

7.3 Brittle Failure

It is relatively straightforward to design plastics artefacts to avoid ductile failure, using procedures described in the previous section. In practice, ductile materials often fail in a brittle manner, which becomes much more difficult to predict from a theoretical standpoint. Brittle fracture is a low energy process characterized by failure at low strain, with little or no permanent deformation. Components contain small, crack-like defects which can act as stress concentration features; these microcracks grow under load and may eventually lead to a catastrophic failure.

Having examined some concepts of yielding which promote ductile failure in plastics, it is now appropriate to examine the opposite end of the failure mode spectrum by suitable revision of the principles of *brittle fracture* in solid plastics. By using separate approaches to study failure, we do not attempt to imply that failure diagnosis, and subsequent analysis, is always based initially upon a simple choice of theory according to the mode of failure; on the contrary, it has already been stated that the mechanical properties of many commercial grades of plastics are characterized by a high level of sensitivity to a significant number of variables. These variables might represent microstructural aspects of the polymer, effects of processing and/or design, or elements associated with the test methods adopted; whatever the most prominent influences on failure may be, the fact remains that the whole issue of mechanical failure analysis (and subsequent use of appropriate theory to predict failure) is clouded by the fact that *ductile-brittle transitions* are so often encountered by plastics materials.

Many engineering materials other than polymers suffer from the effects of brittleness, so that the analytical science of brittle failure analysis is well-developed, and is generally referred to in terms of *Fracture Mechanics,* which seeks to determine:

(1) how residual strength changes with defect size;
(2) what is a tolerable flaw in a manufactured component;
(3) what is the critical crack length at the expected service load;
(4) how often should crack inspections be performed in service.

Since the majority of effort has been directed at linear-elastic materials, the term *Linear-Elastic Fracture Mechanics, (LEFM)* is often used. We have already seen in Chapter 6 that the mechanical behaviour of most plastics materials is both non-linear and time-dependent; immediately, some of the difficulties associated with the application of LEFM theory to describe fracture processes in plastics become clear. Inevitably therefore, an increasing element of effort is necessary to establish the validity of the fracture mechanics approach. In particular, there are some factors associated with plastics which require special attention:

- Brittleness is often seen in normally-ductile materials;
- There are numerous factors which can change the mode of failure, and influence mechanical behaviour on either side of a ductile-brittle transition;
- Since many of these factors relate (directly, or indirectly) to the manufacturing processes used for plastics, difficulties are often encountered when attempting to relate fracture data from "idealized", laboratory-prepared specimens, to real products;

- Even when brittle fracture of plastics is deemed to have occurred, we should always attempt to qualify the influence of yielding (at the tip of propagating defects) on the overall response.

At this stage, it may be useful to refer to some notable review papers on polymer fracture[7.9–7.11], and to textbooks which are dedicated to the analysis of brittle failure and fracture mechanics in polymers[7.2,7.12–7.14].

Our approach in this section is to reintroduce and summarize the principles associated with various elements of fracture mechanics, discuss the criteria which underpin their general application to studies of brittleness in plastics, and finally, to refer to some examples of test methods and fracture behaviour associated with plastics products.

7.3.1 The Importance of Defects in Brittle Failure

All fundamental studies of mechanical strength in polymers have shown that:

- the measured fracture strength is always substantially lower than would be predicted from theoretical calculations of rupture forces for primary, or intermolecular bonds;
- the extent to which these observations hold true depends upon the physical size of test-specimens, with larger components generally behaving in a more brittle, and less predictable manner.

It is recognized that the origin of these effects, and indeed the whole basis of fracture mechanics, is centred upon the presence of *crack-like defects* in engineering materials. The rationale for analysis then becomes concerned with the degree of *stress intensity* created by the presence of any inherent flaws in a material and whether, or at what point, this is likely to increase to an apparently critical level (thereby promoting catastrophic failure), as a result of *crack growth* or increased application of stress. We are generally less concerned with the origin and nature of defects in materials, since these occur in an uncontrolled manner during all fabrication techniques. Furthermore, we may consider fracture mechanics as a "worst-case design tool"; ie. we should assume that defects will be present, then formulate a test methodology to study what their effect will be on the in-service performance of a component under stress. A simultaneous design programme to avoid ductile failure ought then to introduce a degree of reliability for each mode of failure.

Whilst these arguments may be applied with equal validity for all classes of engineering materials, there are some additional factors which are of relevance to the presence of flaws and cracks in plastics. Some examples of macroscopic defects associated with processing and component design are described and illustrated in a later section of this Chapter (7.6.3). However, not all defects (or other forms of stress concentration) created during manufacture are evident until studies at the microscopic, or sub-microscopic level are carried out; an additional factor of great relevance is therefore the scale of defects appropriate to a specific polymer and/or product. Kausch[7.14] has presented a paper which describes the origin and role of defects in the mechanical behaviour of solid plastics. These studies are based upon the heterogeneous structural nature of plastics, and suggest some specific features which are important within this context:

- Polymer degradation and chain scission;
- Local differences in orientation (chain anisotropy), crystallinity, crosslinking or additive concentration;
- Other forms of flaws, inclusions or stress-enhancing features (surface defects, crazing, contaminants, residual stress).

Additional examples of *macroscopic defects* in plastics products which may lead to stress concentrations include:

- Sharp internal corners and changes in cross-section (where a transition to plane strain loading; see Section 7.3.3.2, may occur);
- Pigment, or filler-agglomerates;
- Incomplete, or imperfect joints;
- Flash, or witness marks on the gates of injection mouldings.

It is therefore important and extremely desirable to specify the typical size of defects which may be present in components fabricated from plastics; indeed, it is not uncommon to use an *inherent flaw size* (measured on a routine basis by ultrasonic techniques, for example) as part of a quality assurance programme for certain types of product. In addition, many controlled type-tests and development programmes make use of externally-machined notches to concentrate the stress sufficiently to generate failure parameters appropriate to the brittle mode. The theories upon which such studies are based are summarized in subsequent sections.

7.3.2 Brittle Fracture in Solid Plastics

In this section, our intention is to revise some of the principles and developments of fracture theory, in order to establish a basis from which we may diverge to consider some of the specific characteristics of plastics, in the context of brittle failure.

7.3.2.1 Early Fracture Theories

The pioneering research on brittle fracture is often attributed to the work of GRIFFITH[7.15] in the early part of the twentieth century, in which it was recognized that:

- Brittle fracture can be directly related to the presence of cracks or defects in a material, with failure originating from the most severe of these;
- When fracture occurs, the energy required to increase the length of a defect by creating new surfaces must be balanced by a corresponding decrease in potential energy in a loaded sample;
- Fracture strength is inversely related to the length of any existing notch or defect.

The second of these statements forms the basis of the *energy-balance* approach to fracture, which is discussed in the next section; the other points were later exemplified by an important relationship which was able to account for the proposals made regarding the dependence of failure stress (σ_f) on defect length (a):

$$\sigma_f = \left(2\frac{E\gamma}{\pi a}\right)^{1/2} \tag{7.4}$$

where γ and E represent the surface energy and elastic modulus of the material.

Equation 7.4 is significant to all subsequent fracture theories, yet since Griffith's work was carried out on energy-elastic, and inherently brittle materials (mainly glass), a discrepancy arose when it became apparent that the measured values of surface energy for polymers (determined by relationships between failure strength and defect length) were far in excess of what were known, by separate experimentation, to be true values. Clearly, an independent, and additional mode of energy dissipation is evident during fracture of plastics; this is also true for other materials (steels, for example) and may be attributed to the development of local yielding, or other forms of *plastic deformation,* at the tip of a growing defect.

When dealing with fracture in materials which may show appreciable signs of ductility, it is sensible therefore to replace the surface energy (γ) with an alternative term which can account for the total work of fracture, including the contributions from plastic yielding: this parameter is usually referred to as the *strain energy release rate (G),* which is deemed to have reached a critical level (hence, critical strain energy release rate, G_C) when fracture occurs. Using similar terminology to equation 7.4:

$$\sigma_f = \left(\frac{EG_C}{\pi a}\right)^{1/2} \tag{7.5}$$

G_C represents the amount of energy required to create unit new surface area, as a crack grows during the fracture process; it therefore takes the units J/m^2.

Subsequent work by IRWIN[7.16] relates the fracture process to the intensity of stress created at the crack tip. For example, in an infinite sheet containing a central defect of length 2a, the *stress intensity factor (K)* is given by:

$$K = \sigma(\pi a)^{1/2} \tag{7.6}$$

Fracture will therefore occur when K reaches a specific level, the *critical stress intensity factor* (K_C). In terms of the other fracture parameters:

$$(K_C)^2 = (\sigma_f)^2 \cdot \pi a = EG_C \tag{7.7}$$

This development in the representation of brittle fracture is extremely valuable to the design engineer; as long as a solution for stress intensity factor can be determined (ie. appropriate to the specimen, notch geometry and chosen loading mode; see 7.3.2.3, below), we have a failure criterion which may be applied to many practical situations where brittle fracture is regarded as a possible consequence of loading.

In the following sections, we examine both the energy basis of fracture, and the stress-intensity approach in more detail.

7.3.2.2 Energy Balance Approach

Crack propagation during fracture can be represented by an energy balance; in qualitative terms, this states that the overall level of energy required to create new surfaces is balanced by the loss in stored elastic energy (which, for any loaded body containing a stress concentration, must itself be concentrated within the vicinity of the defect tip) in the specimen:

$$-\frac{dU}{da} = \gamma\frac{dA}{da} \tag{7.8}$$

$-dU$ represents an incremental loss in potential energy, and dA is the corresponding increase in surface area created by crack growth.

WILLIAMS[7.13] has worked from first principles to develop solutions for G based upon changes in energy during the fracture process. The representation is based on changes in the following incremental energy functions:

dU_1 – Applied energy;
dU_2 – Dissipated energy;
dU_3 – Stored strain energy;
dU_4 – Kinetic energy.

Hence, for any fracture process:

$$dU_1 - dU_2 = dU_3 + dU_4 \tag{7.9}$$

The terms in this expression can be differentiated with respect to the increase in fracture area (A), so that G, the strain energy release rate (per unit area) can be defined by:

$$G = \frac{dU_1}{dA} - \frac{dU_3}{dA} \tag{7.10}$$

According to the mode and geometry of loading, and the stress-strain response of the polymer, solutions for G may be obtained on the basis of this principle, thereby allowing experimental measurement of fracture energy to be determined; several such solutions are given by WILLIAMS[7.13].

7.3.2.3 Stress Intensity Approach

When a sharp defect is present in a stressed body, the stress concentration factor is related to the remote stress (σ) multiplied by the square root of defect length:

$$K = f \cdot [\sigma(a)^{1/2}]$$

The criterion for failure by unstable crack growth is defined by the critical values of K and G. Hence, $\sigma \rightarrow \sigma_f$ as:

$$K \rightarrow K_C$$
$$G \rightarrow G_C$$

These *brittle failure criteria* are analogous, and we therefore use K_C and G_C as material fracture parameters; in particular, K_C, the critical stress intensity factor, has assumed the more definitive term *fracture toughness*, which is expressed in units of $N/m^{3/2}$. It can be shown that K_C and G_C are related by:

$$(K_C)^2 = E \cdot G_C \tag{7.11}$$

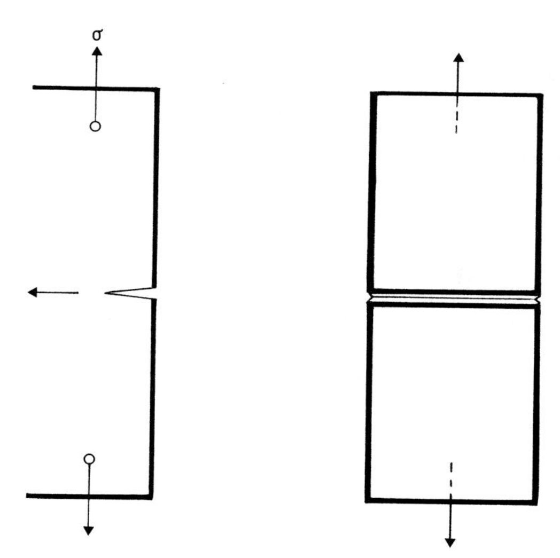

Figure 7.5 Mode I fracture testing (crack opening mode).

Additional subscripts are often used with these fracture parameters, according to the loading geometry and crack propagation mode; for example, K_{IC} represents fracture toughness under "Mode I" conditions of crack growth (Figure 7.5).

Using this approach, we need to specify the dependence of stress intensity factor upon crack type, size and specimen geometry, such that we can evaluate K_C directly from a measurement of failure stress (σ_f). A geometric calibration parameter (Y) is often used to relate stress intensity to remote stress and defect length, so that fracture toughness is evaluated by:

$$K_C = \sigma_f Y(a)^{1/2} \tag{7.12}$$

Y is a function of the size and shape of crack(s) within the component and can be determined according to the intended mode of experimentation; reference is given to a compendium of stress intensity factors[7.17], in which solutions for Y (and hence K) are presented for a multitude of practical cases. In practice, we can measure failure stress (σ_f) as a function of defect length (a), so that by rearranging Equation 7.12, we can see that a plot of $(\sigma_f)^2 \cdot Y^2$ versus $1/a$ will provide a solution for fracture toughness (K_C).

7.3.2.4 Crack-Opening Displacement

An additional property which has helped to rationalize the brittle fracture behaviour of ductile materials is the *crack opening displacement (COD)* parameter. Since plastics materials are generally characterized by relatively low values of yield stress, crack-tip ductility during fracture is of great importance to the overall failure process. It would therefore appear sensible to relate crack extension to the development of a *critical yield strain,* within the vicinity of a stress concentration. For example, in Mode I crack growth (Figure 7.6) we can see how the attainment of a critical strain at the crack tip can be represented by a related (vertical) crack-face

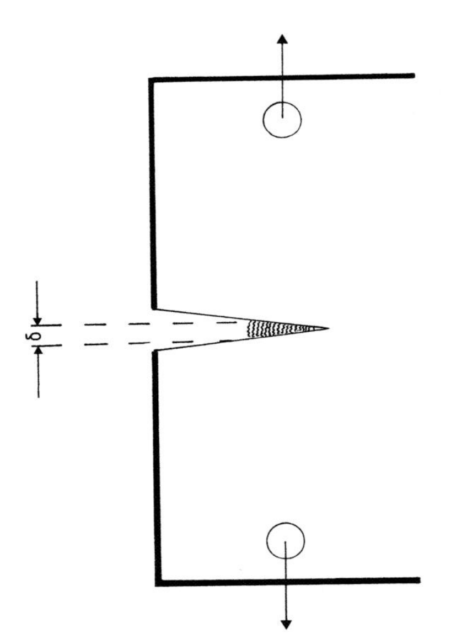

Figure 7.6 Schematic definition of crack opening displacement (δ) in Mode I loading.

displacement; hence "COD", (δ). This parameter can be related to other fracture properties by[7.13]:

$$\delta = \frac{(K_I)^2}{\sigma E} = \frac{G_I}{\sigma} \tag{7.13}$$

Local crack extension will occur when δ reaches a critical value (δ_C) so that in a short-term fracture test:

$$\delta = \delta_C \qquad \text{when} \quad \sigma = \sigma_f$$

However, whilst the attainment of δ_C implies that crack growth will occur, it does not always represent the point at which the fracture process becomes unstable; the crack extension may represent a metastable phenomenon, as part of a "slow crack growth" experiment (see below).

Some critical COD values have been quoted by FERGUSON et al.[7.18]; these are generally of the order of hundreds of microns, and include 0.43 mm for HIPS and 0.75 mm for PC.

7.3.2.5 Time-Dependent Crack Growth

As we have seen, stress intensity factor is a function of both applied stress and the length of any defects in the fracture specimen. The criterion for fracture is that unstable (rapid) crack growth will occur if K reaches a critical value. Another situation of interest emerges, if we examine the kinetics of time-dependent *subcritical crack growth,* for continuous loading where $K < K_C$.

It is useful to be able to predict crack growth rate, and ultimate failure time under such conditions, for which an empirical relationship between crack growth velocity da/dt and stress intensity factor (K_I) is required. An appropriate power law model has been seen to satisfy the behaviour of several plastics[7.19], and taking the example of Mode I loading, this can be quoted in the following form:

$$\dot{a} = \frac{da}{dt} = C \cdot (K_I)^m \tag{7.14}$$

where C and m are coefficients which describe crack growth data appropriate to specific materials and test conditions. Substituting for K_I (from an expression of the form given in 7.12) and rearranging, we can integrate as follows:

$$\int_0^t dt = \frac{1}{C(Y\sigma)^m} \int_a^{a_t} \frac{da}{(a)^{m/2}} \tag{7.15}$$

Since the defect length is a at zero time, and increases to a_t at any time t, we have:

$$\Big[t\Big]_0^t = \frac{1}{C(Y\sigma)^m} \cdot \left[\frac{2a^{1-(m/2)}}{2-m}\right]_a^{a_t}$$

This gives a solution for t, the time taken under constant load (during subcritical crack growth) for the defect to grow to an extended length a_t:

$$t = \frac{2}{C(2-m)(Y\sigma)^m} \cdot [(a_t)^{1-(m/2)} - a^{1-(m/2)}] \tag{7.16}$$

This expression holds true in any given crack growth regime where the coefficients C and m are constant. However, for the total time-dependent fracture process, there may be several such regimes, with $\dot{a}$ generally increasing with loading time; supporting evidence is often derived from microscopic analysis of the fracture surface, which reveals different topographical features in each region of crack growth.

7.3.3 Using LEFM for Plastics – Some Special Considerations

In the previous section, the principles of LEFM were presented, and the relevant material properties appropriate to brittle fracture were defined. Since the widespread introduction of plastics components in engineering applications has occurred, the development of fracture theory for other strong solids has matured considerably; this has allowed these principles to be adopted for plastics materials with some degree of confidence. However, there are some points which should be re-emphasized, which relate to the likely validity of applying LEFM to analyse the apparent brittleness in polymers:

- Because of the relatively low yield stress values of many plastics, plastic deformation at the crack tip is far more likely to occur;
- Whilst a small degree of dissipative energy can be accommodated in the overall work of fracture, it is obvious that as this assumes greater significance, there is a much greater possibility that a fracture mechanics approach will lose its general validity;

- Plastics properties such as fracture toughness and yield stress are dependent on many variables appropriate to fracture testing. Even for a given material therefore, the test conditions necessary to ensure validity are quite restricted.

These comments hold special significance for plastics materials which are normally ductile; ie. the constraints associated with designing valid fracture experiments become more severe. In summary, LEFM works best for plastics which are characterized by a low capacity for energy absorption (eg. in standard tests evaluating impact, or environmental stress cracking resistance), and a high yield stress; these include thermoplastics such as PS and PMMA, and unreinforced thermosetting plastics. It is generally difficult to find fracture data for commercial polymers which may be used for purposes of direct comparison; because of the constraints which influence the validity of the approach, the mode of stressing and test specimens used are subject to wide variation.

Fracture toughness values (in $MN/m^{3/2}$) for plastics generally lie between 0.5 (for unreinforced thermosets) to higher levels (1.0 – 3.0) representing a typical range for thermoplastics, although higher values have been recorded (eg. for highly reinforced plastics). A summary of numerical fracture data (K_C, G_C) has been compiled by KAUSCH[7.2], in which the test conditions used and the original publications whence the data were accessed are also quoted.

In summary, the major problem in applying fracture theory to plastics is to assess the extent to which the plastic deformation zone at the crack tip influences the result; the remainder of this section addresses this point in more detail.

7.3.3.1 The Plastic Zone

When a specimen containing a defect of appreciable size is loaded, the stress concentration rises in the vicinity of the crack; clearly, there must be a limit on the increase in stress which can be sustained, so that in practice, yielding occurs at the crack tip when the stress concentration is sufficiently high (Figure 7.7). The energy absorbed in this yielding process is directly related to the volume of plastically deformed material. It is usual to express the yield zone in terms of its radius (r_Y) at the crack tip; this is related to the other fracture properties by[7.10,7.13]:

$$r_Y = \frac{1}{2\pi} \cdot \left(\frac{K_C}{\sigma_Y} \right)^2 \tag{7.17}$$

Computation of the plastic zone size enables the likelihood of brittle fracture to be estimated. To control the effect of crack-tip yielding and generate valid fracture data, the specimen thickness (b) should always be significantly greater than r_Y; the empirical criterion quoted in (7.18) is often used:

$$r_Y < \frac{b}{4} \tag{7.18}$$

Experience has shown that the fracture properties of many polymers lie close to this limit of validity, for samples produced by conventional moulding techniques. This result emphasizes some important points:

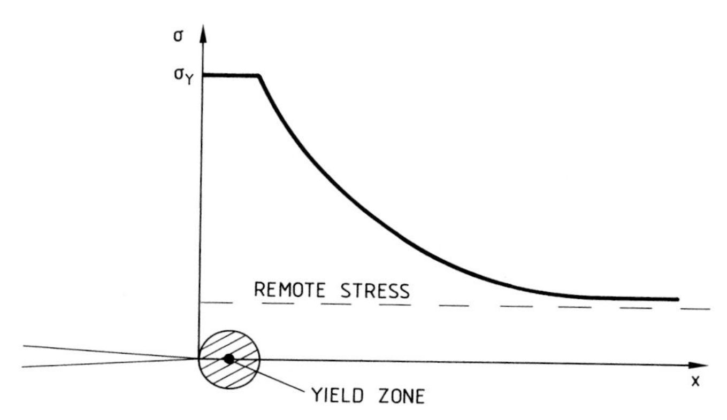

Figure 7.7 Stress concentration (σ) at the tip of a defect under load: as the distance (x) from the defect increases, the stress tends towards the nominal, 'remote stress', whilst yielding occurs close to the crack tip when the yield stress (σ_Y) is exceeded.

- It is often difficult to fabricate specimens to an appropriately high thickness using most melt processing techniques (where uniform cooling rates and microstructural homogeneity are notoriously difficult to achieve);
- A bimodal distribution of fracture energy data is not uncommon, in circumstances where a slight change in test conditions is sufficient to promote greater yielding: for example, complete brittle failures and ductile "hinge-breaks" are often seen within the same sample population.

7.3.3.2 Plane Strain Fracture

Having emphasized the potential influence of sample thickness, it is useful to verify the underlying cause for its prominent effects. As sample thickness is progressively increased, the general effect on total energy absorption (U) and measured fracture toughness (K_C) is illustrated in Figure 7.8. We are generally interested in the worst-case scenario, so that the behaviour of plastics in relatively thick sections is the desirable area in which to design experiments, in order to obtain *geometry-independent fracture* data.

This transition in fracture behaviour can be explained by movement from *plane stress* ("thin" samples) to *plane strain* conditions (Figure 7.9). In relatively thick specimens, the term "plane strain" suggests that the component of deformation in the thickness direction is suppressed, and ε_y reduces to zero. The stress system at the crack tip becomes *triaxial* (σ_y is finite), thereby contributing to an increased yield stress, which then favours the brittle failure mode. The constrained contraction effect in plane strain fracture requires that the expression relating fracture properties (7.11) be modified:

$$(K_C)^2 = \frac{EG_C}{(1-\nu^2)} \qquad (7.19)$$

(where ν is Poisson's Ratio).

This approach can lead to the definition of other *validity criteria* for quoting fracture parameters; indeed, we have already defined conditions for the size of the

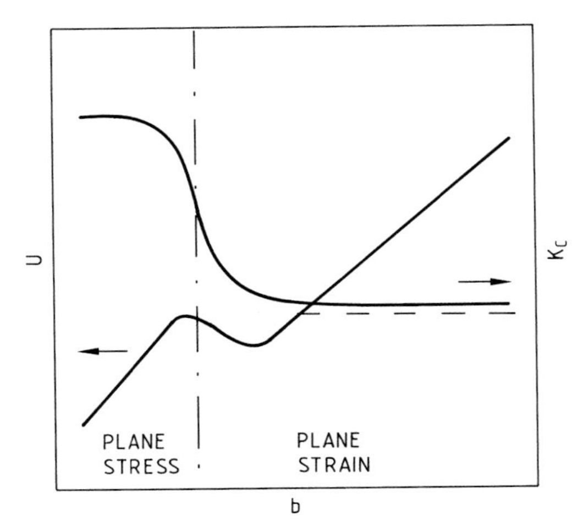

Figure 7.8 General effects of specimen thickness (b) on fracture energy (U) and fracture toughness (K_c), showing the transition from plane stress to plane strain failure.

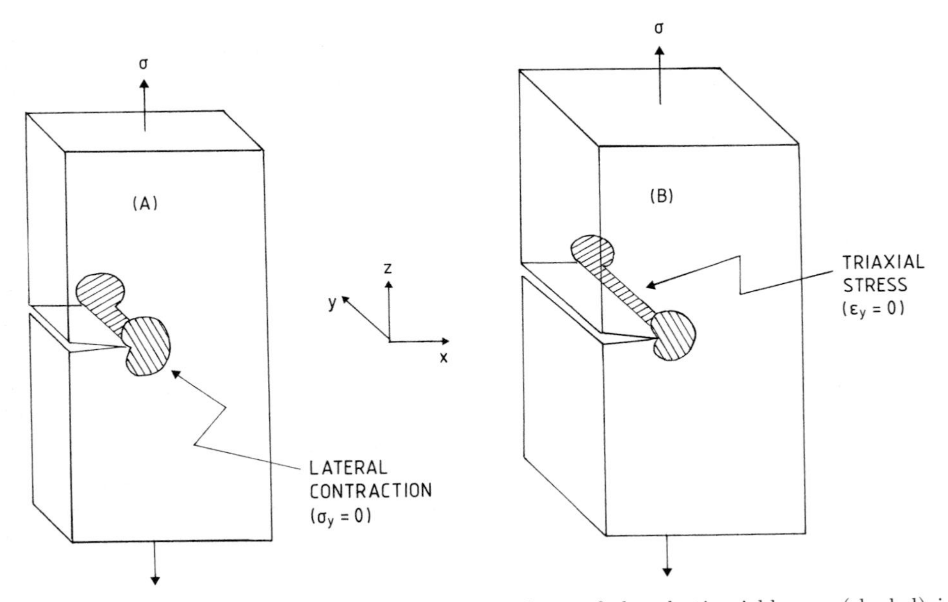

Figure 7.9 Schematic diagrams showing size and shape of the plastic yield zone (shaded) in fracture test specimens differing in thickness.

(A) Plane stress failure with minimal edge constraint, so that ε_y is finite and $\sigma_y = 0$;
(B) Plane strain failure in a thicker sample; with the exception of relatively small zones close to each surface, $\varepsilon_y = 0$ and the stress distribution is triaxial since σ_y is finite.

plastic zone, relative to specimen thickness (b) (Equations 7.17, 7.18). In addition, the minimum width (w) of a specimen required to restrict gross yielding may be estimated by[7.20]:

$$\frac{(K_C)^2}{(\sigma_Y)^2 w} = 0.64Y^2 \cdot \frac{a}{w} \cdot \left(1 - \frac{a}{w}\right)^4 \tag{7.20}$$

The left-side of this expression reaches its optimum value when w assumes its minimum value, ie. to minimize the effects of crack-tip yielding. This allows a minimum value of specimen width to be estimated. For example, using a sample span length to depth (L/w) ratio of 4[7.20]:

$$\frac{(K_C)^2}{(\sigma_Y)^2 w_{\min}} = 0.16$$

i.e

$$w_{\min} = 6.25\left(\frac{K_C}{\sigma_Y}\right)^2 \tag{7.21}$$

7.3.3.3 The J-Method

There have been some other notable developments in fracture testing, with the general objective of obtaining data on tougher plastics without necessarily resorting to excessively thick specimens or extreme conditions of experimentation (eg. using "unrepresentative" low temperatures). An example is the *J-Method*[7.13, 7.21], in which deep-notched samples are loaded to failure. Although the deformation mode represents total yielding, the method is deemed to succeed in characterizing toughness since the stress state is confined to plane strain conditions.

The parameter J is defined in a similar manner to strain energy release rate (G), in terms of energy per unit fracture propagation area; hence J_C is the critical level at which fracture occurs. A suggested test validity criterion with respect to sample thickness (b) and uncracked ligament area $(w - a)$ is quoted by Hashemi and Williams[7.21]:

$$b, \quad (w - a) > 25 \cdot \frac{J_C}{\sigma_Y} \tag{7.22}$$

In effect, this method allows fracture data to be obtained from materials which exhibit non-linear load-deflection characteristics even under high stress concentration. With respect to experimentation, the application of this method should be assured for physically smaller samples, relative to equivalent requirements for the generation of valid G_C data. However, the technique has still to reach a stage of maturity; it has been suggested[7.22], that some analytical modifications may be required when applying the J-method to certain classes of tough thermoplastics.

7.3.3.4 Influences on Fracture Parameters of Polymers

The fracture parameters of relevance to plastics are:

σ_Y – Yield stress
E – Elastic modulus
K_C – Critical stress intensity factor; fracture toughness
G_C – Critical strain energy release rate; work of fracture
J_C – Fracture energy function (J-method).

Some of these are interrelated, appropriate to particular test methods, by expressions which have been presented earlier (eg. Equation 7.19). In earlier sections, we have examined the dependence of various mechanical properties of plastics materials on a number of variables associated with the material formulation, and the test method. Some of these include:

- temperature and strain rate;
- external "environmental" effects (humidity; exposure to UV-radiation; presence of solvents, or surface-active agents);
- orientation of molecules, or fibrous reinforcements;
- molecular weight; effects of blending, copolymerization and additives.

Clearly, we have a situation where many fracture properties will depend on a number of simultaneous, or competing effects, when independent changes are made to an experiment. For example, if a variation is made which directly reduces yield stress, greater plastic deformation will generally occur, thereby increasing the apparent level of toughness. However, certain additives such as inorganic fillers may concentrate stress in plastics, but simultaneously, the resistance to crack propagation in the filled matrix may decrease, resulting in a more brittle product. When initiating research aimed at studying the effect of a change in input variables (especially material formulation changes), it is imperative to establish if the measurements represent a bone-fide change in fracture behaviour, or if a change in failure mode of the material has simply modified the validity of the approach.

Many of these issues are obviously polymer-specific; the reader is therefore referred to the text by WILLIAMS[7.13], in which a summary of previous research on several classes of plastics has been presented.

7.3.3.5 Test Methods

The fracture properties of plastics (fracture toughness (K_C) or strain energy release rate (G_C)) are determined by appropriate test techniques whereby cracks of prescribed shape and dimensions are machined into specimens of finite size. The defect-containing sample is loaded, and failure data are recorded. However, there exists a multitude of techniques (differing in sample/crack shape, loading mode and other test conditions) which have been used for this purpose. Ultimately, the final choice of methodology depends upon the required information, and its intended application; however, the following principles of a fracture test programme should be adopted:

(1) Introduce a defect of appropriate shape and dimensions;
(2) On selected samples, check that the actual crack dimensions are correct and consistent (especially crack-tip radius and the defect length, relative to the overall specimen dimensions). A microscope is an invaluable tool for this purpose;
(3) Condition all samples to the required temperature and humidity, for an appropriate period before testing;
(4) Select and control test variables (temperature, strain rate, machine calibration etc.) and attempt to eliminate any possible contributions to erroneous data (sample twisting, friction losses);
(5) Carry out the test and measure maximum load and total energy absorption (required for K_C and G_C, respectively);

(6) Obtain the appropriate solutions for stress intensity factor (K) and geometric calibration parameter (Y), where applicable;
(7) Check the validity of the data (as described earlier in this section), appropriate to the load-extension data, and to any microstructural observations on the fracture surfaces;
(8) If necessary, change test conditions (4, above) or notch dimensions to ensure that the test is generating valid, plane strain fracture data.

Many different test methods have been used in the research programmes which have been referred to in the texts and papers cited; solutions for G and K are listed for some of these methods by PASCOE[7.23]. Ultimately, the choice of technique depends upon whether the objective is to determine:

- Material characteristics – in which case a "standard" methodology is recommended;
- Design data appropriate to brittle fracture – for which K_C and/or G_C data are necessary;
- Effects of specific forms of embrittlement – impact, fatigue, environment-induced.

Another potential source of embrittlement arises when components are joined together during fabrication by melt-phase welding, or by using a suitable adhesive. Inevitably, the material immediately adjacent to the joint represents a possible plane of weakness, along which brittle crack growth may subsequently occur. An extensive list of test techniques which may be adopted for this special form of brittle fracture has been compiled by KINLOCH[7.24].

7.3.4 Examples of Fracture Mechanics Applied to Plastics Products

7.3.4.1 PVC Pressure Pipe – BS 3505

In order to attain an additional level of assurance against premature brittle fracture in cold-water PVC pressure pipe, the revision of BS 3505 in 1986 included the introduction of a *fracture toughness* specification[7.25]. The requirement is that all pipes of size 3 (89mm external diameter) or greater shall withstand for not less than 15 minutes without breaking of cracking at the notch, the test force corresponding to a true fracture toughness of not less than 3.25 $MN/m^{3/2}$. The test specimens are taken from internally-notched split-rings, machined from extruded pipe sections (Figure 7.10).

A fracture mechanics approach works well for a notch-sensitive material such as unplasticized PVC, and helps to provide some assurance against brittle fracture, in parallel to other tests which are based upon the ductile failure mode. The true fracture toughness (K_{IC}, which is in accordance with the dimensions of the test-piece) is given by:

$$K_{IC} = \left(-2x^2 \cdot \ln \left(\cos \frac{K_C}{x} \right) \right)^{1/2} \tag{7.23}$$

where $x = 32.56(e_n)^{1/2}$ (for wall thickness e_n at the notch, assuming the notch depth (a) is one quarter of the pipe wall thickness, for a material of tensile strength

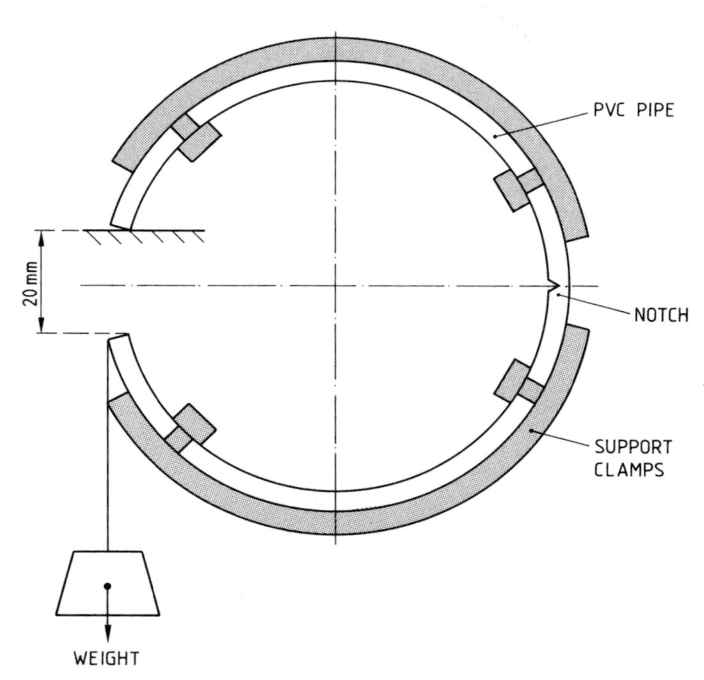

Figure 7.10 Notched split-ring test geometry for the evaluation of fracture toughness in UPVC pressure pipes, according to BS 3505.

50 MN/m^2), and fracture toughness K_C is expressed as:

$$K_C = \frac{3YM}{b(e_n)^{3/2}} \tag{7.24}$$

Y is the geometric calibration parameter ($Y = 1.91$ for $a/e_n = 0.25$), M is the applied bending moment and b is the width of the ring-section at the position of the notch.

This standard test method also contains other tests based upon additional mechanical properties:

(1) Impact resistance – to a falling, hemispherically-ended striker of appropriate mass (chosen according to the nominal pipe size), released from 2 metres at 20 °C;
(2) Creep rupture (hydrostatic pressure) resistance – time-dependent tests at both 20 °C and 60 °C.

Other previous research on fracture properties of PVC pipe is also referred to: MARSHALL and BIRCH have investigated the effect of PVC gelation on time-dependent fracture toughness[7.26], whilst MOORE et al.[7.27, 7.28] have examined the validity of a fracture mechanics approach in detail, with reference to other independent variables of importance, such as molecular weight, processing level and age-in-service.

7.3.4.2 Fracture of Impact Modified PVC Profiles

Although the inherent toughness of plastics such as rigid PVC can be greatly improved by the addition of an *impact modifier,* experience has shown that when premature failure occurs (for example, in imperfect welds, or by abuse experienced during transportation or improper installation of PVC window sections), the failure mode is inevitably brittle. A fracture mechanics analysis on this type of polymer is therefore still of considerable relevance, yet is made more problematic in practice by the lower yield stress of impact modified plastics promoting the tendency to observe gross yielding in "fracture" tests.

For example, the thickness-dependence of notched impact strength in modified PVC window profiles has been demonstrated[7.29]; a transition to a *plane stress* state, as impact strength increases below a specific profile thickness level, is thought to be responsible for this behaviour.

In order to investigate Charpy impact properties appropriate to the brittle mode, a Trade Standard[7.30] has specified the use of a sharper notch (of tip radius 0.1mm), for testing at ambient temperature; this test is based on a "pass-fail" principle, for which the target impact strength is 12 kJ/m^2.

Experience has shown that when testing some of the more ductile specimens, a mixed-mode response can still be obtained, whereby a proportion of the sample population exhibit *"hinge-failures",* caused by crack blunting induced by gross yielding. Figure 7.11 illustrates a typical fibrillar morphology of highly-strained, cold-drawn PVC, close to the position of defect-termination. This effect results in some unusually high impact strength values, giving a bimodal distribution of data appropriate to both failure modes, showing that these particular test conditions lie close to a ductile-brittle transition.

The use of LEFM provides a means of eliminating data which reflect the presence of such a transition. By Charpy testing with a "dead-sharp" notch at a lower temperature, it is possible to investigate the direct effect of changing independent variables on fracture properties, without the secondary (but often dominating) influence of geometric variables such as specimen thickness[7.31]. Figure 7.12 shows that whilst the Charpy impact strength of the profiles (according to the BPF/GGF test[7.30]) appears to vary in a non-systematic manner with extrusion temperature (specimen thickness is modified by such changes, due to differences in die swell behaviour, and the required levels of drawdown appropriate to profile sizing), the geometry-independent G_C data are relatively insensitive to the morphological changes induced by processing.

This example demonstrates how to eliminate the effects of thickness (created by plane stress conditions at the crack tip), during a pendulum-type impact test. The fracture mechanics approach to analyzing plastics' impact characteristics will be discussed further, in Section 7.5.2.

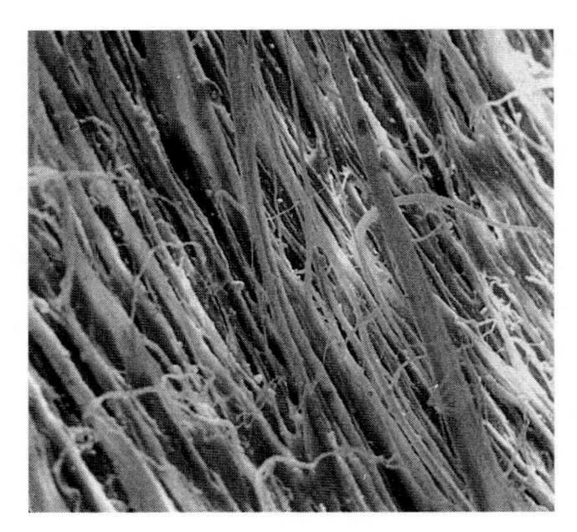

Figure 7.11 Scanning electron micrograph showing the fibrillar morphology of drawn PVC, at the root of a 'hinge-break' Charpy impact test specimen. (Reproduced by permission of Elsevier Applied Science Publishers Ltd.)

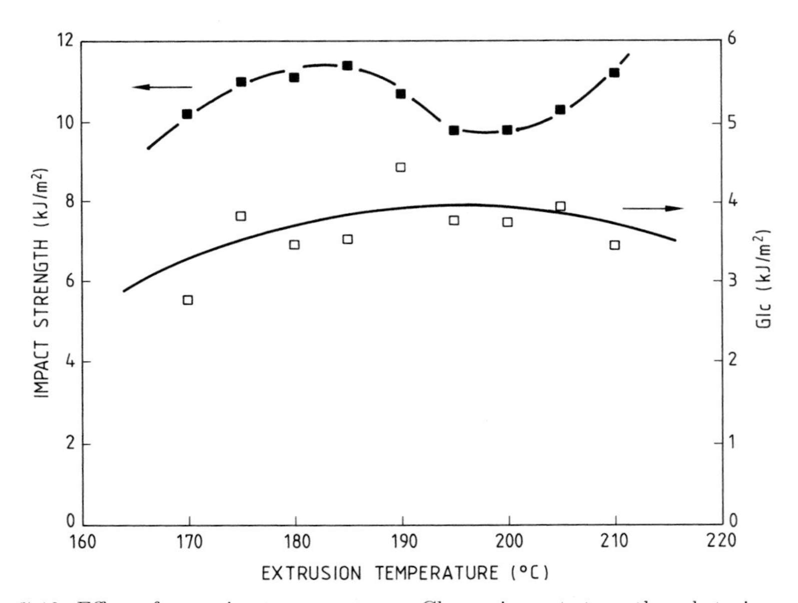

Figure 7.12 Effect of extrusion temperature on Charpy impact strength and strain energy release rate (G_{IC}) in impact-modified PVC.

7.4 Failure of Plastics Composites

7.4.1 Continuous Fibre-Reinforced Plastics Composites

Fibre reinforced plastics are much stronger in the fibre direction than transversely and thus their failure strength will be a function of the relative orientation of the applied load (Figure 7.13). The designer must, therefore, consider all 3 components of stress to determine the failure criterion.

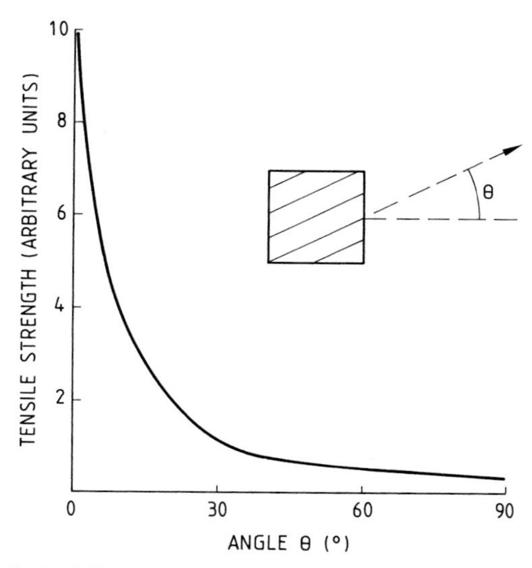

Figure 7.13 Off-axis tensile strength of unidirectional, reinforced plastic laminate.

It is usual[7.32, 7.33] to design composite structures on the basis of *first ply failure (FPF)*, rather than by catastrophic failure of the material. The main advantages of the FPF approach are that it allows for an adequate safety factor to be an integral part of the design stress, and maintains the integrity and stiffness of the component. The failure of the first ply will be followed by other ply failures until ultimate failure occurs.

Criteria based on maximum stress or maximum strain have also been used to predict composite failure. These techniques do not allow for any interactions between the various stress components and tend to overestimate the strength of composites[7.32]. It is preferable to use a method which allows for these interactions, based on the Von Mises approach for conventional isotropic materials. The relationship can be expressed as follows (using the terminology of TSAI and HAHN[7.32]):

$$\left(\frac{\sigma_x}{X}\right)^2 - \frac{\sigma_x \sigma_y}{X^2} + \left(\frac{\sigma_y}{Y}\right)^2 + \left(\frac{\tau_{xy}}{S}\right)^2 = k \tag{7.25}$$

σ_x, σ_y, τ_{xy} are the stress components on the principal axes, and X, Y, S are the longitudinal, transverse and shear strength properties (respectively).

Failure will occur within a ply when k ≤ 1. The approach is conservative, and k is effectively a safety factor which can be used in design.

The procedure for a laminate containing plies oriented in different directions is as follows:

- For the selected ply, transform the applied stresses (σ_1, σ_2, τ_{12}) to the stress components in the principal axes (σ_x, σ_y, τ_{xy}):
- Substitute into equation 7.25 to calculate k;
- Repeat these steps for each ply.

The lowest value of k represents the weakest layer; if $k \leq 1$, then the ply will fail. It should be appreciated that this does not imply catastrophic failure of the laminate.

The failure envelope of a *cross ply GRP laminate* is shown in Figure 7.14; the curve represents the locus of first ply failure. Numerous examples of this technique are discussed by TSAI and HAHN[7.32].

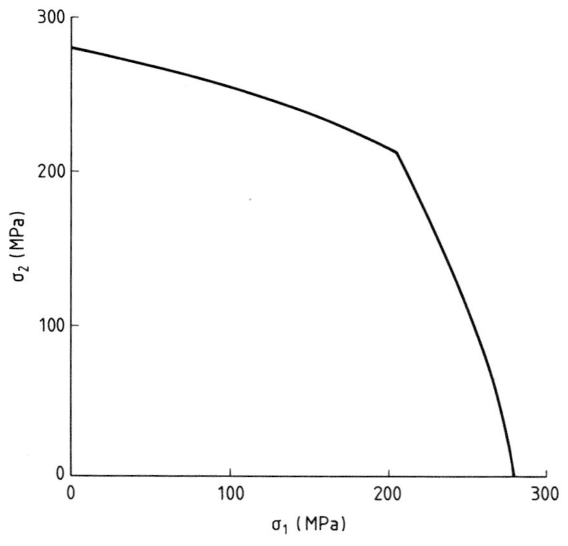

Figure 7.14 First-ply failure envelope for crossply [0/90]s carbon fibre reinforced plastics (CFRP) laminate, subject to biaxial tensile stress.

In practice, many fabricated laminates suffer from the effects of damage in the outer fibres *(gel coat cracking)* when put under continuous load; the stress concentrations associated with such defects can then be responsible for premature (low-strain) failure, especially if the failure mechanism is environment-related, such as occurs when GRP is loaded in an acidic atmosphere[7.34]. This has led to the term: *stress corrosion* of glass reinforced plastics composites and represents, in effect, a form of *environmental stress cracking.* Mechanisms of environmental crack growth in reinforced plastics are inevitably quite complex, and lie outside the scope of this treatment.

The reader is referred to the monograph by HULL[7.35] for further details on the strength and failure characteristics of plastics composite materials.

7.4.2 Short-Fibre Reinforced Plastics Composites

Analyzing the failure strength and fracture properties of plastics materials reinforced by the addition of short fibres has become a complex problem, for the following reasons:

- Typical fibre length may be of subcritical size, thereby limiting the validity of, or requiring modifications to, the theory of continuous fibre composites;
- Stress concentrations at fibre-ends become far more significant as the average fibre length decreases;
- Fibre breakage occurs during melt processing, so that a fibre length distribution will exist in the product. This distribution will be very different to the nominal (initial) fibre length distribution, and will vary with processing conditions during compounding and subsequent moulding;
- Preferred fibre orientation occurs in moulded products. Once again, the magnitude of this effect will be process-dependent, but the influence of anisotropy is rarely insignificant, since the alignment of the reinforcing fibres has a very important bearing on the ultimate failure properties of fibre-reinforced composites.

Fibre length (and length distribution) and *fibre alignment* (and therefore the extent of anisotropy through the thickness of moulded products) are of critical importance to engineers and designers who wish to predict the failure strength of short-fibre reinforced plastics. In turn, the mode of loading (tension, high-velocity impact, notched flexure etc.) must be specified before the effect of such changes can be predicted with confidence.

(A) Fibre length. In the previous Chapter (Sections 6.6.2.2, 6.6.2.4), we described how a simple rule of mixtures may be used to predict the strength of continuous fibre composites. However, in situations where short fibres are used as reinforcing materials, values of failure stress predicted by Equation 6.34 (for example) will inevitably represent overestimates. FOLKES[7.36] has described how the average stress carried by the fibres at failure ($\overline{\sigma}_F$) is much lower than the inherent fracture stress of the reinforcing material (σ_F) , even if the fibres are greater than critical length:

$$\frac{\overline{\sigma}_F}{\sigma_F} = 1 - \frac{L_C}{2L} \tag{7.26}$$

For fibres of subcritical size:

$$\frac{\overline{\sigma}_F}{\sigma_F} = \tau_I \frac{L}{2R} \tag{7.27}$$

L_C is the *critical fibre length,* and τ_I is the fibre-matrix *interfacial shear strength.* The values of σ_F may then be substituted into expressions of the type given in 6.34, to allow more realistic estimates of composite failure strength to be made; that is, by accounting for the attrition in fibre length which occurs during the processing of these materials.

(B) Fibre orientation. The effects of off-axis loading in fibre composites were discussed in the context of stiffness in Section 6.6.2.3. When attempting to extend the appreciation of anisotropic effects to failure characteristics, the point of overriding importance is to be able to specify an appropriate *failure mode.*

HULL[7.35] and FOLKES[7.36] each refer to the original paper by STOWELL and LIU[7.37], which described three distinct failure mechanisms appropriate to loading at different angles (θ) to the direction of fibre orientation.

Tensile fibre fracture occurs at very low values of θ; at intermediate angles ($10° < \theta < 60°$) the failure mode is dominated by excessive *shear stress* at the fibre-matrix interface (inducing failure by *delamination*), and as θ approaches 90°, the fibres simply act to concentrate the stress in the matrix material, which is itself subject to ultimate failure without any significant degree of reinforcement being imparted to the composite material.

Expressions relating the dependence of failure stress on θ have been derived[7.37], and the approach can be extended further to specify failure stress in some special cases[7.36]. Another more qualitative, but practically-relevant point to consider however, is the fact that when short-fibre composites are fabricated, the thermomechanical history will impose a complex orientation distribution of fibres[7.38], making strength evaluation a more complex issue.

(C) Energy-absorbing characteristics. The requirements for high toughness in short-fibre composites do not necessarily comply with the demand for high intrinsic strength and stiffness. This arises due to two additional, and highly-significant mechanisms of energy-absorption: *interfacial debonding* between the fibre and matrix, and *frictional pull-out energy* required to remove reinforcing fibres from their original positions in the composite, as crack growth occurs.

The energy component associated with this specific aspect of failure (U_f) assumes its maximum level when the average fibre length is of critical size (L_C)[7.36], and is given by:

$$U_f = V_F \tau_I \frac{(L_C)^2}{12D} \tag{7.28}$$

V_F is the volume fraction of fibres and D is the fibre diameter.

To obtain the maximum energy-absorption, the role of the *fibre-matrix interface* is of particular note. Although an increase in τ_I will be reflected by greater toughness, this is only true in circumstances where debonding and pull-out represent the dominant failure mechanism; that is, where fibre length is below L_C.

If the envisaged applications for short-fibre plastics composites are likely to require a high degree of toughness, then the effects of fibre-attrition during melt processing may often be tolerated without undue concern; clearly, this example reflects a case of compromise between the mechanical properties which are sought.

(D) Effects of processing. In addition to the effects on fibre length distribution and alignment, there are other aspects of processing which influence composite properties:

- **Degradation and interfacial characteristics.** Some evidence has been gained to suggest that the impact, and tensile strength characteristics of reinforced PETP are diminished in hot and humid environments, not only by hydrolysis, but also because of a change in the interfacial strength and the associated failure mode[7.39];
- **Influence of weld-lines.** Not unexpectedly, the morphology of internal welds in injection moulded short-fibre composites is complex; in consequence, the relationships between processing and attainable mechanical strength are difficult

to predict. Of particular concern is the phenomenon of *transverse fibre orientation,* within the vicinity of a weld.

The data in Figures 7.15 and 7.16 illustrate some of these effects. In 7.15, the dependence of ultimate tensile failure stress (σ_f) of short glass-fibre reinforced

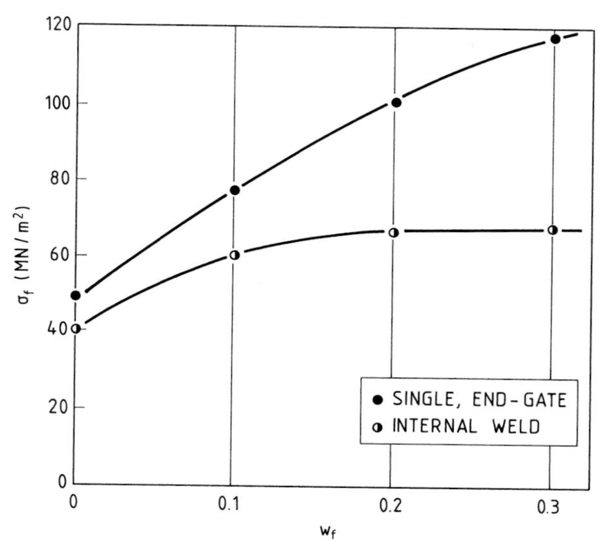

Figure 7.15 Dependence of tensile failure stress (σ_f) on glass-fibre concentration (weight fraction, w_f), for injection moulded PETP, showing the effect of internal flow-welds.

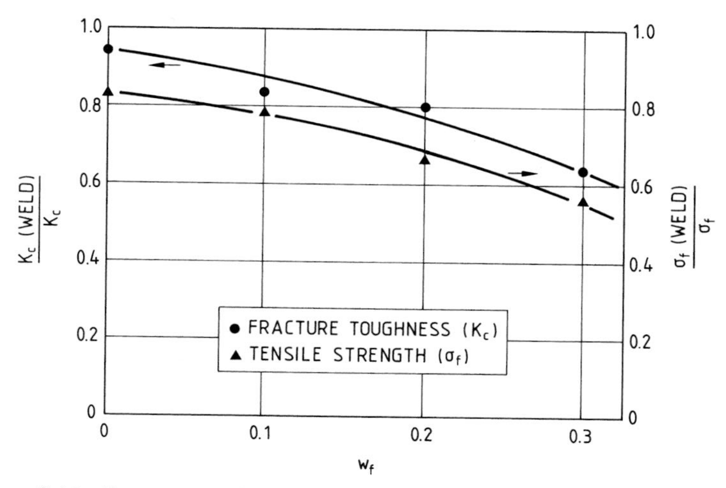

Figure 7.16 Fracture toughness (K_c) and tensile strength (σ_f) data for injection moulded compounds of short glass-fibre reinforced PETP. On each scale, the properties of samples containing an internal weld are plotted (relative to the equivalent properties from single, end-gated mouldings), against weight fraction of reinforcement.

PETP on fibre content (in the range 0 – 30 %, by weight) is evident: the failure strength of specimens which do not contain an internal weld varies directly with fibre concentration, but this is not true in the weld-containing samples, injection moulded using twin, end-gates. This type of mould filling mechanism was referred to in Chapter 5, and generally results in a drop in reinforcing efficiency due to *transverse fibre alignment* at the weld-interface.

Fracture toughness is also modified in a similar manner by the presence of the internal weld (Figure 7.16): although fracture toughness of all samples was shown to increase with fibre content, the K_C values (derived from testing double-edge notched (DEN) specimens in tension, with the notch-tip coinciding with the weld-plane) of welded specimens (relative to unwelded) are significantly decreased when the degree of reinforcement increases. A similar decrease in normalized tensile strength is also evident.

7.5 Some Specific Modes of Failure in Plastics

So far in this Chapter, we have reviewed some concepts of ductile failure and brittle fracture in plastics, and highlighted some of the factors which promote the tendency to observe brittleness in an otherwise ductile polymer. However, experience has shown that the stress system, timescale and external loading conditions each play an important part in the mechanical performance of plastics products. In consequence, it is possible to group together and distinguish some individual aspects of failure which represent predominantly, though not exclusively, brittle fracture modes.

7.5.1 Creep Rupture

When failure occurs over an extended period of time under static loading conditions, *creep rupture* or *static fatigue* are terms which are often used to describe this failure mode. It is indicative of the viscoelastic nature of plastics' mechanical properties, though it is important to separate two distinct types of behaviour, as described below.

(1) Ductile failure. Failure occurs by *gross yielding,* which is usually evident by the extensive deformation in the vicinity of the rupture point (eg. Figure 7.3) and by *stress whitening.* As the applied stress decreases, the time to failure becomes extended; it is true to suggest that this failure mode is simply indicative of a time-dependent yield stress characteristic in plastics.

(2) Brittle failure. Failure occurs as a result of *unstable crack growth,* with little or no macroscopic evidence of yielding, which is confined to a small zone at the crack tip. Main crack initiation takes place from a pre-existing defect within the component, and failure is then determined by the kinetics of slow crack growth (Section 7.3.2.5) appropriate to the test conditions.

It is wrong to suggest that the long-term creep rupture characteristics of any plastics material will be either wholly-ductile, or wholly-brittle; in practice, the observed failure mode will depend upon test conditions (stress system, temperature, test media) and not least, upon time. This point is taken further in the next section.

7.5.1.1 Prediction of Creep Rupture Failure

A particular commercial sector which is specifically concerned with creep rupture performance is the *plastics pressure pipe* industry. The main problem which is faced is the necessity to ensure that failure can be avoided over what may be extensive pipe lifetimes of up to 50 years. Obviously, it is not practicable to run proof tests which even approach such lifetimes, so that a great deal of effort has been put into modelling creep rupture data so that a design stress can be chosen on the basis of extrapolation.

The usual format for presentation of creep rupture test data is by relating an internal pressure, or stress function (an independent variable, but one which is usually plotted on the ordinate) to the logarithm of failure time (measured as a dependent variable). An example of these data, for HDPE pipes tested according

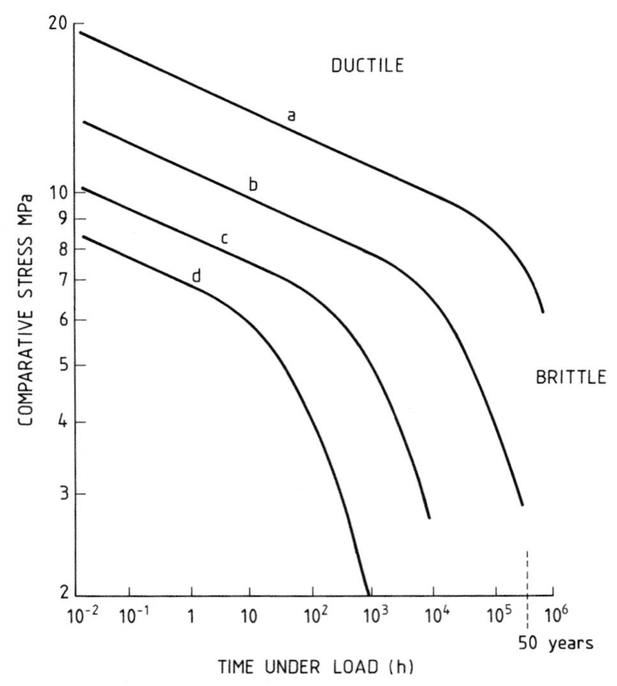

Figure 7.17 Creep rupture data for HDPE (type 1) pipe according to DIN 8075: (a) 20 °C; (b) 40 °C; (c) 60 °C; (d) 80 °C. (Reproduced by permission from Hoechst UK Limited, Polymers Division.)

to DIN 8075, is illustrated in Figure 7.17, where the stress function is a pipe hoop stress, varied by changing internal water pressure. Several points are of interest:

- In this case, a log-log representation (see below) is used to model individual failure points;
- The effect of increasing temperature is to reduce failure time at a given internal pressure, though the gradient $\frac{d \log \sigma}{d \log t}$ is similar for a given mode of failure;
- The most striking, and important point which emerges from such data is the change in failure mode which gives rise to a change of gradient in the representation, with a transition to brittle fracture occurring over shorter lifetimes, at elevated temperature.

It is this latter observation which has necessitated the development of predictive models for pipe failure times. Clearly, there are two competing failure modes which must be taken into account, and potentially the most damaging of these is the risk of brittle fracture taking place in the "extrapolation period" of the component lifetime.

BRAGAW[7.40] has presented an analysis of models for the prediction of the lifetime of pipe systems, based upon activated rate-process theory of failure under creep loading conditions, which can be expressed in general terms by an Arrhenius equation of the form:

$$\frac{df}{dt} = A \cdot \exp\left(- \frac{B\sigma}{CT} \right)$$

df/dt is failure rate, σ is applied stress, T is absolute temperature and A, B and C are constants.

Several mathematical models have been used to relate failure time (t_f) to internal pressure (P), and/or temperature (T). These have been reviewed[7.40, 7.41], and on the basis of the degree of fit which is evident from test data on polyethylene pipe, the following representation is deemed most appropriate to model the combined effects of temperature and pressure level:

$$\log t_f = A_0 + \frac{A_1}{T} + A_2 \cdot (\log P) \tag{7.29}$$

A_0, A_1 and A_2 are constants, obtained by statistical regression from practical data.

For a thin-wall pipe, nominal hoop stress (σ_H) is twice the axial stress (σ_A) and is related to internal pressure (P) and pipe dimensions (external diameter D and thickness b) by:

$$\sigma_H = 2\sigma_A = \frac{PD}{2b} \tag{7.30}$$

Since hoop stress is directly proportional to pressure, σ_H can be substituted for pressure in Equation 7.29; hence the use of a log-log representation to relate hoop stress to failure time.

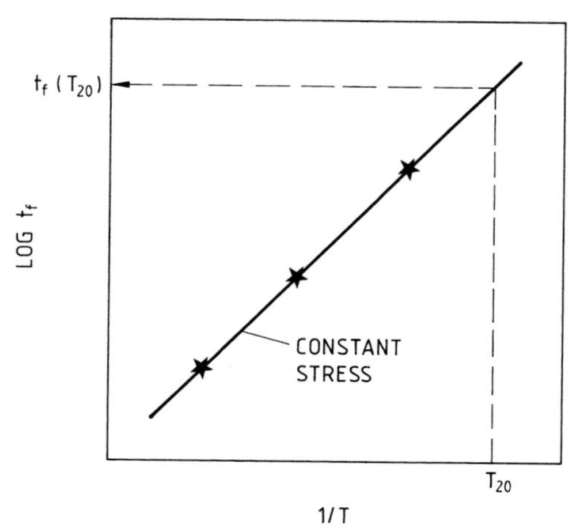

Figure 7.18 Arrhenius plot relating failure time (t_f) to the inverse of absolute temperature $\frac{1}{T}$, showing how extrapolation can be used to predict brittle fracture at low temperature (eg. 20 °C, as shown).

However, repeated reference must also be made to the mode of failure, since the coefficients in an empirical model such as 7.29 will depend upon whether ductile yielding or brittle crack growth is responsible for failure. For example, if we wish to use a set of mixed-mode failure data (Figure 7.17) to estimate brittle failure time at the lowest test temperature at a stress of (say) 5 MN/m^2 (ie. which is not possible directly from the medium-term test data), we can read-off failure time versus temperature data points. These usually satisfy an Arrhenius relationship, so that extrapolation may be carried out on a linear plot of log t_f against $\frac{1}{T}$ to obtain the brittle-mode failure time (Figure 7.18).Some national standards based upon the evaluation of creep rupture characteristics of plastics products are cited[7.42, 7.43].

7.5.1.2 Creep Rupture Resistance at Elevated Temperature

The onset of brittle failure over extended loading times at elevated temperature has limited the applicability of some plastics for pressure pipe applications where constant exposure to high temperature is envisaged. For example, the maximum operating conditions for pipes in the UK are:

Cold water	10 bars (1.0 MN/m^2) pressure at 20 °C;
Hot water	2 bars (0.2 MN/m^2) pressure at 70 °C;
Central heating	2 bars (0.2 MN/m^2) pressure at 85 °C.

In addition, the effects of freezing conditions, and short excursions to temperatures as high as 105 °C must be considered.

PVC and polyethylene pipe materials offer an excellent range of properties at relatively low material cost, for the distribution of water and gas. However, enhanced creep rupture properties are necessary for pressure applications at higher temperature. Greater retention of short-term properties is required, and the transition from ductile to brittle failure under creep loading should not be of equivalent concern. Plastics materials offering such characteristics include polybutylene (PB)[7.44], crosslinked polyethylene (XLPE) (see below) and chlorinated PVC (CPVC).

The use of *crosslinking* to enhance the mechanical fracture resistance of PE is of special importance. The presence of primary bonds between adjacent molecular segments results in an improvement in several characteristics, such as fracture toughness, environmental stress cracking resistance and (as appropriate to applications for hot-water distribution) long-term creep rupture performance at elevated temperature[7.45 – 7.47]. Fine tuning of attainable properties, appropriate to end-application, can be achieved by changing the degree of crosslinking (or *gel content)* and subsequent crystallinity, according to the chosen chemical mechanism of crosslinking. These include the use of organic peroxide catalysts to initiate a free-radical crosslinking process[7.48], crosslinking using silane groups created by a reaction between a PE graft copolymer and a catalyst masterbatch in the presence of moisture[7.46], and by means of irradiation[7.47 – 7.49].

7.5.2 Impact Properties

A plastics product is more likely to fail when it is subjected to an impact blow, in comparison to the same force being applied more slowly. In molecular terms, the material has greater opportunity to undergo compensating molecular motion *(relaxation)* in the second case. In consequence of this potentially damaging aspect of viscoelasticity, a great deal of effort has been expended in generating *impact tests* which generate quantitative data appropriate to material characteristics, but equally, the information should also reflect the likely behaviour of real products subjected to impact abuse in service. In this section, we examine some typical test methods used to evaluate the response of plastics materials or products to impact loads, and show how the principles of fracture can be extended to impact test applications.

7.5.2.1 Test Methods and Principles

In all these techniques, we seek to establish the unit amount of *impact energy* which can be sustained by plastics materials subjected to high strain rate loading conditions. Since the ability to absorb energy is dependent upon the shape, size and in particular, on the thickness of the component, it is important that all such tests be carefully standardized such that as far as possible, an *impact strength* parameter (for example, energy absorbed per unit area of failure) can be obtained. Clearly, this cannot be feasible for many fabricated products, where the thickness may necessarily be subject to wide variations (for example, in a range of moulded tanks of different capacity), or where the thickness distribution changes during production runs (extrusion blow moulding being a representative case in point). These effects have given rise to the development of different families of impact tests, where the measurement principle may reflect a demand for either an inherent material property, or a characteristic of the product, taking on-board not only the material characteristics, but also the effects of processing, design and dimensional fluctuation.

(A) Standard pendulum-type tests. Pendulum tests represent a traditional, but still commonly-utilized impact test methodology, whereby a test specimen (notched or unnotched) is held or clamped in a fixed position, and is subject to a high-velocity blow imparted by a pendulum impactor. The sample geometry, and the mechanism by which the load is applied (hence the stress system) can be chosen appropriate to the data requirements. Three such examples are illustrated in Figure 7.19: the Charpy test is a high-speed three-point flexure test, the Izod method is based on a notched cantilever geometry, and tensile-impact represents a high strain rate version of the conventional tensile test.

• **Test Principle.** Prior to testing, the system is calibrated such that without any positive resistance to the swing, the pendulum comes to rest at a height calibrated to "zero energy". During the impact test itself, the energy absorbed by the sample is equal to, and measured by the loss in potential energy suffered by the pendulum. This arbitrary measurement of energy can be converted into an impact strength measurement by dividing by either the ligament area of the specimen (Charpy; tensile-impact), or by the specimen width (Izod).

A simple energy balance for the swing of a pendulum of mass m, initially released from a height h_1 and coming to rest at h_2 gives:

$$mg(h_1 - h_2) = U_I + U_L$$

where U_I is the absorbed impact energy, and U_L represents energy losses such as:

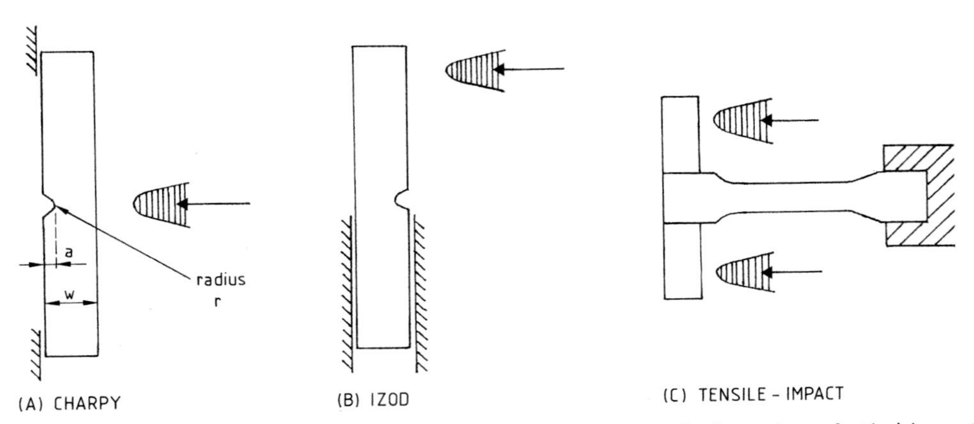

Figure 7.19 Test geometry, loading mechanism and clamping mode for various plastics' impact tests.

- the kinetic energy of broken fragments of specimens;
- resistance due to friction in the pendulum bearings;
- off-axis loading.

Since the notched impact strength, S_I, is impact energy per unit ligament area, we have:

$$S_I = \frac{mg(h_1 - h_2) - U_L}{b(w - a)} \tag{7.31}$$

Some impact data of this type for PP are presented in Figure 7.20. Although such test data are quantitative, and relatively easy to obtain, the major disadvantages are associated with the following factors:

• **Notch sensitivity.** The total energy absorbed in an impact test may be superficially broken down into contributions representing *crack initiation,* and *crack propagation.* Clearly, the energy requirement for main crack initiation will be reduced, often substantially, if the stress distribution is made triaxial, and much more concentrated, by the introduction of a notch into the test specimen. Since the impact fracture characteristics of certain plastics materials are much more prone to *notch-sensitivity* than others (these include rigid PVC, and some of the so-called engineering thermoplastics such as PC, POM and PA resins), it follows that attempts to rationalize plastics materials in terms of impact resistance can only be successful if the notches in all such tests are of identical size and shape. Expressed another way, the ranking of plastics in terms of the apparent impact toughness evaluated by the pendulum techniques depends entirely upon the magnitude of stress concentration induced by the notch; it is not appropriate therefore, to attempt to use this type of information in anything but a comparative manner on competitive grades of plastics.

However, the use of notches in standard impact tests has become commonplace; it can be shown[7.13] that the concentration of stress (σ_C) at the tip of a round-ended

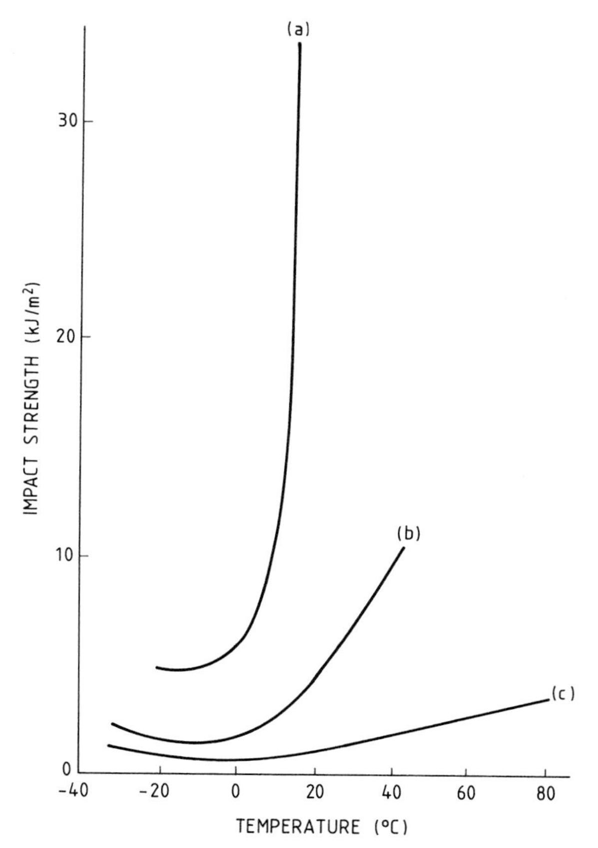

Figure 7.20 Temperature dependent impact strength data for PP homopolymer: (a) un-notched; (b) notch radius 1 mm; (c) notch radius 0.25 mm. (Reproduced with permission from ICI Chemicals and Polymers Ltd.)

notch may be expressed in terms of a *stress concentration factor (SCF):*

$$SCF = \frac{\sigma_C}{\sigma} = 1 + \left(\frac{a}{r}\right)^{1/2} \tag{7.32}$$

σ_C/σ is the ratio of concentrated, to remote (nominal) stress, and a,r represent the notch length and tip radius, respectively.

HAWORTH and WALSH[7.39] have reported the dependence of the impact properties of fibre-reinforced PETP resins, as a function of stress concentration. It must be noted that this stress concentration parameter is of limited applicability, since:

(1) the implication that an infinite stress is developed as defects become extremely sharp is clearly erroneous. In such a case, local yielding would occur at the crack tip;
(2) correlations between brittle failure stress and notch geometry are made more complex by changes in the dominant mode of failure.

• **Temperature sensitivity.** Since the ductility and the brittle strength of plastics are both temperature-dependent, it is inevitable that the fracture resistance under high-speed loading is itself very sensitive to test temperature. An increase in temperature, like a decrease in impact velocity, will allow greater opportunity for molecular relaxation, thereby promoting a higher measured impact resistance (especially if the degree of yielding is significant). In accordance with such transitions between ductile and brittle behaviour, temperature may be used as a manipulative variable to examine failure characteristics in either mode.

PP is a good example of a commodity polymer which can be associated with "low temperature brittleness", under impact loads. PP copolymers have a much lower glass transition temperature (of around $-60\,°\mathrm{C}$) in comparison to conventional PP homopolymers, and in consequence, the onset of embrittlement occurs at a significantly lower temperature, giving designers greater freedom to engineer products with increased assurance against impact failure. It should also be appreciated that PP copolymers owe their enhanced fracture resistance to the reduced yield stress, which is usually evident for material grades of similar molecular weight, under equivalent test conditions.

• **Sensitivity to specimen dimensions and processing history.** Alongside other parameters such as fracture toughness, impact strength is also geometry-dependent, and is fraught with interpretative difficulties if the test specimens are prepared by different fabrication techniques (for example, by injection moulding (oriented) or by compression moulding (isotropic), or even by altering processing conditions. These issues, although long-recognized, still impose severe problems for Standards' Authorities in their attempts to rationalize specifications for impact strength of plastics materials.

(B) Falling-weight impact (FWI) tests. A general lack of correlation between pendulum test data and impact-related failure characteristics in plastics products has resulted in the development of less sophisticated, and more qualitative methods related to the fabricated article. Pendulum impact test data represent material properties specific to the conditions of test, and to standard sample geometries; experience has shown that such data should not be used to evaluate product properties, nor can they necessarily be indicative of valid linear elastic fracture mechanics properties. In contrast, data from any of a family of drop tests (or *falling-weight impact* methods) may be used to assess the likely effects of such loading conditions experienced in-service.

• **Test Principle.** A missile of given mass (m), shape and size is released, and falls by gravity onto the test specimen from a predetermined height (h). A "pass-fail" criterion is used to examine the impact-failure process, and the degree of imposed energy is calculated accordingly (mgh).

Test variables include striker mass, impact velocity (hence strain rate, determined according to drop height), stress intensity associated with the striking-head (hemispherical-indentor, corner-cube etc.), and other more orthodox parameters such as temperature. The testing principle is relevant to a wide range of products, including extruded pipes, sheet and PVC window profiles, biaxially oriented PE or PETP films, and blow moulded HDPE drums; hollow-form components are often tested by filling with a liquid medium, then dropping onto a hard, flat surface.

Some variations of the basic principle are often adopted in order to obtain a mean, or median failure energy, using a "pass-fail" criterion with drop height (or mass) as an active test variable. For example, BS 2782 Methods 306 B/C[7.50] advocate the use of fall heights of 300 mm or 600 mm (respectively), for impact testing sheet or moulded specimens. When failure occurs, the mass is decreased by a standard increment, until the energy is insufficient for failure; the increment trend is then used in the opposite sense, and at least 20 samples must be tested, per run. The drop-weight impact energy is then calculated from the failure population, and represents conditions under which a 50 % probability of failure may be anticipated.

Figure 7.21 illustrates some typical data from this technique determined by drop-tests on containers stretch blow moulded from PETP.

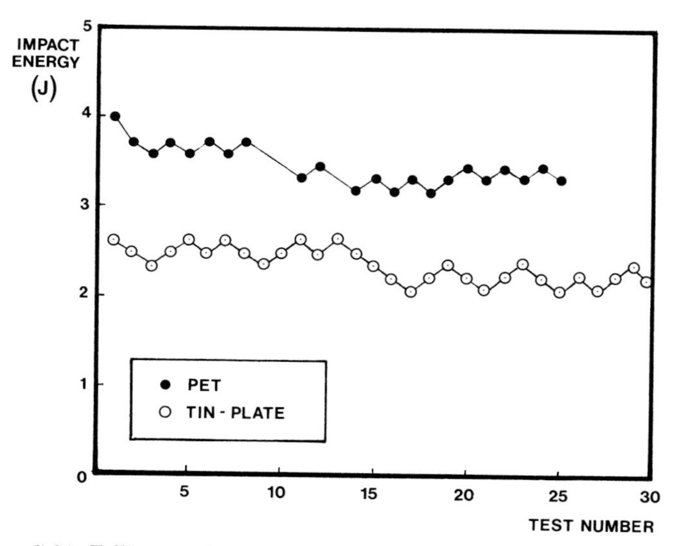

Figure 7.21 Falling-weight impact test data for biaxially oriented PETP, and tin-plate containers (varying increments of corner-cube mass, at a fixed drop-height of 300 mm and test temperature 23 °C).

(C) Instrumented impact testing. The past decade has witnessed a significant expansion in the application of instrumented impact testing[7.51–7.53]. A transducer is fitted to the impactor head, which measures the force created by the resistance of the specimen to the high-velocity blow, throughout the few milliseconds usually associated with the impact event. These force-time data are fed through a charge amplifier, stored on a transient recorder, and can be examined on an oscilloscope.

• **Test Principle.** The total potential energy available at the test initiation point (mgh) is chosen such that conditions of *excess energy* are apparent; that is, the energy absorbed during fracture is much lower than the total kinetic energy of the missile when contact is made. Using the assumption of constant impact-head velocity during failure under these conditions, we are able to transpose the readout from force-time to force-deflection, allowing subsequent computations of stress, strain and impact energy to be made throughout impact. In effect, the advance in instrumentation allows the determination of data to describe a high-speed flexure test.

These principles are illustrated by some IFWI data for commodity thermoplastics, in Figure 7.22. Some of the parameters which may be evaluated are illustrated on this figure. The relevance of each of these depends upon the manner in which the test specimen breaks; for example, if low-strain brittle fracture occurs, the total failure energy (U_f) approximates to the energy at peak force (U_p). In the subsequent section, we describe how instrumented impact data may be used to generate fracture mechanics parameters.

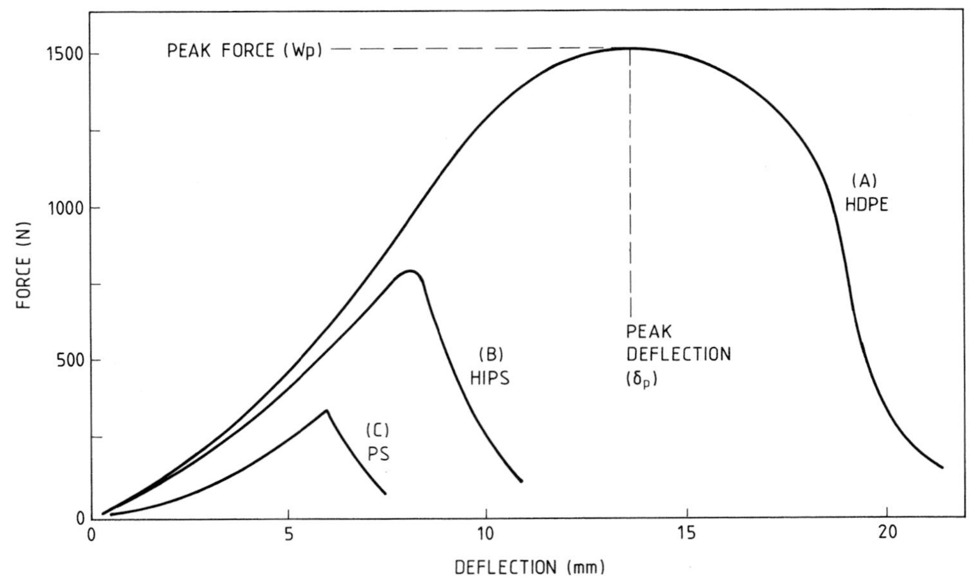

Figure 7.22 Instrumented impact test data for some commodity thermoplastics materials, derived at 3 ms^{-1} with a hemispherically-ended missile of mass 25 kg.
Force-deflection curves for
(A) HDPE (extrusion blow moulded container);
(B) Rubber-modified PS (injection moulded);
(C) Unmodified PS (injection moulded).

Although this principle can be utilized in both pendulum-type and drop-weight impact tests, it is the latter group which have proved of most direct use to the evaluation of impact toughness, and fracture properties of fabricated products; hence the term *Instrumented Falling Weight Impact (IFWI)* techniques[7.53]. Test variables include:

- **Impact energy and velocity.** These variables are usually chosen to generate excess-energy conditions; if changes in applied energy are required, it may be pertinent to vary mass, rather than velocity, to avoid effects due to strain-rate variation.

- **Temperature.** Changes in impact failure mode can be established by systematic variation in temperature, either by the use of a suitable environmental chamber, or by pre-conditioning.
- **Specimen geometry.** A specific problem associated with product evaluation is the effect of specimen thickness, especially in instances where the component shape does not lend itself easily to stress analysis.

A useful calibration experiment in such circumstances is to measure and model the dependence of impact parameters on specimen thickness; peak force (W_p) and peak energy (U_p) are related to thickness (b) by:

$$(W_p, U_p) \ \alpha \ (b)^n \tag{7.33}$$

For example, experiments on injection moulded HIPS[7.54] have supported the validity of a power law relationship between impact parameters and thickness; in the case of HIPS, the representation for each of the parameters quoted was described by an exponent (n) of 1.5. In another study of formulation effects on the IFWI properties of welded PVC profiles[7.55], it was demonstrated that the variations in thickness must be accounted for before the specific influences of compound additives can be identified.

In Table 7.1, we present some data from low-temperature, instrumented falling-weight impact tests on competitive materials for car bumper applications, derived at a test velocity of 3 ms^{-1}; these data have been normalized to 1mm thickness, to account for the effect of thickness variations between the different systems.

Table 7.1 IFWI Data for Car Fenders (−26 °C)

Material	Thickness (mm)	W_p (N)	U_p (J)	U_f (J)
PU	3.7	583	5.75	9.45
Modified PBTP	3.4	820	2.89	5.20
PP Copolymer	2.6	1183	10.1	16.5
GRP	3.4	1324	4.1	14.5
PBTP/PC Blend	3.2	2500	21.0	29.5
HDPE	4.1	1526	13.2	23.0

7.5.2.2 Fracture Mechanics Approach to Impact Failure

It is prudent therefore, to design against impact failure on the basis of fracture mechanics[7.13, 7.31, 7.56, 7.57]. The effects of stress concentration, strain rate and temperature are crucial to the success of this technique, especially the last, as a result of the sensitivity of yield stress to temperature.

Some of the shortcomings associated with using the traditional, pendulum-type impact methods, particularly with respect to the rationalization of failure modes and data presentation, can be overcome by systematic use of the fundamental fracture mechanics principles described earlier. This methodology was first proposed by Plati and Williams[7.56], and will be summarized here.

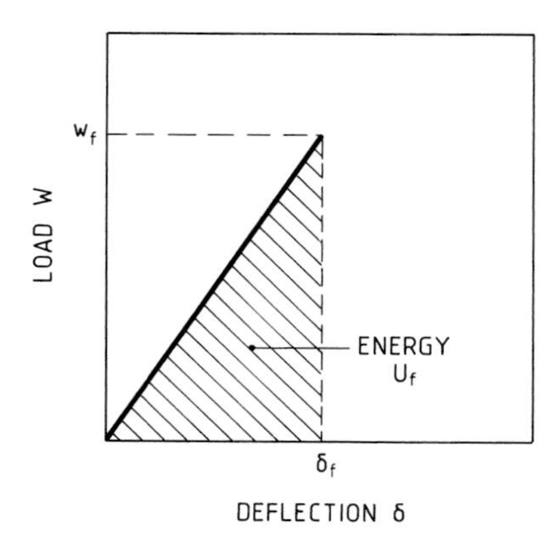

Figure 7.23 Load-deflection-energy relationships for a linear-elastic material.

We start on the assumption that at the test conditions used, the polymer must exhibit linear elastic behaviour, giving a load (W) – deflection (δ) relationship of the type shown in Figure 7.23, thereby allowing the total energy absorption at failure (U_f) to be estimated by:

$$U_f = W_f \frac{\delta_f}{2} = J \frac{(W_f)^2}{2} \tag{7.34}$$

where the specimen compliance, $J = \frac{\delta}{W}$.

When impact energy U reaches a critical value (U_f), strain energy release rate, G assumes its own critical value (G_C), so that:

$$G_C = \frac{1}{b} \cdot \frac{dU}{da}$$

Rearranging the expression for G_C, using Equation 7.34, we obtain:

$$G_C = \frac{W_f^2}{2b} \cdot \frac{dJ}{da} \tag{7.35}$$

If an accurate value of failure energy (U_f) can be measured, we combine Equations 7.34 and 7.35 to obtain:

$$U_f = G_C \cdot bw\phi \tag{7.36}$$

ϕ is a dimensionless factor related to compliance and sample dimensions by:

$$\phi = J \cdot \left[\frac{dJ}{d(a/w)} \right]^{-1}$$

ϕ can be determined for a range of test geometries; PLATI and WILLIAMS[7.56] have presented solutions of ϕ appropriate to Charpy and Izod impact tests, and have found that the LEFM approach helps to rationalize impact properties of all but highly ductile samples. The usual method of approach is to vary specimen and/or notch dimensions, and to plot the measured values of U_f against bw ϕ; the gradient of this relationship will give a solution for G_C.

The technique may be extended to allow calculations of fracture toughness K_C to be made by use of data from an instrumented test. In a Charpy (notched, 3-point bend) test for example, the peak force W_f is measured and used to calculate remote failure stress (σ_f) by:

$$\sigma_f = 3\frac{W_f L}{2bw^2}$$

Fracture toughness and modulus (E) are then evaluated as described in a previous Section, using Equations 7.11, 7.12.

The Charpy impact fracture data presented in Section 7.3.4.2 were derived in this way; this approach can, of course, be extended to examine the influence of temperature, rate effects and various material/compound variables on fracture mechanics parameters. A recent study on the fracture properties of mineral-filled PP compounds has been reported by Chen et al.[7.58], in which G_C was shown to be relatively insensitive to the addition of 10 % $CaCO_3$ fillers, across the temperature range −40 to +40 °C.

7.5.3 Dynamic Fatigue Characteristics

This mode of failure, often referred to simply as *fatigue,* is a well-known failure mechanism for metals products: the cracking of welded Liberty ships in the 1940's and the failure of Comet aircraft in the 1950's were each established as fatigue failures. The term 'fatigue' refers to situations where a fluctuating, usually *cyclic load* is continually applied to engineering solids.

Reference was made in Chapter 6 (Section 6.4.6) to the dynamic mechanical properties of plastics, measured under conditions of cyclic stress or strain. The distinctions between conventional *dynamic mechanical tests* (used to evaluate information on the low-strain viscoelastic properties of plastics, since stress, strain and time vary simultaneously), and *fatigue characteristics* (a failure phenomenon) are made in terms of the stress magnitude and frequency of loading; the former is significantly higher in the case of dynamic fatigue. Experience has shown that plastics also suffer from this route to failure, although (in comparison with metals) perhaps less frequently, not so spectacularly, and by two distinct routes:

(A) Unstable crack growth – brittle fracture. Classical fatigue failure occurs as a result of defect propagation under high-frequency stress, leading to a sudden, catastrophic failure in the brittle mode. As such, it can be evaluated on the basis of a fracture mechanics approach.

(B) Thermal fatigue – ductile yielding failure. Under rapid-cycling conditions where significant viscous motion occurs, the build-up of heat in low-conductivity plastics leads to situations where the applied stress may exceed the yield stress of the polymer (ie. at elevated temperature) within the locality of the temperature rise. In this case ductile failure occurs, exemplified by gross plastic deformation in the region of rupture, so that a failure analysis based upon the mechanics of crack-growth becomes irrelevant.

This mechanism of fatigue failure effectively limits the frequency and amplitude of stress which can be tolerated in investigations of dynamic fatigue failure by unstable crack growth. Clearly, it is not easy to accelerate fatigue testing appropriate to the brittle mode; this factor has long been recognized, and has hindered many attempts to study the fatigue behaviour of engineering thermoplastics.

7.5.3.1 Analysis of Fatigue by Fracture Mechanics

The pioneering work in the application of fracture theory to fatigue loading is generally attributed to PARIS et al.[7.59]; hence the *Paris law* is used to relate the rate of fatigue crack growth (increase in defect length per loading cycle, $\frac{da}{dn}$) to the alternating stress intensity factor ($\triangle K$) developed by the magnitude of applied stress:

$$\frac{da}{dn} = A(\triangle K)^m \tag{7.37}$$

A and m are power law coefficients (constant for a material under specified conditions), and $\triangle K$ is given by:

$$\triangle K = K_1 - K_2 = (\sigma_1 - \sigma_2)Y(a)^{1/2} \tag{7.38}$$

where subscripts 1 and 2 refer to the maximum, and minimum values of stress (and hence stress intensity) in the fatigue loading programme.

However, there are some factors which may detract from the validity of a Paris power law representation. These include:

- the original defect size (pre-cracked samples should be used); and
- the degree of crack-tip yielding (which should be strictly controlled).

Also, as pointed out by BUCKNALL[7.60] following some fatigue studies on HDPE, crack growth rate is sensitive to the ratio σ_2/σ_1, and cannot be adequately predicted if the loading pattern is subject to change. The measured values for exponent m in this study were all close to 4, which appears to be typical for many classes of polymers, where m-values generally lie in the range 3 – 5.

Using a procedure analogous to that described earlier for slow crack-growth under constant stress (Section 7.3.2.5), we can evaluate the time-dependent (hence cycle-dependent) fatigue behaviour of plastics. Taking the example of crack-opening conditions, the alternating stress intensity factor ($\triangle K_I$) is given by:

$$\triangle K_I = Y \triangle \sigma (a)^{1/2}$$

Making a substitution for K into the Paris law, and rearranging, we can form an integral as follows:

$$\frac{1}{A(Y\triangle\sigma)^m} \int_a^{a_t} \frac{da}{a^{m/2}} = \int_0^{n_t} dn \tag{7.39}$$

If the defect length increases from a (at zero time) to a_t (over any number of cycles, n_t), we can solve the integral to give:

$$n_t = \frac{2}{A(Y\triangle\sigma)^m(2-m)} \cdot [(a_t)^{1-(m/2)} - a^{1-(m/2)}] \tag{7.40}$$

In the important situation where defect length reaches a critical size, unstable crack growth is evident and the number of fatigue cycles can be estimated accordingly, if prior knowledge of Paris law coefficients (A and m), specimen geometry and stress system (σ) are obtained.

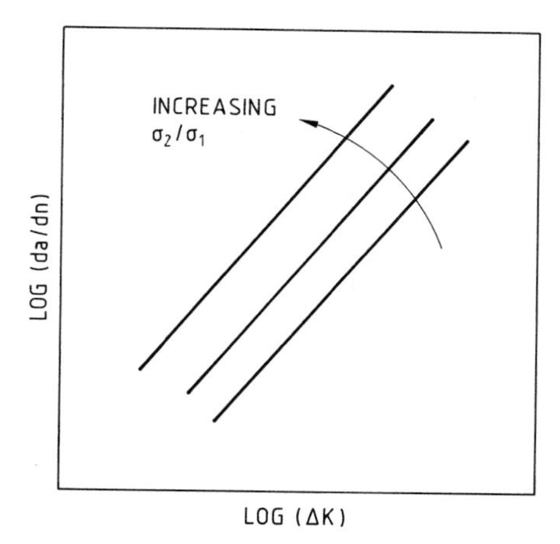

Figure 7.24 Log-log representation of Paris law for plastics fatigue properties. (After BUCKNALL[7.60]; see text for definitions.)

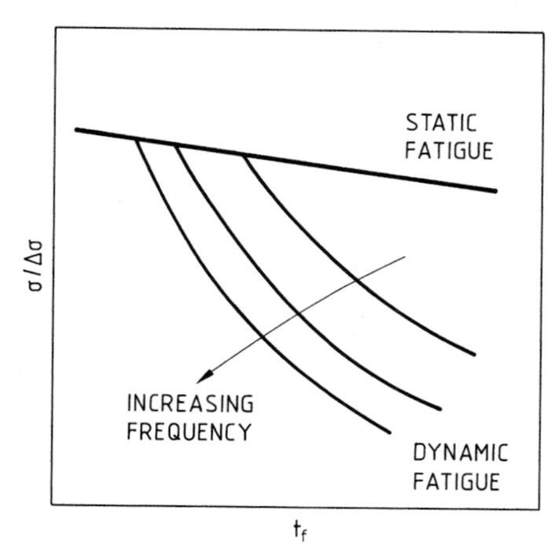

Figure 7.25 Failure data for static fatigue (stress (σ) versus failure time (t_f)), and for dynamic fatigue (stress amplitude ($\triangle\sigma$) versus failure time) in plastics materials.

Fatigue data are usually expressed in two distinct ways. In Figure 7.24, the Paris power law is represented by a log-log plot of da/dn versus $\triangle K$. An alternative approach is to use an *S-N representation,* illustrating the dependence of "cycles to failure" against stress (or stress amplitude). However, this might be misleading in situations where the frequency-dependence of fatigue is investigated; Figure 7.25 shows how an increase in frequency is associated with a decreased failure time, under circumstances which may have shown greater fatigue cycles to failure. This plot also compares dynamic fatigue failure with static fatigue (creep rupture characteristics) in the ductile mode.

7.5.3.2 Fatigue Loading and Viscous Heating

The observation of a local temperature rise in plastics as a result of inelastic deformation under cyclic loading promotes a tendency towards thermal fatigue failure, as a result of the drop in yield stress at higher temperature. WILLIAMS[7.13] has analyzed energy dissipation in such circumstances, and quotes the following relationship between temperature (T) and time (t) under fatigue loading conditions:

$$T = T_0 + \left[n\pi L q^2 (\triangle\sigma)^2 \cdot \frac{\tan\delta}{hE_2} \right] \cdot \left[1 - \exp\left(-\frac{ht}{L\varrho C_P} \right) \right] \tag{7.41}$$

The test variables are initial temperature (T_0), number of cycles (n), sample volume/surface area ratio (L), stress ampliture ($\triangle\sigma$) and q, a factor which depends upon stressing mode ($q = 1$ in direct tension). Material properties which influence the extent of heat dissipation include the dynamic loss tangent ($\tan\delta$) and loss modulus (E_2), surface heat transfer coefficient (h), density (ϱ) and specific heat capacity (C_P).

This model describes not only the viscoelastic character of plastics under fatigue loading, it also accounts for the temperature losses sustained by the specimen due to surface convection effects.

Equation 7.41 can be used to estimate the maximum temperature rise before the onset of thermal fatigue failure, which can then be used to define limiting conditions for test frequency and stress amplitude. For PMMA[7.61] tested at 50 Hz (50 s^{-1}), the limiting stress amplitude to avoid thermal fatigue is of the order 22 MN/m^2.

7.5.3.3 Some Practical Fatigue Studies on Engineering Plastics

HERTZBERG and MANSON[7.62] have compiled a thorough review of fatigue effects in engineering plastics materials. More recent studies by BOWMAN and co-workers on the fatigue characteristics of HDPE pressure pipe systems have demonstrated the sensitivity of component lifetime to the structure of injection moulded fittings[7.63], and to the occurrence of microstructural heterogeneities (voids, contaminant particles) in extruded pipe-sections[7.64]. The work by BUCKNALL cited earlier[7.60] was also concerned with the fatigue performance of HDPE materials for gas distribution pipe systems.

7.5.4 Environmental Stress Cracking Resistance

A specific source of embrittlement in plastics materials which has frequently posed unanticipated problems in service is the phenomenon of *environmental stress cracking (ESC)*. This failure mechanism is defined by any conditions in which an external (or residual) stress is imposed on a specimen which is in contact with an external (usually a liquid or vapour) environment; it is the combination of the stress and the liquid medium which gives rise to premature failure, often seen under conditions which would otherwise appear to offer little threat to the material if either the stress, or the environment were eliminated.

7.5.4.1 Definition and Mechanism of ESC

ESC is not, therefore, a form of direct chemical attack (see also Section 10.3.1), and is not associated directly with chain scission. Its effects have been particularly, though by no means exclusively detrimental to the performance of amorphous polymers stressed in the presence of organic solvents. However, many of the early investigations on environment-related failure were based upon polyethylenes; HOWARD[7.65] describes the environmental stress cracking of polyethylene in terms of the time-dependent brittle failure under a tensile stress appreciably below the limits of its short-term strength. Failure occurs by *surface-initiated brittle fracture* of a specimen or part, especially under polyaxial stress, in contact with a medium in the absence of which fracture does not occur under the same conditions of stress. Combinations of external and/or internal stress may be involved, and the sensitizing medium may be gaseous, liquid, semi-solid or solid.

Early work on the ESC performance of polyethylene showed that the resistance to premature failure may be enhanced by:

- increasing molecular weight and/or decreasing molecular weight distribution;
- refinement of crystalline texture;
- decreasing the degree of crystallinity (in the instance where crystalline texture is not controlled).

Clearly, the task of achieving adequate resistance to environmental failure is related not only to the plastics' material specification, but also to the thermal history during processing, which has a strong influence on the developing crystalline microstructure.

Subsequent experience has shown that in comparison to families of glassy thermoplastics, semi-crystalline polymers generally suffer less-frequently from the effects of ESC, though there is evidence that when polyolefins are stressed in contact with detergents, certain machine oils and/or other surface-active agents, especially at elevated temperature, premature failure by ESC may occur.

• **Mechanism of ESC failure.** The visible damage which occurs by ESC may vary considerably. In some cases a single defect, once initiated, can propagate rapidly through the sample; in other situations *macroscopic crazing* occurs, sometimes visible as local whitening in the region of maximum stress, rendering the component unfit for use but without necessarily leading to brittle cracking. This has resulted in the promotion of an additional term: *environmental stress crazing*. However, these terms should not be used interchangeably, based purely upon subjective evidence of the general appearance of damage zones of ESC failures. Indeed, the whole mechanism of premature failure by environmental contact is based on the assumption that the active medium penetrates any microscopic defects introduced by external stress, and is able to interact with, and diffuse quickly into yielded (eg. crazed and dilated) material at the crack-tip, thereby creating a greater stress concentration, and accelerating the progression of the damage zone. Fracture will occur if and when the localized load-bearing characteristics of the crazed material become inadequately resistant to the increasing stress concentration.

Although the precise mechanism, or mechanisms of ESC are still not fully understood, this sequence of events would appear to account for the observed transition towards more rapid crack propagation, increased brittleness and reduced failure times which are characteristic of ESC phenomena in plastics materials. Some of these points are taken further, and are illustrated with typical micrographs of ESC fracture surfaces (in ABS components) in Section 7.6.3.3.

7.5.4.2 Test Methods and ESC Failure

As the introduction of plastics materials into engineering applications has expanded, there have emerged numerous additional combinations of plastics and external media which can lead to ESC; sometimes failure occurs under severe or complex loading, yet on other occasions it is the presence of an *internal stress* which is sufficient to initiate brittle fracture. With this in mind, it is inevitable that the inventory of ESC test methods has become extremely diverse. Some techniques are very specific to certain plastics and external media, and often, the damage assessment is rather subjective, and is rarely quantitative. Typical independent ESC test variables include:

- **Stress system.** External stresses (in tension, flexure, torsion, or created by internal pressure in containers and hollow components) may be applied or on occasions, moulded components are left without external stress, so that the effects of fabrication conditions on internal stress and ESC can be determined. When external loading is applied, constant-stress (creep) or constant deformation (stress relaxation) modes may be used.
- **Temperature.** Constant temperature, or intermittent/systematic variation in temperature represent viable alternatives. Elevated temperature may promote the diffusion rate, yet plastics may be inherently more brittle at lower temperatures.
- **Specimen dimensions.** Under ideal conditions, dimensions should be held consistent. Where this is not possible, measurement of dimensions should be taken since thickness (for example) will determine the magnitude of applied stress in a constant deformation test.
- **Processing history.** The effects of internal stress and orientation, induced as a result of processing by different methods (or by using various fabrication conditions) should be carefully controlled, and cross-interpretation of data generated on specimens having different processing histories should be attempted only with extreme caution.
- **Environment.** Concentration of external media, exposure times and conditions of contact (eg. total immersion in excess liquid, or a single-dip, allowing vapour evaporation to occur) can each be adjusted.

The measured parameters from such tests are many and varied; these could include changes in weight and dimensions (which characterize physical interactions that may, or may not lead to brittleness under the specified conditions), visible signs of damage (crazing and whitening, cracks, sample delamination) and failure data (time to craze or fracture; critical strain – see below) appropriate to groups of selected independent variables.

The decision to opt for constant stress, or constant strain loading conditions is one which has always attracted particular attention. The former techniques (Figure 7.26) are based upon creep loading, and are therefore more appropriate to envisaged service conditions for thermoplastics. Comparisons between environment-related failures and conventional creep rupture performance can then be made, if control samples are used in the experiment. The format of data representation suggested in Section 7.5.1.1 may be used as a starting point for modelling ESC failure of plastics under constant stress; a change in failure mechanism is usually evident both macroscopically, and from the failure data plots obtained.

One of the earliest ESC standard tests is the so-called Bell Telephone technique[7.66], in which bent strips of material containing a linear defect (the specimens are of standard dimensions; 38 × 13 × 3 mm) are totally immersed in a chemical medium (based on Igepal® – alkyl aryl polyethylene glycol) before being examined for visible signs of permanent damage (Figure 7.27 B). Other, more sophisticated techniques, also based upon a constant deformation principle are related to the existence of a *critical strain parameter* (ε_C) in plastics materials, which describes the minimum degree of deformation at which ESC failure may be anticipated. A *flexural relaxometer* apparatus of the type depicted in Figure 7.27 (A) may be used to relate failure time to bending deformation, and determine critical strains for chosen polymer-environment combinations.

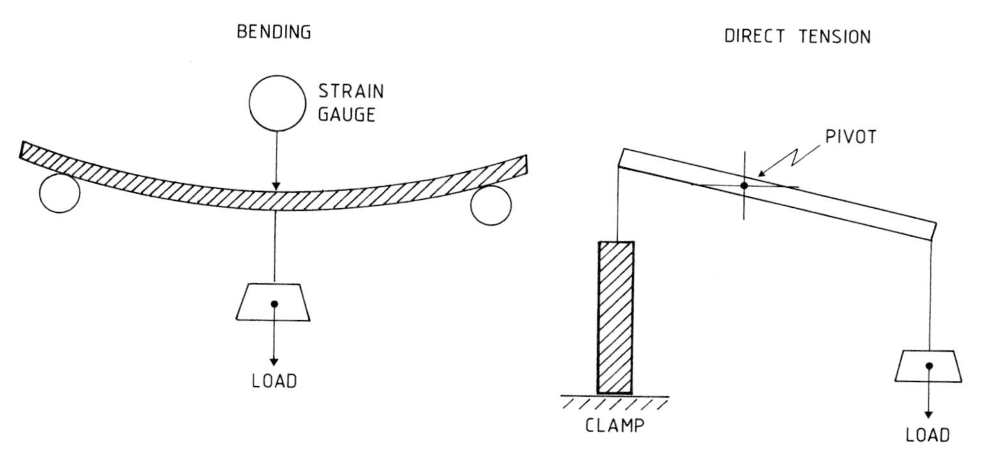

Figure 7.26 Environmental stress cracking tests in the constant stress (creep) mode.

HENRY[7.67] has described how force measurements can be translated into stress relaxation data (Figure 7.28) in order to determine ε_C; a significant increase in relaxation rate (eg. at ε_3) occurs when crazing takes place, thus indicating that the critical strain has been exceeded. Results have shown that for SAN exposed to alcohols and other solvents[7.67], ε_C lies between 0.1 and 1.0 %; it increases markedly with rubber content, and is also sensitive to styrene-acrylonitrile ratio and the solubility parameter of the liquid. However, ε_C was shown to be relatively insensitive to molecular weight. FAULKNER[7.68] extended this approach to show that the critical strains for injection moulded ABS in methanol, though also strongly influenced by rubber content, were significantly increased when flexed in the flow direction, parallel to the principal direction of residual (uniaxial) molecular orientation.

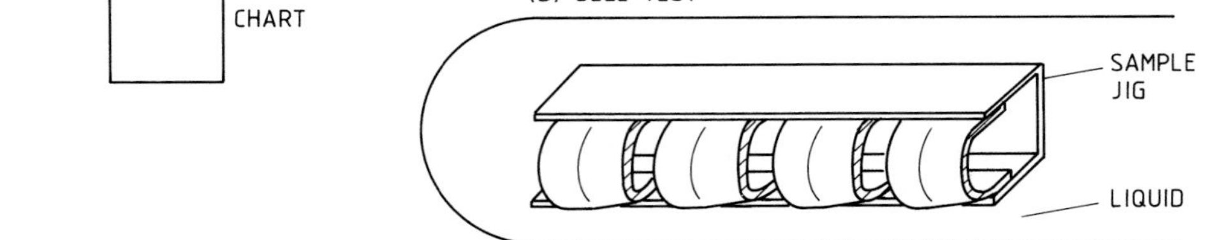

Figure 7.27 Environmental stress cracking tests in the constant deformation mode: (A) Flexural relaxometer; (B) Bell test. (Note the imposed defects on the outer surface of each sample).

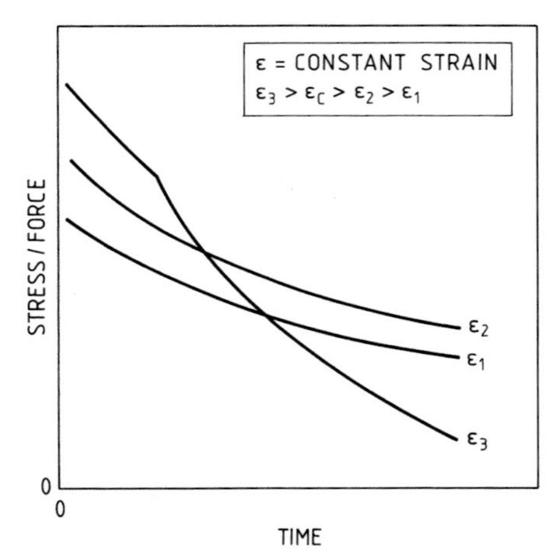

Figure 7.28 Data format from constant-deformation ESC test; the critical strain (ε_c) lies between ε_2 and ε_3 (after HENRY[7.67]).

This latter result is consistent with some constant-deformation ESC data derived in our own laboratories on PC with high-octane petrol (Figure 7.29), where decreases in failure time (at a given strain) were observed for specimens injection moulded at successively higher mould temperatures, where the extent of molecular orientation, the surface component of (compressive) residual stress and the degree

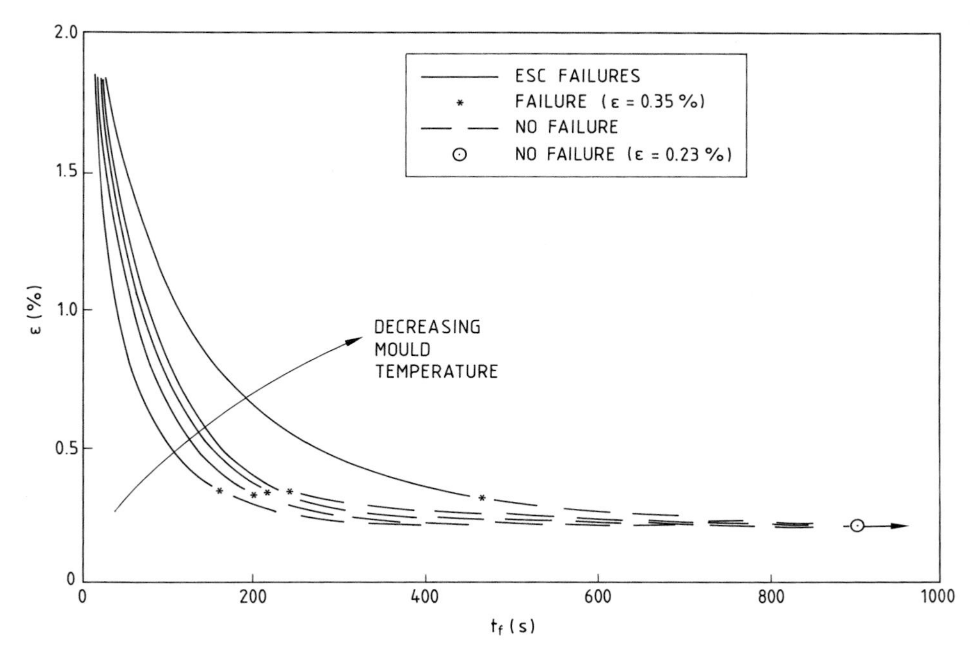

Figure 7.29 ESC data for polycarbonate (Lexan® 101; GE Plastics) in 'four-star' petrol: dependence of failure time (t_f) on applied flexural strain (ε). Mould temperature was varied between 75 and 115 °C (as shown), in increments of 10 °C; individual data points have been omitted, for purposes of clarity.

of free volume are each diminished. The resolved critical strain level for PC in petrol is of the order 0.3 %.

A comprehensive review of all published test methods for environmental stress cracking resistance of plastics (to 1975) has been compiled by TITOW[7.69]. A subsequent publication by ORTHMANN[7.70] traces the chronological developments in ESC tests, with respect to appropriate national standard techniques.

7.5.4.3 Application of Fracture Mechanics to ESC

The acceptance of the critical strain approach to study environmental effects in plastics has helped to quantify the most important test variables and compound constituents which influence ESC resistance in linear thermoplastics. In particular, the theories of solvent crazing and zone growth are well-developed, and ESC has been reviewed thoroughly in this context by KAMBOUR[7.71], and by WILLIAMS[7.13].

However, there has been much less success in accounting for ESC resistance and crack growth kinetics in terms of conventional LEFM parameters, in a manner which might account for the physical interactions between the polymer and the environment. Some of this inertia must be associated with the imprecise nature of the damage process, whereby cracks of critical size are propagated from solvent-crazed material.

7.6 Evaluation and Investigations of Failure Behaviour

7.6.1 Introduction and Review of Test Methods

In this section, our broad intention is to define how product failure may be evaluated, with reference to some examples of premature failure in plastics products. Generally, the examples which are illustrated represent the not uncommon phenomenon of brittle fracture in polymers which are normally ductile. In consequence, a highlighted feature of this section is the demonstrated use of *microscopic techniques* to investigate the nature of any transition from ductile to brittle behaviour. It must be stated that whilst this philosophy is often used in a retrospective manner to investigate the pre-occurrence of failure in products, it should not be considered restricted to this somewhat negative role. Instead, it is our belief that microscopy is a valuable tool at the product development stage, so that any aspects of polymer microstructure which may be likely to initiate brittle fracture can be identified, and any remedial action (modifications to tooling, component design or material specification) carried out.

A glance through any compilation of Standards covering the mechanical properties of plastics reveals an extensive, but confusing array of methods associated with failure, which can be grouped as follows:

(A) Simple tensile, flexural or compressive tests;
(B) Short-term fracture tests;
(C) Long-term strength, such as creep rupture or subcritical crack growth;
(D) Simple impact – Charpy, Izod, notched tensile-impact, and falling weight;
(E) Instrumented impact;
(F) Dynamic fatigue;
(G) Environmental stress cracking; and
(H) Abrasion and wear.

Most of these test procedures have value in the context of investigating failure, with the exception of compression techniques, in which a ductile response in materials which might be brittle under most other stress systems is often induced. Some of the test geometries have already been encountered within Chapters 6 and 7 and, whilst obviously the details of the methods are important, and can be found in International Standards (ISO) Publications, or in British Standards, (BS 2782, or BS 4618), it is the varied response of plastics materials under load which concerns us here.

The two most common tests for failure are the *tensile test* and the 3-point, or 4-point *flexural tests,* the latter being used for stiffer, or more brittle materials. From these starting points, the severity of a given technique may be increased in several different ways, by appropriate and systematic variation of:

- stress concentrating factors;
- increasing the rate of testing;
- decreasing temperature;

- cyclic stressing, and
- an increasingly severe environment.

These factors individually and collectively increase the chance of brittle fracture, as do several materials' parameters; their contributions may be quantified on the basis of a fracture toughness measurement. The remainder of this chapter will be concerned with considering how these factors arise in manufacture but first, the methods available for examining failures are surveyed.

7.6.2 Investigations of Failure

A very wide variety of techniques has contributed to our understanding of failure, the use of many of which will be exemplified in the next several pages. However, a group of methods based upon *microscopy* has proved to be very powerful. These include a number of special microscopical techniques:

(A) microtomy: specimen preparation, particularly of thin sections;
(B) visual inspection (especially of fracture surfaces, or thin sections);
(C) light microscopy for improved resolution;
(D) other techniques "enabled" by the light microscope:
 - study of phase objects,
 - differential interference methods, and
 - hot stage microscopy.

(E) scanning electron microscopy, (SEM), for improved depth of focus and surface topography; and
(F) SEM-based dispersive X-ray analysis.

The reader is referred to texts by HEMSLEY[7.72], and by SAWYER and GRUBB[7.73] for more detailed information on the use of microscopic techniques in plastics failure analysis.

7.6.3 Factors Inducing Premature Failure

By reference to tests enumerated in Section 7.6.1, some idea can be gained of the general failure behaviour; departure from expectation may be analyzed under the headings given in Section 7.1. Some of these topics have already been referred to in Chapter 1; experience has shown that their repetition in the present context is justified by their importance.

7.6.3.1 Materials Factors

Perhaps the simplest reason for premature failure in a plastics component is the use of polymer of insufficient chain length, (too low molecular weight). This might arise by:

- starting with polymer of inadequate molecular weight, in order to comply with constraints imposed by melt phase processing. For example, low molecular weight (high melt flow index) grades are essential for adequate shear flow in

rotational moulding, and are often preferred in injection moulding in order to reduce pressure drop for mould filling (see Chapter 5);
- degradation of "satisfactory" polymer by subjecting it to an excessively severe thermal treatment (or an excessive melt phase residence time), thereby causing degradation;
- a specific chemical decomposition mechanism, such as hydrolysis (Chapter 10);
- incorporating undue amounts of recycled material, or insufficient stabilizer.

This is one of the easiest causes of premature failure to investigate. Two routes are available: first, to measure a molecular weight characteristic (usually in solution or alternatively, a melt viscosity technique), or second, to determine the fracture toughness or impact strength, showing that it is unacceptably low.

One example will suffice: PETP is notoriously sensitive to absorbed moisture, which readily induces hydrolysis at processing temperatures. Thus, the toughness of plastics based on PETP is considerably reduced when processing material containing only trace quantities of moisture[7.39].

In our experience, possibly the greatest incidence of failure is concerned with the unsatisfactory *dispersion of additives,* especially pigments. A pigment's main function is to colour a product, but the effect on other properties may be serious, especially if the pigment is introduced to the parent polymer by the addition of a small quantity of a concentrated *masterbatch.* Furthermore, masterbatch formulation, and the method of incorporating the pigment may be the source of yet more problems, as we shall see later.

One of the most problem-laden areas is the colouring of rotational mouldings, a process which, as was noted in Chapter 5, subjects the polymer melt to minimal shear; that is, there is little opportunity for *dispersive mixing* in the melt stage of the shaping process. Thus the problem of inadequate *pigment dispersion* is added to that of poor compaction and homogenization. The result is that many coloured rotational mouldings have poorly dispersed pigment, which is generally to be found on the surface of the polymer powder, and leads to products of low strength and low toughness.

An example is a rotational moulding in LDPE (overall dimensions $1.25 \times 1.25 \times 0.5$ m), which failed under mild impact blows. Thin sections of the moulding taken in the vicinity of the crack can be readily prepared by *sledge microtome,* and low magnification with common light is sufficient to reveal that the pigment dispersion is far from satisfactory (Figure 7.30). A more acceptable dispersion is shown in Figure 7.31. Sometimes there is the added complication that the pigment, or a constituent of it, nucleates crystallization of the polymer matrix, frequently polyethylene (Figure 7.32). The uneven dispersion of the nucleant contributes to a very irregular texture, which may then have an undesirable influence on crack propagation resistance. This was also demonstrated in Figure 5.35 by the *row nucleation* effect induced by pigmentation.

The only certain remedy to prevent poor pigment dispersion and/or uneven pigment distribution in rotational mouldings is to prepare the powder from compound which has been melt-mixed. However, this is not attractive commercially, and other means of improving pigment incorporation have been sought. The success of pigment dilution with polymer (a type of masterbatch technique) depends on

Figure 7.30 Rotational moulding in LDPE showing poor pigment dispersion.

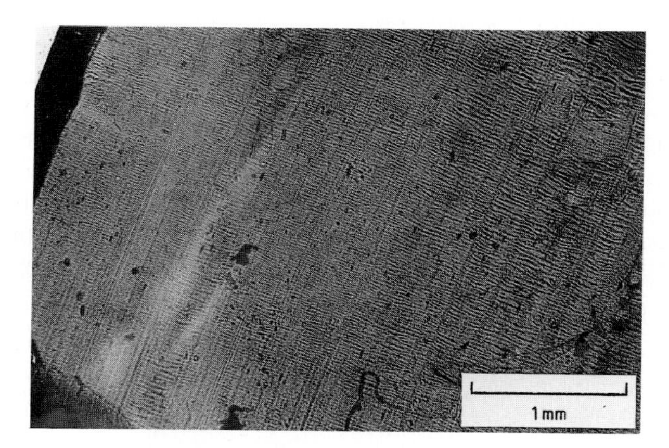

Figure 7.31 Rotational moulding in LDPE showing improved pigment dispersion.

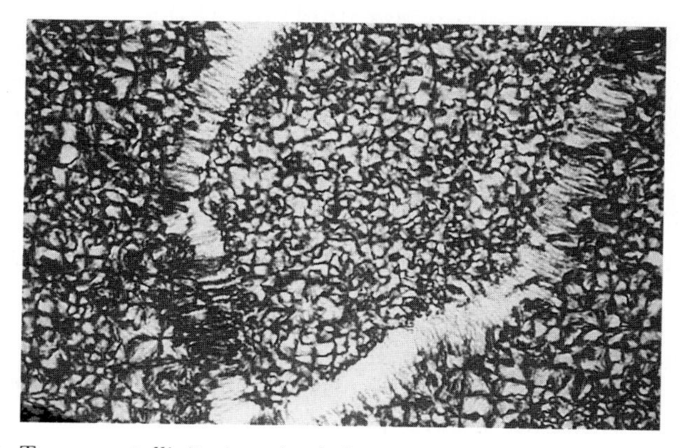

Figure 7.32 Trans-crystallinity in polyethylene, nucleated by poorly-dispersed pigment.

the nature of the pigment, and on the viscosity of each constituent; it proves much easier to mix a low viscosity masterbatch in a polymer of higher molecular weight. The masterbatch itself should be based on the same parent polymer, and should preferably be added frequently, ie. avoiding the effects of excessive pigment concentrates. The results often vary from unacceptable to very satisfactory, as illustrated in Figures 7.30 and 7.31, respectively.

Another class of additive which increases the chance of failure is filler, especially near-spherical particles which have only limited reaction at the interface with the plastics matrix. PP is embrittled by substantial amounts of clay filler, especially if the dispersion is not optimized.

Masterbatch techniques for the incorporation of additives may overcome some problems, but will undoubtedly create others. Polymers are generally immiscible, so that the process of mixing a concentrated dispersion of the additive, and adding small quantities of this masterbatch to the matrix polymer is fraught with possible problems. It is common practice that the masterbatch is made in the most convenient polymer for the manufacturer, a convenience which may lead to cheapness, but which can also lead to failure in the final product. The masterbatch makes two contributions to failure: poorly mixed polymer interferes with the development of texture in the matrix polymer, and regions of high pigment concentration deteriorate the strength and toughness. LDPE is frequently used as a carrier polymer for masterbatch additions to polyolefins, including HDPE, LLDPE, LDPE and PP, despite exhibiting a lack of miscibility with most of them. Figure 7.33 shows a section of a large blow moulded container in PP, coloured with a masterbatch based on LDPE, which failed unexpectedly on low temperature impact. The crack initiated in the *pinch off* region, where very poor dispersion, both of pigment and carrier polymer, was evident.

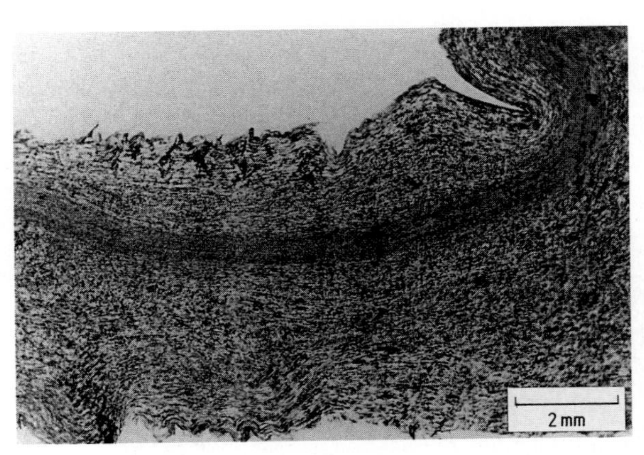

Figure 7.33 Poor distribution of LDPE masterbatch carrier polymer in polypropylene blow moulding.

The immiscibility of many polymers discussed above is used to advantage in improving the impact performance of many plastics; compare HIPS with PS, and ABS with SAN. In such cases, attention is paid to the *interface* between the matrix resin and the dispersed rubber, where the compatibility is frequently improved by grafting. Without this treatment, however, even formally similar polymers such as LDPE and LLDPE are not strictly miscible; a blend of such materials may lead to textural features which can hasten failure in critical regions of the product.

7.6.3.2 Processing, Fabrication and Design

Very large numbers of factors contribute to failure under this heading; we cannot imply that all such factors can always be specified, but the examples given within this section should help to separate the origin of some influential variables arising from the manufacture of the plastics product.

(A) Design. This involves two considerations, the design of the product, and the means of achieving the design. Features which create *stress concentrations* are aspects of a poor design technique for plastics, as with metals; thus, sharp corners and abrupt changes in section thickness may both contribute to early failure.

Corners should be generously radiused and thickness changes, if really essential, should take place over an extended distance. The presence of an abnormally thick section in a component is undesirable for other, secondary reasons:

- it may be associated with a transition towards triaxial stress (plane strain) when loaded;
- there may be difficulties associated with moulding and cooling, resulting in internal voids, surface sink marks, or excessive residual stress.

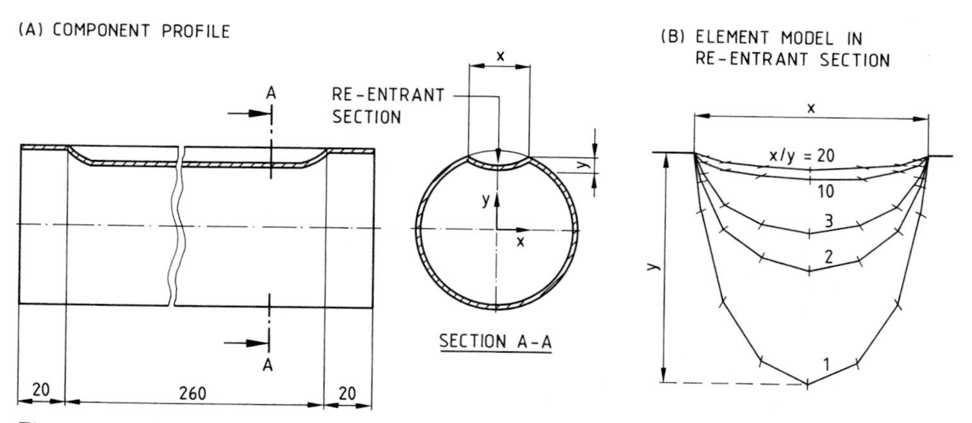

Figure 7.34 Stress concentration in a re-entrant (concave) section of a plastics pressure vessel, analyzed by finite element analysis.

(A) Component drawing, with re-entrant section width (x) and depth (y);
(B) Profile of 8-element concave models used to investigate the effect of section depth ($x : y$ ratios between 20 : 1 and 1 : 1).

There is now much greater opportunity for flexibility in design of plastics products, in particular to determine by computer simulation, based on *finite element analysis (FEA),* features which lead to stress concentration. This approach has been applied to car coolant reservoir tanks, where a re-entrant section of the tank was shown to be a significant stress concentrator; this feature coincided with the region of brittle failure on testing. Figure 7.34 shows how this effect was simulated by FEA of a pressure vessel. Using this approach, the effects of section depth and element thickness on the degree of stress concentration were computed; the results are given in Table 7.2, and show that geometric shape is the most important factor in modelling concave sections.

Table 7.2 Stress Concentration in Concave Sections

Shape of Concave Section	Thickness (mm)	x/y (Figure 7.34)	Stress Concentration Factor
2-Element V	3	2	10
2-Element V	6	2	4
2-Element V	9	2	2
2-Element V	12	2	1.3
8-Element Concave	6	0.5	27
8-Element Concave	6	1	14
8-Element Concave	6	2	2.9
8-Element Concave	6	3	2.6
8-Element Concave	6	10	2.1
8-Element Concave	6	20	1.3

(B) Processing. This provides the greatest opportunity for inducing failure in products which purport to be strong and tough. Since virtually none of the factors is obvious to visual inspection, then additional Quality Control, or Quality Assurance procedures must be adopted. The following factors are examples of those which have been found to affect performance.

• **Melt Processing Temperature.** Too high a processing temperature will lead to significant degradation, with deterioration in mechanical properties and an increase in brittleness arising from lower molecular weight polymer, and/or the presence of decomposition products such as gas inclusions.

Processing temperature too low may result in *unmelted granules* in the product. An example of this effect is given in Figure 7.35, which is a moulding in PA6, moulded at 220 °C, a few degrees below the fusion temperature of this material: (A) is taken in common light, whilst in (B), the sample is examined between crossed polars. Increasing the melt temperature by 20 °C eliminated this problem completely.

An example of *oxidative degradation* contributing to failure is afforded by a rotational moulding which had been seriously overheated at the base. This polyethylene unit failed in service by brittle fracture; subsequent examination of the material in the failed region by *infra-red analysis* revealed a high level of oxidation, indicative of considerable degradation, and the consequent deteriorated mechanical properties.

• **Blowing Pressure.** In the cooling stage of the blow moulding process, effective cooling requires that the moulding remains in good thermal contact with the cool

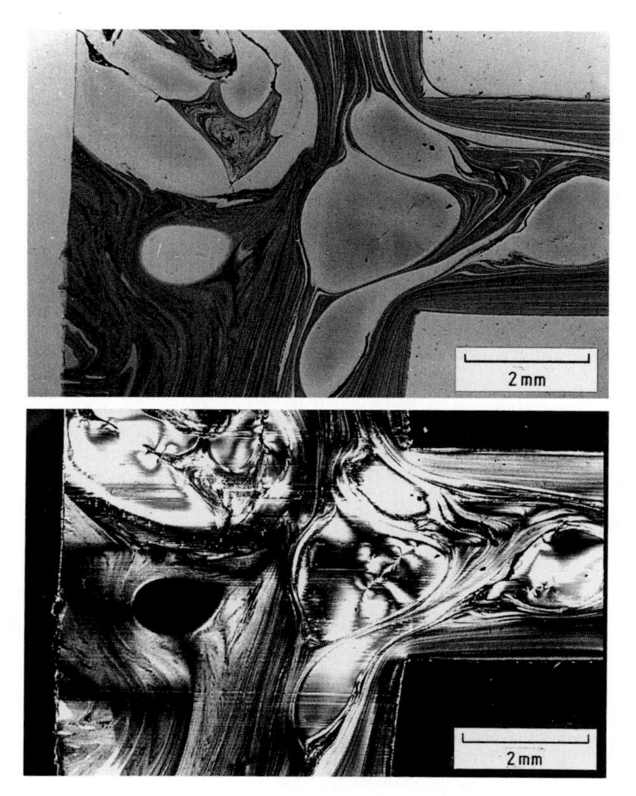

Figure 7.35 Unmelted granules of PA6 in injection moulding.
Sections viewed: (A) with common light, and (B) through crossed polars.

metal mould. To achieve this, a higher pressure is needed than is required to blow the shape originally, to counteract thermal contraction and the additional shrinkage associated with crystallization of highly-crystalline plastics such as HDPE polymer. A reduced rate of cooling is shown in Figure 7.36, with the lower pressure leading to loss of contact with the mould. This is undesirable economically, but technically, the slower cooling encourages cracks to develop at the inner surface (Figure 1.18), increasing the chance of premature failure when under external stress.

• **Injection Moulding Capacity.** The moulding capacity of an injection press is often expressed as the maximum mass of polystyrene which the machine can deliver in a single shot. This can be misleading, since it is the *shot volume* which is constant; the maximum shot weights corresponding to various polymers are different, and vary according to their densities. Thus a manufacturer required to make a moulding in PP must recalculate the capacity of his press by multiplying the "PS capacity" by the ratio of the respective densities (density should strictly be taken appropriate to the injection pressure and temperature). This means that the shot size for PP is some 19 % lower than the quoted capacity in "PS units", and if the moulder persists in moulding a shot size in PP beyond the capabilities of the machine, this will result in *unmelted granules* of PP entering the moulding. This creates a hiatus

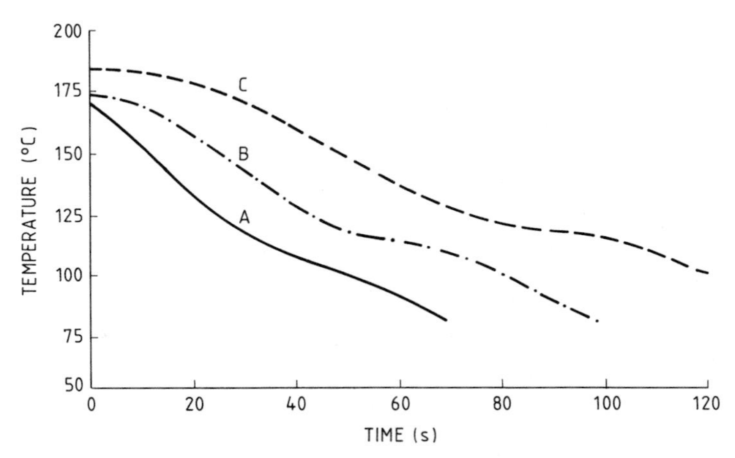

Figure 7.36 Rate of cooling on the inner surface of HDPE blow moulding as a function of pressure: (A) 2.43 bar; (B) 1.67 bar; (C) 1.11 bar. At low pressure levels, the component loses contact with the mould because of thermal shrinkage effects.

in the crystalline texture, and a region of incipient failure, Figure 7.37; transferring the job to a machine of increased shot weight and greater melting capacity removed the problem completely.

Figure 7.37 Gate region of an injection moulding in PP, processed on a machine of insufficient capacity (moulding weight in excess of the shot size), allowing unmelted polymer granules to enter the moulding.

- **Unsuitable Mould Cooling Regime.** In shaping thermoplastics by injection or blow moulding, it is usual that the mould temperature be as cold as practicable, since this reduces the cycle time (Chapter 5). Similarly, in cooling extrudates, the manufacturer wishes to cool the product as rapidly as possible, in order to optimize production line speeds. These efforts to hasten the cooling rate may enhance the chance of failure. Quenching blow mouldings has been shown to lead to *morphological interfaces,* which are regions where crack initiation is facilitated by stresses generated by the change in texture[7.74]. Similarly, the rapid chilling of thick extrudates may set up excessive *residual stress,* leading to extensive internal voiding, or to premature failure under load. This is avoided by imposing programmed slow cooling; a very good example is provided by the extrusion of LDPE covered submarine telephone cable[7.75]. The environmental stress cracking data on polycarbonate (in flexure) shown earlier (Figure 7.29) are exceptional in this respect; we may speculate that the improved performance at lower mould temperature is related to a higher degree of compressive residual stress, at the outer moulding surface.

• **Contamination.** This may be adventitious (where the contamination is frequently particulate), or intentional (blends with other polymers), or the addition of masterbatch based on an alien polymer. Its importance has already been raised in Chapter 1. A particle contaminant is a stress raiser, which can lead to early failure, yet assigning the origin of such contamination is often difficult, involving identification and investigation of how it entered the product. Identification may be on the basis of morphological features observed with a light or scanning electron microscope (SEM), or more frequently, by X-ray dispersive analysis, coupled with observations by SEM.

The blending or alloying of polymers is a popular theme currently, extended into masterbatch techniques for incorporation of additives. In such systems *miscibility* and/or *dispersion* are extremely important. Miscibility is, however, rare amongst polymers, and even more so amongst crystalline polymers; blends which are not miscible can lead to regions of morphological weakness with premature failure. This has been referred to earlier; Figure 7.33 shows poor dispersion of LDPE masterbatch in a polypropylene blow moulding.

• **Internal Welds and Pinch-off Effects.** The importance of these features is greatest when they are combined with other factors which are deleterious to service; thus, an internal weld, or pinch-off in blow moulding becomes a danger region in a product with poor pigment, or other polymer, dispersion. Problems associated with internal welds have been investigated by CRAWFORD[7.76], whilst the pinch-off as a weakening feature has been discussed by SHELLEY[7.77]. Poor pigment and polymer dispersion, in the region of the pinch-off in the base of a large container, initiated the crack which propagated across the whole base, Figure 7.38. With greater attention paid both to dispersion and to the pinch-off, the feature can assume an innocuous role in the texture.

Regions of an extrusion blow moulded product where the material is required to seal to itself are shown in Figure 7.39; clearly, the technique as carried out leads to an incipient crack.

• **Welding Processes.** Welding techniques are important in the fabrication of plastics products; there is a wide range of methods available to the designer, and the successful implementation of these methods has greatly extended the applications range for plastics products[7.78]. However, polymer microstructure is always affected

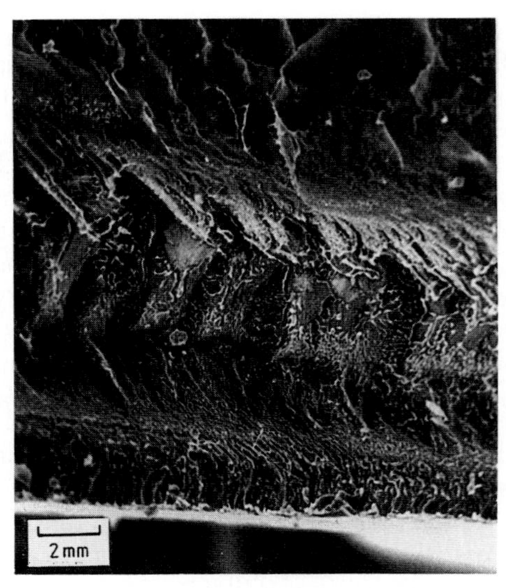

Figure 7.38 Poor mixing of masterbatch in the matrix polymer can lead to failures in the pinch-off region of a blow moulding.

Figure 7.39 Regions in an L-ring drum blow moulding where the polymer is required to seal to itself.

by the thermomechanical history associated with any welding process and almost without exception, this may lead to deterioration in the mechanical properties which determine the durability of a product.

Hot plate welding, friction welding and the use of adhesives are obvious and very common routes to join plastics components. In all of these methods, poor technique can be a detrimental factor, and is exemplified in the welding of additional features to a HDPE container by the hot plate method. The temptation is always to use heat soaking times which are far too short, and to compensate by using an excessively high temperature: this could result in thermal degradation. A properly formed weld

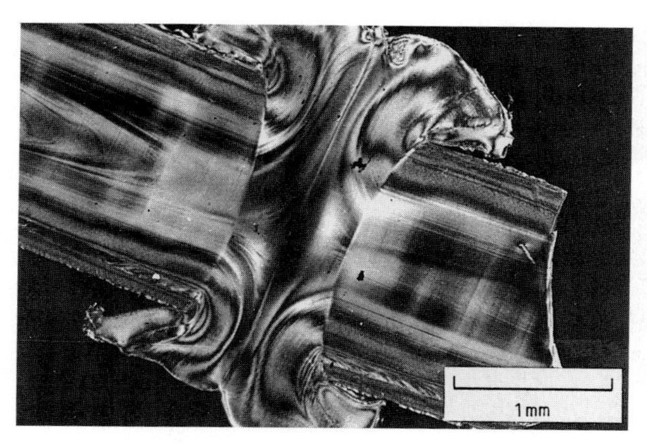

Figure 7.40 Crossed polars micrograph of a satisfactory hot-plate weld in HDPE; note however, the high orientation of the injection moulded parts.

is shown in Figure 7.40, taken between crossed polars. Several features are of note:

(1) The welding bead accounts for about one half of the material melted by the hot plate; any material which has been excessively heated should be deposited into the bead.
(2) The weld zone is distinguished from the unheated polymer by a change in texture (including the phenomenon of transverse molecular orientation), with the internal boundary constituting a line of weakness whence many failures are initiated.
(3) The positions where this interface meets the component surface are particular points of weakness, and stress concentrations are particularly severe in situations where the weld bead is not removed. Even when the bead is sanded or machined, great care must be taken to ensure that the effects of the remaining surface defects can be controlled.

Frequently, components to be welded to a main structure are injection moulded; it is important that they should be of similar material to the major component and should also be moulded sympathetically, without undue residual orientation (which would relax, causing shrinkage and alignment problems, on subsequent welding).

The problem of *interfacial mis-alignment* is highlighted in Figure 7.41; in this case, the PVC profiles were welded under conditions of excessive clamping pressure. When loaded, the stress distribution in improperly-fabricated joints becomes concentrated and more complex, and a transition towards embrittlement is almost inevitable.

• **Orientation and Residual Stress.** For stresses applied in the direction of orientation, the mechanical properties are often enhanced, whilst in the transverse direction there is deterioration which can lead to failure. The benefits and deficiencies of orientation can also be experienced in filled compounds, especially when anisotropic fillers, including fibres, are used. An example of *molecular orientation,* in injection moulded PC, is shown in Figure 7.42.

Figure 7.41 Section through a hot-plate welded PVC profile, showing the effect of interfacial misalignment.

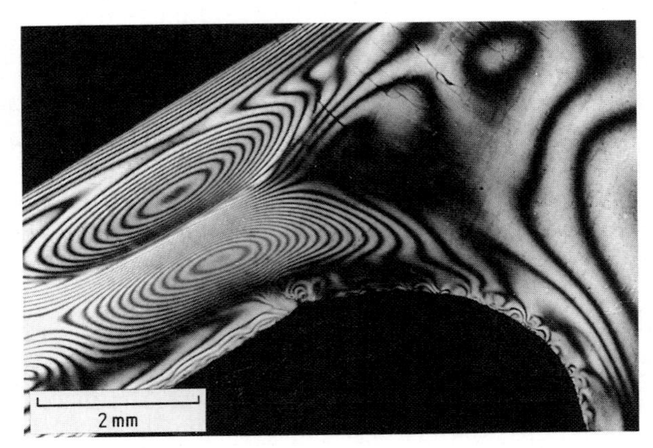

Figure 7.42 Residual orientation in a polycarbonate moulding, viewed through crossed polars.

The deformation and fracture behaviour of oriented PS has been investigated for both uniaxially-drawn sheets[7.79], and for injection moulded samples[7.80]. These studies have shown that tensile fracture properties in particular are very sensitive to the direction and extent of orientation; injection moulding introduces further complexities, due to the non-uniform distribution of orientation which usually exists in such components.

• **Problems with Gates in Injection Moulding.** The gate of a mould must be one of the regions where confused texture might be anticipated, and most mouldings fall far short of the optimum. The difficulty is inherent in the function of the gate, as it allows more material to enter the mould to compensate for shrinkage on cooling. It is almost inevitable that such material will have had a different thermal history to that already in the mould. Further, the movement of polymer will generate a morphology which is different, even for optimum functioning of the gate, and subsequent separation of the moulding from the material residing in the feed channels often leaves a defective and unsightly witness mark on the component surface.

Under workshop conditions and under the pressure of minimizing cycle time, less satisfactory textures will be encountered. Examples have been shown in Figures 1.16 and 7.37; with such confused texture it is perhaps not surprising that premature failure often occurs by crack initiation at the gate. It is therefore of some importance that the mould designer recognizes that gate location should be chosen in accordance with the objective of avoiding likely areas of stress concentration under load.

7.6.3.3 Environmental Factors

Included in this section are temperature, and thermal history, and effects on failure properties due to exposure to moisture and other external media; the chemical effects of these will be treated more thoroughly in Chapter 10.

• **Temperature.** The exposure of plastics to low temperature, with the imposition of mechanical stress, may evoke a much more brittle response than is experienced at ambient temperature; for example, PP homopolymer is much more likely to fail at 0 °C than at 23 °C. This has been demonstrated by impact data for unnotched PP, Figure 7.20, in which the temperature decrease has a profound effect[7.81].

Another aspect of temperature exposure is excursion to very high temperature for a short time, or treatment at somewhat lower temperatures for longer times, especially in the presence of air or oxygen. *Physical ageing* effects are very important in polymers[7.82], especially for amorphous plastics where small-scale molecular motions occur, thereby reducing free volume; these effects appear to suppress molecular relaxations in stressed plastics, and are linked to embrittlement effects, as experienced in PC, for example.

• **Moisture.** The absorption of moisture by a plastics material generally softens – and toughens – in the short term; polyamides are typical plastics showing this behaviour. In the longer term, and especially at moderately elevated temperatures, water is frequently degradative, leading to premature failure, as exemplified by the hydrolysis of thermoplastic polyesters[7.39], and to stress corrosion of glass-reinforced laminates based on polyester resin[7.34].

• **Environmental stress cracking.** This is a failure mechanism requiring the combined action of stress and an active environment, with neither being effective separately. ESC mechanisms, test methods and data have been discussed previously, in Section 7.5.4.

The phenomenon is exemplified by coolant reservoir tanks for motor vehicles, manufactured in impact-modified PP. The combined effect of pressure, high temperature and aqueous ethylene glycol causes characteristic cracking, Figure 1.19. As with many failure phenomena, performance is improved as the molecular weight is increased, provided that there is not a catastrophic increase in the residual strain in the product, ie. induced by processing, as a result of higher melt viscosity.

Extrusion blow moulding is a process which requires high molecular weight polymer and is the preferred method of shaping for this application. Injection moulding is inferior in three respects:

(1) injection moulding requires "easy-flow" (low molecular weight) polymer;
(2) orientation and residual stress are inevitable; and
(3) the tank, injection moulded in two halves, needs to be welded.

Each of these encourages failure by environmental stress cracking, in the presence of ethylene glycol.

Styrene polymers, including high-toughness ABS resins, are vulnerable to this type of failure, sometimes in seemingly innocuous environments; the cracking shown in Figure 7.43 was induced in a stressed bar of ABS, merely by sticking a self-adhesive label over the stressed component. The failure in Figure 7.44 was caused by solvent sticking an ABS plug into an ABS moulding: tetrahydrofurane was the solvent employed.

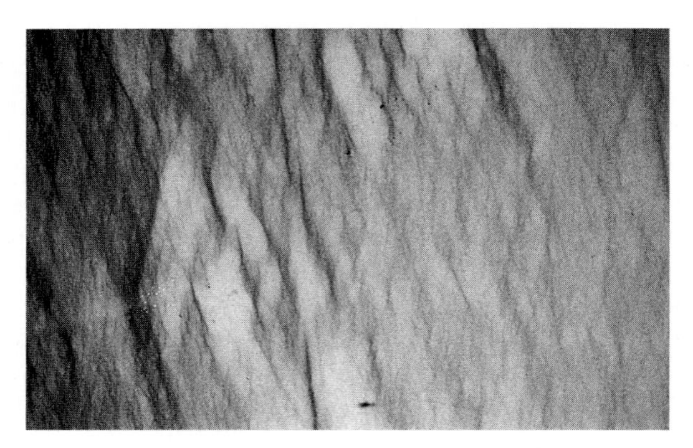

Figure 7.43 Environmental stress cracking of ABS, accelerated by a constituent of a self-adhesive label.

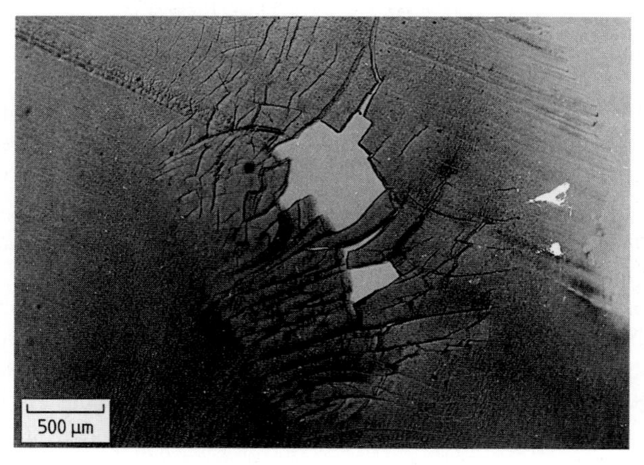

Figure 7.44 Environmental stress cracking of an ABS product, under the influence of tetrahydrofurane, introduced via an adhesive.

7.6.3.4 Applied Stresses

The manner in which a stress is applied, and the timescale of loading, affect the response of a plastics product; a number of separate cases can be considered, many aspects of which have been covered in detail in section 7.5, and will be summarized further:

(A) Impact loading. High velocity impact loads often lead to premature failure in plastics components, due to the material's inability to relax and conform to the applied stress. This situation is made worse if the stress system is triaxial, and if stress concentration is apparent. Analysis of impact failure by fracture mechanics is desirable.

(B) Time under load. Long term tests on HDPE pipe under internal pressure first drew attention to unexpected failures at long times, especially at elevated temperatures. These failures were unexpected for two reasons: first, they occurred at a time/stress combination at significantly shorter times than extrapolation would indicate; secondly, the failures were brittle in nature, occurring at low elongation. This behaviour has now been recognized as general for some families of plastics materials; typical creep rupture data for HDPE[7.83] have been shown in Figure 7.17.

(C) Dynamic fatigue. Component failure under high frequency, oscillating stress is known as fatigue; plastics structures, in common with many other engineering materials, are prone to premature brittle failure by this mechanism. As discussed in Section 7.5.3, analysis of fatigue fracture can be quite complex, but one redeeming feature of this failure mode is the relative ease with which its occurrence can sometimes be recognized by microscopic analysis; Figure 7.45 illustrates this point adequately.

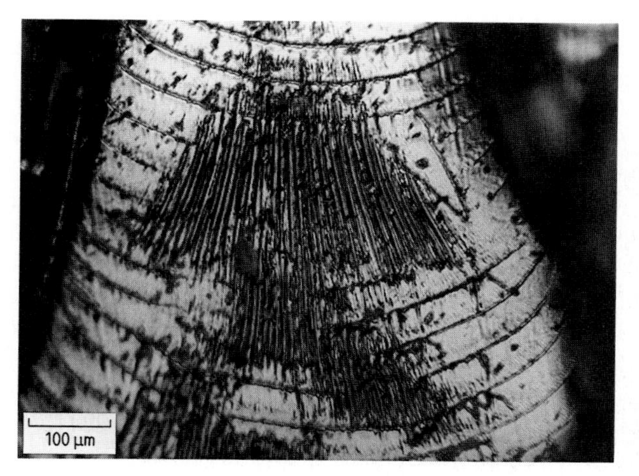

Figure 7.45 Fatigue failure in a polycarbonate moulding; note the steady progression of the crack.

(D) Complex stress systems. The response of a component depends on the severity of stress which is applied; this has two aspects, the nature of the stress and its magnitude. The stress system may be prone to give ductile behaviour (eg. plastics under compressive load), or likely to evoke a brittle response; in the ultimate, triaxial tension.

(E) Abuse of product. A product is usually manufactured with some application in mind, so that at the design stage some decision is made about the severity of treatment expected in service. The design does not usually give guarantees for good performance, without reference to the use or abuse imposed on the product. Exemplifying, a metal food-can would not survive attack with a large hammer, nor would we expect it to. It will however, perform adequately when subjected to the treatment suffered by cans in normal service use. Similar experience is available for plastics, and components can be developed which are well suited to the application, but they will not necessarily survive malicious attack.

In summary, test procedures and the interpretation of results should be based on a knowledge of:

(1) the initial properties required for a particular application;
(2) the various factors and mechanisms which can cause deterioration; and
(3) materials, and environmental factors which affect the resistance to deterioration.

References

7.1 BROSTOW, W. and CORNELIUSSEN, R.D., EDS., *Failure of Plastics,* Hanser, Munich, (1986).

7.2 KAUSCH, H.H.: *Polymer Fracture,* Springer-Verlag, New York, (1978).

7.3 WARD, I.M.: *Mechanical Properties of Solid Polymers,* Wiley, New York, (1971).

7.4 BOWDEN, P.B.: *The Yield Behaviour of Glassy Polymers,* in *The Physics of Glassy Polymers,* HAWARD R.N., Ed., Applied Science, London, (1973).

7.5 BROWN, N.: *Yield Behaviour of Polymers,* in *Failure of Plastics,* BROSTOW W. and CORNELIUSSEN R.D., Eds., Hanser, Munich, (1986).

7.6 WHITNEY, W. and ANDREWS, R.D.: *Yielding of Glassy Polymers: Volume Effects,* J. Polym. Sci: Polym. Symp., **C16,** (1967), 2981.

7.7 BOWDEN, P.B. and JUKES, J.A.: *The Plastic Flow of Isotropic Polymers,* J. Mat. Sci., **7,** (1972), 52.

7.8 BROWN, N.: *A Theory of Yielding of Amorphous Polymers at Low Temperature – A Molecular Viewpoint,* J. Mat. Sci., **18,** (1983), 2241.

7.9 ANDREWS, E.H.: *A Generalised Theory of Fracture Mechanics,* J. Mat. Sci., **9,** (1974), 887.

7.10 WILLIAMS, J.G.: *Fracture Mechanics of Polymers,* Polym. Eng. Sci., **17,** (1977), 144.

7.11 WILLIAMS, J.G.: *Fracture Mechanics of Non-Metallic Materials,* Phil. Trans. Roy. Soc. Lond., **A299,** (1981), 59.

7.12 KINLOCH, A.J. and YOUNG, R.J.: *Fracture Behaviour of Polymers,* Applied Science, London, (1983).

7.13 WILLIAMS, J.G.: *Fracture Mechanics of Polymers,* Ellis Horwood, Chichester, UK, (1984).

7.14 KAUSCH, H.H.: *The Nature of Defects and Their Role in Large Deformation and Fracture of Engineering Thermoplastics,* in *Interrelations between Processing, Structure and Properties of Polymeric Materials,* SEFERIS J.C. and THEOCARIS, P.S., Eds., Elsevier, Amsterdam, (1984).

7.15 GRIFFITH, A.A.: *The Phenomena of Rupture and Flow in Solids,* Phil. Trans. Roy. Soc. Lond., **A221,** (1920), 163.

7.16 IRWIN, G.R.: *Fracture,* Encyclopaedia of Physics **6,** 551, Springer-Verlag, Berlin, (1958).

7.17 ROOKE, D.P. and CARTWRIGHT, D. J.: *Compendium of Stress Intensity Factors,* HMSO, London, (1976).

7.18 FERGUSON, R.J., MARSHALL, G.P. and WILLIAMS, J.G.: *The Fracture of Rubber-Modified Polystyrene,* Polymer **14,** (1973), 451.

7.19 HERTZBERG, R.W.: *Deformation and Fracture Mechanics of Engineering Materials,* John Wiley, New York, (1976).

7.20 CHAN, M.K.V. and WILLIAMS, J.G.: *Plane Strain Fracture Toughness Testing of High Density Polyethylene,* Polym. Eng. Sci., **21,** (1981), 1019.

7.21 HASHEMI, S. and WILLIAMS, J.G.: *Fracture Characterisation of Tough Polymers using the J Method,* Polym. Eng. Sci., **26,** (1986), 760.

7.22 NARISAWA, I. and TAKEMORI, M.T.: *Fracture Toughness of Impact-Modified Polymers Based on the J-Integral,* Polym. Eng. Sci., **29,** (1989), 671.

7.23 PASCOE, K.J.: *General Fracture Mechanics,* in: *Failure of Plastics,* BROSTOW, W. and CORNELIUSSEN, R.D., Eds., Hanser, Munich, (1986).

7.24 KINLOCH, A.J.: *Review - The Science of Adhesion Part 2; Mechanics and Mechanisms of Failure,* J. Mat. Sci., **17,** (1982), 617.

7.25 BS 3505, (1986), *Unplasticised Polyvinyl Chloride (PVC-U) Pressure Pipes for Cold Potable Water,* British Standards Institution, London.

7.26 MARSHALL, G.P. and BIRCH, M.W.: *Design for Toughness in Polymers. 3 – Criteria for High Toughness in UPVC Pressure Pipes,* Plast. Rubb. Proc. Appl., **2,** (1982), 369.

7.27 MOORE, D.R., STEPHENSON, R.C. and WHALE, M.: *Some Factors Affecting Toughness in UPVC Pipe Materials,* Plast. Rubb. Proc. Appl., **3,** (1983), 53.

7.28 MOORE, D.R., PREDIGER, R. and STEPHENSON, R.C.: *Relevance and Application of Fracture Toughness Measurements for UPVC,* Plast. Rubb. Proc. Appl., **5,** (1985), 335.

7.29 TEMPELS, R.: *Rigid PVC Impact Behaviour – Influence of Specimen Thickness,* Kunststoffe **80,** (1990), 24/58.

7.30 BPF/GGF Publication, *A Trade Standard for UPVC Windows,* 2nd Ed., British Plastics Federation, London, (1989).

7.31 CALVERT, D.J., HAWORTH, B. and STEPHENSON, R.C.: *The Use of Fracture Mechanics to Describe the Impact Strength of PVC Window Profiles,* Presented at PRI International Conference PVC '90, Brighton, UK, (1990).

7.32 TSAI, S.W. and HAHN, H.T.: *Introduction to Composite Materials,* Technomic Publishing, Westport, USA, (1980).
7.33 TETLOW, R.: in *Carbon Fibres in Engineering,* LANGLEY M., Ed., McGraw-Hill, New York, (1973).
7.34 MARSHALL, G.P. and HARRISON, D.: *Design for Toughness in Polymers. 2 – Environmental Stress Corrosion of Chemically-Resistant Polyester Resins and Glass Reinforced Laminates,* Plast. Rubb. Proc. Appl., **2,** (1982), 269.
7.35 HULL, D.: *Introduction to Composite Materials,* Cambridge University Press, Cambridge, UK, (1981).
7.36 FOLKES, M.J.: *Short Fibre Reinforced Thermoplastics,* Research Studies Press, John Wiley, Chichester, UK, (1982).
7.37 STOWELL, E.Z. and LIU, T.S.: *On the Mechanical Behaviour of Fibre-Reinforced Crystalline Materials,* J. Mech. Phys. Solids., **9,** (1961), 242.
7.38 BRIGHT, P.F. and DARLINGTON, M.W.: *Factors Influencing Fibre Orientation and Mechanical Properties in Fibre Reinforced Thermoplastics Injection Mouldings,* Plast. Rubb. Proc. Appl., **1,** (1981), 139.
7.39 HAWORTH, B. and WALSH, G.M.: *Effects of Hydrolysis on the Physical Properties of Short Glass-Fibre Reinforced Poly(Ethylene Terephthalate),* Brit. Polym. J., **17,** (1985), 69.
7.40 BRAGAW, C.G.: *Prediction of Service Life of Polyethylene Gas Piping Systems,* Proc., 7th Plastic Fuel Gas Pipe Symposium, American Gas Association, New Orleans, USA, (1980).
7.41 BRAGAW, C.G.: *The Forecast of Polyethylene Pipe and Fitting Burst Life using Rate Process Theory,* Presented at PRI International Conference, Plastics Pipes V, York, UK, (1982).
7.42 ASTM D2837 (1976), *Obtaining Hydrostatic Design Basis for Thermoplastic Pipe Materials,* American Society for Testing Materials.
7.43 DIN 8075, Parts 1/2, *HDPE Type 1/2 Pipes; General Quality Requirements and Tests,* German National Standard.
7.44 KEMP, S.G.: *Polybutylene – Selected Material and Property Aspects,* Plast. Rubb. Proc. Appl., **3,** (1983), 169.
7.45 ENGEL, T.: *Forging and Crosslinking of Thermoplastics,* Plast. and Polym., **38,** (1970), 174.
7.46 SCOTT, H.G. and HUMPHRIES, J.F.: *Novel Crosslinking Method for Polyethylene,* Mod. Plast., **50,** (1973), 82.
7.47 JENKINS, H. and KELLER, A.: *Radiation-Induced Changes in Physical Properties of Bulk Polyethylene 1. Effect of Crystallisation Conditions,* J. Macromol. Sci., **B11,** (1975), 301.
7.48 OTANI, K.: *Crosslinking Process of Polyethylene,* Jap. Plast., **8,** (1974), 18.
7.49 ZYBALL, A.: *Irradiation Crosslinking of Polyethylene in the Presence of Polymerisable Additives,* Kunststoffe, **67,** (1977), 16/461.
7.50 BS2782 (1970) Methods 306B/C, *Impact Strength (Falling Weight Method with Sheet Specimens),* British Standards Institution, London.
7.51 GUTTERIDGE, P.A., HOOLEY, C.J., MOORE, D.R., TURNER, S. and WILLIAMS, M.J.: *A Versatile System of Impact Tests and Data,* Kunststoffe, **72,** (1982), 9/543.

7.52 TURNER, S., REED, P.E. and MONEY, M.: *Flexed Plate Impact Testing – Some Effects of Specimen Geometry,* Plast. Rubb. Proc. Appl., **4,** (1984), 369.

7.53 JONES, D.P., LEACH, D.C. and MOORE, D.R.: *The Application of Instrumented Falling Weight Impact Techniques to the Study of Toughness in Thermoplastics,* Plast. Rubb. Proc. Appl., **6,** (1986), 67.

7.54 REED, P.E. and TURNER, S.: *Flexed Plate Impact Testing II. The Behaviour of Toughened Polystyrene,* Plast. Rubb. Proc. Appl., **5,** (1985), 109.

7.55 HAWORTH, B., LAW, T.C., SYKES, R.A. and STEPHENSON, R.C.: *Factors Determining the Weld Strength of Window Corners Fabricated from Impact Modified PVC Profiles,* Plast. Rubb. Proc. Appl., **9,** (1988), 81.

7.56 PLATI, E. and WILLIAMS, J.G.: *The Determination of the Fracture Parameters for Polymers in Impact,* Polym. Eng. Sci, **15,** (1975), 470.

7.57 MCCRUM, N.G., BUCKLEY, C.P. and BUCKNALL, C.B.: *Principles of Polymer Engineering,* Oxford University Press, Oxford, UK, (1988).

7.58 CHEN, L.S., MAI, Y.M. and COTTERELL, B.: *Impact Fracture Energy of Mineral-Filled Polypropylene,* Polym. Eng. Sci., **29,** (1989), 505.

7.59 PARIS, P.C., GOMEZ, M.P. and ANDERSON, W. E.: *A Rational Analytic Theory of Fatigue,* Trend. Eng., **13,** (1961), 9.

7.60 BUCKNALL, C.B. and DUMPLETON, P.: *Factors Affecting Fatigue Crack Growth in HDPE,* Polym. Eng. Sci., **5,** (1985), 343.

7.61 CONSTABLE, I., WILLIAMS, J.G. and BURNS, D. J.: *Fatigue and Cyclic Thermal Softening of Thermoplastics,* J. Mech. Eng. Sci., **12,** (1970), 20.

7.62 HERTZBERG, R.W. and MANSON, J.A.: *Fatigue of Engineering Plastics,* Academic Press, New York, (1980).

7.63 BARKER, M.B., BOWMAN, J. and BEVIS, M.J.: *Fatigue Properties of High Density Polyethylene Pipe Systems,* J. Mat. Sci: Lett., **15,** (1980), 265.

7.64 BARKER, M.B., BOWMAN, J. and BEVIS, M.J.: *The Performance and Causes of Failure of Polyethylene Pipes subjected to Constant and Fluctuating Internal Pressure Loadings,* J. Mat. Sci, **18,** (1983), 1095.

7.65 HOWARD, J.B.: *Why do Plastics Stress-Crack?* Polym. Eng. Sci., **5,** (1965), 125.

7.66 ASTM D1693 (1970), *Environmental Stress Cracking of Ethylene Plastics,* American Society for Testing Materials.

7.67 HENRY, L.F.: *Prediction and Evaluation of the Susceptibilities of Glassy Thermoplastics to Environmental Stress Cracking,* Polym. Eng. Sci., **14,** (1974), 167.

7.68 FAULKNER, D.L.: *Anisotropic Aspects of the Environmental Stress Cracking of ABS and SAN Copolymers in Methanol,* Polym. Eng. Sci., **24,** (1984), 1174.

7.69 TITOW, W.V.: *A Review of Methods for the Testing and Study of Environmental Stress Failure in Thermoplastics,* Plast. and Polym., **43,** (1975), 98.

7.70 ORTHMANN, H.J.: *Environmental Stress Cracking of Thermoplastics,* Kunststoffe, **73,** (1983), 17/96.

7.71 KAMBOUR, R.P.: *A Review of Crazing and Fracture in Thermoplastics,* J. Polym. Sci: Macromol. Rev., **7,** (1973), 1.

7.72 HEMSLEY, D.A.: *The Light Microscopy of Synthetic Polymers,* RMS Series 07, Oxford Science Series, (1982).

7.73 SAWYER, L.C. and GRUBB, D.T.: *Polymer Microscopy,* Chapman and Hall, London, (1987).

7.74 GARG, S.K. and BIRLEY, A.W.: *Effect of Morphology and Microstructural Interfaces on the Properties of Linear Polyethylene Blow Mouldings,* Plast. Rubb. Proc. Appl., **2,** (1982), 105.

7.75 BAGULEY, E.: *Polyethylene Submarine Telephone Cables,* PI Trans. and J., **32,** (1964), J120.

7.76 CRAWFORD, R.J., KLEWPATINOND, V. and BENHAM, P.P.: *The Influence of Injection Moulding Conditions on the Impact Strength of a Thermoplastic,* Plast. and Rubb: Proc., **4,** (1979), 151.

7.77 SHELLEY, R.M.: *Development of HDPE Fuel Tanks,* PhD. Thesis, Loughborough University of Technology, (1987).

7.78 WATSON, M.N. (Ed.), *Joining Plastics in Production,* Welding Institute, Cambridge, UK, (1988).

7.79 HULL, D. and HOARE, L.: *Deformation and Fracture of Oriented Polystyrene,* Plast. Rubb: Mat. Appl., **1,** (1976), 212.

7.80 HULL, D. and HOARE, L.: *The Effect of Orientation on the Mechanical Properties of Injection Moulded Polystyrene,* Polym. Eng. Sci., **17,** 204, (1977).

7.81 ICI Technical Literature: *Propathene Data for Design,* Technical Service Note PP110, ICI Chemicals and Polymers Ltd., Wilton, UK.

7.82 STRUIK, L.C.E.: *Physical Aging in Amorphous Polymers and Other Materials,* Elsevier, Amsterdam, (1978).

7.83 Hoechst Plastics Technical Literature: *Pipes; Hostalen GM 5010 T2 and Hostalen GM 7040 G,* Hoechst-Celanese Plastics, Frankfurt am Main, Germany.

Chapter 8
Electrical Properties

8.1 Introduction

Polymers are inherently *electrical insulators,* as a consequence of their structures of covalently bonded discrete molecules, a quality which has been long exploited for restricting access and affording environmental protection to conductors carrying electrical power. Plastics are also used to restrain high electric fields.

Early insulation was almost inevitably derived from natural products, often vegetable in origin, although mica also found widespread application. A much-favoured insulant was gutta percha (trans-polyisoprene, an isomer of natural rubber), the properties of which encouraged manufacturers of cables to invest in plantations of gutta-producing trees. With the commercial introduction of phenol-formaldehyde plastics in about 1910, it was inevitable that the electrical industry should be in the forefront of its development. Also, further expansion of the plastics industry occurred in the 1940's by the replacement of rubber insulation with PVC.

The growing plastics industry was thus nurtured by the powerful cable and electrical goods industry, and fed on the technology of that industry, providing a firm engineering base for the measurement and presentation of the electrical properties of plastics. These methods have been generally accepted, and provide a sound basis for the measurement of these properties.

8.1.1 Some Electrical Test Methods

The electrical properties of plastics are determined by:

(1) chemical structure;
(2) physical texture and morphology;
(3) residues from polymerization, impurities and additives.

Species in the last category contribute significantly to the electrical properties of plastics products, to a greater extent than for any other range of characteristics. As with mechanical properties, responses to electrical stimuli divide into *low stress phenomena,* which are not particularly dependent on the electrical stress level (or potential) and are thus similar to the mechanical moduli of elasticity, and the behaviour at high stresses, where failure results in both the mechanical and electrical cases. The similarity between low stress *dielectric measurements* and dynamic mechanical data is very close; both depend on chemical and physical structure, and both have been developed into convenient instruments for investigating such structures.

Electrical test equipment often used for plastics materials includes:

- the dielectric spectrometer, and
- the dielectric thermal analyzer.

The former, an early model of which was described by HYDE[8.1], scans the frequency response of permittivity and Tan δ at a series of temperatures, often by Fourier Transform of the "current versus time" response to a step change in voltage. The latter monitors Tan δ at a constant frequency as the temperature is increased linearly with time, and is a further extension of the family of differential thermal methods already considered in Chapters 1 and 6.

However, in view of the importance of *permittivity* and *power factor,* and their dependence on frequency and temperature, the development of instruments and methodology to measure these quantities has been prolific. Depending largely on frequency, but taking accuracy into account, the following instruments and techniques are relevant. The electrode system may be a two or three terminal device, the latter using a guard ring; both may be used with liquid immersion techniques, the electrode system being filled with a liquid of known permittivity, and the sample under investigation being immersed in the liquid. With matched permittivities of liquid and sample, the permittivity can be determined to high accuracy, since the effects of fringing fields are constant.

(A) Fourier transform techniques

- *DC step response;* 10^{-4} to 10^{6} Hz
- *Time domain spectroscopy;* 10^{4} to 10^{10} Hz.

(B) Limited frequency range techniques

Various methods cover the range 10^{-3} to 10^{10} Hz; the most important of these are summarized below.

- *Schering bridge*[8.2] (10^{1} to 10^{6} Hz) has been much used in polymer investigations. It is a capacitance bridge, which for precise work is operated with a Wagner earth[8.3].
- *Transformer ratio arms bridge* (10^{1} to 10^{6} Hz); whilst conventional bridges, such as the Schering, are based on the Wheatstone bridge principle, where the impedances of the ratio arms are resistance-capacity networks, Lynch, in particular, has developed a bridge in which the ratio arms are closely-coupled transformers[8.4–8.6].
- *Microwave bridges* (10 to 70 GHz)[8.7].
- *Transmission lines* (0.3 to 40 GHz)[8.8].

(C) Resonance techniques from 10^{4} to 10^{11} Hz

These techniques depend on the restoration of resonance in an appropriate inductance-capacitance-resistance, or microwave cavity circuit, when resonance has been disturbed by insertion of a sample. The following variants are commonly used:

- *Hartshorn and Ward apparatus* (0.1 to 30 MHz); the method was first described in 1936[8.9], with improvements being made by BARRIE[8.10], and by REDDISH et al.[8.11]. The last workers have developed a high resolution instrument, with an uncertainty in loss angle of $+/-0.6\mu rad$ at $80\mu rad$, measured at 6 MHz. Permittivity can be measured to 1 % accuracy levels.

- *Q – meter measurements* (0.1 to 250 MHz); a special electrode system, in conjunction with a commercial Q-meter, has enabled Hill to make measurements to $+/-0.5\mu rad$ in the frequency range 1 to 70 MHz[8.12].
- *Re-entrant cavity* (0.1 to 1 GHz); this is the high frequency equivalent of the Hartshorn and Ward apparatus, and has been analyzed by PARRY[8.13].
- *Microwave cavity measurements* (9 to 70 GHz)[8.14].

Selection from these methods allows measurements to be made to the desired levels of accuracy, at the frequencies required.

In the following sections of this Chapter, we review the electrical properties of plastics materials at low stress, at high stress (where *electrical breakdown* is more likely), static charge characteristics, and we also introduce the concept of "electrically-conductive" plastics composite materials, before reviewing some applications of plastics products in the electrical engineering sector.

8.2 Electrical Properties at Low Stress

The recommended method of presentation of data for insulating materials is in contour plots of the time- and temperature-dependent permittivity, (dielectric constant), and dielectric dissipation factor (Tan δ)[8.15,8.16]; some examples of these data for acetal copolymer are given in Figures 8.1 and 8.2, respectively, and volume resistivity data for PES are illustrated in Figure 8.3. The quantities portrayed are defined in the following terms.

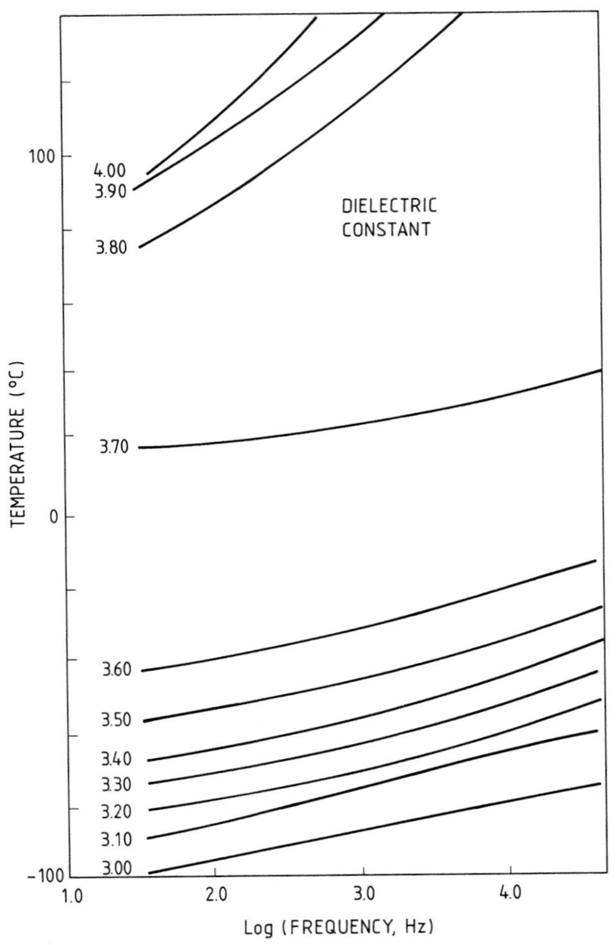

Figure 8.1 Contour map of dielectric constant (relative permittivity), on axes of temperature and (log) frequency, for POM copolymer. (Courtesy of ICI).

Relative permittivity, (ε_R) is the ratio of the capacitance (C) of a given configuration of electrodes with the plastics material as the dielectric medium, to the capacitance

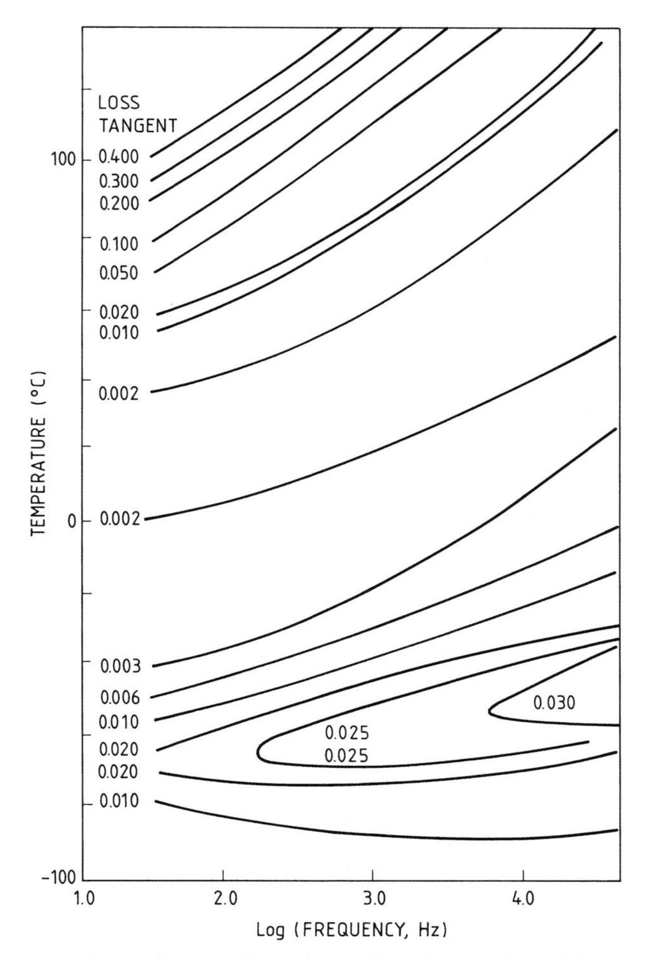

Figure 8.2 Contour map of loss tangent (which is equivalent to power factor, at low values), on axes of temperature and (log) frequency, for POM copolymer. (Courtesy of ICI).

(C_V) of the same configuration of electrodes with vacuum as the dielectric.

$$\varepsilon_R = \frac{C}{C_V} \tag{8.1}$$

Dielectric loss angle, (δ) of an insulating material is the angle by which the phase difference between the applied voltage and the resulting current deviates from $\pi/2$ radians when the dielectric of the capacitor consists entirely of the dielectric material. Thus the electrical properties of plastics in an alternating field, like mechanical properties under dynamic stress, are controlled by the effects of viscoelasticity and can be characterized by the lag between the applied stimulus and the response variable.

Dielectric dissipation factor, loss tangent, is the ratio of the electrical power dissipated in a material to the total power circulating in the circuit. It is the tangent of the

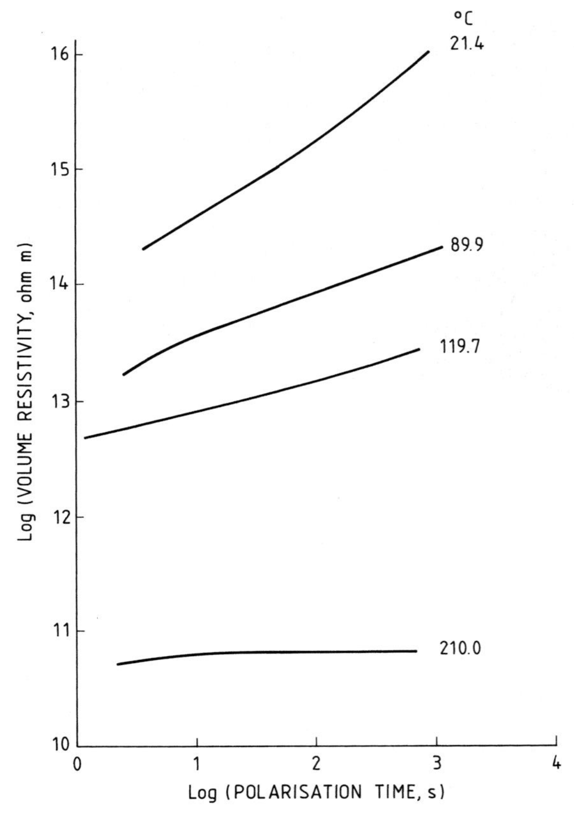

Figure 8.3 Effects of temperature (°C) and time on the apparent volume resistivity of PES. (Courtesy of ICI Advanced Materials).

loss angle (δ), and is therefore directly analogous to the Tan δ function relevant to dynamic mechanical testing (see Section 6.4.6), which describes the relationship between storage and loss moduli. Relative permittivity (ε_R) is described by the complex permittivity function of the material (ε^*) relative to the permittivity of a vacuum (ε_V), which may be expressed in a format equivalent to that quoted earlier (Equation 6.26) for complex modulus:

$$\varepsilon_R = \frac{\varepsilon^*}{\varepsilon_V} \tag{8.2}$$

where ε_V, the complex electrical compliance function (permittivity), is given by:

$$\varepsilon^* = \varepsilon_1 + i\varepsilon_2 \tag{8.3}$$

Loss index, ε_2 of an insulating material is the product of its dissipation factor and the relative permittivity, so that:

$$\text{Tan}\,\delta = \frac{\varepsilon_2}{\varepsilon_1} \tag{8.4}$$

As the dissipation factor of a plastics material may be in the range Tan $\delta = 0.000005$ to 0.3, it is common practice to use angular units for the lower parts of the range.

Microradians (μrad) are often used, since as $\delta \rightarrow 0$, Tan $\delta \rightarrow \delta$ (in radians).

($1 \mu rad = 10^{-6}$ in Tan δ units).

Details of experimental methods are given elsewhere[8.16].

Volume Resistance is the ratio of the direct voltage applied to the electrodes in contact with the test material, to the steady-state direct current flowing between them. Currents flowing along the surface of the specimen, and polarization effects at the electrodes, must be excluded. The *volume resistivity* (ϱ_V) is taken as the volume resistance reduced to a cubical unit volume. The SI unit of ϱ_V is the ohm.m ($\Omega \cdot m$), although the ohm.cm ($\Omega \cdot cm$) is used frequently. For more detail, see Reference 8.17. Almost invariably for low-conductivity plastics (where volume resistivity is often as high as $10^{16} \Omega m$), the value is dependent on the time of electrification, with the *apparent volume resistivity* increasing with increasing time of electrification, typically by an order of magnitude between measurement times of 10 – 100 s.

Surface Resistivity (ϱ_S) is defined as the ratio of the electrical field strength to the current density in a surface layer of an insulating material. Practically, the measurement of this quantity is associated with thin samples, and usually presents some difficulties; the state of a polymer surface is often characterized by other methods. Test techniques for surface resistivity are also given in Reference 8.17.

Illustrative electrical properties data, for a number of polymers, are given in Table 8.1.

Table 8.1 Electrical Properties of Some Commercial Plastics

LDPE	PP	PTFE	POM	PES	PPVC	ABS	PA66	
							WET	DRY
Relative Permittivity (1kHz), ε_R								
2.28	2.15	2.05	3.70	3.50	$\simeq 7-8$	2.70	3.3	5.0
Dielectric Dissipation Factor (1kHz), (Tan δ)								
30	1000	$15-20 (\mu rad)$						
			0.002	0.003	$\simeq 0.3$	0.006	0.025	0.2
Volume Resistivity (60s.), $\log \varrho_V$ (ϱ_V in $\Omega \cdot$ cm)								
17.5	17.5	18	14	17	$9-10$	15	15	12

There is an *electrical-mechanical analogy* which is often used to compare these classes of properties in plastics materials. The viscoelastic nature of polymers creates similarities in the material response, both for mechanical and electrical stimuli. For example, under dynamic excitation, the independent and measured variables move out of phase (stress and strain in mechanical tests; voltage and current in electrical tests) and complex notation is used to describe the overall response, as shown in Equation 8.2. Permittivity is therefore a measure of compliance to an electrical potential (load).

Hence in a dynamic electric field, the reorientation of permanent dipoles may not keep pace with the changes in applied load. In this way, we might therefore expect mechanical and dielectric loss peaks to coincide, since similar molecular motions are likely to be involved.

8.2.1 Molecular Origins of Permittivity and Power Factor

The interaction of an electric field with a polymer is concerned with the *electrically active elements* in the structure, comprising electrons, atoms and atom-atom linkages which are unbalanced electrically. An additional loss mechanism associated with interfacial processes is often observed in multiphase systems; additives, such as antioxidants, are frequently the origins of such losses.

- **Electronic polarization** is the slight disturbance of the electrons with respect to the nucleus by the applied field. Since the applied field is weak compared with the field associated with the nucleus, electronic polarization is rarely encountered in a consideration of electrical properties. It is relevant, however, to electromagnetic radiation of very high frequency; we shall meet it in the next chapter as a factor affecting the *refractive index.*

- **Atomic polarization** has its origins in the distortion of the arrangement of atomic nuclei by the applied field. Its magnitude is usually quite small, and its contribution to the dielectric properties of polymers is insignificant. It will not be considered further.

- **Orientational polarization** occurs when the applied electric field interacts with the permanent electric dipoles, which arise from asymmetry of charge in the molecules. For any one species of dipole, $\tan\delta$ will vary with frequency as shown in Figure 8.2; the frequency of the $\tan\delta$ maximum is known as the *relaxation frequency.* The number and amplitude of the individual loss processes depends on the number of dipolar species present, and their possible separate or cooperative motions.

Figure 8.2 illustrates two areas in which dielectric losses for polyacetal are high, for the following reasons:

(1) A conduction process (low frequency, high temperature region);
(2) A dipolar relaxation mechanism (loss maximum at low temperature).

POM is therefore considered to be a relatively low electrical loss plastics material, and it is convenient that power factor assumes a low, and relatively consistent level between -50 and $+50$ °C.

- **Interfacial polarization.** Effects on dielectric properties which are associated with material discontinuities are usually termed *Maxwell-Wagner Effects,* after the first workers to examine them theoretically[8.18]. Such processes usually occur at frequencies less than 1 MHz, and arise when one of the components, usually the polymer, behaves essentially as a dielectric, and the other as a conductor. Loss results from charge redistribution with the periodic reversal of the alternating field; the relaxation frequency is determined by the properties of the components, and by their physical distribution.

8.2.2 Permittivity and Power Factor of Commercial Polymers

The simplest materials are the non-polar polymers, exemplified by PE, for which the Clausius-Mosotti relationship for electrical permittivity is:

$$\frac{\varepsilon_R - 1}{\varepsilon_R + 2} = K\varrho \tag{8.5}$$

K is a material constant, and ϱ is density.

Further, the experimental value of K (0.327, for PE) lies in close agreement with the theoretical value of 0.326. The permittivity datum for polyethylene is 2.276 at a density of 920 kg/m^3. This is very close to the square of the *refractive index* (a property also measured as a response to electromagnetic waves), for the same density; this relationship was first pointed out by Maxwell. Applying it to Equation 8.3 derives a quantity called the *molar refraction* (R_M):

$$R_M = \frac{n^2 - 1}{n^2 + 2} \cdot \frac{M}{\varrho} = \frac{N_A \alpha}{3\varepsilon_0} \tag{8.6}$$

Equation 8.6 is known as the Lorenz-Lorentz relationship; in it, M is the molecular weight (strictly of the repeat unit) and ϱ is the density. N_A is Avogadro's number, ε_0 is another constant (the permittivity of free space) and α is the polarisability.

This method provides a route for calculating the *molecular polarisability* from a macroscopic quantity which can readily be measured to high accuracy. If the polarisability is indeed measured, via the refractive index referred to visible light, then its origin is solely electronic.

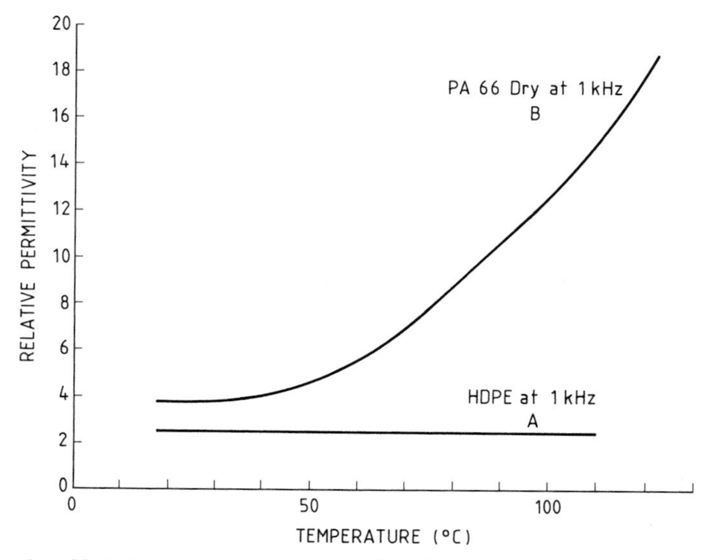

Figure 8.4 Variation of relative permittivity with temperature for non-polar (HDPE) and polar (PA66) polymers.

Additivity is a useful characteristic of molar refraction, so that to a first approximation at least, the molar refraction of a molecule can be predicted by summing the contributions of its various parts. For a non-polar polymer, therefore, it is possible to calculate the permittivity from a knowledge of the refractive index, ($\varepsilon_R = n^2$) or from first principles via the molar refraction, calculated from a table of bond refractions.

It should be noted that the method fails when the molecule contains delocalized electrons, such as conjugated double bonds. The permittivity of non-polar polymers is relatively independent of exciting frequency, but that of polar polymers decreases with increasing frequency and increases with increasing temperature, (Figure 8.4). With decreasing frequency or increasing temperature, ε_R increases as each succeeding polarization mechanism adds its contribution; ε_R reaches its highest value at zero frequency[8.19].

The *dielectric thermal analyser, DETA,* facilitates the measurement of Tan δ as a function of temperature, and is available commercially,[8.20]. The equipment operates in the frequency range 20 Hz to 100 kHz.

8.2.3 Volume Resistivity

The usual experimental arrangement for measuring resistivity is to determine the current which results from the application of a potential difference across the specimen. Current response versus time is obtained, and the data are transformed into resistivity versus time curves (Figure 8.5), which show marked time dependence.

Very few plastics show true (ohmic) resistance at room temperature; this is more likely to be encountered at higher temperatures, especially in the presence of ionic impurities or additives. The results in Figure 8.5 show true conductivity at 88 °C. By contrast, the apparent resistivity of PTFE over the time of electrification reaches very high values, with no hint of equilibrium conductivity, Figure 8.6.

The *one-minute volume resistivity,* the ratio of potential to current obtained one minute after electrifying the specimen, is frequently quoted for plastics materials. This is an arbitrarily chosen time, with no special significance, except its convenience for laboratory measurements; the datum is but one point taken from a set of curves such as those given in Figure 8.5. More detail of the measurement of volume resistivity is given in Reference 8.16.

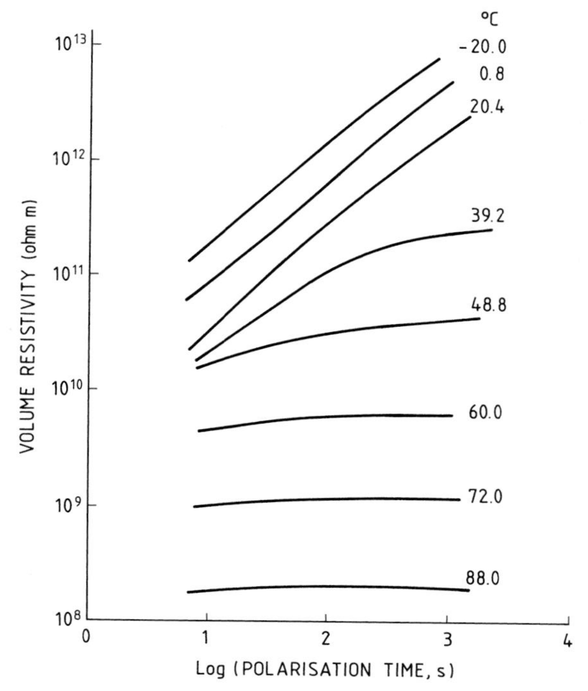

Figure 8.5 Variation of apparent volume resistivity with time of electrification, at various temperatures (°C), for PVC.

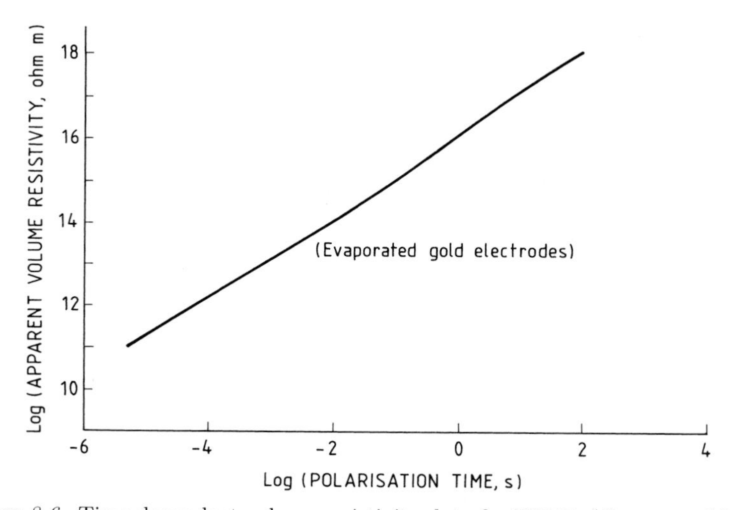

Figure 8.6 Time-dependent volume resistivity data for PTFE. (Courtesy of ICI).

8.2.4 Surface Resistivity

Definition of this term has been given earlier in the chapter, but any such simple definition hides a number of difficulties. Firstly, a surface must be a layer with a finite thickness, so that, inevitably, there will be a measurement component related to volume resistivity. More problems arise in ensuring the authenticity of the surface, since there is often a tendency for *additive migration* to occur in solid plastics (or for surface deposition of "defective" low molecular weight fragments), and in addition, it is difficult to minimize the contact resistance between the electrodes and the specimen. Finally, environmental factors, especially humidity, affect the accuracy and validity of surface resistivity data.

Test methods for surface resistivity are available in the Standards literature[8.17] and in compilations of data.

8.3 Electrical Properties at High Stress: Electric Breakdown

In Section 8.2, the electrical properties examined could be regarded as true materials' properties, albeit that they depended on time (or frequency) and temperature; they are virtually independent of the geometry of the system.

However, as the electrical potential is increased, the geometry of the sample and electrodes gains in importance, and it is only possible to report the performance of particular systems. Other factors affecting the electric strength are given elsewhere[8.21, 8.22], and include:

(1) frequency, waveform and time of application of the voltage;
(2) ambient temperature, pressure and humidity;
(3) gaseous inclusions, moisture or other contamination; and
(4) the electrical and thermal characteristics of the medium.

Thus, whilst engineers can utilize the low stress properties directly in design, at high stresses the results of standard tests only serve as an approximate guide to the choice of material and to the design of a component.

Various phenomena may be involved in breakdown. Although, like mechanical failure, it is generally associated with imperfections, it is useful, nevertheless, to conceive a material property, *dielectric strength,* which is defined as the potential at breakdown, divided by the thickness of the insulator. The test voltage is defined as the peak value divided by 1.414; in other words, it is the maximum field strength which the dielectric can withstand without failure. The intrinsic electric strength of many polymers is very high, frequently in excess of 100 MV/m, but is an elusive property; factors which detract from high performance include:

- thermal breakdown – the increase in temperature as current is passed induces higher electrical conductivity, and a runaway breakdown effect is feasible;
- internal, or external gas discharge – electrical conductivity is increased when ionic species (impurities) are present;
- surface deterioration – the presence of moisture and other impurities leads to preferential conduction effects, and may result in surface damage known as tracking;
- electronic breakdown – this is created by a small number of free electronic carriers in the bulk plastic material;
- electromechanical failure – loss of intrinsic strength occurs at elevated temperature, ie. close to primary transitions, as with mechanical failure.

Some of these modes of electrical failure are discussed in more detail in the following sections.

8.3.1 Thermal Breakdown

An alternating field applied across a dielectric will generate heat by viscous relaxation processes, the amount of heat will depend on the dissipation factor. With a constant potential, there may be sufficient conductivity for the *ohmic (resistance) heating* to be significant. This latter contribution increases with increasing temperature as the conductivity increases, so that the possibility of thermal runaway exists. The contribution due to relaxation processes may decrease with increasing temperature, thereby helping to stabilize the system. According to the heat balance equation quoted by BLYTHE[8.21], the power dissipated per unit volume is equal to the rate of increase in enthalpy, plus the rate of heat loss.

Approximate numerical solutions have been obtained for this equation, the main feature being identification of a *critical voltage,* above which the temperature increases indefinitely. Practically, this may terminate useful life by softening/melting, or by chemical degradation. Thermal breakdown is most likely to be encountered at elevated service temperatures, where conduction may be important, and also at high frequencies, where the dissipation of electrical energy into heat (via *dipolar relaxation* mechanisms) can be large. Once again, we can refer to an analogous mechanical failure characteristic, in this case thermal fatigue failure.

8.3.2 Gas Discharges

The breakdown voltage of air, approximately 3 MV/m, is very much lower than that of most polymers; thus, on application of a high potential to a solid polymer, discharges will occur at any gas inclusions in the polymer, including voids. These latter are known as *internal discharges;* when these are formed, the high temperature generated within the discharge space can readily induce local degradation in the surrounding polymer.

To achieve long life in a high voltage insulator, gas discharges must not occur under the working conditions. The potential at which discharges start is an important characteristic for polymers to be used as high voltage insulators; it is known as the *discharge inception voltage, (DIV).*

The design of high voltage insulation systems which do not fail by discharges at the edges of the electrodes is important. In practice, the cause of many failures is deterioration in the surface quality. Contaminants, such as salt and moisture, allow increased conduction over the surface; this may lead to *tracking,* the appearance of conducting paths over the surface, due to the combined effects of *electric stress* and *electrolytic contamination.* Standard tests[8.22, 8.23] reproduce conditions which favour this type of high voltage failure; for example, a *comparative tracking index* indicates a plastics material's resistance to tracking under electrical load, in terms of a maximum sustainable voltage.

8.3.3 Electrical Failure of Plastics

The formation of conductive tracks appears to be easiest with polymers containing benzene rings, although PVC is another polymer which performs poorly under these conditions. Nylon resins generally possess an excellent range of electrical properties[8.24], though it should be noted that moisture absorption in polyamides can significantly reduce dielectric strength; however, the equivalent characteristics of PBTP are not affected by moisture ingress to the same extent.

The tracking resistance of plastics may be improved by suppressing surface carbonisation in certain mineral-filled formulations. For example, aluminium trihydrate (ATH) undergoes an endothermic transition at elevated temperature, thereby cooling the plastics matrix. However, additions of filler may also modify or reduce other electrical properties (volume resistivity; dielectric strength) due to their ionic nature and the presence of absorbed water in the bulk material.

The dielectric strengths of some engineering thermoplastics materials are given in Table 8.2. It should be appreciated that these data are dependent upon specimen thickness (with resistance to breakdown increasing for thinner samples), temperature and humidity. Reference should always be given to the precise detail associated with such standard test methods.

Table 8.2 Dielectric Strength Data for Thermoplastics

Material	Method	Thickness (mm)	Dielectric Strength (MV/m)
PA6	IEC 243	3.0	10 – 14
PA66	IEC 243	3.0	10 – 14
PA66*	IEC 243	3.0	8 – 11
PETP*	IEC 243	–	11 – 12
POM	ASTM D149	0.13	82.8
POM	ASTM D149	2.3	19.7
POM	DIN 53481	0.2	60 – 70
PC	ASTM D149	3.2	15 – 20
PC/ABS	DIN 53481	2.3	24
ABS	DIN 53481	1.0	24 – 32
PPS	ASTM D149	3.2	14 – 22
PTFE	DIN 53481	0.125	80
PTFE	DIN 53481	0.3	40

* Data for short glass-fibre reinforced grades

In addition to the methods quoted above, IEC 112[8.23] covers measurements up to 600V on materials with relatively poor tracking resistance. High voltage-low current discharges are covered in IEC 587: *Test method for evaluating resistance to tracking and erosion of electrical moulding material used under severe ambient conditions.*

An extensive review of dielectric breakdown mechanisms in plastics has been compiled by Ku and Liepins[8.18]. Typical electrical failure behaviour is described in terms of the effects of chemical nature, molecular weight, microstructure, morphology and additives such as plasticizers and fillers. Previous research is comprehensively reviewed by these authors, and will not be repeated here.

8.4 Static Charge

8.4.1 Origin and Effects of Electrostatic Charge

A region of high electrical charge in any material will decay at a rate determined by the permittivity and the volume resistivity. For metals, resistivity is low, and charge decays very rapidly but for polymers, which generally have very high resistivities, the charges may be retained over very long timescales. Charge may reside anywhere in a material, but the mechanism of acquiring charge dictates that it is most commonly encountered on *surfaces,* where it arises by contact between dissimilar materials, and is enhanced by friction, (rubbing). Charging may also be effected by *ion bombardment* of a surface, such as in the *Corona Discharge* treatment of PE films or PETP bottles, to improve their printing performance or coatability.

Electrical breakdown of the air frequently limits the charge density which can be accumulated; since this occurs at about 3 MV/m, it is practically difficult to achieve better than 1 % of the polymer's insulation capability if large air gaps are present.

The presence of electrostatic charge is recognized by mutual attraction or repulsion with other surfaces, possibly with small discharges (sparks), and frequently with marked attraction of dust; characteristic dust patterns form according to the static charge distribution on the component surface, an effect which is undesirable in many plastics appliances in domestic and business environments. In packaging, static charge accumulation is one of the most serious problems. The situation is alleviated by including *antistatic agents* ("antistats") in the formulation; sometimes mixed systems give beneficial effects but above all, the polymer surface must be treated to optimise the performance of the system.

The flow behaviour of polymer powders may be affected by electrostatics, either adversely (for mixed charges), or beneficially (for charges of the same sign). The nuisance value of electrostatic charge can be compensated by exploitation of the ability of polymers to retain charge in applications such as electrostatic powder coatings.

8.4.2 Measurement of Charge

Such measurements should be based on techniques which do not depend on the measurement of current; the gold-leaf electroscope was an early instrument in this field, but has been replaced by more robust instruments giving quantitative results. The use of a probe, which is positioned distant from the surface to be measured, is the basis of modern techniques; an A.C. signal is generated by a *field mill,* a rotating earthed vane which chops the signal induced at the probe. The signal is amplified and analyzed.

The decay of charge on a HDPE moulding containing a typical antistatic agent is illustrated in Figures 8.7 (A and B), the former being measured 24 hours after moulding, and the latter six days later. The difference is very significant, and is explained by the additive taking some time to reach the surface, and thereby become

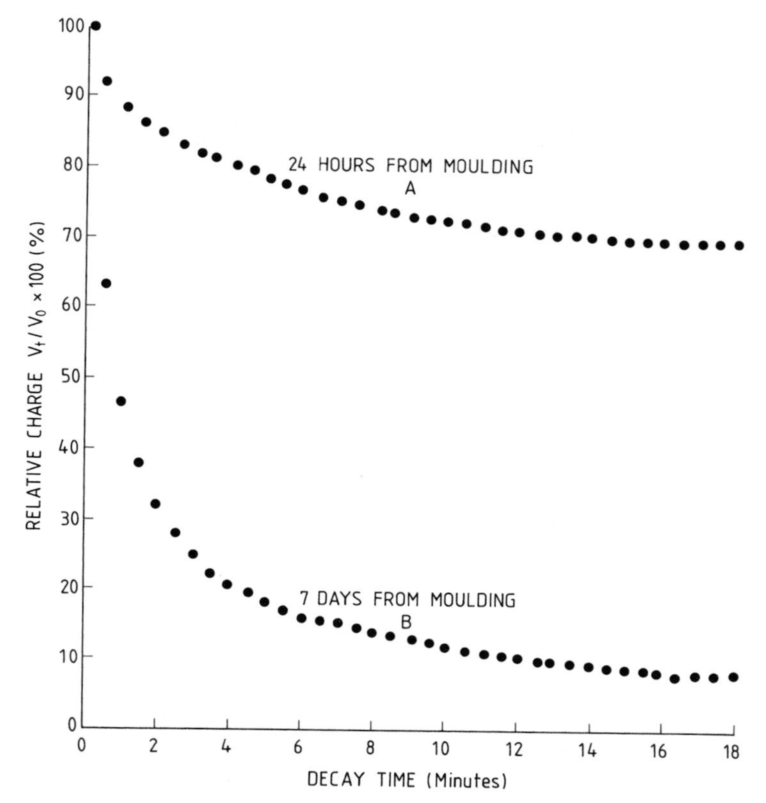

Figure 8.7 Electrical charge decay curves for PE moulding containing antistatic agent: (A) 24 hours after moulding; (B) 6 days later.

effective. A suitable instrument has been developed and is marketed by John Chubb Instruments Limited.

Two factors contribute to the electrostatic behaviour of a polymer surface: the ability to accept charge, and the rate of charge decay. Such interacting phenomena are best accommodated in an *index,* or a *rating.* Such a classification considerably aids the comparison of antistatic additives.

8.5 Electrically Conductive Plastics

Polymers are generally considered to be strong electrical insulators, a fact which is reflected by the diverse range of applications for which plastics are used, in which insulation and low electrical loss are important considerations (wire covering, power cables, electrically-powered component housings). However, the degree of resistance to an electrical potential varies considerably, and a typical range of *volume resistivity* data for commercial plastics materials covers several decades (between 10^8 and 10^{18} Ωm, but typically $10^{10} - 10^{14}\Omega m$), according to the polarity of the molecular species in the repeat units, and the additives present in the compound. When these characteristics are considered alongside the other advantages of plastics products (high specific stiffness, toughness and easy-shaping properties), the reason for increasing demand for high-performance plastics parts in electrical engineering becomes evident.

However, in some circumstances it is desirable to combine the advantages of mouldability and mechanical properties with other properties which may require higher electrical conductivity. These might include some of the following applications[8.25]:

- antistatic products (floor-coverings, conveyor belts and tubes, stackable containers, playground equipment, and in the packaging of sensitive electronic components), in which the accumulation of surface electrical charges must be eliminated, or carefully controlled;
- moulded instrument housings (HIPS, PVC, modified-PPO) and telecommunications equipment, which require shielding to electromagnetic interference (EMI) effects;
- heat sinks (cast thermosetting plastics moulds; thermoplastics bearings and moving parts);
- polymer electrode materials;
- various components for audio-visual applications.

Plastics and hydrocarbon intermediates such as waxes are also seeing increasing use as *flow vehicles* for the mass-production manufacture of conductive, or magnetic powder metallic components, for example by injection moulding. In short, there is tremendous scope for using *electrically-conductive plastics* in applications for which metals, or other inorganic materials have traditionally been used.

In this section, we give a brief review of how these properties may be realized, firstly by the use of electrically-conductive additives and secondly, by advanced polymer synthesis techniques.

8.5.1 Electrical Conductivity afforded through Additives

The concept of easily-shaped, *conductive mouldable thermoplastics* is very attractive to design engineers, for applications as electrically- resistive heat sources, antistatic products and parts requiring EMI shielding. So long as the additives can be effectively dispersed in the dielectric plastics matrix (which generally presents few problems, since the effect of fillers on melt viscosity does not preclude the use of conventional continuous compounding methods), formulation studies may be carried out to obtain the optimum levels of property enhancement, per unit material cost.

The inventory of commercial plastics compounds for such purposes is increasing rapidly, as more effective additives are developed, and as the fundamental relationships between additive constituents, polymer morphology and electrical conductivity are understood. Some of the groups of conductive additives which are being utilized and developed for such purposes are reviewed here.

(A) Carbon fillers. Carbon black is an exceptionally useful additive in polymer compounds, as a result of its reinforcing capability (particularly with elastomers), its effectiveness as a colorant, good processing characteristics, low cost and general availability. Carbon is also a very effective conductor, and the addition of higher quantities of carbon black filler can considerably reduce volume resistivity in plastics, but to an extent which can be manipulated very effectively by the microstructure developed in the fabricated parts.

The volume resistivity data in Figure 8.8 show how the electrical conductivity of filled plastics is improved when a *core-shell morphology* is developed in the product. This may be achieved by processing polymer particles having an electrically-conductive coating; control of mechanical work during subsequent processing is desirable, to maintain a network of conductive pathways through the sample.

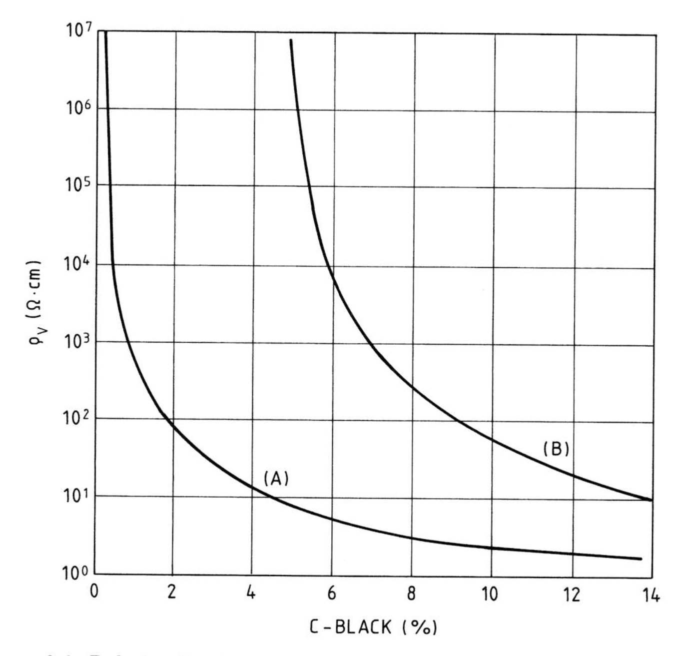

Figure 8.8 Relationship between volume resistivity (ϱ_v) of plastics and weight percentage carbon black content for 'core-shell' morphology (A) and for a uniform filler particle distribution (B). (Data reproduced by permission of BASF[8.25])

Other forms of carbon are also used in plastics including carbon fibres, in situations where exceptional mechanical stiffness may be the most important property specification.

(B) Metallic fillers. The concept of utilizing high-strength, high aspect ratio fibres (glass, carbon) in plastics materials to produce composites offering exceptional mechanical properties is now well-established, and has proved extremely successful. Other physical properties may also be enhanced in a similar manner, by the introduction of heterogeneous species having desirable properties, appropriate to the end-application. An example of significant interest is the family of plastics composites modified by the addition of metallic powders, flakes or fibres, typically aluminium.

Figure 8.9 compares the relative electrically-conducting effectiveness of aluminium-alloy flake, and carbon black[8.26] in filled plastics; the *critical volume fraction* (V_C) of filler required for conductivity is lower, in the case of the metallic system. Additional work by BIGG on aluminium-filled PP has shown that V_C is reduced further by increasing the aspect ratio of the fibres, or plate-like flake reinforcement material[8.27].

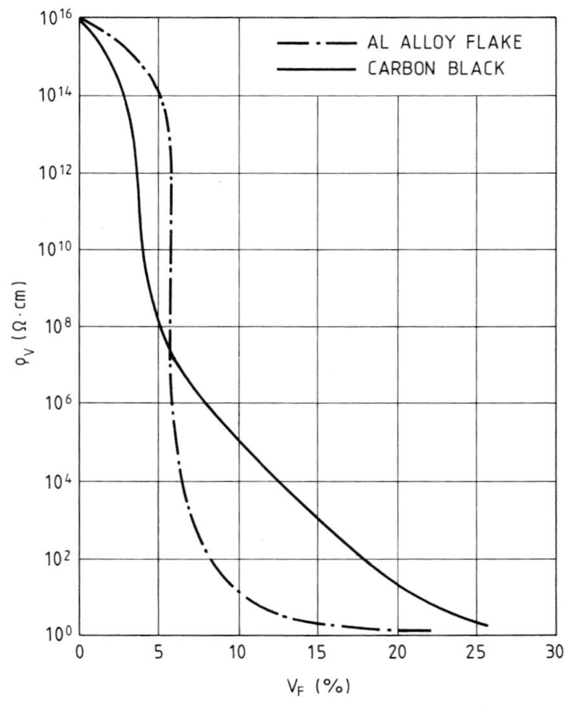

Figure 8.9 Comparison of aluminium alloy flake (Transmet^{TM} K-102) and carbon black, as electrically-conductive additives in plastics. (Data reproduced by permission of Transmet Corporation[8.26].)

EMI shielding characteristics in plastics are also realized by the addition of conductive metallic fillers, with a more effective barrier to electromagnetic waves obtained if part thickness, volume fraction of filler or electrical conductivity are increased. In moulded parts, preferential orientation of the high aspect ratio fillers occurs in the flow field, resulting in anisotropic conductivity, or shielding properties; thus the electrical efficiency of the composite part depends upon processing history[8.25].

8.5.2 Intrinsically Conductive Polymers

Electrical conductivity is realized by the presence of *chain unsaturation* and *electron delocalisation* effects. Much research effort and interest has therefore been devoted towards the development of polymers with intrinsic electrical conduction characteristics, brought about by the presence of conjugated groups, and by doping techniques.

KU and LIEPINS[8.18] have reviewed the previous literature in this field, and describe the fundamental concepts of conduction, measurement techniques, and the principles of conductive-polymer synthesis. They define *electrically-conductive polymers* as those whose conductivity exceeds 0.01 $(\Omega m)^{-1}$, and make the following distinctions in synthesis techniques used to prepare such *synmetals:*

- Pyrolysis – (PAN fibres)
- Ziegler-Natta catalysis – (Polyacetylenes)
- Electrochemical synthesis – (Polypyrrole film)
- Condensation polymerisation – (Polyphenylenes; polyquinolines).

Doping can be used to modify the conductivity characteristics further, especially in polyacetylenes, and by electrochemical synthesis techniques.

For more information on conductive polymers, the reader is referred to dedicated texts and review articles[8.18, 8.28, 8.29]; some current commercial applications are also described in technical company literature[eg. 8.25].

8.6 Electrical Applications of Plastics

As we have seen earlier in this Chapter, plastics have generally good electrical properties; indeed, some are so outstanding that they are essential components in critical applications. Thus, submarine telecommunications cables spanning the Atlantic and Pacific oceans would not be possible without the high quality PE used as insulant. Similarly, high voltage (power factor correction) capacitors, which are essential for the electricity supply grid, have been reduced in size and cost by being based on oriented PP film. However, the contribution of plastics in these prestigious applications has been reported elsewhere. We shall be concerned here with less exciting applications, where qualities in addition to the electrical properties are the reasons for their use; these include processing characteristics (such that small and intricate shapes can be manufactured economically), mechanical properties (resistance to impact loads, stiffness at elevated temperature), dimensional stability and retention of electrical (and other) properties in hot and/or moist environments, over long periods of time.

8.6.1 Switches, Microswitches and Connectors

Dimensional stability, high strength and adequate electrical insulation and tracking resistance are the qualities required in these applications, together with the ability to be shaped easily (usually by injection moulding) into single components of complex shape; glass-reinforced polyamide 66 satisfies these demands, and offers the added advantage that the terminals of small components can be soldered without damaging the casing. This illustrates that, although PA 66 is not regarded primarily as a plastic for the electrical industry, the combination of properties makes it a first choice candidate for these applications. Another candidate whose properties are less affected by moisture is saturated (thermoplastic) polyester; PBTP is preferred, and is again capable of glass reinforcement.

Amongst some relatively new developments in thermoplastics polymer alloys, PC/ABS blends are chosen for connectors and switch casings where high impact strength, excellent surface quality and low mould shrinkage are other important requirements which are fulfilled by this amorphous material. Since resistivity and dielectric strength are relatively unaffected by the influence of external moisture, this material is suitable for waterproof switches[8.30].

A particular application in which a very accurate level of electrical conductivity is required is an injection moulded chip carrier: insulation is necessary to prevent short circuits between contacts (during assembly and in use), but equally, any static electrical discharge must occur relatively quickly[8.25].

8.6.2 Insulant for Domestic Wiring

Again, following the theme that plastics offering the best electrical properties are not necessarily chosen for a wide range of applications, the use of plasticized PVC as the electrical insulator for domestic light and power cables is firmly established, in spite of the indifferent electrical properties; ie. electrical conductivity will increase with increasing plasticizer content. Note that the usual ester plasticizers also change the non-flammable character of PVC, allowing the material to burn, the plasticizer being the flammable component.

In this application, the electrical properties are adequate, but the most important selection criteria are the easy processing by extrusion (wire-coating), attractive raw material price, and the ability to tailor the material formulation to generate specific types of properties.

8.6.3 Insulant in British Rail Signal System

To enable the position of a train to be established, the monitoring circuit depends on isolating the rails from the concrete sleepers. PA 66 pads between rail and sleeper provide sufficient electrical insulation and mechanical stiffness, whilst tolerating the impact forces generated by the passage of a high speed train.

8.6.4 Some Applications for Polyethersulphone (PES)

Any variations in the electrical properties of PES up to temperatures as high as 200 °C are minimal. Other qualities offered by this polymer, in addition to adequate electrical insulation, include stiffness and exceptional creep resistance at elevated temperature, and resistance to degradation[8.31]. Applications based on these and other properties include:

- a heating element holder in a hand drier endures a working temperature of 100 °C, with fault peaking to 200 °C;
- a moulded circuit board in PES is likely to be cheaper and easier to manufacture (by injection moulding) than epoxy/glass, has a higher temperature rating, is inherently fire retardant and maintains low D.C. loss under harsh conditions: furthermore, the board can be flexible and thermoforming is also possible.

8.6.5 Polyimide Films in Microelectronics

Polymer dielectrics are rapidly gaining importance as substitutes for inorganic (ceramic-based) insulating layers between metallic conductors in microelectronic systems. These applications are dominated by polyimide films[8.32], which offer the following characteristics:

- high electrical resistivity, breakdown strength and low dielectric loss;
- stability to temperatures as high as 450 °C;
- coating of substrates with thin films is feasible, and patterning can be achieved in lithography;
- topographic features can be transformed into a single plane by polyimide films;
- polyimide has excellent chemical resistance.

Polymers are finding additional applications in microelectronics, such as microresists, as injection moulded components for chip encapsulation and as printed circuit boards. These rapidly developing, and exciting applications for plastics in the electronics industry have been thoroughly reviewed by SAONE and MARTYNENKO[8.33].

References

8.1 HYDE, P.J.: *Wide-frequency Range Dielectric Spectrometer,* Proc. I.E.E., **117,** (1970), 1891.

8.2 STOUT, M.B.: *Basic Electrical Measurements,* 2nd Ed., Prentice Hall, Englewood Cliffs NJ, USA, (1960).

8.3 FIELD, R.F.: *Dielectric Measuring Techniques: Permittivity,* in *Dielectric Materials and Applications,* Von HIPPEL, A.R.,Ed., MIT Press, New York, (1954).

8.4 LYNCH, A.C.: *A Bridge Network for the Precise Measurement of Direct Capacity,* Proc. I.E.E. Part B, **104,** (1957), 363.

8.5 LYNCH, A.C.: *Measurement of the Dielectric Properties of Low Loss Materials,* Proc. I.E.E., **112,** (1965), 426.

8.6 BS 4542, (1970), *Determination of the Loss Tangent and Permittivity of Electrically Insulating Materials in Sheet Form,* British Standards Institution, London.

8.7 LYNCH, A.C. and AYERS, S.: *Measurement of Small Dielectric Loss at Microwave Frequencies,* Proc. I.E.E., **119,** (1972), 767.

8.8 ROBERTS, S. and VON HIPPEL, A.R.: *A New Method for Measuring Dielectric Constant and Loss in the Region of Centimetre Waves,* J. Appl. Phys., **17,** (1946), 610.

8.9 HARTSHORN, L. and WARD, W.H.: *The Measurement of Permittivity and Power at Frequencies from 10^4 to 10^8,* J.I.E.E., **79,** (1936), 567.

8.10 BARRIE, I.T.: *Measurement of Very Low Dielectric Losses at Radio Frequencies,* Proc. I.E.E., **112,** (1965), 408.

8.11 REDDISH, W., BISHOP, A., BUCKINGHAM, K.A. and HYDE, P.J.: *Precise Measurement of Dielectric Properties at Radio Frequencies,* Proc. I.E.E., **118,** (1971), 255.

8.12 HILL, G.J.: *Dielectric Materials, Measurements and Applications,* I.E.E. Conference Publicaton No. 129., I.E.E., London, (1975).

8.13 PARRY, J.V.L.: *The Measurement of Permittivity and Power Factor of Dielectrics at Frequencies from 300 MHz to 600 MHz,* Proc. I.E.E., Part III, **98,** (1951), 303.

8.14 MORENO, T.: *Microwave Transmission Design Data,* Dover Publications, New York, (1958).

8.15 BS 4618 (1970), *Recommendations for the Presentation of Plastics Design Data: Permittivity and Loss Tangent,* Parts 2.1 and 2.2, British Standards Institution, London.

8.16 I.E.C. 250 (1969), *Recommended Methods for the Determination of the Permittivity and the Dielectric Dissipation Factor of Electrical Insulating Materials at Power, Audio and Radio Frequencies, including Metre Wavelengths,* Bureau Central de la Commission Electrotechnique Internationale, Geneva.

8.17 I.E.C. 93 (1980), *Methods of Test for Volume Resistivity and Surface Resistivity of Solid Electrical Insulating Materials,* (Address as above).

8.18 KU, C.C. and LIEPINS, R.: *Electrical Properties of Polymers; Chemical Principles,* Hanser, Munich, (1987).

8.19 REDDISH, W.: *Dielectric Study of the Transition Temperature Regions for Polyvinyl Chloride and Chlorinated Polyvinyl Chloride,* J. Polym. Sci: Polym. Symp., **C14,** (1966), 123.

8.20 Polymer Laboratories Technical Literature: *PL DETA – Dielectric Thermal Analyser,* Polymer Laboratories Ltd., Loughborough, U.K.

8.21 BLYTHE, A.R.: *Electrical Properties of Polymers,* Cambridge Solid State Science Series, Cambridge University Press, (1979).

8.22 I.E.C. 243 (1967), *Recommended Methods of Test for Electric Strength of Solid Insulating Materials at Power Frequencies,* (Address as in Reference 8.16).

8.23 I.E.C. 112 (1979), *Method for Determining the Comparative and the Proof Tracking Indices of Solid Insulating Materials under Moist Conditions,* (Address as in Reference 8.16).

8.24 BASF Technical Literature: *Ultramid® Nylon Resins (PA) – Product line, Properties, Processing,* BASF Aktiengesellschaft, Ludwigshafen, West Germany, (1984).

8.25 BASF Technical Literature: *II – Electrically Conductive Plastics;* in *BASF Plastics, Research and Development,* BASF Aktiengesellschaft, Ludwigshafen, West Germany (1986).

8.26 Transmet Corporation Technical Literature: *Transmet^{TM}, for Conductive Plastics,* Transmet Corporation, Columbus, Ohio, USA (1980).

8.27 BIGG, D.M.: *Mechanical and Conductive Properties of Metal Fibre-Filled Polymer Composites,* Composites, **10,** (1979), 95.

8.28 DUKE, C.B. and GIBSON, H.W.: *Conductive Polymers,* in *Encyclopaedia of Chemical Technology,* Grayson M, Ed., John Wiley, New York, (1982).

8.29 BIGG, D.M.: *Conductive Polymeric Compositions,* Polym. Eng. Sci., **17,** (1977), 892.

8.30 Bayer Technical Literature: *Bayblend T Thermoplastic Blend,* Bayer AG, KL Division, Leverkusen, West Germany, (1979).

8.31 ICI Technical Literature: *Victrex PES Data for Design, Unreinforced Grades,* Technical Service Note VX101, ICI Advanced Materials, Wilton, UK, (1984).

8.32 MITTAL, K.L. Ed., *Polyimides, Synthesis, Characterization and Applications,* Vols I and II, Plenum, New York, (1984).

8.33 SAONE, D.S. and MARTYNENKO, Z.: *Polymers in Microelectronics; Fundamentals and Applications,* Elsevier, Amsterdam, (1989).

Chapter 9
Optical Properties

Many diverse phenomena are discussed in this Chapter, sharing the common characteristic of being responses to *electromagnetic radiation* within or close to the wavelength range 400 – 800 nm, that is, visible light. This is not to deny that important effects result from radiation outside this band: *ultra violet (UV)* radiation of shorter wavelength has a considerable degradative effect on many polymers, whilst the *infrared (IR)* region is particularly fruitful for information concerning the structure of polymers. The former effect will be considered in Chapter 10, and both UV and IR spectroscopic techniques will be commented upon later in this section. The scope of the present discussion includes the following, and generally conforms to the appropriate British Standard[9.1] recommendation:

(1) Reflection

- reflective losses at a change of refractive index;
- direct reflection factor versus angle of incidence;
- variation of gloss with angle of incidence and solid angle of receptor; and
- total reflection factor.

(2) Refraction

- variation of refraction with wavelength (dispersion curve);
- total internal reflection (light pipes and fibre optics);
- refractive index versus temperature;
- birefringence: orientation and stress-optical effects.

(3) Light scattering

- transparency – origins of loss of clarity (resolution), and of loss of contrast;
- specular transmittance, (direct transmission factor);
- turbidity (scattering coefficient) and absorption coefficient;
- attenuation (extinction coefficient);
- forward scattered fraction (haze);
- surface scattering (and oiling out of surfaces);
- clarity measurements by loss in resolution and by instrumental methods.

(4) Light transfer

- total transmission factor (total transmittance) and hiding power;
- pigmentation and light transmission.

(5) Absorption of light

- selective absorption (colour) and colour measurement;
- problems associated with colourants;
- autofluorescence effects in plastics.

These properties are relevant to the application of plastics in many fields, including:

(a) optical instruments (camera lenses, light pipes etc.);
(b) bottles and containers (especially for potable liquids);
(c) packaging films and sheet;
(d) lamp (lantern) covers; and
(e) coloured products.

• SPECTROSCOPIC TECHNIQUES

Additionally, since the interaction of radiation with the structure of the polymer results in a diversity of responses, these can be analyzed to gain information concerning the structure of the material. This facility has largely been ignored, with the exception of one particular area: the use of *infrared spectrophotometry* to determine chain structures. Somewhat less use has been made of *ultraviolet spectroscopy*.

(A) Infra-Red Spectroscopy

Stretching, deformation and vibration of bonds in organic molecules give rise to specific absorptions of infrared radiation, the absorption being proportional to the number of active groups[9.2]. The wavelength range covered is 2.5 – 50 μm, or as usually expressed, in frequency units 5000 – 200 cm^{-1} (ie, pulses per cm, Kaysers). Radiation absorption occurs at specific frequencies for particular bonds, so that a polymer with a variety of bonds has a multiplicity of absorptions, which represents a "fingerprint" for the material. This is illustrated by reference to high density polyethylene (HDPE), which has a very simple infrared spectrum (Figure 9.1). Since this polymer is almost entirely composed of methylene groups, the absorption

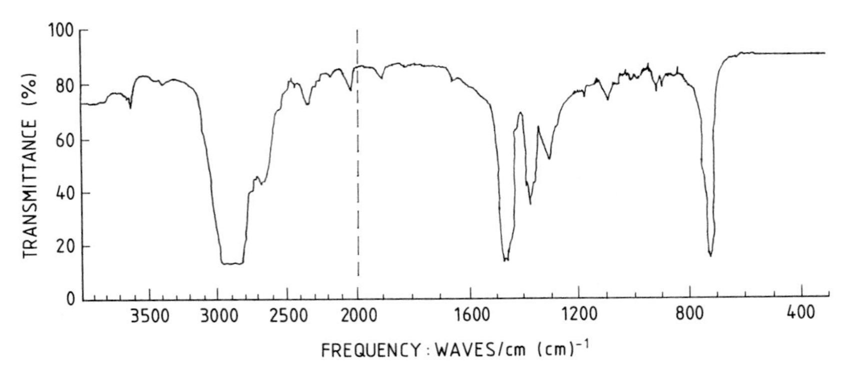

Figure 9.1 Infra-red spectrum of HDPE.

peaks can be identified with the deformation of that group. Note that there is a "doublet" at 720 – 740 cm^{-1}; the two peaks are ascribed to motion in the crystalline and noncrystalline regions of the polymer.

IR spectroscopy is a powerful and accurate technique and has been used widely in the characterization of polymers for some considerable time. The text by HASLAM et al.[9.3] is a notable contribution in this field, as a result of its emphasis on the application of IR techniques to the identification and analysis of plastics. It should also be noted that since many plastics absorb strongly in this range, IR absorption is widely used for heating plastics, especially in heat-softening processes such as thermoforming and stretch blow moulding (see also Chapter 5).

(B) Raman Spectroscopy

This technique is another example of vibrational spectroscopy in polymer technology; in this case, the resultant transitions between energy levels which result from the vibration of primary bonds are measured in response to a concentrated beam of monochromatic radiation in the visible region. When the radiation interacts with the polymer sample, a small fraction is scattered with a changed wavelength, a phenomenon known as *Raman scattering.* It is the analysis of the intensity of peaks at which a change of wavelength occurs (the "Raman shift"), which gives rise to the characterization principle in this technique. Since the degree of Raman scattering is rather low, high intensity sources of visible radiation are generally required, and the sample should preferably be optically clear; lasers have therefore been used for this purpose.

For further information on these vibrational spectroscopic techniques and their application to polymer characterization, the reader is referred to other, more specialized texts[9.4, 9.5].

(C) Ultra-Violet Spectroscopy

The preferential absorption of UV or visible light can generally be related to specific types of primary chemical bonds in a polymer system. For example, π-electrons present as part of chain unsaturation will lead to absorption of energy in the visible range; this is the reason why degraded PVC changes colour (from pale yellow, progressively to orange, brown and ultimately, black) as the degree of chain unsaturation increases. In Figure 9.2, UV spectra are shown for samples of PVC which have been welded at different temperatures; the greater absorption, and the shift to higher wavelengths for specimens welded at 250 °C is directly attributed to the increased concentration and length of conjugated (double-bond) sequences in PVC, created by thermal degradation[9.6, 9.7]. The sensitivity of this technique is emphasized by the fact that there is no visible sign of surface discoloration on the products.

Although some useful information on specific chemical groups may be gained from UV spectroscopy, the technique is not used as extensively in polymer characterization as (say) infrared, because a complete characterization of chemical species cannot be achieved. However, ultraviolet illumination finds application in *fluorescence microscopy;* for example, PVC degradation causes the polymer to fluoresce when irradiated with UV radiation (fluorescence is the remission of radiation at lower frequency; see Section 9.6). Additionally, many antioxidants fluoresce strongly, allowing their location within the polymer to be determined.

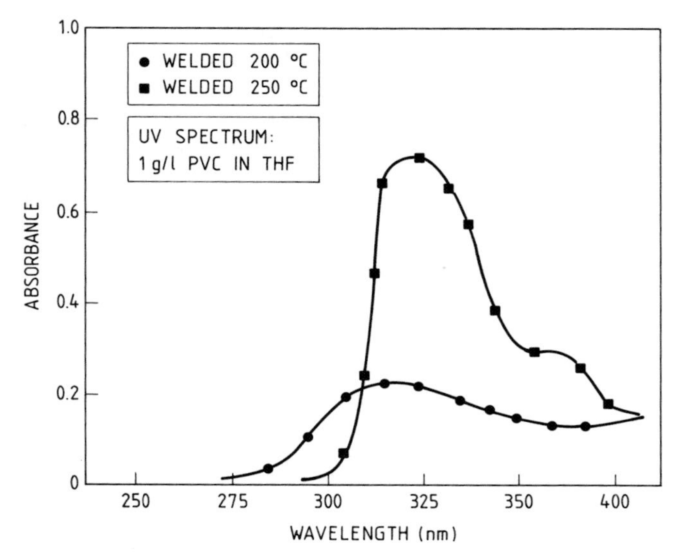

Figure 9.2 UV spectra for rigid PVC samples hot-plate welded at different temperatures.

9.1 Reflection

Simple reflection at a surface follows the laws of geometric optics, ie. the angle of reflection equals the angle of incidence. The quantity of radiation reflected depends on the refractive index difference for the materials which constitute the surface, on the quality of the surface (with respect to flatness and to damage from scratches etc.) and on the angle of incidence. The lowest reflective loss ($\triangle$ R) is for normal incidence, being:

$$\triangle R = \frac{(n_1 - n_2)^2}{(n_1 + n_2)^2} \tag{9.1}$$

n_1 and n_2 are the refractive indices of the two media; for air and PMMA, the refractive indices are 1.00 and 1.50 (approximately), leading to reflection of some 4 % of the incident normal radiation at both the air-PMMA and the PMMA-air interfaces. The reflection is of a similar magnitude, independent of which refractive index is the larger.

The *direct reflection factor (R)* is defined by the ratio of light flux reflected geometrically, ϕ_R, to the incident light flux, ϕ. A curve showing the relationship between R and the angle of incidence is the preferred presentation and is shown for PMMA in Figure 9.3.

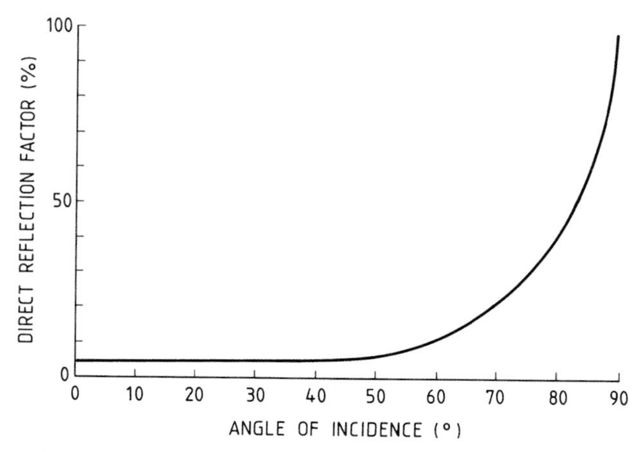

Figure 9.3 Direct reflection factor versus angle of incidence, for PMMA sheet at 20 °C. (Reproduced from Perspex handbook, by courtesy of ICI).

The reflection of light at a surface, which defines the quality of that surface, depends not only on the light reflected geometrically, but extends to an appreciable solid angle around the reflection direction. This visual impression of a surface is monitored by the measurement of *gloss*[9.8, 9.9] (Figure 9.4), the numerical value of which depends on the design of the glossmeter, especially the solid angle at the receptor, and on the angle of incidence, which should be stated. Standards specify the angles to be used (eg. 20°, 45° or 60° and 80°).

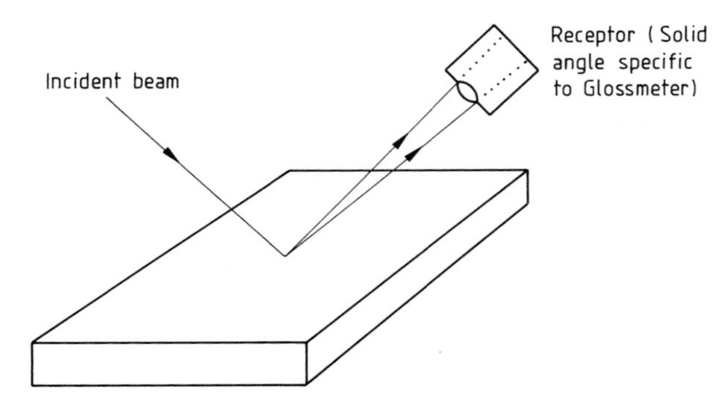

Figure 9.4 Principle for the measurement of gloss in sheet plastics.

Simple surface reflection (gloss) is adversely affected by:

- flow defects in extrusion;
- phenomena such as surface crystallization;
- the surfaces of plastics mouldings replicating an imperfect finish on a mould cavity;
- scratches and erosion incurred in service;
- the presence of fillers, especially those of large size, including fibres.

Structural foam components are often associated with a characteristic surface swirl pattern, which is formed by the collapse of small bubbles as the foaming melt is cooled, ie. upon contact with an injection mould, or a sizing die in extrusion. The quality of the surface is determined by the degree and overall pattern of reflected light, and therefore depends critically on the foam processing conditions.

The *total reflection factor* (R_T) is defined as the ratio of the total energy reflected or scattered backwards ϕ_{sc}^b, to the energy of the incident radiation (ϕ):

$$R_T = \frac{\phi_{sc}^b}{\phi} \tag{9.2}$$

The surface reflection of monochromatic light can sometimes be used as a measurement principle for other properties. For example, WHITE and co-workers[9.10, 9.11] have developed a laser reflection technique to measure the curvature of moulded plastics specimens in order to characterize through-thickness profiles of residual stress by the layer-removal method.

9.2 Refraction

9.2.1 Direct Refraction

Refraction, or the change in direction of a light beam as it passes from one medium to another, is caused by a change in velocity of the light, and consequently its wavelength, since the frequency remains unchanged. The *refractive index* (n) is defined for a specified wavelength in vacuum, (or a specific frequency), as the ratio of the velocity in vacuum (V_V) to the velocity (V) in the material:

$$n = \frac{V_V}{V} \tag{9.3}$$

Refractive indices are frequently quoted against air as standard, since the velocities in vacuum and air are very similar; the quantity is measured as the ratio of the sin of the angle of incidence to the sin of the angle of refraction.

The *dispersion curve* gives the variation of refractive index with the wavelength of the light; an example of these data is given for PMMA, in Figure 9.5.

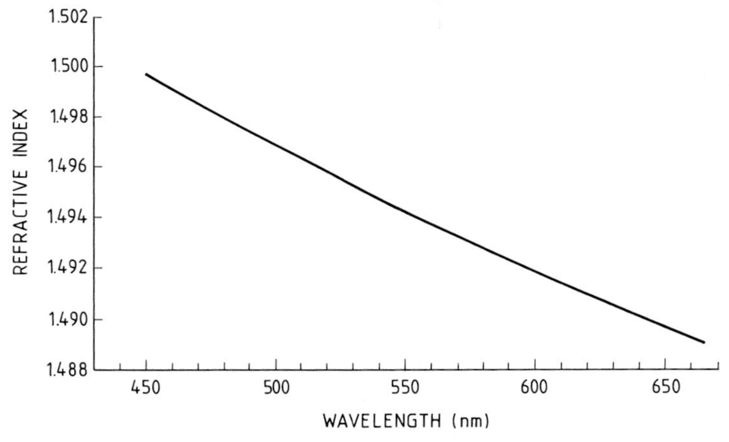

Figure 9.5 Dispersion curve of refractive index versus wavelength of electromagnetic radiation, for PMMA sheet at 20 °C. (Reproduced from Perspex handbook, by courtesy of ICI).

An important application for transparent plastics (and glasses) is in light pipes and optical fibres. Light inside the "pipe", a solid cylinder of material, is totally internally reflected if it strikes the surface at an angle of incidence greater than $\mathrm{Sin}^{-1}(1/n)$ (where n is the refractive index), assuming the other material is air. For smooth surfaces, there is little loss of light at the repeated reflections and, for PMMA in particular, little absorption, so that light can be "piped" over short distances with little loss in intensity. *Optical fibres,* bundles of small diameter light pipes, depend on similar principles, but these are flexible, and are currently undergoing increasing application in communications.

The variation of refractive index with temperature, for a specific wavelength, is usually presented as a curve (see Figure 9.6). The decreasing refractive index with increasing temperature is of similar magnitude to the decrease in density. The slope of the curve in Figure 9.6, the *temperature coefficient of refractive index,* is sometimes quoted to express the temperature dependence of this optical characteristic.

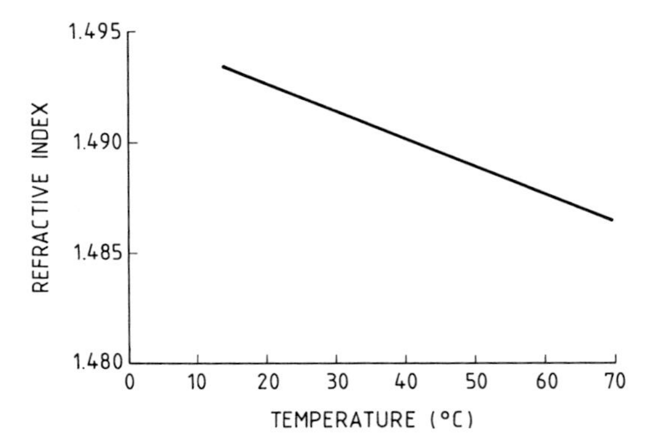

Figure 9.6 Variation of refractive index with temperature: PMMA sheet illuminated at 587.3 nm (Reproduced by courtesy of ICI).

9.2.2 Birefringence

9.2.2.1 Definition and Measurement

The phenomenon of *double refraction* in plastics materials, or *birefringence,* is a consequence of *optical anisotropy* and arises due to a polymer chain having different refractive indices along and across the main chain axis; such a chain is therefore said to be optically anisotropic. For chains distributed randomly, there is no resultant anisotropy, but if they are oriented preferentially in one direction (uniaxial), or in a plane (planar), birefringence may be observed. A ray of polarized light incident on the material gives rise to two light components, vibrating in planes which are mutually perpendicular. The velocities of these correspond to the refractive indices, which are conventionally referred to the three principal directions in the material. *Birefringence* is defined as the maximum algebraic difference between refractive indices measured in two orthogonal directions (x and y, say), and provides a convenient way of monitoring molecular orientation, or a state of stress in fabricated plastics parts. Hence:

$$\triangle n_{XY} = n_X - n_Y \tag{9.4}$$

Birefringence is usually measured directly using polarized electromagnetic radiation; two measurement techniques are summarized here.

- **Polarized Light**

A birefringent sample is placed between crossed polars, at 45° to the polarization axes, and the difference in velocity between the "fast" and "slow" waves (light wave components vibrating parallel to, and perpendicular to the main-chain axis in the

x-y plane) gives rise to a *constructive interference* effect when the light emerges from the specimen (Figure 9.7). The *optical path difference* (or *relative retardation, D)* can be measured by its characteristic colour; if the incident light is polychromatic, the retardation can be identified on Newton's scale using a *Michel-Levy Chart,* or by an optical compensator[9.5] (ie. a device with precalibrated and adjustable birefringence characteristics). If filters are used and monochromatic light is incident on the sample, the relative retardation is identified by measuring the order of dark interference fringes (see below); since retardation, birefringence and applied stress are usually all in direct proportion, the use of monochromatic light is favoured in *photoelastic stress analysis* (Section 9.2.2.3).

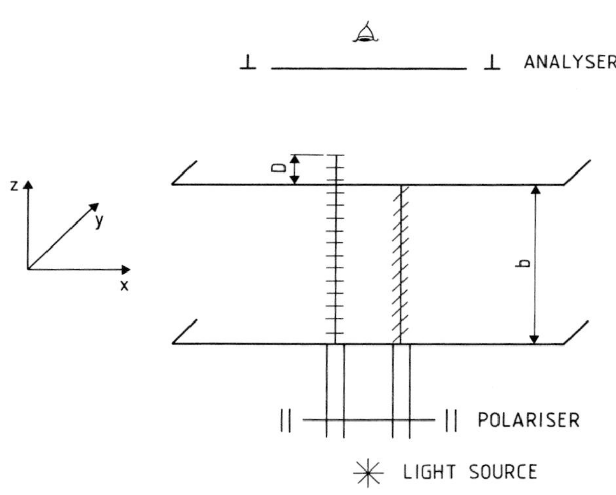

Figure 9.7 Polarized light technique to measure relative retardation (D) in a specimen of thickness (b), in order to determine birefringence in the x-y plane ($\triangle n_{xy}$).

Once the optical path difference is known, it can be related to birefringence by:

$$D = b \cdot \left[\frac{V_A}{V_X} - \frac{V_A}{V_Y} \right] = b \cdot [n_X - n_Y] = b \cdot \triangle n_{XY} \tag{9.5}$$

V is the velocity of light in air (V_A), and in the x and y-directions in the polymer (V_X, V_Y).

Thus, it is suggested that birefringence is directly proportional to optical path difference. In practice however, this is only true where structural regularity is apparent throughout the entire thickness of the specimen; the usual case concerning fabricated plastics is quite different, so that the total birefringence must be considered as the integral sum of retardation contributions from optically anisotropic material throughout the thickness:

$$\triangle n = \int_0^b dn = \int_0^b \frac{dD}{db} \tag{9.6}$$

Through-thickness birefringence profiles are therefore obtained only by careful experimentation involving the machining of thin sections, and subsequent measurement of optical path difference on the remaining specimen; data from this type of technique have been derived for various injection moulded amorphous plastics[9.12–9.14]. The importance of this type of characterization study was emphasized by the findings of WHITE et al.[9.12], who demonstrated the relationship between premature crazing in stressed PS and the existence of positive birefringence (transverse orientation).

- **Spectrophotometric Technique**

When the magnitude of birefringence is quite high, it often proves difficult to provide a sufficiently accurate method of compensation. In these circumstances, a spectrophotometer is often used[9.15], in which the intensity (I) of monochromatic polarized light transmitted through a highly birefringent specimen is compared to the equivalent intensity from a split (reference) beam, over a spectral distribution of input wavelengths (Figure 9.8). The phase difference (θ) between the components of light emerging from the sample is dependent on wavelength and birefringence, and is given by:

$$\theta = \frac{2\pi}{\lambda} b(\triangle n_{XY}) \tag{9.7}$$

m is an integer appropriate to the conditions at which constructive interference occurs, such that:

$$m = \frac{\theta}{2\pi} = \frac{Db}{\lambda}.$$

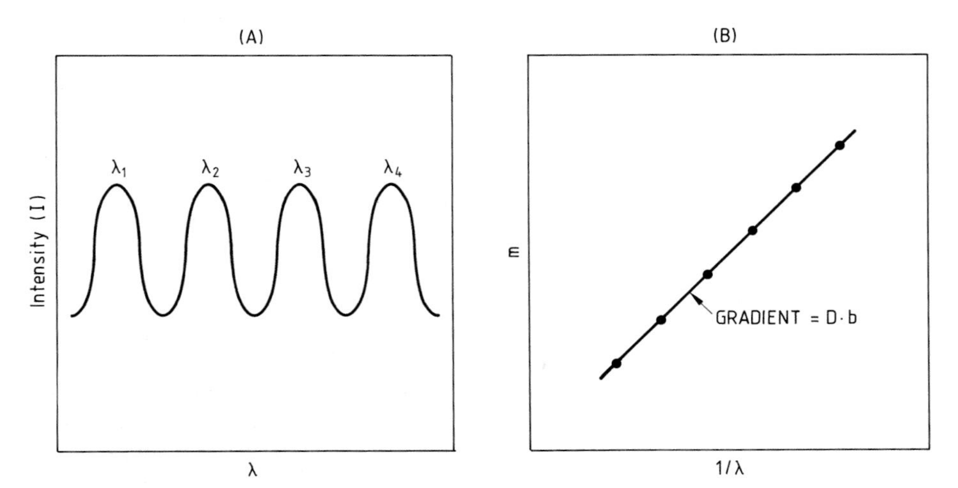

Figure 9.8 Spectrophotometric technique to measure birefringence: (A) Intensity (I) versus wavelength (λ); (B) Integer (m) versus reciprocal wavelength $\frac{1}{\lambda}$.

If a spectrometer is used to scan over a wide range of wavelengths, then the gradient of a plot between m and λ (Figure 9.8) can be used to evaluate optical path difference, allowing birefringence to be calculated using the expression stated earlier (Equation 9.5).

Dispersions of particles in an otherwise isotropic polymer can exhibit birefringence if the particles, although isotropic, are nonspherical, and are preferentially oriented, or have a different refractive index to the matrix material; this effect is known as *form birefringence.* A similar result may occur if the sample is associated with a heterogeneous domain-type microstructure, as occurs in some thermoplastic elastomers.

9.2.2.2 Birefringence and Molecular Orientation

The degree of molecular orientation induced by processing is an important factor in the morphology of plastics parts, which modifies the mechanical properties in the bulk of the material. Orientation originates as a result of an applied stress during processing, and is rendered stable if relaxation is suppressed by rapid cooling; several types of orientation were classified in an earlier section (6.6.1), and the phenomenon was also highlighted, in a processing context, throughout Chapter 5.

Since the polarisability of covalent bonds in plastics materials is direction dependent, birefringence will inevitably be controlled, to some extent, by any *preferred orientation* of the main-chain polymer backbone; birefringence is therefore known to be a strong function of molecular orientation in fabricated plastics, and many studies have been carried out to relate the development of birefringence to some primary processing variables such as:

- melt temperature and shear stress (injection moulding);
- strain rate and draw ratio (fibre spinning).

However, it is also known that the relaxation of side-groups (without any equivalent changes in main-chain orientation) can also induce significant changes in measured levels of birefringence[9.16].

(A) Intrinsic Orientation Birefringence

It is generally found that when the development of orientational birefringence is measured against a variable such as draw ratio, which is responsible for its creation, the birefringence first increases positively, then tends towards a limiting value: an *intrinsic orientation birefringence,* the value of which depends upon the chemical groups of which the repeat unit consists.

When polymer molecules are oriented, the predominant array direction of primary bonds (for any given volume of material) will change, thereby modifying the interactions with polarized light as a result of optical anisotropy. Hence the orientation birefringence can be modelled by an idealized rubbery network, in which it is assumed that molecular segments are aligned along two (only) orthogonal axes. Birefringence may be estimated according to the principal polarisabilities of the chain segments (p_X and p_Y, parallel and perpendicular to the chain-orientation axis), as a function of the applied extension ratio (λ_X):

$$\Delta n_{XY} = \frac{2\pi}{45} \cdot \frac{n^2 + 2}{n} \cdot N_C \cdot (p_X - p_Y) \cdot \left(\lambda_X^2 - \frac{1}{\lambda_X} \right) \tag{9.8}$$

n is the mean refractive index and N_C represents the number of network chains per unit volume.

Birefringence is therefore strongly influenced by the extent of deformation which induces chain alignment, and is also dependent upon the difference in polarisability, parallel and perpendicular to the oriented molecular chains.

(B) Optical Properties and Orientation

An optically anisotropic polymer is one in which the transmission velocity of light is direction-dependent, ie. the refractive index varies according to the vibrational direction of the light. If we confine the discussion to conditions in which the birefringence arises only by molecular orientation, it follows that the complete, 3-dimensional characterization of orientation can only be achieved by measuring refractive index along all corresponding principal axes.

An *optical indicatrix* can be used to characterize the appropriate optical properties in this way[9.17]. An indicatrix surface is defined by the extremities of an infinite number of vectors (in 3-dimensional space) drawn from a unique point, with the length of the vectors representing the refractive index of light vibrating parallel to it. Hence the indicatrix for an optically isotropic material is represented by a sphere, and a uniaxially-oriented sample is described by an ellipse (Figure 9.9); in the latter case, the value of n_Y may be greater, or less than n_X, thereby determining the *sign of birefringence.* For commercial monofilament products such as polyamides and polyesters, $n_Y > n_X$ and the fibres have positive birefringence; in the case of acrylics (PMMA), the birefringence is negative.

When using birefringence measurements to characterize molecular orientation in plastics, it is important to realize that a single, in-the-plane measurement determines only the difference in principal refractive indices within the plane. For example, a zero measurement of Δn_{XY} might represent:

- isotropic material;
- equibiaxial orientation within the x-y plane.

It is also conceivable that a similar result would be obtained if the incident light were coaxial with the orientation direction in a uniaxially oriented sample (see Table 9.1). It is necessary therefore, to take birefringence measurements about two mutually perpendicular planes within the material, in order to characterize all three principal refractive indices.

Table 9.1 Orientation, Refractive Index and Birefringence in Plastics

Orientation	Refractive indices	Birefringence		
		(Δn_{xy})	(Δn_{xz})	(Δn_{yz})
Isotropic	$n_X = n_Y = n_Z$	0	0	0
Uniaxial (x)	$n_X \neq n_Y = n_Z$	Finite	Finite	0
Biaxial	$n_X \neq n_Y \neq n_Z$	Finite	Finite	Finite
Equi-biaxial (x, y)	$n_X = n_Y \neq n_Z$	0	Finite	Finite

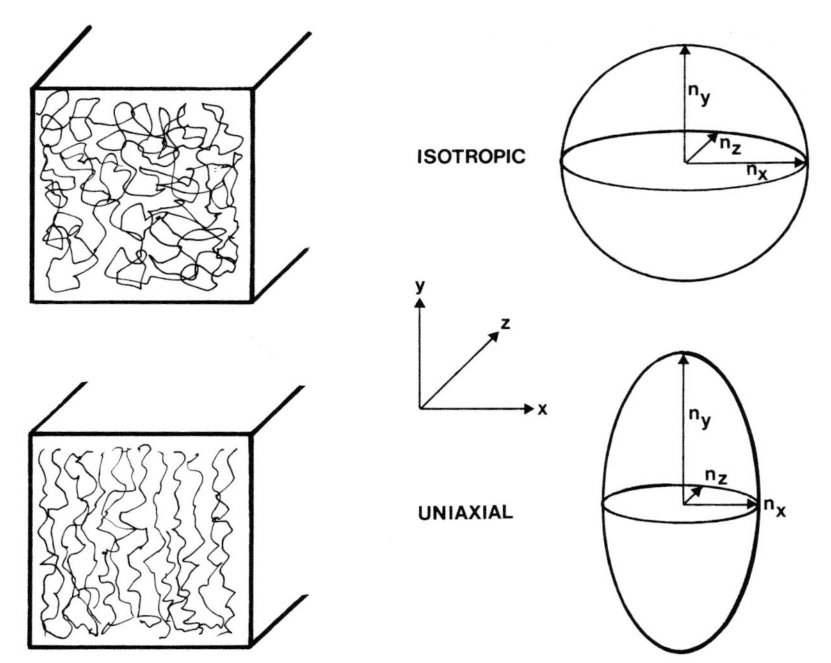

Figure 9.9 Indicatrix representation of isotropic and uniaxially-oriented plastics (n_i is refractive index in the i-direction).

(C) Orientation Functions

• Uniaxial Deformation and Orientation

The interrelationships between processing conditions, orientation and physical properties of plastics are very complex. Since orientation is a directional property, it is desirable to relate the preferential alignment of polymer chains to some well-defined and controllable axes which are characteristics of processing. As an example, consider an element of material which has been uniaxially stretched: if the angle between the polymer chains (c-axis) and the deformation direction is θ_C, then the birefringence is related to an *orientation function*, given by:

$$f_H = \frac{3\mathrm{Cos}^2\theta_C - 1}{2} \tag{9.9}$$

This orientation parameter is often referred to as the Herman's orientation function; the term $\mathrm{Cos}^2\theta_C$ represents the average value of θ_C for all the structural entities within the sample. Several important implications arise from this approach:

Isotropic material	$\mathrm{Cos}^2\theta_C = \frac{1}{3}$	$f_H = 0$
Perfect uniaxial orientation	$\mathrm{Cos}^2\theta_C = 1$	$f_H = 1$
Perpendicular to fibre-axis	$\mathrm{Cos}^2\theta_C = 0$	$f_H = -\frac{1}{2}$

- **Biaxial Deformation and Orientation**

A similar approach has been developed to model the direction and magnitude of biaxial orientation in fabricated plastics[9.18, 9.19]. In this case, orientation is characterized by the angles θ_{CX} (between the c-axis and the first deformation axis), and θ_{CY}, the equivalent angle with respect to the second (perpendicular) strain axis. The *biaxial orientation functions* are given by:

$$f^B_{CX} = 2\mathrm{Cos}^2\theta_{CX} + \mathrm{Cos}^2\theta_{CY} - 1 \tag{9.10}$$

$$f^B_{CY} = 2\,\mathrm{Cos}^2\theta_{CY} + \mathrm{Cos}^2\theta_{CX} - 1 \tag{9.11}$$

Once more, some special cases should be emphasized:

Isotropic material	$f^B_{CX} = f^B_{CY} = 0$
Uniaxial orientation (x-axis)	$f^B_{CX} = 1; \quad f^B_{CY} = 0$
Uniaxial orientation (y-axis)	$f^B_{CX} = 0; \quad f^B_{CY} = 1$
Uniaxial orientation (z-axis)	$f^B_{CX} = f^B_{CY} = -1$
Equi-biaxial orientation (xy-plane)	$f^B_{CX} = f^B_{CY} \neq 0$

These principles, and any actual measurements for non-ideal samples, are best expressed graphically, using a plot of f^B_{CY} versus f^B_{CX} (Figure 9.10), on which the special examples described above are evident. All states of biaxial molecular orientation can be described adequately by the appropriate coordinates within the isosceles orientation triangle.

Using this approach, it is feasible to relate the state of biaxial orientation to the manufacturing conditions by which it is introduced, as long as relevant measurements can be made. According to WHITE and SPRUIELL[9.19–9.21], the orientation functions for amorphous plastics are related to birefringence as follows:

$$f^B_{CX} = \frac{n_X - n_Z}{\Delta n'} \qquad f^B_{CY} = \frac{n_Y - n_Z}{\Delta n'} \tag{9.12}$$

$\Delta n'$ is the intrinsic orientation birefringence of the polymer.

For semicrystalline plastics, Equations 9.10 and 9.11 may again be used, but in a suitably transposed form, in terms of the principal crystallographic axes (a, b and c).

These functions may also be obtained from wide-angle X-Ray pole figures; this technique has been used to generate such orientation data in previous studies on biaxially oriented films of HDPE[9.22, 9.23] and PETP[9.24].

9.2.2.3 Birefringence and Stress

Birefringence also results when stress resides in, or is applied to a polymer; the *stress optical coefficient (C)* is the ratio of the birefringence (Δn) in a material and

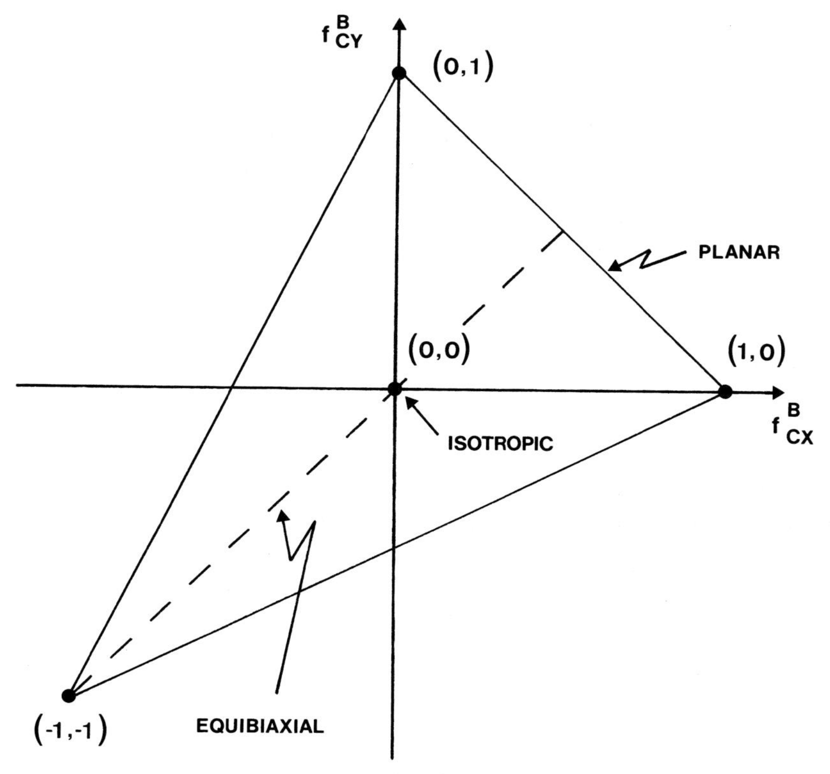

Figure 9.10 Representation of uniaxial, or biaxial molecular orientation by the isosceles 'orientation triangle'. See the text for the definitions of the orientation functions f^B_{CX}, f^B_{CY}. (After WHITE and SPRUIELL[9.19–9.21].)

the difference in principal stress (within the same plane (x-y, say)), normal to the incident light which produces it:

$$\Delta n_{XY} = n_X - n_Y = C_{XY} \cdot (\sigma_X - \sigma_Y) \tag{9.13}$$

The stress optical coefficient is a most important quantity in *photoelastic stress analysis;* the unit is the Brewster, equal to 10^{-12} m^2/N. High values of C indicate sensitive stress-optical materials, which are very useful in characterizing the direction and magnitude of stress (and stress distribution) in photoelastic experiments. Cast sheets of translucent epoxy resin are often used as photoelastic model materials; this polymer has a stress-optical coefficient of around 50×10^{-12} m^2/N. Most other commercial polymers have positive stress-optical coefficients; however, plastics such as PS and PMMA are less sensitive to this effect, and may, in certain circumstances, exhibit negative stress-optical coefficients, ie. the lower, in-plane refractive index is parallel to the maximum principal stress. It is therefore very important to specify the sign of birefringence, in order to determine the nature of the applied stress.

Even in situations where changes in birefringence in plastics are induced by an externally-applied stress, the effect of molecular orientation cannot be overlooked, since the stress-optical coefficient is itself orientation dependent[9.25]. The coefficient (C) of both thermoplastics and thermosetting resins is also dependent upon temperature[9.26, 9.27]. It is therefore always desirable to use isothermal test conditions in photoelastic measurements, and to use isotropic specimens of consistent thermal history.

An additional problem concerned with this latter point is the separation of birefringence effects originating from both molecular orientation and a residual stress system; since the aspects of processing which promote one of these phenomena may also produce the other (notably in injection moulding of amorphous thermoplastics) it is inevitable that these respective contributions to birefringence are often confused, and prove difficult to separate. BROUTMAN and KRISHNAKUMAR[9.28] have described a quick, qualitative method to address this problem in moulded polycarbonate, in which a saw-cut is introduced into a specific location of the testpiece; any disturbance of the birefringence pattern indicates the predominance of double refraction effects due to a physical (internal) stress. A more thorough, quantitative technique, involving the analysis of birefringence in successively-machined thin sections, has also been reported by SAFFELL and WINDLE[9.29].

Further reference to photoelastic stress analysis, and the use of optically-anisotropic plastics in such experiments, can be found in the text by KUSKE and ROBERTSON[9.30].

9.3 Light Scattering

The scattering of light as it passes through a material which is essentially transparent leads to two effects: *aberration* of the image (loss of clarity), and reduction in *contrast* in the image. The former is associated with scattering at very small angles, whilst the latter occurs with scattering at high angles: both are aspects of *transparency,* and both are caused by variation of refractive index within the material. Small size of the dispersed phase leads to high angle scattering, and large size is associated with low angle scattering; the intensity of scattering is also dependent on the mismatch in refractive index.

Polymers which contain a large proportion of scatterers are opaque, as, for example:

- high impact polystyrene, which contains a significant quantity of rubber particles;
- polypropylene, polyethylene or nylon, which are highly crystalline.

If the particles are smaller than the wavelength of the radiation, scattering is reduced progressively; thus PETP products which have been crystallized thermally into the *spherulitic* form are opaque, whereas containers or films which consist of much smaller (and highly oriented) *chain-extended crystallites* usually offer a high degree of optical clarity. Light scattering from solid polymers can, therefore, be a basis for the study of the texture of the materials.

9.3.1 Light Transmission and Transparency

Transparency may be quantified by considering a narrow beam of light of specified frequency falling normally on to the plastics material. A fraction of the light is reflected, and some is scattered by imperfections at the surface; this latter component is usually eliminated from the measurement by "oiling-out" the surface with a liquid of similar refractive index to the polymer. A frequent practical arrangement has the matching refractive index liquid in a cell in the optical system; the absorbing and scattering characteristics can then be determined by a "specimen-in, specimen-out" comparison.

The *direct transmission factor (T),* also known as the *specular transmittance,* is defined by the ratio:

$$T = \frac{\phi_{\text{undev}}}{\phi} \tag{9.14}$$

where ϕ is the flux in the absence of a plastics specimen, and ϕ_{undev} represents the undeviated flux when the sample is included in the measurement cell.

The difference between these two quantities represents the sum of flux absorbed (ϕ_A), and the scattered component (ϕ_{sc}):

$$\phi_{\text{undev}} = \phi - \phi_A - \phi_{sc} \tag{9.15}$$

For comparatively low concentrations of the scattering species, and for relatively thin specimens (of less than 3mm thickness), the relationship between the direct transmission factor and the thickness (b), can be approximated by:

$$T = \exp[-b(\sigma + K)] \tag{9.16}$$

σ is a measure of the light scattered, and is known as the *scattering coefficient,* or *turbidity.* It is the fraction of the light flux scattered in passing through unit thickness of the material; it has the dimensions of inverse length, usually expressed in cm^{-1}. The wavelength of the radiaton in vacuum or air must be recorded. K is the fraction of light energy absorbed in passing through unit thickness of the material; it has similar dimensions to σ. The sum $(\sigma + K)$ is known as the *attenuation* or *extinction coefficient,* and has similar dimensions to its components.

The scattered radiation is the sum of that scattered forwards (ϕ^f_{sc}), between 0 and 90°, measured from the direction of the incident beam, and that scattered backwards. The total flux transmitted (ϕ_T), is given by:

$$\phi_T = \phi_{\text{undev}} + \phi^f_{sc} \tag{9.17}$$

The *forward scattered fraction,* χ^f_{sc}, is:

$$\chi^f_{sc} = \frac{\phi^f_{sc}}{\phi_T} \tag{9.18}$$

9.3.2 Haze and Clarity

The lower contrast when viewing an object through a scattering medium is due to the light scattered forwards at high angles, which is conveniently expressed as the *forward scattered fraction* (χ^f_{sc}), and is often termed *haze* (multiangle scatter). In standard tests[9.31], haze is determined as the proportion of light scattered between two arbitrary angles set by the instrument, compared with the total transmitted light, ϕ_T. The limiting angles on standard hazemeters are frequently 2.5° and 90 °, so that the haze determined on such an equipment is lower than the forward scattered fraction by the scattering between 0° and 2.5°. For some materials, notably PETP film, this discrepancy may be substantial and may lead to a lack of correlation between standard measurements and visual effects.

The data in Figure 9.11 demonstrate the dependence of haze (measured according to ASTM D1003[9.31]) of PETP bottles on the container wall crystallinity, which was altered by variation of thermal history during processing. A deterioration in the surface quality, resulting in increased light scatter, appears to occur only when a specific level of crystallinity has been achieved. Other experiments showed that the light scattering could be drastically reduced by immersing samples in a liquid of similar refractive index (cassia oil), which suggests that haze is related to *surface light scattering* caused by the size and/or irregularity of crystallites, possibly

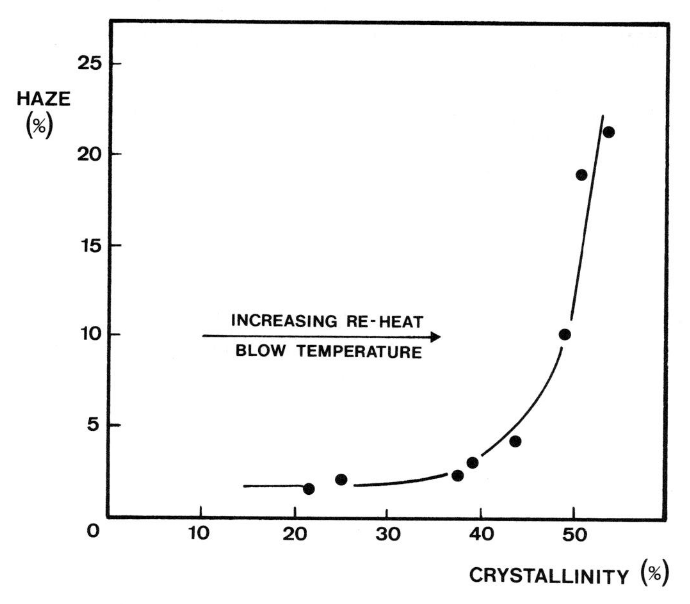

Figure 9.11 The dependence of haze (measured according to ASTM D1003) on crystallinity (measured by thermal analysis) for PETP containers stretch blow moulded at different reheat temperatures.

exaggerated by imperfect replication of the mould surface (ie. as the more crystalline product tends to shrink to a greater extent, during the cooling phase).

Clarity is the capacity of the sample for allowing details in the object to be resolved in the image; it is perfect only when the scattering is zero, and is dependent on the angular distribution of the scattered intensity, and on the geometry of the optical system, particularly the distance between the object and the plastics sample through which it is being viewed. Loss of clarity is assessed by measuring the angular separation of two points which can just be resolved in the image, in comparison with the object viewed directly, for a magnification of 1 in the optical system.

It has been suggested that the use of *optical transfer functions* would allow more objective measurement of scattering phenomena[9.32]. However, since the understanding of scattering by polymers is not increased thereby, the subject will not be discussed further here.

9.4 Light Transfer

In many applications, the ability to transmit light to a specific level is extremely important, in addition to other characteristics such as surface gloss, colour shade and hiding power. These phenomena are discussed in subsequent sections.

9.4.1 Transmission and Hiding Power

The *total transmission factor* (τ), or *total transmittance,* is relevant to the application of plastics as lamp or lantern covers, and is defined as the ratio of the total flux transmitted (ϕ_τ) to the incident flux (ϕ), which is assumed to be a narrow beam normal to the surface of the sample:

$$\tau = \frac{\phi_\tau}{\phi} \tag{9.19}$$

The thickness of the sample and the nature and type of illuminant must each be recorded: white light is preferred since this is employed in the majority of applications.

A further requirement in this application is that the opacity of the plastics material be sufficient to "hide" the filament of the light source. This is measured as the *hiding power* and depends upon:

(1) the refractive index difference between matrix and dispersed phase, frequently a white filler; and
(2) the size and concentration of the dispersed phase.

9.4.2 Pigmentation and Light Intensity

In applications for colour-tinted sheet plastics products such as roofing panels, reference should be made to transmission spectra to cover the entire visible and near-UV regions, since the extent of light transmission will be very sensitive to the absorption characteristics of the pigmentation system.

For a given incident beam, the transmission intensity (I) of radiation through a plastics medium is also dependent upon thickness and the concentration of the absorbing species. This is most frequently described by the Beer-Lambert law:

$$I = I_0 \cdot e^{-kCb} \tag{9.20}$$

I is the light intensity after passing through a specimen thickness b, I_0 is incident intensity, k is a coefficient relating to the efficiency of the absorbing medium, and C is its concentration.

A number of spectrophotometers function on the basis of this principle, such as IR thickness gauges for rigid, clear packaging components. If the appropriate absorption spectra of a given material are known, such that the wavelength and intensity of the incident radiation can be specified, photometric techniques can be developed to measure and relate the intensity of transmitted light to the thickness and/or concentration of absorbant. This is exemplified by the data in Figure 9.12, in which a linear relationship is obtained between $\log(I)$ and the predetermined concentration of green masterbatch in a series of thin-walled packaging components, injection moulded in HDPE. Light scattering also occurs due to the presence of crystallites in certain plastics; clearly, this is a possible source of error, in this type of experiment.

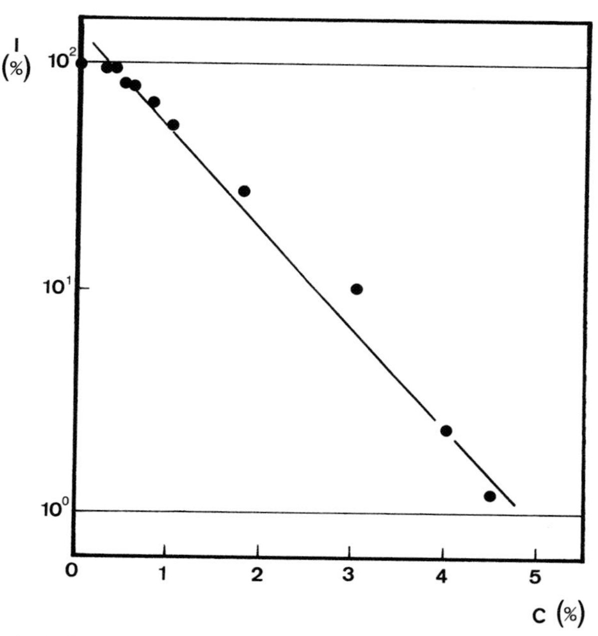

Figure 9.12 Transmitted light intensity (I, expressed relative to the intensity through an otherwise similar, unpigmented component, under equivalent conditions) versus masterbatch concentration (C) for pigmented, thin-walled injection mouldings in HDPE.

9.5 Absorption of Light

This area is concerned with selective response within the visible region, 400 – 800 nm in the light reflecting, transmitting or diffusing characteristics of isotropic materials, which do not fluoresce or phosphoresce. Many of the standards are due to C.I.E. (Commission Internationale d'Eclairage), and depend on the response being analyzed into the behaviour in three spectral regions, characterized by the tristimulus values. Instruments to make such measurements include spectrophotometers and colorimeters, designed to measure the tristimulus response directly[9.33]. Reference to colour measurement in plastics is also made in Chapter 10 (Section 10.2.2), in the context of polymer degradation.

9.5.1 Colour Measurement Phenomena

- *Spectral metamerism* is exhibited by colours which show a change in the closeness of match when the spectral distribution of the illuminant is altered. Two colours which match under one illuminant may not match under a different one; such colours differ in the distribution of reflected or transmitted energy, but under certain illuminants give similar tristimulus values. The phenomenon may also be a consequence of observers with colour-matching capabilities very different to those of the standard observer.
- *Geometrical metamerism* is the relative variation in the spectral reflection and tristimulus values which is found when a pair of surface colours, which match under given conditions, are illuminated or viewed under a different geometry.

9.5.2 Some Problems Associated with Colourants in Plastics

The desire to manufacture self-coloured products from plastics and eliminate thereby the need for painting (or surface coating) is based on their inherent optical properties and the favourable economics. However, the inclusion of pigments and dyes in a plastics formulation can lead to several unforseen problems.

Some pigments are hard, for example ferric oxide, and their inclusion in a moulding powder can result in very rapid wear of the gate region when such formulations are injection moulded, a problem which did not arise when compression moulding the same materials. Failure of the product may result from imperfect dispersion of colourant, especially a particulate pigment; furthermore, poor dispersion leads to inefficient colouration. Pigment accumulated along the flow lines invites a crack to be formed in that region.

A possible side effect of the dispersion of pigments and dyes in crystallizable polymers is to nucleate the crystallization process[9.34]. Polypropylene is particularly susceptible to the effect, for which some appropriate data are illustrated in Table 9.2; all the samples were cooled at 10 °C per minute. Most of the coloured materials showed nucleation to some extent, as manifested in a higher crystallization temperature on cooling from the melt state. Whilst most of the colourants nucleate PP into the usual stable structure, *α-form-crystallinity*[9.35, 9.36], certain dyestuffs nucleate a different species of crystallite, which differs from the α-form in melting temperature and mechanical properties. *β-form-crystallinity*[9.36, 9.37], as induced by quinacridone red, leads to lower yield strength coupled with very much increased elongation at break under tensile stress; consequently, the low-strain modulus is much reduced, in comparison with α-form PP.

Table 9.2 The Crystal-Nucleating Effect of Colourants in Polypropylene

Polymer description	Melting point(s) T_m (°C)	Crystallisation temp. T_X (°C)
• **Natural homopolymer A: MFI (2.16 kg, 230 °C) = 1.8 g/10min.**		
After extrusion	165.9	118.1
With colourants:		
Quinacridone Violet	165.5	131.8
Quinacridone Red	147.8, 165.8	128.4
Phthalocyanine Green	166.9	127.0
Chromophtal Blue	165.1	127.6
Chromophtal Orange	164.4	124.1
Chromophtal Yellow	165.6	123.8
Chromophtal Red	164.8	119.8
• **Natural homopolymer B: MFI (2.16kg, 230 °C) = 0.9 g/10min.**		
Raw material	166.2	111.9
After extrusion	165.1	121.1
With colourants:		
Quinacridone Red	146.9, 164.7	131.0
Quinacridone Violet	165.4	131.5

The data presented in this Table were obtained by Differential Thermal Analysis (DTA); T_m is the peak temperature of the *melting endotherm* on the heating cycle, and T_X is the equivalent peak temperature of the *recrystallization exotherm,* on cooling from the melt-state. A wide range of nucleating capability is illustrated in these data; the greatest nucleating activity leads to crystallization at the highest temperature in the cooling mode, and to the development of the finest texture in products melt processed in pigmented grades of PP (see also Chapter 1). However, the melting and recrystallization behaviour of β-form crystallites in PP is also dependent upon shear history[9.38], and cooling rate[9.39]. The two melting temperatures for quinacridone red are attributed to the β crystalline form (lower melting temperature), which after melting recrystallizes to the α-form, with the higher melting temperature[9.40].

The sensitivity of the micromorphology and mechanical properties of PP copolymer to the presence of pigments, recycled material (hence molecular weight) and other compound additives has been demonstrated by BEVIS et al.[9.40,9.41]. Clearly, the evolution and development of commercial plastics as materials offering attractive, self-colouring characteristics has unfolded some complex side issues affecting other physical properties. However, these need not pose significant problems to products in service, so long as the influence of the colourant on the component microstructure can be closely characterized and understood.

9.6 Autofluorescence in Plastics

9.6.1 Origin of fluorescence

Fluorescence is the phenomenon whereby molecules or characteristic chemical groups absorb incident energy, then re-emit a proportion of this as photons of light, at a lower frequency. Many substances such as dyestuffs will absorb ultraviolet radiation, which is re-emitted within the visible range, thus providing the characteristic *fluorescent colours,* for which the additives are generally specified.Whilst there exists a multitude of organic species which are known to fluoresce[9.42], the *autofluorescence* of plastics is encountered far less frequently. This should be appreciated in the context of the chemical groups which are responsible for the origin of luminescence effects; notably aromatic groups, main-chain unsaturation, and other chromophoric side-groups such as those containing $C = O$ and $C \equiv N$ bonds. The origin of fluorescence lies with conjugated bond systems, in which the delocalized π-electrons are able to absorb energy with relative ease.

9.6.2 Measurement Principles

The use of fluorescence microscopy to characterize polymers was first reported by HEMSLEY et al.[9.43]. Rigid PVC was the material used in this study, which tends to fluoresce with greater intensity as the concentration of polyene sequences, formed as a result of dehydrochlorination, increases following processing.

With this technique, it is vital to determine some information on the fluorescent species, so that the experimental conditions can be chosen to satisfy the UV-absorption and emission (fluorescence) spectra, for any given material.

- **UV Absorption Spectra**

The response of plastics compounds to UV radiation should be predetermined, so that the wavelength of an appropriate monochromatic source can be chosen in a strongly-absorbing region. A high pressure mercury vapour lamp (having a strong emission peak wavelength of 365 nm) has been found to be suitable for some examples of plastics[9.43].

- **Fluorescence Emission Spectra**

The occurrence of autofluorescence is often quite easy to detect, using a standard scientific photometer. If the complete emission spectrum can be determined, it is possible to detect and characterize changes in properties which are responsible for a change in measured fluorescence intensity; these might include the extent of polymer decomposition, the composition of copolymers, the concentration of specific additives and (dependent on the technique used), specimen thickness.

9.6.3 Application to Plastics

(A) PVC Technology

Due to its commercial importance coupled with its indifferent stability, PVC has been the subject plastics material for many developmental studies involving autofluorescence. Reference has already been made to the pioneering work on fluorescence microscopy of PVC powder blends and extrudates[9.43], and similar techniques have also been used to characterize the microstructure of hotplate welded PVC profiles[9.44] (see also Chapter 10).

(B) Coated and Multilayer Products

If differences in the fluorescence emission characteristics of composite plastics products can be specified, techniques can be developed which allow the thickness of individual layers to be determined, usually with a high degree of accuracy. For example, this method can be applied to sections from thin-wall packaging films and containers, in order that the thickness of an expensive, low-permeability resin is adequate and uniformly-distributed, without being excessive; clearly, the possibility to obtain data on coating thickness distribution carries important advantages over gravimetric techniques.

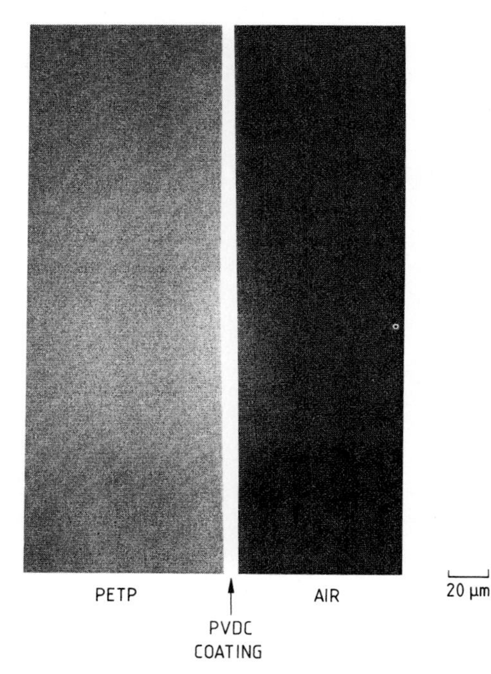

Figure 9.13 UV-fluorescence micrograph showing a dip-coated PVDC layer (of about $8\mu m$ thickness) on a PETP beer-container[9.45, 9.46]. (Reproduced with permission from Elsevier Applied Science Publishers Ltd.)

Despite having similar emission spectra, PETP and PVDC can easily be distinguished in sections subjected to UV light (Figure 9.13). If this technique is carried out on a microscope, it becomes relatively straight forward to measure the PVDC latex coating thickness, using an eyepiece with movable crosswires, precalibrated using a graticule. Figure 9.14 shows typical thickness data for PETP beer bottles coated with between 1 and 3 layers of PVDC.

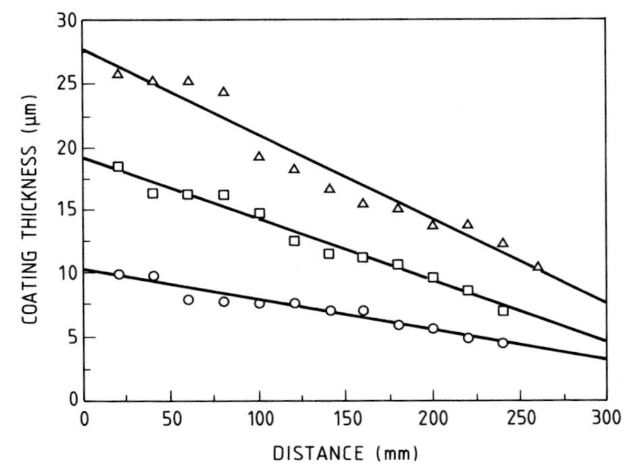

Figure 9.14 PVDC coating thickness data, on blow moulded PETP bottles, as measured by fluorescence microscopy: (○) standard, single-coat application, (□) two dip-coating cycles, (△) three coating cycles[9.45, 9.46]. (Reproduced with permission from Elsevier Applied Science Publishers Ltd.)

References

9.1 BS 4618, (1968), *Recommendations for the Presentation of Plastics Design Data: Optical Properties,* Part 5.3, British Standards Institution, London.

9.2 HUMMEL, D.O., and SCHOLL, J.: *Atlas of Polymer and Plastics Analysis,* Carl Hanser, and VCH Publications, Munich, (1989).

9.3 HASLAM, J., WILLIS, H.A., and SQUIRREL, D.C.M.: *Identification and Analysis of Plastics,* Heyden, London, (1980).

9.4 SIESLER, H.W., and HOLLAND-MORITZ, K.: *Infrared and Raman Spectroscopy of Polymers,* Marcel Dekker, New York, (1980).

9.5 CAMPBELL, D. and WHITE, J.R.: *Polymer Characterization Physical Techniques,* Chapman & Hall, London, (1989).

9.6 RAO, C.N.R.: *Ultra-Violet and Visible Spectroscopy,* Butterworth, London, (1961).

9.7 OWEN, E.D. (Ed.): *Degradation and Stabilisation of PVC,* Applied Science, London, (1984).

9.8 ASTM D523 (1985): *Test for Specular Gloss,* American Society for Testing Materials, Philadelphia, USA.

9.9 ASTM D2457 (1970): *Specular Gloss of Plastic Film,* American Society for Testing Materials, Philadelphia, USA.

9.10 COXON, L.D., and WHITE, J.R.: *Residual Stresses and Aging in Injection Moulded Polypropylene,* Polym. Eng. Sci., **20,** (1980), 230.

9.11 HAWORTH, B., HINDLE, C.S., SANDILANDS, G.J., and WHITE, J.R.: *Assessment of Internal Stresses in Injection Moulded Thermoplastics,* Plast. Rubb. Proc. Appl., **2,** (1982), 59.

9.12 WHITE, E.F.T., MURPHY, B.M., and HAWARD, R. N.: *The Effect of Orientation on the Internal Crazing of Polystyrene,* J. Polym. Sci; Polym. Lett., **7,** (1969), 157.

9.13 CUCKSON, I.M., HAWORTH, B., SANDILANDS, G.J., and WHITE, J.R.: *Internal Stress Assessment of Thick-Section Injection Mouldings,* Intern. J. Polym. Mat., **9,** (1981), 21.

9.14 HAWORTH, B., SANDILANDS, G.J., and WHITE, J.R.: *Characterisation of Injection Mouldings,* Plast. Rubb. Intern., **5,** (1980), 109.

9.15 YANG, H.H., CHOUINARD, M.P., and LINGG, W.J.: *Birefringence of Highly Oriented Fibres,* J. Polym. Sci; Polym. Phys., **20,** (1982), 981.

9.16 QAYYUM, M.M., and WHITE, J.R.: *Time-Dependent Birefringence in Glassy Polymers during Stress Relaxation and Recovery,* Polymer, **23,** (1982), 129.

9.17 BLOSS, F.D.: *An Introduction to the Methods of Optical Crystallography,* Holt, Reinhart and Winston, New York, (1961).

9.18 STEIN, R.S.: *The X-Ray Diffraction, Birefringence and Infrared Dichroism of Stretched Polyethylene. III. Biaxial Orientation,* J. Polym. Sci., **31,** (1958), 335.

9.19 WHITE, J.L., and SPRUIELL, J.E.: *Specification of Biaxial Orientation in Amorphous and Crystalline Polymers,* Polym. Eng. Sci., **21,** (1981), 859.

9.20 WHITE, J.L., and SPRUIELL, J.E.: *The Specification of Orientation and its Development in Polymer Processing,* Polym. Eng. Sci., **23,** (1983), 247.

9.21 CAKMAK, M., SPRUIELL, J.E., and WHITE, J.L.: *A Basic Study of Orientation in Poly(Ethylene Terephthalate) Stretch-Blow Moulded Bottles,* Polym. Eng. Sci., **24,** (1984), 1390.

9.22 CHOI, K.J., SPRUIELL, J.E., and WHITE, J.L.: *Orientation and Morphology of High-Density Polyethylene Film Produced by the Tubular Blowing Method and its Relationship to Processing Conditions,* J. Polym. Sci; Polym. Phys., **20,** (1982), 27.

9.23 GILBERT, M., HEMSLEY, D.A., and PATEL, S.R.: *Effect of Processing Conditions on the Orientation and Properties of Polyethylene Blown Film,* Brit. Polym. J., **19,** (1987), 9.

9.24 CAKMAK, M., SPRUIELL, J.E., WHITE, J.L., and LIN, J.S.: *Small Angle and Wide Angle X-Ray Pole Figure Studies on Simultaneous Biaxially Stretched Poly(Ethylene Terephthalate) (PET) Films,* Polym. Eng. Sci., **27,** (1987), 893.

9.25 ANDREWS, R.D., and RUDD, J.F.: *Photoelastic Properties of Polystyrene in the Glassy State. 1. Effect of Molecular Orientation,* J. Appl. Phys., **28,** (1957), 1091.

9.26 RUDD, J.F., and GURNEE, E.F.: *Photoelastic Properties of Polystyrene in the Glassy State. 2. Effect of Temperature,* J. Appl. Phys., **28,** (1957), 1096.

9.27 KHAYYAT, F.A., and STANLEY, P.: *The Dependence of the Mechanical, Physical and Optical Properties of Araldite CT200/HT907 on Temperature over the Range-10 to 70 °C,* J. Phys. D; Appl. Phys., **11,** (1978), 1237.

9.28 BROUTMAN, L.J., and KRISHNAKUMAR, S.M.: *Cold Rolling of Polymers. 2. Toughness Enhancement in Amorphous Polycarbonates,* Polym. Eng. Sci., **14,** (1974), 249.

9.29 SAFFELL, J.R., and WINDLE, A.H.: *The Influence of Thermal History on Internal Stress Distributions in Sheets of PMMA and Polycarbonate,* J. Appl. Polym. Sci., **25,** (1980), 1118.

9.30 KUSKE, A., and ROBERTSON, G.: *Photoelastic Stress Analysis,* Wiley Interscience, London, (1974).

9.31 ASTM D1003 (1977): *Haze and Luminous Transmittance of Transparent Plastics,* American Society for Testing Materials, Philadelphia, USA.

9.32 WILLMAITH, F.M.: *Transparency, Translucency and Gloss* in *Optical Properties of Polymers,* MEETEN, G.H.: Ed., Elsevier Applied Science, London, (1986).

9.33 BS 3900: *Tristimulus Measurements; Direct Measurements Part 8 (1986): Determination of Colour and Colour Differences. Part 9 (1986): Measurement. Part 10 (1986): Calculation,* British Standards Institution, London.

9.34 BINSBERGEN, F.L., and DE LANGE, B.G.M.: *Heterogeneous Nucleation in the Crystallisation of Polyolefins: Part 2 Kinetics of Crystallisation of Nucleated Polypropylene,* Polymer, **11,** (1970), 309.

9.35 PADDEN, F.J., and KEITH, H.D.: *Spherulitic Crystallisation in Polypropylene,* J. Appl. Phys., **30,** (1959), 1479.

9.36 CHUA, J.O., and GRYTE, C.C.: *Studies on the α and β-Forms of Isotactic Polypropylene by Crystallisation in a Temperature Gradient,* J. Polym. Sci; Polym. Phys., **15,** (1977), 641.

9.37 Turner-Jones, A., and Cobbold, A.J.: *The β Crystalline Form of Isotactic Polypropylene,* J. Polym. Sci; Polym. Lett., **B6,** (1968), 539.

9.38 Oragaun, H., Hubeny, H., and Mushnik, H.: *Shear Induced β-Form-Crystallisation in Isotactic Polypropylene,* J. Polym. Sci; Polym. Phys., **15,** (1977), 1779.

9.39 Fujiwara, Y., Goto, T., and Yamashita, Y.: *Comparison of Premelting and Recrystallisation Behaviour of β-Phase Isotactic Polypropylene by Heating and Cooling at Different Rates,* Polymer, **28,** (1987), 1253.

9.40 Murphy, M.W., Thomas, K., and Bevis, M.J.: *Relationship Between Injection Moulding Conditions, Micromorphology and Impact Properties of Polypropylene: A Typical Commercial Grade,* Plast. Rubb. Proc. Appl., **9,** (1988), 3.

9.41 Williams, D., and Bevis, M.J.: *The Effect of Recycled Plastic and Compound Additives on the Properties of an Injection Moulded Polypropylene Copolymer,* J. Mat. Sci., **15,** (1980), 2834; 2843.

9.42 Unilever Education Booklets: Advanced Series, *8. Fluorescence,* Unilever Education Section, London, (1975).

9.43 Hemsley, D.A., Higgs, R.P., and Miadonye, A.: *UV Fluorescence Microscopy in the Study of PVC Powders and Extrudates,* Polym. Comm., **24,** (1983), 103.

9.44 Haworth, B., Law, T.C., Sykes, R.A., and Stephenson, R.C.: *Factors Determining the Weld Strength of Window Corners Fabricated From Impact Modified UPVC Profiles,* Plast. Rubb. Proc. Appl., **9,** (1988), 81.

9.45 Robinson, T.M.: *Factors Determining the Adhesion and Barrier Properties of PVdC – Coated PET Beverage Containers,* PhD Thesis, Loughborough University of Technology, 1990.

9.46 Harworth, B., and Robinson, T.M.: *The Measurement of Thin PVdC Coatings on PET Substrates Using Fluorescence Microscopy,* Polym. Testing, **10**, (1991), 205.

Chapter 10
Other Physical and Environmental Properties

In the first of these categories can be found phenomena such as *permeation* and the *absorption* (or *desorption)* of fluids (particularly water) and gases/vapours into plastics. These properties often determine the suitability of a given material for a specific application: for example, plastics packaging for foodstuffs must be resistant to the ingress of oxygen and/or atmospheric moisture, so that there is no quality deterioration in the food over the recommended shelf-life of the product.

Chemical mechanisms of *polymer decomposition* are strictly outside the scope of this book; however, since the exploitation of physical properties in plastics components is occasionally limited by such effects, some of the more important elements and consequences of degradation are included. It has been noted earlier that most commercial plastics contain appropriate chemical additives, and that some are blends or alloys: the *chemical resistance* and *fire properties* depend upon the nature and proportions of all the constituents within the plastics compound.

10.1 Physical Interactions with Plastics

10.1.1 Solubility

10.1.1.1 Some General Concepts

The molecular chains in thermoplastics are held together only by secondary interchain forces, not by primary chemical bonds as in cross-linked systems.

Separation of polymer molecules by an external chemical species is, therefore, not qualitatively different from the solution of an organic substance of lower molecular weight in a solvent. Generally, we can state that:

- low-polarity plastics are dissolved by solvents of similar solubility parameter;
- non-polar solvents dissolve hydrocarbon polymers, and polar solvents dissolve polar polymers;
- the extent of dissolution is a function of diffusion-rate, and is therefore temperature-dependent;

Solubility parameters are quoted for some common solvents (paraffins, ethers, ketones, alcohols, chlorinated, and other common solvent liquids), together with some equivalent data for semi-crystalline, and amorphous plastics, in Table 10.1 (Data from Ref.10.1).

Table 10.1 Solubility Parameters $(MJ/m^3)^{0.5}$ for Polymers & Solvents

Polymers		Solvents			
PTFE	12.6	n-pentane	12.8	Di-ethyl ether	15.1
PE	16.3	n-hexane	14.9	Di-methyl ether	18.0
PP	16.3	n-octane	15.5		
POM	22.6			Turpentine	16.5
PA66	27.8	M.B.K.*	17.6	CCl_4	17.6
		M.P.K.*	17.8	Xylene	18.0
		Acetone	20.4	Toluene	18.2
PEMA	18.3			Benzene	18.7
PMMA	18.7	n-octanol	21.0	Chloroform	19.0
PS	18.7	n-hexanol	21.8		
PVC	19.4	n-butanol	23.2	Trichloroethylene	19.0
PC	19.4	n-propanol	24.2	Tetrachloroethylene	19.2
PET	21.8	Ethanol	26.0	Methylene dichloride	19.8
		Methanol	29.6	Ethylene dichloride	20.0
		Phenol	29.6		

* M.B.K. – Methyl *n*-butyl ketone;
M.P.K. – Methyl *n*-propyl ketone

However, it must be appreciated that the simple rules on polymer solubility which have been quoted are often broken when crystallinity is present; even a relatively low degree of molecular order usually requires that a further degree of interaction is necessary to activate dissolution. For example, PVC is considered to be more solvent resistant than polycarbonate; this is probably related to the modest degree of crystalline order which is apparent in parts fabricated from rigid PVC.

More detailed consideration of polymer solutions can be found in standard texts[10.1, 10.2].

Component designers often associate polymer solubility with penetrating *liquids,* so that emphasis is often placed upon the potential likelihood of part-failure by complete dissolution, by chemical attack, or by other forms of degradation arising from the interaction; such issues are discussed elsewhere in the present chapter (Sections 10.2.3, 10.3).

However, the solubility of plastics to diffusing *gases* generally occurs at much faster rates (due to the smaller physical size of the penetrating molecules), and presents problems for the precise specification of packaging products, whose performance is always designed to a prescribed shelf-life.

An initial estimate of the *dissolved* equilibrium concentration (C_D) of small gaseous molecules (of partial pressure p) within a polymer barrier can be obtained using *Henry's Law:*

$$C_D = Sp \tag{10.1}$$

(where S is the relevant gas solubility constant).

In many practical situations, however, this simple rule cannot accurately describe gas solubility in plastics; although S is usually considered to be a direct linear function of the volumetric proportion of the amorphous phase, commercially fabricated products may exhibit sharp *crystallinity gradients* across the barrier through which sorption takes place. Also, sorption is strongly temperature-dependent, and non-ideal behaviour is very common in the *glassy-phase* (ie. below T_g) of amorphous thermoplastics.

10.1.1.2 Dual Sorption Model

The deviation of practical sorption data from predictions based upon Henry's law (particularly when considering gases such as carbon dioxide and ethane), have led to the development of a two-mode mechanism (the *dual sorption model)* to describe sorption in plastics materials[10.3 – 10.5]; using this model, the total effective gas concentration (C^*) is given by:

$$C^* = C_D + C_H \tag{10.2}$$

C_H is the concentration of gas molecules assumed to be adsorbed into "holes", ie. in *microvoids* which are always present in the morphology of amorphous, or semi-crystalline plastics. In effect, the dual sorption model implies that where such defects occur, Henry's Law will yield an underestimate of gas solubility.

Since C_D is inversely related to crystallinity, and C_H depends upon the affinity and degree of saturation of the sorbed gas molecules, the overall gas concentration, according to the dual sorption model, is summarized as:

$$C^* = S_{\text{MAX}} \cdot (1 - V_X) \cdot p + C_H^* \cdot \frac{bP}{1 + bP} \tag{10.3}$$

where S_{MAX} is Henry's law solubility constant for the amorphous matrix, V_X is the fraction of crystallinity, C_H^* represents the hole-saturation constant and b is another parameter (constant for a given combination of polymer and sorbant), which represents the affinity for sorption into microvoids.

Hopfenberg and Stannett[10.6] have summarized the important elements of dual sorption in glassy polymers and conclude that, at low pressure, C_H generally dominates the overall contribution to sorption, but as pressure increases, C_H gradually tends towards its saturation value (C_H^*), and conventional dissolution (characterized by C_D) becomes more important.

This model carries some relevance to many practical applications for plastics; it helps account for the fact that, contrary to elementary theory, sorption of a specific gas may not always be in direct (inverse) proportion to the degree of crystallinity. It tells us that the perfection, or *homogeneity* of the crystalline morphology is equally important to the in-service performance of the product.

10.1.2 Barrier Properties: Sorption, Diffusion and Permeation

The low densities of plastics materials arise as a consequence of their open structures; it is perhaps not surprising that small molecules can pass through the structure without affecting, or being affected by the diffusion process. Water vapour, oxygen and carbon dioxide are examples of such small molecules. The barrier properties of plastics are important in many applications, notably in packaging. These depend on:

(1) the chemical nature of the barrier material and the fluid;
(2) the concentration of the fluid, or partial pressure of the gas;
(3) environment temperature, with respect to the transition temperature of the barrier;
(4) permeation time, and the development of steady-state conditions;
(5) area of barrier material; and
(6) thickness of barrier.

In general, the gas barrier characteristics of plastics may be expressed by reference to three coefficients:

- Solubility Constant, S (as discussed above);
- Permeability Coefficient, P;
- Diffusion Coefficient, D.

These are related by[10.7]:

$$P = D \cdot S \tag{10.4}$$

Permeation of plastics materials is therefore a function of both the *solubility* of the sorbant gas, and also the *diffusion* rate across the barrier. The following section describes in more detail how these properties are related.

10.1.2.1 The Physics of Mass Transport

Diffusion is the term given to a transport process by which molecules move in a direction from a high-concentration to a low-concentration site, at a rate which depends upon the initial concentration gradient across the barrier, in the appropriate direction. Fick's laws are used to express diffusion in mathematical terms:

$$J = -D\frac{dC}{dx} \tag{10.5}$$

Equation 10.5 is *Fick's first law,* which relates the steady-state mass transfer rate (per unit area), J, to the concentration gradient of diffusant in the x-direction dC/dx, by the *diffusion coefficient, D*. Diffusion is a thermally-activated rate process, so that coefficient D is strongly temperature-dependent. This relationship is strictly only applicable to steady-state diffusion conditions. For many practical situations involving the diffusion of small gaseous molecules through plastics membranes, Fick's laws may be used to model the transport process, which is then termed *"Fickian Diffusion"*.

Integration of Fick's first law gives an expression relating flux (J) to the diffusion coefficient (D) and gas concentrations C_2 and C_1, at either face of a gas barrier of thickness x, by:

$$J = \frac{1}{x}\int_{C_1}^{C_2} D \cdot dC = D \cdot \frac{C_2 - C_1}{x} \tag{10.6}$$

Similarly, for a given plastics barrier/fluid system at equilibrium, the steady-state flux due to *permeation* is given by:

$$J = \frac{1}{x}\int_{p_1}^{p_2} P \cdot dp = P \cdot \frac{p_2 - p_1}{x} \tag{10.7}$$

where p_1 and p_2 are the partial pressures of the permeating species on either side of the barrier, and P is the permeability coefficient.

From these expressions, the total amount of permeating species (Q), per unit time (t) and cross-section (A) can be evaluated thus (see also Figure 10.1):

$$Q = DAt \cdot \frac{C_2 - C_1}{x} = PAt \cdot \frac{p_2 - p_1}{x} \tag{10.8}$$

The units of diffusion coefficient (D) are $m^2\,s^{-1}$; however, the chosen units for permeability coefficient are often conflicting, and vary between laboratories according to the materials and types of experimentation being used. Examples of such variations include:

(1)	SI Unit:	$mol \cdot m \cdot N^{-1} \cdot s^{-1}$
(2)	"European" Unit:	$cm^3 \cdot mm \cdot m^{-2} \cdot s^{-1} \cdot cm(Hg)^{-1}$
(3)	"American" Units:	$cm^3 \cdot mil \cdot (100in^2)^{-1} \cdot day^{-1} \cdot atm^{-1}$
		$cm^3 \cdot \mu m \cdot m^{-2} \cdot day^{-1} \cdot bar^{-1}$

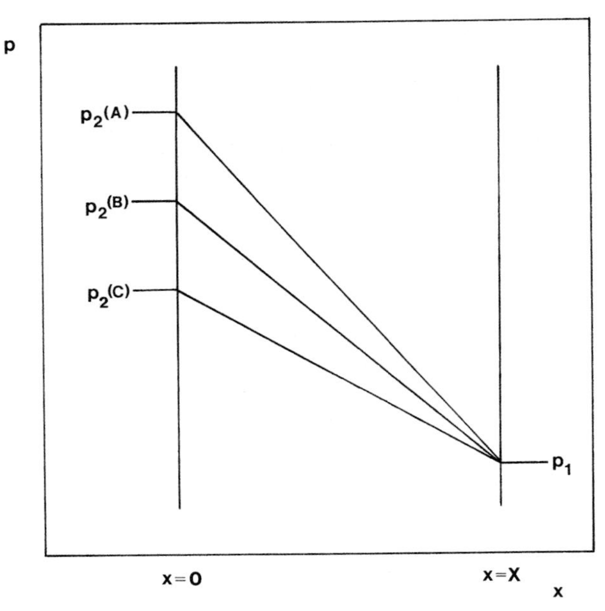

Figure 10.1 Steady-state permeation through a barrier of thickness x for a range of gas pressures ($p_2(A)$, $p_2(B)$ and $p_2(C)$) at the incident face. The gradient is related to the ratio of permeant transport rate to permeation coefficient, (Q/P), for a barrier of given geometry.

Fick's second law can be used to describe the time-dependent change in gas concentration. For example, considering unidirectional diffusion in the positive-x direction, and assuming that D is independent of gas concentration:

$$\frac{\delta C}{\delta t} = \frac{\delta}{\delta x}\left(D \cdot \frac{\delta C}{\delta x}\right) = D \cdot \frac{\delta^2 C}{\delta x^2} \tag{10.9}$$

10.1.2.2 Application to Plastics Barrier Materials

Permeability is generally a steady state concept, applicable to thin barriers, as found in packaging. For other systems, where greater thickness of barrier materials is involved, the establishment of steady state conditions may take a considerable time, relative to the lifetime of the component. It may then be necessary to analyse the system in greater detail, invoking the solubility and diffusion coefficients.

Cellular plastics must also be considered separately, because of their structural inhomogeneity, and their high transmission rates to permeating vapours and fluids; details of suitable test techniques for plastics foams are given in the appropriate Standards (eg. Ref. 10.8).

In practice, a solution for diffusion coefficient (D) from Equation 10.9 is obtained according to the boundary conditions relevant to the application or experiment involved. Such solutions have been derived for a number of dissimilar but related situations, and have been presented in more dedicated texts[10.7,10.9]; a brief reference is given here to applications of significant commercial interest, for which appropriate calculations of barrier properties are usually required.

(A) Thin plastics films (steady pressure-gradient)

CRANK[10.7] has evaluated the appropriate solution to equation 10.9, relating the amount of diffusant, $Q(t)$, to the maximum concentration of the penetrant (C_0, ie. at the incident face of the barrier) for a path length (film thickness) x. Over a relatively long measurement interval, t, this solution reduces to:

$$Q(t) = \frac{D \cdot C_0}{x} \cdot \left(t - \frac{x^2}{6D}\right) \qquad (10.10)$$

Figure 10.2 illustrates this solution, and shows two points of specific interest. First, in the initial non-linear phase, the concentration of the permeating species is increasing *within* the bulk of the film; ie. the process is unsteady. Therefore, equating $Q(t)$ to zero in Equation 10.10 and rearranging, we have:

$$t^* = \frac{x^2}{6D} \qquad (10.11)$$

t^* is known as the *experimental time-lag,* beyond which the data can be used to evaluate valid, steady-state values of diffusion coefficient by measurement of the gradient, $D \cdot \frac{C_0}{x}$.

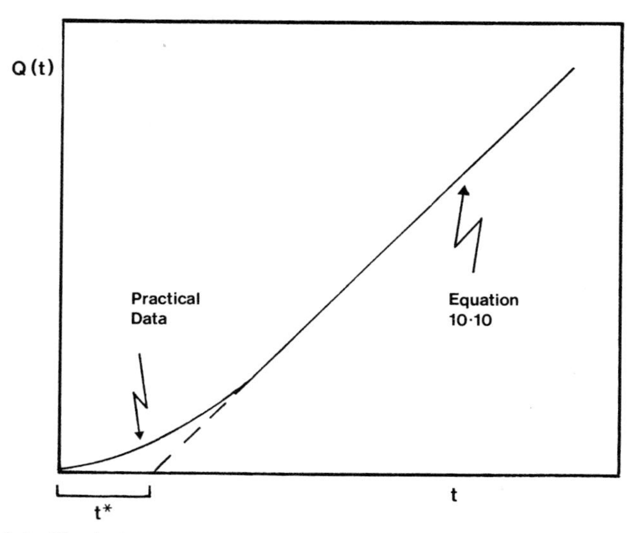

Figure 10.2 Typical permeation behaviour in a thin film, illustrating non-linearity, and the subsequent development of steady conditions following an experimental time-lag t^*.

As long as the experimental timescale, $t > t^*$, this method is applicable to plastics films or sheets; ie. in applications where the concentration gradient of penetrant remains unchanged throughout the experiment. Examples include water vapour penetration into moisture-absorbing foodstuffs through extruded plastics film-wrapping (LLDPE, EVA or PP), and the penetration of saturated organic vapours through blow-moulded containers (HDPE fuel tanks, or PETP cosmetics bottles). In the latter examples, of course, the gas barriers may be appreciably thicker, and the experimental time lag will be considerably increased, as a result.

(B) Diffusion from opposite surfaces

This situation arises when products are initially placed in direct contact with an environment which is rich in diffusant; penetration then occurs towards the mid-plane of the product, often from each of the faces which are normal to the transport direction. The diffusion process is transient; the concentration of diffusant at any position varies with time, with the barrier-layer thickness, and is also determined by the relevant diffusion coefficient for the diffusant and the specific barrier.

Expressions involving through-thickness concentration gradients can be transposed, by integration, into solutions presented in terms of *mass uptake,* which then provides a convenient measurement principle for experimental purposes. For a diffusion process through a barrier of thickness x, characterized by a coefficient D, the mass uptake at any time (t) is defined by:

$$M(t) = M_{\infty}\left[\left(\frac{4}{\pi}\right)^{0.5} \cdot \left(\frac{Dt}{x^2}\right)^{0.5}\right] \tag{10.12}$$

where M_{∞} is the equilibrium mass uptake, for infinite diffusion time. This type of data are often plotted as mass uptake versus square root of time ($M(t)$ vs. $t^{0.5}$), or by the fraction of *saturation* mass uptake $\frac{M(t)}{M_{\infty}}$ versus $\frac{t^{0.5}}{x}$: a linear plot is often considered to be indicative of a *"Fickian"* transport process. However, such transient data must be used with caution when evaluating coefficients of diffusion; note also that Equation 10.12 is invalid when the diffusion rate is retarded, during the approach towards equilibrium conditions.

Figure 10.3 illustrates these points with reference to the moisture uptake of glass-fibre reinforced PETP in hot water at 87 °C; the diffusion rate and the equilibrium mass uptake level appear to vary with composition, but with each grade, linearity is gradually lost over large intervals of time. It was found[10.10] that the approach to "equilibrium" was accelerated by simultaneous chemical hydrolysis of the matrix polymer, associated with corresponding desorption of decomposition products. This was far more prominent in PETP compounds containing an increased content of short glass fibre reinforcement (PETP-B, in Figure 10.3).

By definition, it is possible to equate $M(t)/M_{\infty}$ to 0.5, for a Fickian diffusion process, in order to evaluate a "half-life" with which to characterize the mass uptake process; this leads to a unique relationship between the apparent diffusion coefficient (D), the path length (x) and time (t):

$$D = \frac{0.0494}{(t/x^2)^{0.5}}$$

Therefore, the half-life is given by:

$$t_{1/2} = 0.049 \cdot \frac{x^2}{D} \tag{10.13}$$

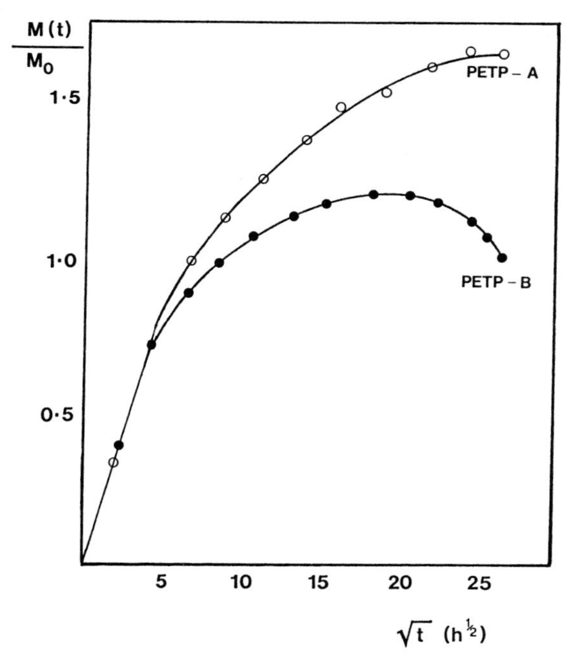

Figure 10.3 Transient-state mass transport in plastics, from an infinite diffusant concentration: time-dependent moisture ingress (expressed as a percentage weight gain, $M(t)/M_O$) in short glass-fibre reinforced PETP is illustrated, together with the effect of fibre loading. (PETP-A and PETP-B contain 10 % and 30 % glass-fibre reinforcement, respectively).
Note the weight loss which becomes apparent in PETP-B during the experiment: this is due to the simultaneous desorption of low-molecular weight decomposition products, formed by hydrolysis. (Reproduced from Ref. 10.10, with permission from the publisher, Society of Chemical Industry, SCI)

These simple techniques provide approximate values of D, evaluated on the basis of simultaneous sorption and desorption processes. On condition that experiments be executed to simulate in-service conditions with a degree of accuracy, diffusion data are often of considerable importance to practical designers. This is particularly true where water is the diffusing species, in applications involving drying, or wetting of plastics raw materials and/or finished products. Some practical examples include:

(1) Feedstock pre-drying. Residual moisture in granular feedstocks of hygroscopic plastics is expelled by evaporation in hopper-driers, prior to melt processing: the minimum drying time depends upon the initial concentration and depth of moisture penetration, the geometric shape of the polymer chips, and the appropriate diffusion coefficient.

(2) Moisture-uptake in products. Most commercial plastics will absorb water to some extent; the rate of absorption is diffusion-dependent, and the final degree of water-uptake is determined by the chemical nature of the individual plastic material (hydrogen bonding, and polycondensation plastics are especially prone to this effect), in conjunction with the concentration gradient across the diffusion interface.

For 3 mm acrylic (PMMA) sheet wetted from one side, D is 0.3×10^{-12} m^2/s, whence the half-life is 17.5 days. However, D is very temperature-dependant: drying (or wetting) at elevated temperatures is far more rapid, and in many cases the saturated water content appears to be little affected by changes in temperature. There are, however, exceptions and limitations to this simple behaviour when the product is *plasticized* by the water (effectively, defined by a drop in T_g) to the rubbery state. PMMA is a good example; it rapidly absorbs water at 100 °C, becomes rubbery, then takes-up more moisture (at increased rates), becoming foamed. These changes are purely physical; they take place without breaking the primary covalent bonds in the polymer chains.

The saturation level of moisture-uptake determines the extent to which the mechanical properties of plastics products are modified by its presence. In general, yield strength and stiffness decrease with moisture content, but toughness may be enhanced since the energy-barriers for relaxation may be reduced if the water molecules are able to penetrate adjacent polymer chains. These changes are obviously time-dependent in practice, and properties data should always be carefully standardized, and therefore used or quoted with care.

Some additional effects associated with water absorption (hydrolysis; blistering of GRP) are addressed elsewhere in this chapter (Sections 10.2.3; 10.4).

10.1.2.3 Some Implications for Plastics Products

Any materials selection exercise for products which are to fulfil a barrier-property criterion, must also include reference to other properties which may, indirectly, affect the performance of fabricated components. A good commercial example of this is the biaxially-oriented, stretch blow moulded container:

(A) Carbonation-retention

There are deemed to be four independent mechanisms by which CO_2 molecules are lost from carbonated beverages packed in, for example, stretch blow moulded PETP[10.11]:

Sorption. CO_2 is adsorbed into the PETP according to the carbonation gas pressure; an increase in sorption is observed at lower temperatures. The dual sorption model referred to in Section 10.1.1.2 describes this process with some accuracy; in addition to conventional dissolution, the presence of microvoids and associated microstructural defects is important with respect to sorption, and also in terms of the transient-state permeation which occurs subsequently.

Creep. A carbonated drinks container can be regarded as a cylindrical pressure vessel operating under a "constant" pressure (therefore constant hoop stress) condition. Like all plastics, oriented PETP suffers from *creep* effects (seen by increases in time-dependent hoop strain), which results in diametric expansion, and increased volumetric capacity in the bottle. Although volumetric expansion is usually limited to levels around 2 %, the effect contributes to a reduction in the internal carbonation pressure. However, this is compensated by the fact that the driving force for creep, the wall hoop stress, is itself diminished by creep effects.

In consequence, the overall effect is assumed to be equilibrated during the early stages of the shelf-life. The effect may be further controlled by increasing the degree of *stress-induced crystallite orientation* in the hoop direction, which effectively enhances creep modulus. This may be achieved at the design stage (by using a more appropriate circumferential stretch ratio), or during the process itself (by using "low-temperature, high strain-rate" preform expansion conditions).

Closure loss. The design of the bottle-seal should be such that closure losses of CO_2 are strictly controlled; for plastics closures, the relaxation of *interfacial stress,* in this context, should not be overlooked.

Permeation. For most carbonated foodstuffs containers of appreciable service-life, CO_2 retention (and therefore the ultimate shelf-life of the product) is primarily determined by gas *permeation,* making this property the most important, though not the only factor of commercial relevance.

The overall effect of these individual mechanisms of CO_2 pressure loss from carbonated drinks containers is illustrated in Figure 10.4; note the enhanced rate of permeation in lower-capacity containers, which arises due to their greater "surface area to volume" ratio.

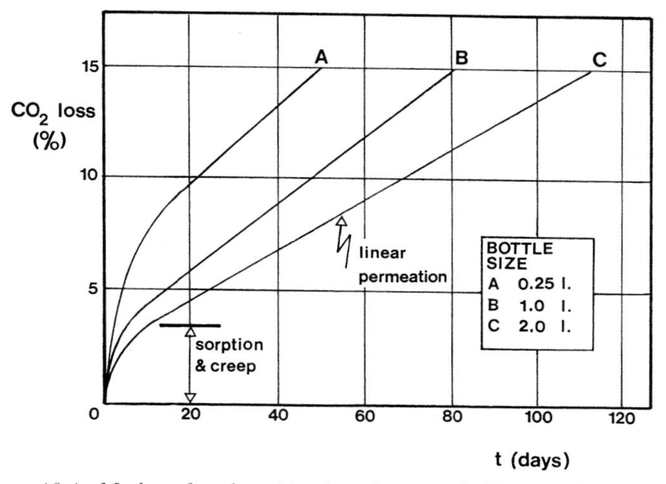

Figure 10.4 Modes of carbonation-loss for stretch blow moulded PETP containers of different capacity (0.25 – 2.0 litres; 4 volumes carbonation at 25 °C). (Data reproduced from Ref. 10.14, with permission from ICI Chemicals and Polymers Ltd.).

For purposes of direct comparison, some permeation resistance data are presented later in this section (Table 10.3).

(B) Oxygen, moisture and chemical barrier resistance

In contrast to the problems associated with CO_2-loss, other types of unpressurized applications are more critically dependent on permeation effects alone, since sorption is at equilibrium, and creep does not occur[10.12].

10.1.2.4 Factors Affecting Permeation Resistance of Plastics

Having stated the importance and practical relevance of permeation, it is of interest to summarize the factors which control this property in commercial products. Permeation resistance is generally affected by[10.12]:

(1) Crystallinity: High molecular chain-order and perfection are highly-desirable, in this context;

(2) Orientation: Orientation is generally considered to be beneficial, yet the direction and type of chain-alignment should be clarified, since stress-induced crystallization often accompanies the development of orientation, and may dominate the effect on barrier properties;

(3) Chain rigidity: A high glass-transition temperature is desirable, since permeation kinetics are facilitated by enhanced chain-mobility in the amorphous phase.

(4) Interchain attraction: Plastics containing high-polarity, or hydrogen bonding species are prominently permeation-resistant. Groups of high cohesive energy density (hydroxyls, halogens, nitrile or esters) are known to increase the flow resistance to penetrating species. This is illustrated in Figure 10.5, by reference to the loss of oxygen transmission resistance in ethylene-vinyl alcohol (EVAL, or sometimes EVOH) copolymers, with increasing ethylene content. However, although strong hydrogen bonding is responsible for the exceptional barrier properties of EVAL in the dry state, it is also responsible for the attraction to moisture, which, when absorbed from humid

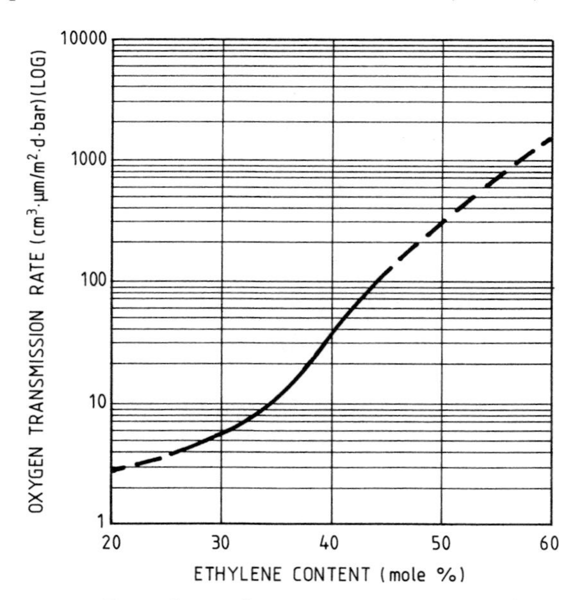

Figure 10.5 Dependence of oxygen transmission rate (at 25 °C, 0 % relative humidity, in accordance with ASTM D3985) on ethylene content, for EVOH copolymers. (Reproduced by permission from Solvay Chemicals (UK) Limited).

environments, induces significant losses to the permeation resistance of the polymer.

(5) Compatible additives: Permeability can be modified by combining different plastics, or by compounding with other chemical additives. However, in order to achieve an improvement, a strong degree of interphase compatibility must exist between the species concerned: in the case of adding inorganic modifiers, a key property which must be attained is interfacial adhesion.

The individual advantages of different families of plastics can be realized by manufacturing products which are coated, surface-modified, or which contain a number of different materials in successive layers.

A convenient mechanism has been proposed by SALAME[10.13] to describe the dependence of plastics' permeation resistance on polymer structure: an empirical *permachor number* (π) may be evaluated for any new polymer, according to:

- the atoms and individual chemical groups upon which the material is based, and their individual Permachor values π_i);
- the cohesive energy density of the polymer chains, and the free volume.

Higher permachor numbers indicate increasing permeation resistance, due to the presence of greater intermolecular forces: some examples are given in Table 10.2. When dealing with semi-crystalline plastics (including oriented materials), some account of the degree of crystallinity must be made; appropriate mechanisms of correction have been suggested[10.13] in order to account for the effects of crystallinity and molecular orientation.

Table 10.2 Example of Permachor Values for Chain-Segments and Polymers (after SALAME[10.13])

Chemical Groups		Incremental Permachor (π_i)
Main-chain:		
	$-CH_2-$	15
	–⟨benzene ring⟩–	60
	–O–	70
	–COO– (ester)	102
	–CONH– (dry/wet)	309/210
Side-chain:		
	$-CH_3$	15
	–⟨benzene ring⟩–	39
	–Cl	108
	–CN	205
	–OH (dry/wet)	255/100

Table 10.2 (cont) Examples of Permachor Values for Chain-Segments and Polymers (after SALAME[10.13])

Polymers*		Permachor (π)
PE (LD; 43 % crystallinity)		25
PE (HD; 74 % crystallinity)		39
PP (60 % crystallinity)		31
PC		31
ABS		39
SAN (25 % AN)		44
PVC (rigid)		61
PETP (quenched)		59
PETP (30 % crystallinity)		65
PETP (45 % crystallinity)		70
PVDC (copolymer; amorphous)		86
PVDC (copolymer; 50 % crystallinity)		97
PAN		110
EVAL (33 % ethylene)	(dry/wet)	97/72**
EVAL (23 % ethylene)	(dry/wet)	117/70***

* All values represent unoriented polymers
** 60 % crystallinity
*** 70 % crystallinity

10.1.2.5 Multi-Layer Packages

It is apparent that there is considerable variation between the barrier properties which are characteristic of different families of plastics materials; this is emphasized by the permeability data presented in Table 10.3.

With prior knowledge of permeation resistance, gas-barrier properties and shelf-life calculations are relatively easy to evaluate, especially for isotropic barriers of an individual material. Inevitably, however, it may not be possible to specify a single plastics material which satisfies permeation and shelf-life criteria, in combination with additional requirements such as mechanical toughness, clarity, and adequate melt-processing characteristics.

In response to these challenges, plastics packaging technology has advanced considerably in recent years. Development of processing techniques based upon *co-extrusion* and *co-injection moulding* has led to widespread commercial exploitation of composite products which combine the exceptional barrier properties of EVAL or PVDC with, for example, the toughness and optical clarity of PETP, or with the processibility and relatively low cost of oriented PP. These products have become known as *multi-layer containers*, or *multi-ply films*.

Table 10.3 Permeation Data for Plastics Materials**

	$P(O_2)$	$P(CO_2)$	$P(H_2O)$
EVAL (dry)	4 – 40	12 – 140	600 – 1800
PVDC (copolymer)	15 – 85	45 – 120	10 – 25
PA (MXD6; oriented)	52	–	1100
PAN	300	640	1600
PETP (oriented)	1600	6650	800
PA6/PA66	2000	4000	4300/1500
UPVC (unoriented)	3100	11400	880
PCTFE	4400	11200	15
PP (oriented)	44000	137500	160
HDPE	53000	191000	145
PS (oriented)	102000	400000	2900
PC	108000	406000	4500
LDPE	178000	595000	560

The data are taken from Reference 10.15; quoted units are:

$P(O_2)\&P(CO_2) : cm^3 \cdot \mu m \cdot m^{-2} \cdot day^{-1} \cdot bar^{-1}$ (ASTM D1434 / D3985)
$P(H_2O) : g \cdot \mu m \cdot m^{-2} \cdot day^{-1}$ (ASTM E96)

** These data should be regarded as typical, rather than absolute. Permeation resistance is exceptionally sensitive to the attainment of precisely controlled test conditions, to small changes in morphology (orientation, crystallinity) and also to initial moisture content, for hygroscopic plastics such as PA, and notably, EVAL.

It is possible to analyse the permeability of the more complex packages by adding the "resistance" to permeation (within each successive barrier layer), using a simple "series" mechanism. The total permeation (Q), under steady-state conditions, is given by:

$$Q = \frac{p_1 - p_2}{\sum(R_i)} \tag{10.14}$$

where p_1 and p_2 are partial gas pressures, and $\sum(R_i)$ is the total resistance to permeation of the package, obtained by summation from each of i individual layers (of thickness x and permeability P) from which it is fabricated:

$$\sum(R_i) = \frac{x_i}{P_i} = \frac{x_1}{P_1} + \frac{x_2}{P_2} + \cdots \tag{10.15}$$

Some examples of multi-layer *film-extrusion* products were quoted in Section 4.2.5.7. High-barrier *plastics containers* are generally blow moulded from co-extruded parisons, or from coinjection moulded preforms, or in contrast, thermoformed from co-extruded sheet. Alternatively, latex coatings such as PVDC may be applied to containers by spraying or by dip-coating, in a separate stage[10.14]. The following products are examples of multi-layer, permeation-resistant plastics packages, with typical layer thicknesses quoted in parentheses:

- PETP/PVDC (400/10μm): clear bottles, for enhanced oxygen-ingress resistance;
- EVAL/Adhesive/PP (20/20/300μm): 3-layer, pasteurisable fruit-juice container;
- PP/Adhesive/EVAL/Adhesive/PP (370/20/20/20/370μm): "hot-fill" and squeezable packages for ketchup, sauces etc.
- PS/Adhesive/EVAL/Adhesive/LDPE: thick sheets, for thermoforming into aseptic, and controlled-atmosphere foodstuffs containers.

The range of different material combinations and available processing techniques for multilayer packages is expanding rapidly, despite the fact that recycling may become a more difficult proposition. For more thorough guidance in this area, the reader is advised to consult up-to-date technical literature from commercial organizations involved in the development of materials[10.14, 10.15] and processing machinery[10.16], specific to this sector of the plastics industry.

10.2 Chemical Degradation of Plastics

Chemical interactions with plastics materials or products may be classified by various mechanisms, according to the nature of *degradation,* the chemical kinetics and the consequences of the reaction.

The most damaging effect of decomposition is the decrease in molecular weight which follows as a consequence; various physical and mechanical properties of plastics parts may be modified, as a result. This effect is potentially most damaging to the in-service product, when the degradation process may not be anticipated and often remains undetected. However, excessive depolymerization is usually apparent by more obvious signs of change, such as *discoloration,* surface modification (*etching*) or by direct chemical attack.

Since chemical reaction rates are accelerated at higher temperatures, a minor amount of degradation often occurs as an inevitable consequence of melt processing; assuming that the reaction can be controlled, the effects upon product performance may be minimal and can be overcome by judicious choice of molecular weight and compound additives. However, assuming significant melt-phase chemical decomposition occurs, components may be considered unacceptable on the basis of *surface discoloration,* or due to the presence of other visual blemishes such as *flow lines* or *splash marks,* which provide evidence that the rheological properties of the resin have been modified by the interaction with degradation products.

In addition, some similar decomposition mechanisms may occur – albeit at significantly-retarded rates – in the solid phase; *hydrolysis* of condensation-polymers, and *UV-weathering* of resins containing unsaturated groups being good examples.

Summarized in the following section are some of the most common modes of chemical decomposition with which commercial plastics are often associated. The exact chemical mechanisms involved are beyond the scope of this text (indeed, in many cases, there is much current debate and disagreement on such topics); our approach therefore concentrates upon the consequences of degradation, in the context of the material, the process and the product, (with examples, as appropriate) and provides only qualitative information on the reaction and its occurrence. We also give a brief reference to *biodegradability,* a property of increasing potential and importance in an environment-conscious society.

10.2.1 Direct Reaction with Oxygen

Many polymers react with oxygen, especially at high temperatures, and are degraded thereby, the reaction frequently being a chain reaction in which polymer radicals add oxygen to yield peroxy radicals; these in turn attack more polymer, forming a hydroperoxide and regenerating the polymer radical. The oxidation process may be accelerated by further decomposition of the hydroperoxide to generate more free radicals, either photochemically or by participation of a transition metal ion, which also might be a catalyst for the generation of radicals from the polymer.

There are three ways of improving the resistance to oxidation; these are exploited commercially by compounding with *antioxidants,* a special class of *antidegradant:*

(1) interfering in the chain reaction by addition of complex amines or phenols which yield stable radicals;
(2) by "scavenging" the hydroperoxides, destroying them harmlessly; and
(3) by eliminating metal ions, usually by forming complexes.

Combinations of these antioxidation mechanisms can lead to *synergistic* effects. However, the ultimate choice of stabilizer system for a given compound is inevitably a difficult decision to make, since the following criteria must be addressed at an early stage of planning:

- The factors which will promote the tendency towards degradation during all stages of processing and post-fabrication (maximum temperature, residence time distribution, viscous heating);
- Equivalent considerations relevant to service-use, such as atmospheric conditions (UV-weathering, moisture absorption, temperature fluctuation, applied or internal stress);
- The effectiveness of the compounding process, as manifested by stabilizer distribution, in addition to the other factors listed above;
- Economics and cost-competitiveness; the addition of even small quantities of expensive additive systems may add a considerable premium to the selling price of "commodity" plastics.

10.2.1.1 Oxidation of Polyolefins

Although polyolefins are relatively inert, oxidation can occur whenever environmental conditions are favourable; in consequence, care must be taken to ensure that the stabilization/antioxidant system is adequately formulated. Some of the important points in this context are summarized below:

(A) Chain Branching. It has been found[10.17] that the susceptibility of polyolefins to oxidise is related to the degree of chain-branching in the polymer, since the propagation kinetics are more favourable (the *rates of reaction* are higher) at the sites of tertiary carbon atoms. A further point which limits the oxidative stability of PP (in comparison to the polyethylenes), is the higher concentration of hydroperoxides which are formed under "equivalent" conditions.

(B) Crystallinity. It is usual to assume that at temperatures well below T_m, the crystalline phase is impermeable to oxygen molecules (see Section 10.1.1). In consequence, however, if the molecules in the amorphous phase of a highly-crystalline polymer (eg. HDPE) are oxidised preferentially as a result[10.18], a sharp drop in toughness may result. This is attributable to the effective loss in molecular weight in the rubbery amorphous phase (which implies that the inter-crystallite *tie molecules* may also be severed), resulting in a decrease in crack propagation resistance, especially under high strain-rate loading conditions, where the inability to relax and absorb significant strain energy is most critical.

(C) Melt Processing. In addition to the relevance of the fundamental research which attempts to link microstructural effects and chain-chemistry to oxidative stability, the importance of the processing phase should never be underestimated. The following points relate to some aspects of processing which can determine, or modify, the oxidation resistance of fabricated parts:

- The *thermomechanical history* associated with shaping, and the subsequent rate of cooling, is largely responsible for determining microstructural features such as crystalline morphology in plastics products;
- Free-oxygen in the vicinity of polymer melts undergoing high shear is known to induce *chain-backbone scission* in polyolefins; this effect will always modify the rheological properties of the polymer in the process under consideration, according to the extent of oxidation (determined by the amount of available oxygen, the melt temperature and the local shear stress), the speed of reaction and the flow requirements downstream of the initiation site. In addition, the severity of oxidative degradation induced during processing may be related to the product's inability to withstand stress, in combination with ageing or weathering effects in service.
- There are many possible chemical mechanisms of melt phase oxidation. Although *chain scission* is an important and an often dominating mode of degradation, it should never be implicitly assumed that it is the only possible effect; for example, *chemical crosslinking* may also occur by a competing mechanism in LDPE[10.19].

Many of these issues, with respect to both the modes of autoxidation in polyolefins, and to the classification of stabilizer/antioxidant systems, are considered in greater detail in appropriate review papers[10.20,10.21].

10.2.2 Photochemical Reactions

For photochemical reaction to occur, it is necessary that two conditions are met:

(1) the radiation must be absorbed by the polymer;
(2) the quantum of energy absorbed must be sufficient to cause the reaction to take place.

Plastics containing carbonyl, or unsaturated groups readily absorb *Ultraviolet* (UV) radiation and are directly *photosensitive*, as a result. In sunlight, although the UV radiation (wavelength range 290 – 400 nm) represents only a fraction of the total energy, it is usually the most damaging component to plastics because polymers are usually transparent to radiations of wavelength exceeding 400 nm.

Due to filtering effects in the upper atmosphere, the highest-energy radiation reaching the earth's surface has a wavelength, $\lambda = 290$ nm; such radiation is not absorbed by hydrocarbon polymers, yet they suffer photochemical degradation. This apparent anomaly is solved by recognizing that the introduction of active groups, *chromophores* occurs during processing, which tend to promote photochemical reactions. The primary reaction is independent of temperature, and ceases when the radiation source is removed; by contrast, the frequently important secondary reactions, including oxidation, are temperature-dependent and can proceed in the absence of further radiation. The incidence of photochemical degradation in plastics is therefore increased by *chain-unsaturation* (eg. in ABS), but may be reduced by incorporation of specific additives:

- Screening additives, such as carbon black, or other pigments;
- Additives which absorb radiation, eg. o-hydroxy benzophenone;
- Additives which transfer energy from activated polymer.

Photosensitisers promote free-radical formation on irradiation and may provide, therefore, a self-destruction method for the elimination of litter, or for the elimination of agricultural mulch film; the cheapest additive of this type is probably ferric stearate.

10.2.2.1 Thermal, Thermo-oxidative and Photodegradation of PVC

Despite the versatility afforded from the excellent range of properties which are offered by rigid and/or plasticized PVC, a major problem associated with this polymer has traditionally been its instability to both temperature and to photo-initiated chemical changes, by *dehydrochlorination*. On decomposing, the presence of evolved hydrogen chloride gas worsens the consequences of the chemical reaction, since HCl is seen as a "toxic" degradation product which has a tendency to attack the surfaces of lower-grade tool steels in processing equipment.

Also, due to the formation of *polyene* (double bond) sequences, which arise during dehydrochlorination, progressive compound *discoloration* (eg. white/transparent PVC tends to pale yellow, then to deeper yellows, orange-red, and finally to deep brown/black when degraded excessively), even at dehydrochlorination levels as low as 0.1-0.2 %, inevitably accompanies thermal decomposition in PVC.

(A) Thermal Degradation. It has recently been proposed that the low thermal stability of PVC is due, at least in part, to the presence of *structural inhomogeneity* in the as-polymerized molecule, which accelerates the rate of decomposition by activating the $C-Cl$ bonds in the system. Such structural irregularities are thought to include initiator residues, chain unsaturation, branches, isomers and oxidation structures. HJERTBERG and SORVIK[10.22] have presented a thorough review of current research in this area, which highlights the development of chain structure during polymerization, and the influence of structure on subsequent thermal stability during processing, where the superimposed influence of environmental oxygen further complicates the range of active mechanisms of PVC degradation.

The deterioration in physical properties which follows thermal degradation is made more severe if the compound has been specifically developed and formulated for a "high-performance" application. For example, the data presented in Figure 10.6 show how the mechanical durability of hot-plate PVC welds, fabricated from window profiles extruded in impact modified PVC (toughened by the incorporation of rubbery polyacrylate particles) can fall dramatically as welding temperature is increased. The higher temperature is associated with a greater extent of degradation, exemplified by the increased degree of *optical fluorescence* within the vicinity of the weld[10.23]. It can also be seen that a greater degree of stabilization (by increasing the concentration of either the barium/cadmium/lead stabilizer, or the lubricant

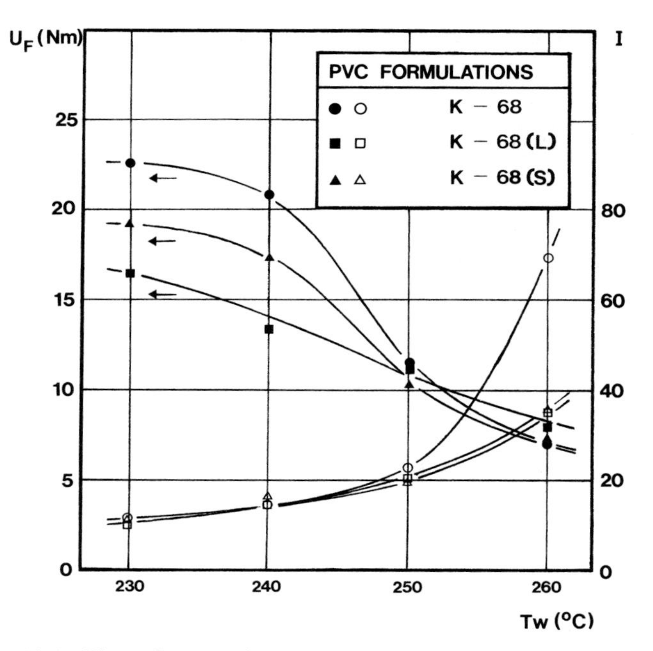

Figure 10.6 The influence of thermal degradation on the mechanical properties of hot-plate welded, impact-modified PVC: failure energy (U_F; closed symbols) of welded corner-sections and interfacial decomposition (expressed as fluorescent light intensity, in arbitrary units; open symbols) are plotted against weld temperature (T_W). PVC compounds include a standard K-68 formulation, and other variants based upon the same formulation, but containing an additional 0.7 parts lead stearate lubricant (K-68(L)), and additional 1.5 parts stabilizer (barium/cadmium/lead system, K-68(S)). (Reproduced from Ref. 10.23, with permission from Elsevier Applied Science Publishers.)

system) retards the development of *polyene* (chain-unsaturation) sequences, but cannot overcome the tendency towards degradation-induced embrittlement.

During the past decades of development of PVC compounds, for applications involving ever-increasing demands of mechanical durability and environmental stability, a considerable amount of expertise has been accumulated to ensure that commercial stabilization techniques (and end-use performance) are optimized. However, it must be stated that the level of understanding is increasing all the time, and since there are inevitable side-issues which stimulate or force changes to the concepts of PVC formulation technology (eg. toxicity effects thought to be associated with lead, or cadmium-based stabilizer systems), investigations on the fundamental basis of PVC degradation – and the mechanisms of chemical stabilization – remain part of an extremely active research area. Some recommended review papers associated with thermal decomposition are given elsewhere[10.22, 10.24–10.26].

(B) Photodegradation. Despite the fact that PVC is nominally free from chain unsaturation, it remains one of the most sensitive plastics to (natural or artificial) *weathering* and *ageing* effects, and to the associated losses in physical properties which arise as a direct consequence: these usually include a range of mechanical properties (notably the resistance to high strain-rate impact loads, and to fracture), surface gloss, colour and interfacial free energy. These observations must be related to existence of chromophoric groups in commercial PVC compounds (residual chain unsaturation, extended polyene sequences and carbonyl groups deposited during polymerization, together with contaminants, and other structural defects referred to in Section (A)), which are able to absorb sufficient energy to *initiate* photodegradation, which is then thought to *autocatalyse* in PVC as the concentration of chromophores increases. Like thermal decomposition, photodegradation is also made more complex by the co-existence of atmospheric oxygen, and the likelihood of parallel *photoxidation*.

The potential effects of UV-weathering demand that PVC formulations be formulated precisely, on the basis of some thorough research which must include environmental exposure. However, where product *colour-retention* is of strict importance, it must be appreciated that not only is the photodegradation (and discoloration) environment-dependent but also, the initial colour and progressive changes in coloration also depend upon the nominal PVC formulation. Clearly, compound formulation, UV-weathering and part-colour are highly interdependent; changes in stabilization type or level cannot be simply equated with degradation resistance alone.

UV-initiated decomposition is a *surface* phenomenon. When it occurs, polyene sequences are produced, and since it is the double-bond conjugations which are responsible for discoloration, these too are concentrated in surface layers of up to 250 μm and can act as a "UV-screen", minimizing the potential threat of further environmental damage. In consequence, the effect of photodegradation on physical properties depends upon the nature and context of the test, relative to the potential influence of the affected part surfaces. For example, a significant reduction in mechanical toughness will generally be evident only if (by the chosen test method) the stress is concentrated at the weathered surface, where the energy required for main-crack initiation is likely to be significantly reduced. It must also be noted that since *crosslink formation* accompanies *chain scission* and the development of polyenes[10.27], the deterioration in properties is not always attributable to a simple reduction in molecular weight; changes in *crosslink density* and *molecular weight distribution* make direct interpretation of test data more difficult. It is thought that chain scission controls properties in the early stages of photoxidative degradation, but crosslinking gradually assumes greater importance as the extent of decomposition becomes more severe.

The effects of *natural weathering* are not always easily classified, since there exists a range of climatic conditions which each has its own influence on polymer property deterioration. Reference must be given to *solar radiation energy,* as appropriate to local climatic conditions; for example, the annual figure for Central Europe is 350 – 400 KJ/m^2, which increases to 500 – 600 KJ/m^2 for the temperate Meditteranean climate, and attains levels up to 800 – 850 KJ/m^2 in arid desert regions such as Arizona or the Persian Gulf. Other climatic conditions of importance include mean (annual) and maximum (seasonal) temperatures, together with air humidity and rainfall; past experience has shown, for example, that solar energy imparted during the cold winter months is not insignificant, and that PVC profiles weathered in a dry environment are damaged more severely than would be the case under equivalent conditions (solar energy and temperature) in tropical climates. In addition, the colour of the plastic profile determines the surface temperature of the component (under given conditions), since light shades tend to reflect a greater proportion of the incident radiation.

Although natural weathering is the most obvious and desirable mode of testing for UV-stability in plastics, it is equally important to develop test techniques which can simulate and accelerate the effects of outdoor exposure. There are many arc-light sources available, but correlation with natural sources is generally poor. Using radiation of spectral energy distribution between 300 – 750 nm from a filtered xenon arc, the "Xenotest" is thought to correlate most directly with the effects of sunlight, and complies with relevant standards, such as DIN 53487. In addition to the possible deterioration of mechanical properties, the effects of surface weathering are manifested in other ways, such as direct colour changes, surface 'chalking' and a reduction in gloss-retention characteristics.

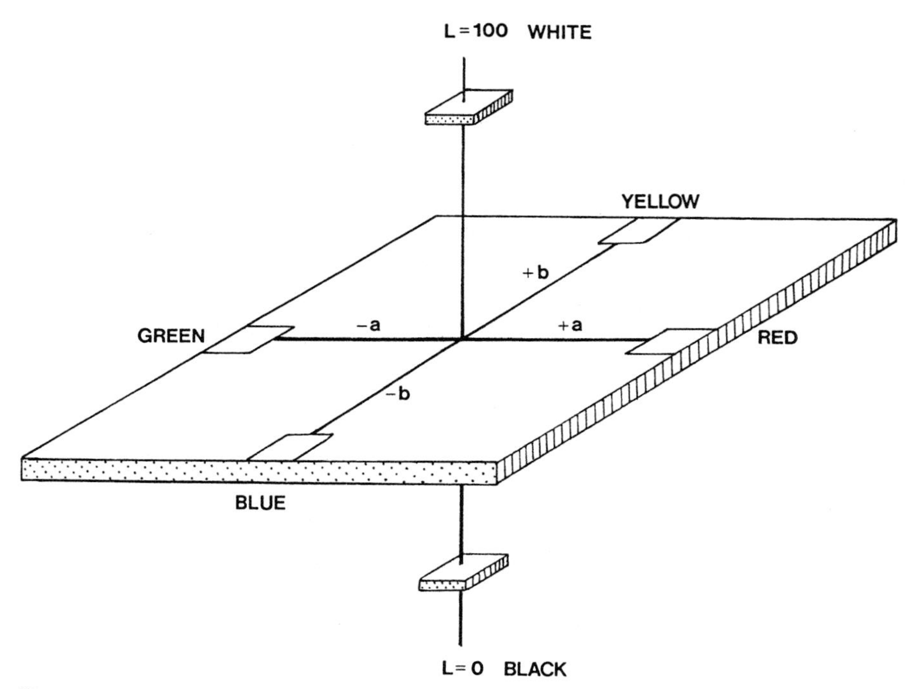

Figure 10.7 3-dimensional analysis of colour, according to the CIE-Lab. system. L (0 – 100) represents 'black to white', a (+/−) is 'red to green' and b (+/−) describes the 'yellow to blue' scale.

Discoloration is particularly damaging to light-coloured, and aesthetically-sensitive PVC products such as films or extruded window profiles; the overall colour change ($\triangle$ E) is generally measured on a 3-dimensional scale of colour (eg. according to DIN 6174; see Figure 10.7), using the following formula:

$$\triangle E = [(\triangle L)^2 + (\triangle a)^2 + (\triangle b)^2]^{0.5} \qquad (10.16)$$

where $\triangle L$, $\triangle a$ and $\triangle b$ represent changes on the white-black, red-green and yellow-blue colour axes (respectively).

In practice, it is often the tendency to observe yellowing which damages the visual appeal of such products. In conequence, a parameter of particular significance is the *Yellowness Index* (YI), which, according to ASTM D1925, is defined by comparison with a magnesium oxide reference, or by a change in yellowness index ($\triangle YI$) following a prescribed weathering treatment.

Two basic mechanisms of surface decomposition are thought to occur in PVC profiles pigmented white with TiO_2 (see Figure 10.8):

(A) Photochemical – surface PVC is attacked;
(B) Photocatalytic – non-optimized grades of additive may catalyse PVC photodegradation preferentially, around the pigment particles.

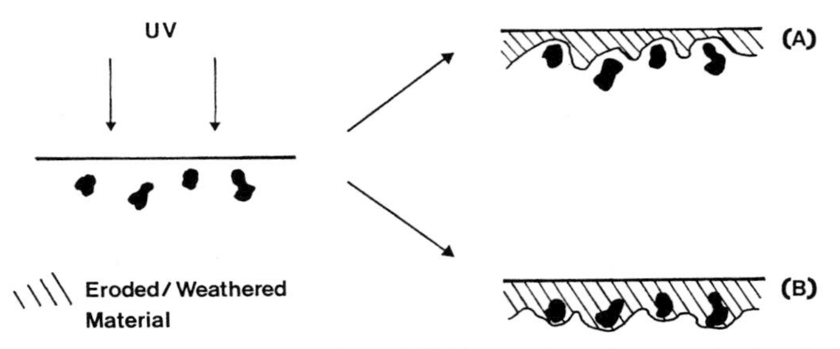

Figure 10.8 Some alternative modes of UV-induced surface weathering in PVC, around particles of TiO_2: (A) Photochemical; (B) Photocatalytic (After GAGNE; Ref. 10.28).

Generally, TiO_2 pigments are very effective co-stabilizers to outdoor weathering for PVC, since their presence reduces the rate of photochemical attack, giving increased colour stability and gloss retention. However, it is imperative that the optimum grade of TiO_2 is chosen (highly-durable, and inorganic-coated rutiles appear to have the best properties), to reduce the rate of photochemical degradation, and to minimize the likelihood of photocatalytic attack[10.28].

Some additional reference citations to photodegradation of PVC, are also quoted here[10.27, 10.29–10.31].

(C) PVC Stabilization. Having examined some of the most common mechanisms and effects of decomposition in PVC, it must be stated clearly that there now exists a vast range of effective chemical *stabilizers* which are available for use in PVC formulations, in order to prevent or minimize the effects of thermal, thermo-oxidative or photodegradation. These include organo-tin compounds, metal compounds and complexes (based upon lead, zinc, cadmium and barium) such as carboxylates, organic phosphites and secondary stabilizers such as epoxies and other organic compounds.

Inevitably, many stabilizers are most cost-effective (and are therefore used) in appropriate *combinations*, according to the modes of stabilization which appear to be most effective, the chemical mechanism(s) of stabilization which are involved, and also depending on additional factors such as processibility (stearates, for example, add a degree of lubrication to rigid PVC compounds), cost and on occasions, toxicity-related factors.

There are some notable reviews on this subject which should be referred to for further information; see especially the text edited by OWEN[10.26], and other works cited therein[eg. 10.32].

10.2.3 Hydrolysis

Some polymers, particularly those based on step growth mechanisms of polymerization, are susceptible to hydrolysis during processing. Examples are saturated polyesters and polyamides, which must be exceptionally dry before being subjected to high temperatures.

Quantities of moisture as small as 100 ppm (0.01 %), have been shown to have a significant effect on the molecular weight of PETP following melt processing (eg. see Figure 5.26). Polyesters are also susceptible to *solid-phase hydrolysis* in service, especially at elevated ("hot-water") temperatures. Individual mechanisms of hydrolytic depolymerization in PETP have been reviewed by ZIMMERMAN[10.33]; practical evidence supports an earlier observation[10.34] that the formation of carboxylic end-groups in PETP, through progressive chain cleavage, is autocatalytic.

Figure 10.9 shows the respective effects of melt phase degradation and solid-phase hydrolysis (in water at 87 °C) on the impact strength of short-fibre reinforced PETP[10.10]; the latter mechanism appears to be more severe, since the fibre-matrix interface is weakened in the presence of water molecules.

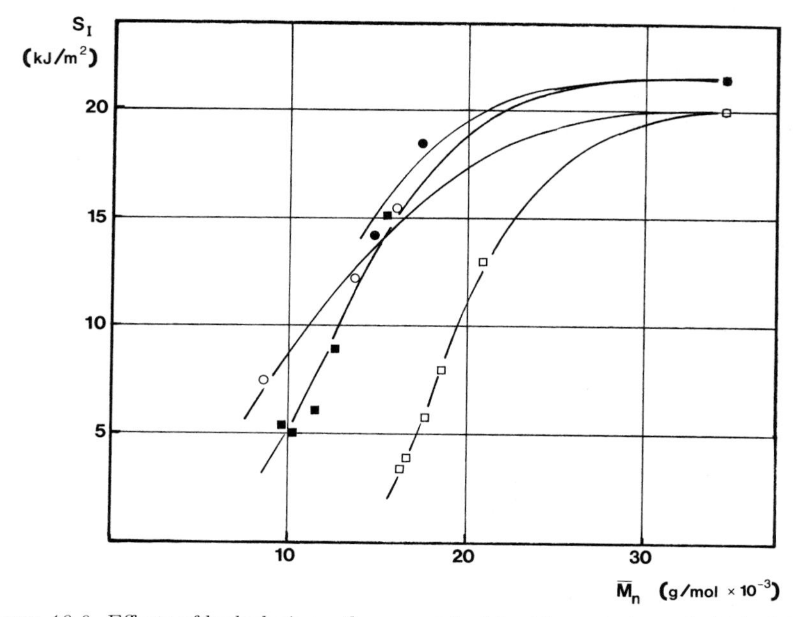

Figure 10.9 Effects of hydrolysis on the un-notched Izod impact strength (S_I) of short glass-fibre reinforced PETP, expressed as a function of number average molecular weight ($\overline{\mathrm{Mn}}$): (○) Fibre loading 10 %, melt-phase degradation; (•) fibre loading 30 %, melt-phase degradation; (□) fibre loading 10 %, solid-phase hydrolysis; (■) fibre-loading 30 %, solid-phase hydrolysis. (Reproduced from Ref. 10.10, with permission from the publisher, Society of Chemical Industry, SCI)

Polyamides are sensitive to aqueous acids, polyesters to aqueous alkalis, but neutral water is also damaging to various plastics materials, either by hydrolysis, or by its ability to penetrate long-chain systems, thereby introducing a physical modification to existing properties. These effects have been covered by some technical publications on thermoplastic polyesters, both unmodified[10.35–10.37], and on fibre-reinforced systems[10.38].

The deteriorating action of water is not confined to thermoplastics: cross-linked polyesters and polyester-based polyurethanes suffer hydrolysis when high temperature and moisture are encountered together. In the latter case, some delay in degradation is obtained by adding a "carbodiimide", (a carboxyl group scavenger), since there is evidence of catalysis by carboxyl groups during hydrolysis.

10.2.4 Biodegradation

Unlike the other mechanisms of degradation which have been described earlier, biodegradation must be considered to be desirable, since it is intended to make better use of this principle with respect to disposable plastics products of a single useage cycle, such as packaging for foodstuffs and agricultural products. Since there are several ways in which plastics can be recycled, or by which energy-recovery can take place (eg. by incineration), most studies on biodegradability are concerned with products which eventually constitute "litter", at the end of their "commercially-useful" life.

A loose definition of *biodegradability* is a material's tendency to suffer degradation by living organisms present in the natural environment (bacteria, fungi). It occurs by reaction between the chemical enzymes secreted by the organism, and the polymer chains or additives which make-up the compound.

According to WIPPLER[10.39], there are three basic stages which constitute biodegradative molecular breakdown:

(1) a change in appearance (discoloration), due to slight chemical change associated with the onset of decomposition: this may be due to the formation of double-bond conjugations in the polymer chain;
(2) a more significant change in a range of physical properties, as a result of a substantial degree of main-chain scission, resulting in a much lower average molecular weight;
(3) visual disappearance of the polymer, when mechanical breakdown *("fragmentation")* is sufficiently advanced to disperse the material to submicroscopic scale.

The majority of commercial plastics are not directly biodegradable. In order to improve the potential biodegradability, a number of criteria must be addressed with respect to both the polymer and the micro-organism:

- The environment in which biodegradation is to take place must be appropriate to the requirements of the mechanism; the organism must be able to flourish, and the natural elements (moisture, sunlight) balanced accordingly;
- A high surface area to mass ratio is required, not only to accelerate the passage of secreted liquids through the medium, but also to assist in fragmentation, once decomposition is initiated;

- The molecules or additives in the compound must possess particular groups of chemicals which can react with the organism;
- The chances of observing significant biodegradation are reduced if the polymer is highly crystalline, or is of high molecular weight.

The route towards the development of further biodegradable plastics materials is concentrated upon *directly-degradable polymers,* and on *biodegradable additives.* The former option is limited by the chemical nature of most existing commercial plastics, and the constraints of cost and market inertia associated with new, unproven alternatives. However, there are some current developments underway, such as plastics based upon Poly(β-hydroxybutyrate) (PHB) (developed by ICI) and Poly(ε-caprolactone) (Union Carbide).

With respect to biodegradable additives, the most effective option appears to be the incorporation of modified (naturally-occurring) materials, such as starches[10.40], although it must be anticipated that some associated disadvantages will arise (reduction in physical properties, and/or unpredictable decomposition whilst in-service) if large quantities of such additives are involved.

Further reference is given to this subject elsewhere[10.39 – 10.42].

10.3 General Chemical Resistance

Many plastics are very resistant to attack by chemicals; indeed, plastics find widespread use in chemical plants, in distribution systems for natural gas and water, and in effluent and sewage disposal networks. It is sometimes simpler to find a solvent which dissolves a polymer than to find a chemical reagent which attacks it.

There is, however, one class of chemical reagent which will attack many groups of plastics: an *oxidizing acid* will generally react even with polyolefins (PE, PP), although some heating may be necessary. Hot, fuming nitric acid reacts readily with polyethylene surfaces[10.43], and preferentially attacks interlamellar regions of the polymer. Another common treatment is *permanganic acid-etching* (application of $KMnO_4$ dissolved in sulphuric acid), which reveals morphological detail in various polyolefins, without penetrating further than $1 - 2\mu m$ below the specimen surface[10.44]. Since both reactants preferentially attack the non-crystalline phase, these reagents etch the surface and thereby develop the detail of the more slowly consumed *crystalline lamellae.* Etching is therefore an aid in the study of polymer *crystalline texture* by Scanning Electron Microscopy (SEM), which has been exemplified in Figure 1.13. This is a SEM micrograph of a section through the thickness of an extrusion blow moulded container, manufactured in HDPE. The micrograph was taken with the specimen tilted, accounting for the elongated aspect of the texture. The surface of the section was first smoothed with a microtome, then treated with permanganate etching reagent (as described), followed by a thorough rinse with water to wash away the debris. After drying, the surface was sputtered with gold, to reduce electron charging of the surface during examination, which would otherwise distort the image.

10.3.1 Types of Chemical Attack

Although the analytical principle described above makes use of the differential chemical resistance of amorphous and crystalline phases of plastics, it is generally true to suggest that, apart from this example, any tendency towards chemical attack (even on a very fine scale) is extremely detrimental towards the physical properties of parts manufactured from the polymer. However, it is often quite difficult to grade the chemical resistance of plastics in a manner which encompasses the different types of response which become apparent in various families of materials, and the severity of chemical attack, as and when it occurs. Furthermore, it is sometimes difficult to distinguish *physical effects* (liquid absorption, diffusion) from *chemical reactions,* without some detailed investigations into the affected areas of the products.

Presented loosely in order of decreasing significance, some examples of chemical interactions with plastics materials include the following:

- **Direct chemical attack:** Usually associated with concentrated, oxidizing acids, though sometimes only seen at elevated temperatures, the effect of direct attack by excessive exposure to a powerful oxidizing reagent is complete dissolution of the polymer.

- **Preferential chemical attack:** This effect is prevalent in semi-crystalline plastics, where the amorphous phase is prone to more rapid chemical attack. However, multiphase polymer systems (blends, copolymers) are also subject to similar effects, according to the chemical nature of each phase, and the degree of compatibility which exists between them. For example, osmium tetroxide (OsO_4) is often used as a staining medium during sample preparation for analysis by optical, or electron microscopy; ruthenium tetroxide (RuO_4) has also been used in this manner, for PC-PBTP (and elastomer-modified) blends[10.45], since it oxidizes the amorphous PC fraction.

- **Environmental stress cracking (ESC):** In strict terms, ESC is not purely a chemical effect; however, there are circumstances in which embrittlement occurs, undoubtedly associated with local chemical decomposition, by a mechanism which would not be possible without the presence of either the stress (whether an externally-applied, or an internal fabrication stress), or the chemical environment.

- **Surface attack:** It is well-known that the surface constituents and characteristics of fabricated plastics may differ considerably from the equivalent properties in the bulk: this can be caused by additive migration, or by inhomogeneities associated with processing, such as unsteady cooling. Consequently, it is possible for chemical reactions to be initiated, but restricted to the surface monolayers of the part. Whilst this type of damage (which may take a considerable time to become apparent) can be optically superficial, it is not unusual to observe a trend towards increased brittleness, since sites are created from which microcracks may propagate more easily, under prolonged external stress.

- **Swelling:** Many organic media are able to penetrate the amorphous phase of plastics materials. If a chemical reaction cannot, or under certain conditions does not occur, the interaction can often be seen as a physical swelling effect, caused by the breakdown of intermolecular bond attraction due to the presence of the migrating species. Since chemical crosslinking tends to reduce the extent of this effect, swelling tests are often used as an empirical measure of the degree of crosslinking in rubbery polymers.

It is inevitable that the severity of chemical attack, in conjunction with any associated physical interactions, tends to be graded on an empirical basis. Materials manufacturers' data sheets often define chemical resistance in an arbitrary manner; this type of information may be digested with confidence only if the precise experimental detail (reagent concentration, temperature, sample type) is defined. For example[10.46], the effect of chemical contact on PA66 resins has been categorized as follows:

"A" No attack. Little or no absorption. Little or no effect on mechanical properties.

"B" Little or no attack. Some absorption causing slight swelling. Slight reduction in mechanical properties at 23 °C.

"C " Some attack. Considerable absorption at 23 °C. Material not suitable for contact unless limited product life is acceptable.

"D" Material decomposes at 23 °C in a short time.

10.3.2 Test Methodology

Due to the difficulties associated with defining the extent of damage in parts which have undergone chemical degradation, it is usual to investigate these types of effects by indirect techniques; ie. to measure some relevant physical properties of test-pieces, which relate the chemical-polymer interaction in useful practical terms.

Typical properties which fall within this context are generally expressed relative to control, or to untested samples, and include the following:

- Weight changes;
- Swelling, and changes in linear dimensions;
- Mechanical properties, such as yield data, and energy-absorbing characteristics;
- Yield of volatiles, and gaseous products.

10.3.3 Chemical Resistance of some Commercial Plastics

Chemical resistance usually diminishes with increasing temperature, since reaction kinetics are rendered more favourable. Amorphous material is generally considered to be more prone to chemical attack, hence the subdivision used below, to illustrate the individual properties of some examples of commercial plastics. All the data have been reproduced from manufacturers' data sheets, as listed in References 10.46 – 10.50.

Semi-crystalline plastics

LDPE - Branched polyethylenes are generally resistant to salts, alkalis and also to many acids, though not oxidizing acids. Halogenated and aromatic compounds are also likely to interact with LDPE. However, the chemical resistance of PE generally increases with density, reflecting the dependence on crystallinity[10.47].

PA66 - Water absorption characteristics are evident, though extensive solid-phase hydrolysis is rare. Although they generally possess good chemical resistance, characteristic of high-crystallinity plastics, PA66 resins are susceptible to attack from concentrated acids, hydrogen-bonding solvents (eg. phenol), and also from halogens such as fluorine, chlorine water and sodium hypochlorite. As with polyethylenes, the chemical resistance of polyamides can also be modified by controlling the degree of crystallinity; the fabricator has the option to enhance crystallinity by increasing the time span which the resin spends in the region between T_m and T_g during the cooling phase of manufacture. This is generally achieved by using a heated moulding tool, and/or by selecting a grade containing a heterogeneous nucleant[10.46].

PEEK - This highly crystalline, linear aromatic polymer has exceptional chemical resistance to a wide range of organic, inorganic and oxidizing reagents; only concentrated sulphuric acid (which induces dissolution) and fuming nitric acid (decomposition), of the commonly-encountered liquids, appear to induce any significant reduction in mechanical properties, following 7-day exposures at ambient temperature[10.48].

Amorphous plastics. Due to the relatively open structure of glassy thermoplastics, it is generally found that this class of polymer is more susceptible to environment-related damage; in particular, there is a much stronger tendency towards swelling, and chemical dissolution (involving a wider range of organic, and inorganic media) than is the case with highly crystalline plastics. Consequently, there are some properties of plastics which can be examined to allow predictions of chemical resistance to be made: these include *solubility parameter* and *polarity,* the importance of which was detailed in Section 10.1.1.

The only practical advantage to be afforded by the tendency towards chemical dissolution is *solvent adhesion,* a principle for bonding plastics which is practically feasible as long as the interaction can be strictly confined to the surface layers associated with the bond interface.

PS Styrene plastics are generally attacked by concentrated acids, but have some chemical resistance when the acid is diluted. The presence of the benzene-ring side groups tends to induce only limited vapour permeation resistance (the solubility parameter is 18.7 $(MJ/m^3)^{0.5}$), and PS is therefore prone to dissolution, swelling and environmental stress cracking by a range of chemicals, notably aromatic and halogenated hydrocarbons, ethers and ketones. The diffusion rate, and also the tendency towards stress cracking, are dependent upon molecular orientation and residual stress in fabricated products. More extensive chemical resistance data-banks are generally provided by manufacturers[10.49].

PVC Having a similar solubility parameter to PS, PMMA, and to PC, rigid PVC would generally be expected to interact with similar families of chemicals. This is true to a point, but PVC (being highly polar, and containing a small degree of crystalline order) is significantly more resistant to inorganic acids, at temperatures up to 60°C. Additional factors which must be considered with PVC compounds include the presence of other polymers and additives within the formulation (the effectiveness of which may be reduced when in contact with aggressive media), and the degree of gelation. For example, poorly-gelled PVC is attacked by acetone at ambient temperature, whereas optimally-processed material is much more resistant; the "acetone test" is often used as an empirical measure of PVC gelation, as part of product "quality control" procedures.

PES The presence of sulphone groups in the chain backbone of amorphous PES results in a relatively high T_g and also, in improved chemical resistance in comparison to some other amorphous plastics. If the free volume can be diminished, for example by high-temperature ageing, the performance can be further enhanced. In summary, PES possesses good resistance to acids, bases and other inorganic solutions, but short-fibre reinforced grades are often recommended for long-term, or elevated temperature exposure, particularly when under external stress[10.50].

Where there is any doubt about the long-term suitability of a particular grade of material in a new application which is associated with exposure to aggressive chemical media, it is always advisable to initiate proof tests which will add supporting evidence to the design and choice of material. Environmental factors such as temperature, fluid concentration and exposure time must be considered alongside product-related variables (material formulation, microstructure, residual stress) in order to produce a more exact and appropriate part-specification.

10.4 Blistering of GRP Laminates

10.4.1 Introduction

Glass fibre reinforced polyester laminates, after prolonged contact with water, sometimes develop disfiguring blemishes, which affect the side of the laminate in contact with water. Since many of the blemishes contain fluid, the phenomenon is known as *blistering of GRP.* A typical blistered laminate is shown in Figure 10.10. Although delamination has been observed in glass-free systems, and also in the absence of a gel coat, it has been found most often in systems based on o-phthalic ester backing resin with chopped strand mat (CSM) reinforcement, and o- or iso-phthalic ester gel coat. The binder for the reinforcement seems to be implicated, since the "powder bound" grade is less prone to blistering than the 'emulsion bound' mat.

Figure 10.10 Blistering typical of a GRP laminate (gelcoat side), after total immersion in water at 40 °C.

Codes of practice have been developed and published, for example by the Reinforced Plastics Group of the British Plastics Federation,[10.51] to provide guidance to fabrication techniques which reduce the chance of blister formation. Even after following optimum procedures, however, there is still a chance of blisters occurring; indeed, it is possible that as many as 30-40 % of boats are affected.

10.4.2 Origin and Nature of Delamination

Recent work[10.52] has indicated that blisters originate in a number of ways; these are additional to the massive delamination which results from poor manufacturing techniques. Whilst contaminants and bubbles generated during casting may contribute to blistering, two factors have been identified as the main origins of blisters: *pre-cracks;* and *fibre-lines.*

These terms require definition: the former are generated during post-curing as *discoidal cracks,* which for clear gelcoats can readily be identified under the

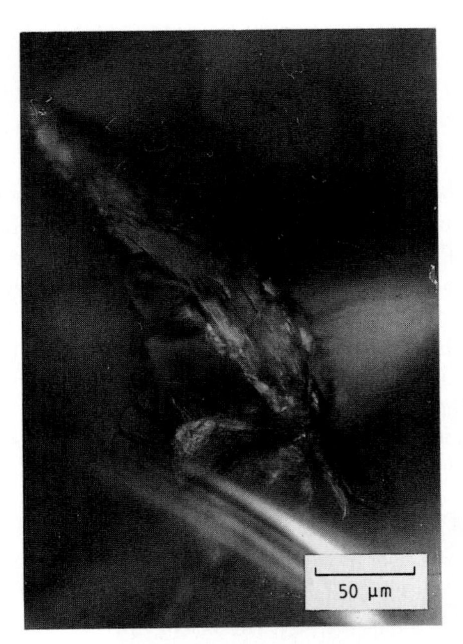

Figure 10.11 Discoidal crack in a GRP laminate; many of these defects appear during post-curing and act as centres for osmotic activity.

microscope; Figure 10.11. Pre-cracks grow more slowly; their number correlates with the birefringence at the gelcoat-backing resin interface, see Table 10.4:

Table 10.4 Pre-crack blistering in GRP Laminates

Sample code Gel/Backing	Maximum birefringence (Δn)	No. of Pre-cracks (per 100 cm^2)
Iso/Ortho	41×10^{-5}	87
Iso/PPG*	54	51
PPG/PPG	36	< 15
NPG/PPG**	321	139
NPG/Ortho	327	133

* PPG is poly(propylene maleate) – See note on isomerisation
** NGP is neopentyl glycol with maleic anhydride/isophthalic acid

We must conclude, therefore, that *stress* is established during the lamination, or curing, which is sufficiently high to cause local failure. Immersion of the laminate allows water molecules to permeate the structure, thereby swelling it. The advancing water front is a perturbation which increases the stress level and creates further discoidal cracks, and other failures associated with stresses around the glass fibres. From the outset, the crack is active in *osmosis,* and there is slow but steady growth.

In measurements of blister growth, it is usual to accelerate the growth rate by immersing the laminate, totally, in distilled water at 40 °C; see Figure 10.12. Pre-cracks develop into circular blisters, which rarely coalesce; in the presence of severe fibre-line blistering, pre-crack blisters may be visible only with difficulty.

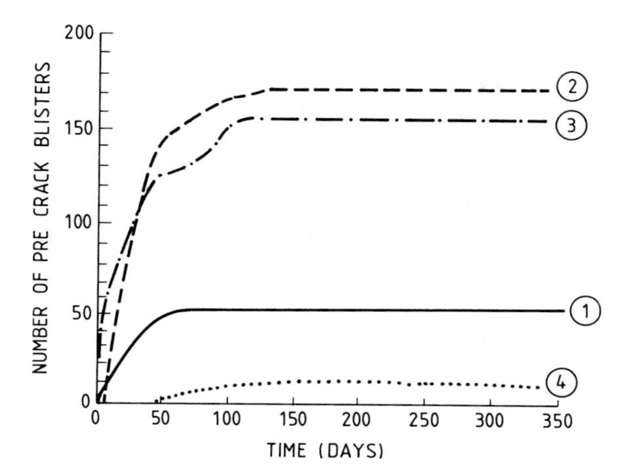

Figure 10.12 Rate of formation of **pre-crack blisters** in distilled water at 40 °C (expressed as number of defects per 100 cm^2):

(1) Isophthalate/Orthophthalate/Em CSM
(2) Isophthalate/Orthophthalate/Po CSM
(3) Isophthalate/Pprop maleate/Em CSM
(4) Neopentyliso/Pprop maleate/Em CSM

(Data reproduced from Ref. 10.52, with permission from the British Plastics Federation).

Fibre-line blisters, as the term implies, are associated with the fibre reinforcement phase; they are first seen as growth along the length of a fibre, giving surface distortion in the laminate. The surface distortion increases with immersion time, presumably being fed by osmosis, with the blisters intersecting and coalescing, so that the total number may decrease with time. The development of fibre-line blisters is shown in Figure 10.13. Attention must be drawn to the much-improved performance of isophthalate or neopentyl glycol gelcoats, when backed by a poly(propylene maleate), which, as will become evident, is largely transformed during polymerization to the trans isomer, (fumarate). Emulsion-bound CSM performs well in these backing resins.

The growth of fibre line blisters again varies considerably for the different laminates. Results for a number of formulations are collected in Figure 10.14, and show the variation in growth rate, from the rapid deterioration of isophthalate gelcoat/orthophthalate/emulsion-bound CSM, to the much superior isophthalate gelcoat/poly(propylene maleate)/emulsion-bound CSM.

The data given in the Figures do not represent all aspects of blistering: in Figure 10.13, the total number of fibre line blisters in curve 2 is greater than for curve 1. However, the size of blisters in powder-bound CSM (Figure 10.14) is very

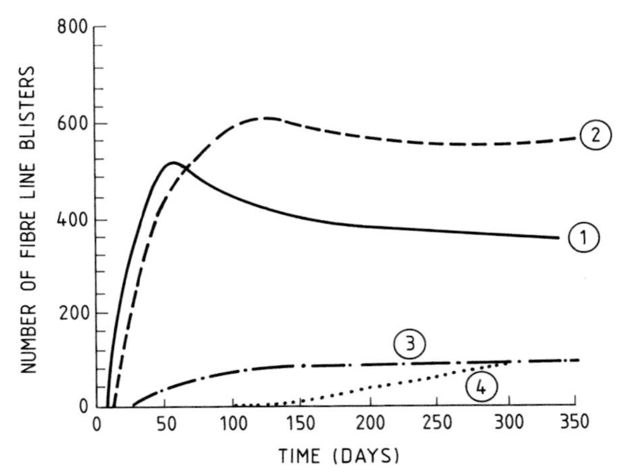

Figure 10.13 Rate of formation of *fibre line blisters;* curves identified as in Figure 10.12; (Data reproduced from Ref. 10.52, with permission from the British Plastics Federation).

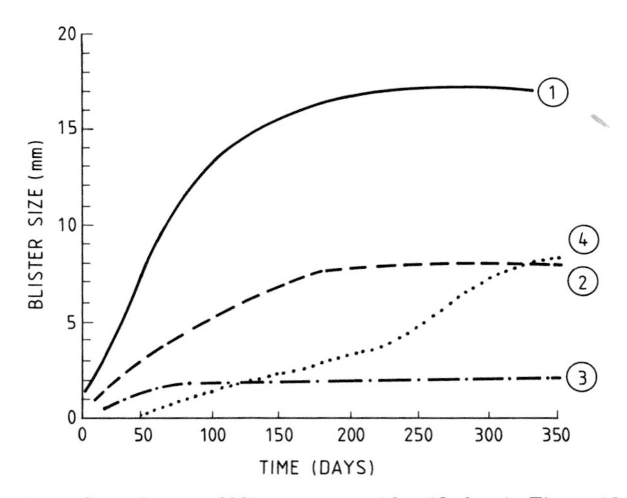

Figure 10.14 Growth rate of blisters; curves identified as in Figure 10.12); (Data reproduced from Ref. 10.52, with permission from the British Plastics Federation).

much smaller than for the emulsion bound CSM. The blisters associated with curve 3 were the smallest encountered.

10.4.3 Influence of the Resin Specification

The same authors have reported the blistering behaviour of a number of other laminates, based on polyesters and on other resins, all of which had been subjected to the accelerated blistering conditions of full immersion in distilled water at 40°C. There is some indication that the propensity to blister is reduced by matching the gel coat to the backing resin, it being argued that the interfacial stress would be reduced by matching the materials, see Table 10.5:

Table 10.5 Blistering Kinetics in GRP Laminates

Composition	Days to blister	
(Gel/Backing/Mat)	(Pre-crack)	(Fibre line)
Iso/Ortho/EmCSM	7	10
Ortho/Ortho/EmCSM	12 – 25	13 – 52
Ortho/Ortho/PoCSM	12 – 25	25 – 45
Iso/Iso/EmCSM	45	60
Iso/Iso/PoCSM	60	100
Iso/Epoxy/EmCSM	**	160
Iso/Epoxy/PoCSM	45	200
Iso/Vinyl/EmCSM	50	74
Iso/Epoxy/PoCSM	50	**

** No blistering evident after 220 days in water at 40°C

In an earlier publication, by BIRLEY, DAWKINS and STRAUSS[10.53], the use of a carbodiimide as a carboxyl scavenger was reported to be beneficial in restricting blistering. It was argued that the ingress of water is facilitated by the presence of hydrolysis products, e.g. of poly(vinyl acetate). The scavenging of the carboxyl groups gives two benefits: removal of a catalyst for the further hydrolysis of PVAC, and removal of the acetic acid, which would be active in osmosis. The addition of 5 % Stabaxol-P to uncured resin, heated for 16 hours at 40°C, reduced the acid value of a typical polyester by 50 %. The addition of the same level of the polycarbodiimide to a laminate retarded the onset of blistering markedly.

In the same paper, it was reported that the use of poly (butyl acrylate) as a binder for the chopped strand mat (CSM) gave a better blistering performance than PVAC; this is explained by the hydrolysis products being less active than those of PVAC, in promoting further degradation of the binder.

10.4.4 Isomerisation of Maleate to Fumarate

In the manufacture of unsaturated polyester resins, it is usual, and cheaper, to introduce the unsaturation via maleic anhydride which, under the conditions of polyesterification, is largely isomerized to the trans-acid (fumaric acid). The analysis can be conveniently carried out by *Nuclear Magnetic Resonance (NMR)* spectroscopy; results obtained over a range of polyesters suggest a conversion of 85 – 90 % is not unusual. This is beneficial to the properties of the final laminate.

10.4.5 Living with Blisters

A number of measures might be taken to reduce the incidence of blistering:

(1) It is necessary to follow good laminating practice with respect to formulations, temperatures and times of the various stages of processing, in order to minimize the development of stress in the system;
(2) Use of a polyester resin which is apparently less permeable to water; for example, neopentyl glycol isophthalate shows good resistance to blistering;
(3) In conventional isophthalate gel coat - orthophthalate backing resin systems, better results are obtained with powder-bound mat compared with emulsion-bound mat;
(4) With emulsion bound mats, some advantage has been observed from the addition of a carboxyl group scavenger, (a polycarbodiimide), to the resin system. Such groups are known to catalyse the hydrolysis of polyesters, including PVAC;
(5) Replacing PVAC binder by poly(butyl acrylate) gives improved resistance to blistering;
(6) Use of a resin based on poly(propylene maleate) in both gel coat and backing resin gives better performance with emulsion bound mat. The improvement does not extend to powder bound mat.

It must be stressed that these possible routes to improved resistance towards blistering are not necessarily additive.

10.5 Flammability and Fire Properties of Plastics

Most plastics are degraded by intense heat to yield smaller molecules; these are often flammable and thus provide fuel in a fire situation. Some plastics resist ignition, whilst others can be lighted easily. Other aspects of fire include *smoke emission* and the release of decomposition products, including toxic gases.

The essential requirements for a fire are heat, fuel and oxygen, as illustrated by the *fire triangle*[10.54] in Figure 10.15. Once combustion has occurred, the course of the fire often accelerates rapidly; the risks involved, and therefore some of the measurements relating to material flammability properties, are summarized below.

- Fire Initiation — Ignition sources; Ignitability; Flammability.
- Fire Propagation — Flashover; Flame Spread; Heat Release Rate.
- Fully Developed Fire — Fire Penetration; Smoke Density, Toxicity and Corrosivity.

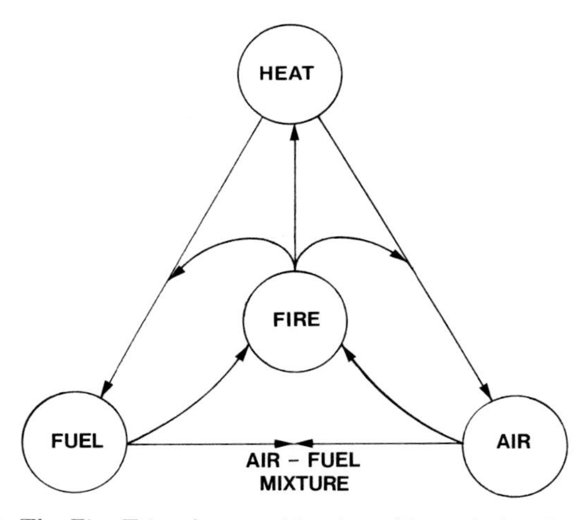

Figure 10.15 The Fire Triangle – combination of heat, fuel and oxygen.

If organic plastics are to act as the source of fuel in a fire, the flammability properties of relevance include ignition and flash points, thermal conductivity, specific heat capacity and the exothermic heat of combustion. Ignition will occur only when the chemical decomposition process induced by external heating produces a sufficient volume of flammable gases; such combustion mechanisms are inevitably quite complex, and involve reactions between radicals formed when the polymer initially decomposes[10.54].

10.5.1 Fire Test Methods

Due to the diverse aspects of flammability and to the large number of relevant variables, tests are notoriously irreproducible and often do not correlate with practical experience; it is probably true to suggest that there is no single test which completely reflects the fire and flammability properties of plastics materials or products. Even if a much more thorough and meaningful test were to emerge, and assuming international agreement was to evolve to support the introduction of such a test, it would require new legislation to fully-implement the procedure. Consequently, it is current practice to attempt to satisfy a range of standards, for purposes of materials development and marketing; towards this objective, a useful document is BS6336[10.55], which features the primary requirement to ensure that any selected fire or flammability tests closely reflect the end-use environment of the product or material under surveillance.

Two of the most commonly adopted flammability test methods are now described in more detail, together with a citation of other techniques relevant to some specific aspects of fire situations.

(1) Critical Oxygen Index. This test (eg. ISO 4589; ASTM D2863), particularly when carried out at elevated temperature, is possibly the most accepted laboratory technique. The procedure establishes the minimum volume concentration of oxygen, admixed with nitrogen, which is required to initiate combustion of the plastic and propagate flame-burning to a pre-specified extent (a flame-spread time, or a distance along the test-sample). Data are usually presented in terms of a *Limiting Oxygen Index (LOI)* value; for most commercial plastics, LOI characteristics lie in the range 15 – 35 %, the most notable exceptions being halogen-containing polymers such as rigid PVC (especially if chlorinated), PVDC and PTFE, which attains exceptional LOI values of over 90 % (see Table 10.7).

(2) Underwriters' Laboratory (UL 94) Rating. Since the flammability of a plastics product is also size and shape-dependent, the UL 94 test takes account of this by simulating the flammability behaviour of moulded products using standard specimen dimensions in conjunction with specified sources of flame ignition. Although this test bears greater resemblance to the anticipated performance of moulded plastics, it should also be understood that other factors (geometric size of product, intensity /time of heating, possibility of heat dissipation) will also affect the in-service result.

The UL 94 methodology can be subdivided further into vertical or horizontal tests (Figure 10.16), the latter also taking flame spread into account. A flame class rating is generally attained by the UL test, relative to the minimum wall thickness which will still maintain the specified rating appropriate to the material; in consequence, data are often presented in terms of a "minimum wall thickness", appropriate to flame class.

(3) Other tests. The following range of test methods has been cited by ICI[10.56] as an appropriate reference list for materials development purposes:

(1) UK - BS476 Parts 6 (fire propagation) and 7 (surface spread of flame test).
(2) USA - ASTM E-184 Tunnel test.
(3) FRG - DIN 4102; DIN 53438.
(4) Italy - DM 26/6/84 CSE/RF2/75A and CSE RF3/77 (building and appliances). Federal Motor Vehicle Safety Standard 302 (FMVSS 302/ISO 3795) UL 94: Vertical burning test (electrical applications).
(5) France - Epidirateur Fire Test.

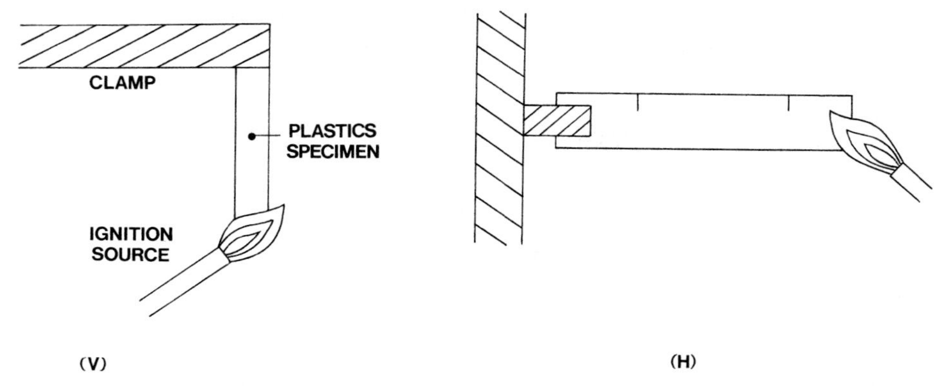

Figure 10.16 Typical arrangements for the Underwriters' Laboratory flammability tests: (V) UL 94V vertical small flame ignitibility (variable sample thickness); (H) UL 94HB horizontal small flame ignitibility/flame spread.

Additional methods include:

ASTM D-1929 – Flash Ignition Temperature; Self-ignition Temperature.
ASTM D-2843 – Smoke Density Test.
ASTM E-662 – Smoke Emission.
ISO 5657 – Radiant Cone Test (Ignition time).
BS6401 – Smoke Density Test at Normal End-Use Thicknesses
BS6425 – Corrosive Fumes.
IEC 695-2-2 – Needle Flame Test.
IEC 754-1 – Corrosive Fumes Test.

Yet more data of significance can be derived from experiments to measure fire spread, heat of combustion, smoke density and carbonaceous residue; the ultimate choice of test method depends upon the degree of relevance to end-use conditions, and the Codes of Practice necessary for purposes of product marketing.

In view of the multitude of test methods and the diversity of influential properties, it is perhaps not surprising that a degree of conflict often exists between results obtained on similar materials by different routes. For example, BATCHELOR[10.57], as part of an overall assessment of fibre-reinforced thermosetting plastics for railway vehicle applications, has reported that an improvement in fire rating to BS 476 can also lead to an increase in smoke emission. The data in Table 10.6 demonstrate this effect, and also show the impressive low-smoke performance of phenolic resins.

Table 10.6 Fire Rating and Smoke Density for Reinforced Plastics

Matrix	BS 476 Classification	Smoke density*
Polyester 1	0	1.60
Polyester 2	1	0.60
Phenolic	0	0.35

* Maximum optical smoke density (arbitrary units), by a method similar to that described by WOOLLEY et al.[10.58].

If extreme care is used, controlled flammability testing can also be used to help identify particular families of plastics materials; qualitative information includes ease of ignition, melting behaviour, colour and severity of the flame, smoke density and odour.

10.5.2 Improving Resistance to Flammability

Resistance to flammability can be enhanced by the incorporation of *flame retardants;* these are additives which reduce flammability by restricting the supply of one or more of the requirements for a fire (heat, fuel or oxygen). Some specific mechanisms by which this can occur include:

(1) the provision of a gaseous blanket which reduces access of air to the flame;
(2) the addition of ingredients which react (usually by evolving water) endothermically; and
(3) incorporation of additives which interfere with the flame reaction.

The market place for flame-retardants is dominated by low-cost aluminium trihydrate (ATH), which decomposes by a strongly endothermic reaction to produce alumina, and water vapour. As a result, the temperature of the local burning zone of the polymer is reduced, and the release of H_2O dilutes the gaseous products produced in the initial stages of decomposition. ATH is very effective in many plastics compounds, and can also be utilized in combination with other flame retardant additives such as antimony oxides, halogen-containing compounds (including chlorinated paraffins) and phosphate esters. As with many other particulate fillers, chemical pre-treatment is possible, in order to optimise the mechanical properties of fire-retardant compounds. For some applications, other cheaper fillers such as calcium carbonate can be added to plastics in combination with (say) ATH, to give an optimum balance of fire properties per unit material cost. KATZ and MILEWSKI[10.59] have written a thorough review of chemical additives in plastics materials, and their role in property enhancement, and TESORO[10.60] has reviewed the chemical modification of polymers for *flame-retardant* compounds.

Reinforced thermosetting resins are often used for structural applications (eg. automotive parts, and panels in other transport-related use) which require high stiffness in combination with fire retardancy. *Hydrated fillers* are commonly added to halogen-containing resins, which are sometimes associated with higher smoke generation, under burning conditions. If non-halogenated resins are used, fire retardancy is more difficult to obtain without a corresponding penalty of higher melt viscosity, associated with higher concentrations of additive. It has been claimed[10.56] that a newly-developed family of thermosetting resins based upon aliphatic monomers (eg. methyl methacrylate), with high levels of hydrated fillers (but containing little, if any halogen-containing fire retardant) provides excellent low-smoke, and low-toxicity characteristics, and is able to maintain the desired low-viscosity flow properties required for fabrication processes such as resin transfer moulding.

10.5.3 General Fire Properties of Some Plastics

The flammability of plastics can vary enormously, from the rapid ignition and explosive behaviour of cellulose trinitrate, to the non-flammability of PTFE. General fire-resistance characteristics of plastics include:

Thermosets:

PF, MF, UF - Difficult to burn in ordinary flames; self-extinguishing.
Polyester-Resin – Burns more readily than phenolic or amino-resins, but flammability can be controlled by formulation.

Thermoplastics:

PTFE	–	Little or no ignition.
PVC	–	Excellent resistance to flammability; however, the self-extinguishing character of rigid PVC compounds is generally lost when plasticizers are added (this depends upon the type and amount of plasticizer).
PA	–	Not easily ignited, but once burning, low-viscosity dripping is evident; characteristic odour of burning hair.
POM	–	Burns readily (with a blue, but scarcely visible flame) and drips; a strong odour of formaldehyde is apparent.
Polyolefins	–	Will burn continuously once ignited, with a characteristic blue-yellow flame and candle-wax odour.
Styrenics	–	Also ignite and burn readily, with a thick, "sooty" smoke; elastomer-toughened grades, HIPS and ABS also evolve an odour of "burning rubber".
CA, CAB	–	Rapid ignition and burning.
CN	–	Extremely flammable and potentially explosive.

Plastics foams[10.61]:

Continuous oxygen ingress is made easier by the open, cellular structures of plastics foams; burning tends to be more rapid than in solid materials. This is a particular problem for polyurethane foams (eg. in furnishings) which is compounded by the emission of toxic gases during combustion. However, it is possible to improve performance by incorporating appropriate chemical additives, and/or fire-retardant interlayers.

A selection of LOI values for plastics, and for other competitive materials, is given in Table 10.7; the data have been reproduced from References 10.62 and 10.63. A LOI value of about 27 – 28 % acts as the divide between materials which will, or will not continue burning in a normal fire environment. Notable plastics whose fire existence is characterized by LOI values exceeding this level include PF (whose thermosetting nature makes it an ideal polymer for flame-resistant laminations),

UPVC, PVDC, PTFE and some high-temperature "engineering" plastics such as polysulphones and polyimides.

Table 10.7 Limiting Oxygen Indices (LOI) of Various Materials (23 °C)

Material	LOI	Material	LOI
Polyacetal	14.9 – 15.7	Nylon 66	24 – 29
		Wood	25.2
Cotton	16.0 – 17.0	Polycarbonate	26 – 28
Natural Rubber	17.2	Neoprene Rubber	26.3
PMMA	17.4	Modacrylic	26.8
Polyethylene	17.4	Nomex	28.5
Polypropylene	17.4	Polysulphone	30 – 32
Polystyrene	17.6 – 18.3	Leather	34.8
Polyacrylonitrile	18.0	Polyimide	36.5
Styrene-Acrylonitrile	18.0	PVDF	43.7
ABS	18.3 – 18.8	PVC (Rigid)	45 – 49
Rayon	18.7 – 18.9	Carbon Black (rod)	59 – 63
Cellulose	19.0	PVDC	60.0
PETP	20.0	PVC (Chlorinated)	60 – 70
PVF	22.6	PTFE	95.0

10.5.4 Halogen Content and PVC

Halogen content is an important aspect of a material's ability to suppress flame-propagation. Halogen-containing materials are relatively difficult to ignite, a fact which is a distinct advantage offered by a tough, structural plastic material like rigid PVC. Although the halides formed during combustion (eg. HCl vapour) contribute to the toxicity of smoke emission from this material, there are other toxic gases which may be released when plastics, or other organic materials are ignited, such as carbon monoxide, and hydrogen cyanide (eg. from polyurethanes, polyamides).

The environmental issues surrounding the performance of plastics materials in fires are often based upon the consequences of ignition in flaming conditions, rather than on the possibility of a significant fire occurring at all. In response, the plastics industry must be able to reassure its users of the positive characteristics which its materials hold. For example, a trade-organization in the UK[10.64] has summarized the combustion properties of PVC as follows (see Table 10.8):

Table 10.8 Combustion Properties of PVC

Fire property	PVC characteristic
Ignitability	Very difficult to ignite with common ignition sources.
Heat Release	The rate of heat release and total heat of combustion are significantly lower than those of most other thermoplastics.
Flame Spread	Limited flame-spread.
Sustained Combustion	PVC forms a protective carbonaceous layer; the emitted HCl acts as a combustion inhibitor.
Smoke Density	Similar to wood when smouldering, but greater under flaming conditions.
Combustion gases:	
Corrosivity	Metallic materials may suffer, but restoration is possible.
Toxicity	HCl is the main gaseous product; the toxic potency of the gases is similar to and not significantly worse than that of many natural and synthetic materials. The build-up of toxic fumes is slow in comparison to rapidly burning materials of a similar toxic potency.
Overall	Resistance to ignition and how well flame is supported and spread are the most significant properties which contribute to fire safety; PVC is particularly good in this respect.

Additional bibliography and journal citations on the subject of fire properties are given in References 10.60 to 10.66.

References

10.1 BRYDSON, J.A.: *Plastics Materials,* 4th Ed., Butterworth Scientific, London, (1982).

10.2 BILLMEYER, F.W.: *Textbook of Polymer Science,* 2nd Ed., John Wiley, New York, (1971).

10.3 BARRER, R.M., BARRIE, J.A., and SLATER, J.: *Sorption and Diffusion in Ethyl Cellulose Part 1: History Dependence of Sorption Isotherms and Permeation Rates,* J. Polym. Sci., **23,** (1957), 315.

10.4 MICHAELS, A.S., VIETH, W.R., and BARRIE, J.A.: *Solution of Gases in Polyethylene Terephthalate,* J. Appl. Phys., **34,** (1963), 1.

10.5 MICHAELS, A.S., VIETH, W.R., and BARRIE, J.A.: *Diffusion of Gases in Polyethylene Terephthalate,* J. Appl. Phys., **34,** (1963), 13.

10.6 HOPFENBERG, H.B., and STANNETT, V.: in: *The Physics of Glassy Polymers,* HAWARD, R.N.: Ed., Applied Science, London, (1973).

10.7 CRANK, J.: *The Mathematics of Diffusion,* Clarendon Press, Oxford, (1975).

10.8 BS 4370, (1988), *Methods of Test for Rigid Cellular Materials; Parts 1–3,* British Standards Institution, London.

10.9 COMYN, J.: Ed., *Polymer Permeability,* Applied Science, London, (1985).

10.10 HAWORTH, B., and WALSH, G.M.: *Effects of Hydrolysis on the Physical Properties of Short Glass-fibre Reinforced Poly(ethyleneterephthalate),* Brit. Polym. J., **17,** (1985), 69.

10.11 MATTHEWS, V., ROBINSON, K., and SYKES, G.: *Melinar® PET – Barrier Performance,* Presented at PIRA Seminar: Developments in Plastics Technology for Packaging, Imperial College London, (1984).

10.12 JONES, K.M., and SYKES, G.: *Melinar® PET - The Route to Improved Barrier,* Presented at PIRA Seminar: PET Containers - An Update, Leatherhead, UK, (1985).

10.13 SALAME, M.: *Prediction of Gas Barrier Properties of High Polymers,* Polym. Eng. Sci., **26,** (1986), 1543.

10.14 ICI Technical Literature: *Melinar® PET - Barrier performance and Viclan® PVdC Coatings,* ICI Chemicals & Polymers, Wilton, UK.

10.15 Solvay Technical Literature: *Clarene® EVOH Resins,* (1987), Solvay & Cie S.A., Brussels, Belgium.

10.16 NEUMANN, E.H.: *Multilayer Bottles by the Co-Injection Technique & Heat Setting of PET Containers, all performed on Single Stage Equipment,* Presented at 9th SPE International Conference: High Performance Plastics Packaging, London, (1988).

10.17 AL-MALAIKA, S., and SCOTT, G.: in: *Degradation & Stabilisation of Polyolefins,* ALLEN, N.S. Ed., Applied Science, London, (1983).

10.18 BILLINGHAM, N.C., and CALVERT, P.D.: in: *Developments in Polymer Stabilisation Vol. 3,* SCOTT, G., Ed., Applied Science, London, (1980).

10.19 CHAKRABORTY, K.B., and SCOTT, G.: *The Effects of Thermal Processing on the Thermal Oxidative and Photo-Oxidative Stability of Low Density Polyethylene,* Eur. Polym. J., **13,** (1977), 73.

10.20 HENMAN, T.J.: in: *Developments in Polymer Stabilisation Vol. 1,* SCOTT, G., Ed., Applied Science, London, (1979).

10.21 AL MALAIKA, S.: *A Critique on Stabilisation Technology in Polyolefins,* Polym. Plast. Technol. Eng., **27,** (1988), 261.

10.22 HJERTBERG, T., and SORVIK, E.M.: in: *Degradation and Stabilisation of PVC,* p.21, OWEN, E.D., Ed., Applied Science, London, (1984).

10.23 HAWORTH, B., LAW, T.C., SYKES, R.A., and STEPHENSON ,R .C.: *Factors Determining the Weld Strength of Window Corners fabricated from Impact-Modified UPVC Profiles,* Plast. & Rubb. Proc. & Appl., **9,** (1988), 81.

10.24 BRAUN, D.: in: *Developments in Polymer Degradation Vol. 3,* GRASSIE, N., Ed., Applied Science, London, (1981).

10.25 MUKHERJEE, A.K., and GUPTA, A.: *Structure and Dehydrochlorination of Poly(Vinyl Chloride),* J. Macromol. Sci. Rev. Macromol. Chem., **C20,** (1981), 309.

10.26 OWEN, E.D.: in: *Degradation & Stabilisation of PVC,* p.197, OWEN, E.D., Ed., Applied Science, London, (1984).

10.27 DECKER C, **ibid,** , p.81.

10.28 GAGNE, B.A.: *The Effect of Titanium Dioxide Pigments and Heat Stabilisers on the Weatherability of White Rigid PVC,* Presented at International Symposium: Current trends in PVC Technology, Loughborough University, UK, (1988).

10.29 OWEN, E.D.: *The Roles of Hydrogen Chloride in the Thermal and Photochemical Degradation of Polyvinyl Chloride,* A.C.S. Symp. Ser., **151,** (1981), 217.

10.30 OWEN, E.D.: in: *Developments in Polymer Photochemistry,* ALLEN, N.S. Ed., Applied Science, London, (1982).

10.31 SZABO, E., and LALLY, R.E.: *World-Wide Weathering of Poly(Vinyl Chloride),* Polym. Eng. Sci., **15,** (1975), 277.

10.32 AYREY, G., HEAD, B.C., and POLLER, R.C.: *The Thermal Dehydrochlorination and Stabilisation of Poly(Vinyl Chloride),* J. Polym. Sci. D. Macromol. Rev., **8,** (1974), 1.

10.33 ZIMMERMAN, H.: in: *Developments in Polymer Degradation Vol. 5,* GRASSIE, N. Ed., Applied Science, London, (1984).

10.34 RAVENS, D.A.S., and WARD, I.M.: *Chemical Reactivity of Polyethylene Terephthalate: Hydrolysis and Esterification Reactions in the Solid Phase,* Trans. Farad. Soc., **57,** (1961), 150.

10.35 KELLEHER, P.G., WENTZ, R.P., and FALCONE, D.R.: *Hydrolysis of Poly (Butylene Terephthalate),* Polym. Eng. Sci., **22,** (1982), 260.

10.36 KELLEHER, P.G., WENTZ, R.P., HELLMAN, M.Y., and GILBERT, E.H., *The Hydrolytic Stability of Glass Fibre Reinforced Poly (Butylene Terephthalate), Poly(Ethylene Terephthalate) and Polycarbonate,* Polym. Eng. Sci., **23,** (1983), 537.

10.37 MARTIN, J.R., and GARDNER, R.J.: *Effect of Long Term Humid Aging on Plastics,* Polym. Eng. Sci., **21,** (1981), 557.

10.38 LHYMN, C., and SCHULTZ, J.M.: *Fracture Behaviour of Collimated Thermoplastic Poly (Ethylene Terephthalate) Reinforced with Short E-Glass Fibre,* J. Mat. Sci., **18,** (1983), 2029.

10.39 WIPPLER, C.: *Degradable Polymers,* A.P.M.E. Report, (1986).
10.40 GRIFFIN, G.J.L.: *Synthetic Polymers and the Living Environment,* Pure. & Appl. Chem., **52,** (1980), 399.
10.41 BREMER, W.P.: *Photodegradable Polyethylene,* Polym. Plast. Technol. Eng., **18,** (1982), 137.
10.42 BAILEY, W.J.: *The Design of New Biodegradable Polymers,* Presented at International Conference on Advances in the Stabilisation and Controlled Degradation of Polymers, Lucerne, Switzerland, (1984).
10.43 PALMER, R.P., and COBBOLD, A.J.: *The Texture of Melt Crystallised Polyethylene as Revealed by Selective Oxidation,* Macromol. Chem., **74,** (1964), 174.
10.44 BASSETT, D.C.: *Principles of Polymer Morphology,* Cambridge University Press, Cambridge, (1981).
10.45 DELIMOY, D., BAILLY, C., DEVAUX, J., and LEGRAS, R.: *Morphological Studies of Polycarbonate-Poly(Butylene Terephthalate) Blends by Transmission Electron Microscopy,* Polym. Eng. Sci., **28,** (1988), 104.
10.46 Celanese Technical Literature: *Celanese Nylon 66: Properties and Processing,* Technical Service Note N102, Hoechst-Celanese Ltd., Watford, UK.
10.47 Hoechst Plastics Technical Literature: *Hostalen LDPE,* Hoechst-Celanese Plastics, Frankfurt am Main, Germany.
10.48 ICI Technical Literature: *Victrex PEEK - A Guide to Grades for Injection Moulding,* Technical Service Note VK2, ICI Advanced Materials, Welwyn Garden City, UK.
10.49 Hoechst Plastics Technical Literature: *Hostyren PS,* Hoechst-Celanese Plastics, Frankfurt am Main, Germany.
10.50 ICI Technical Literature: *Victrex PES: Properties & Introduction to Processing,* Technical Service Note VS2, ICI Advanced Materials, Welwyn Garden City, UK.
10.51 *Repair to Blisters in Glass-Fibre Hulls,* British Plastics Federation Report 244/1, (1979).
10.52 BIRLEY, A.W., and CHEN, F.P.: *Blistering in Glass Fibre Reinforced Polyester Laminates,* Presented at 16th BPF Reinforced Plastics Congress, Blackpool, UK, (1988).
10.53 BIRLEY, A.W., DAWKINS, J.V., and STRAUSS, H. E.: *Blistering in Glass Fibre Reinforced Polyester Laminates,* Presented at 14th BPF Reinforced Plastics Congress, Brighton, UK, (1984).
10.54 TROITZSCH, J.: *International Plastics Flammability Handbook, Principles, Regulations, Testing and Approval,* Hanser Publishers, Munich, (1983).
10.55 BS 6336, (1982), *Development and Presentation of Fire Tests and their use in Hazard Assessment,* British Standards Institution, London.
10.56 SPURR, W.: *A Low Smoke, Low Toxicity System that Adapts Easily,* Plast. Rubb. Week., **1250,** (1988), 15.
10.57 BATCHELOR, J.: *Use of Fibre Reinforced Composites in Modern Railway Vehicles,* Mats. in Engng., **2,** (1981), 172.
10.58 WOOLLEY, W.D., RAFTERY, M.A., AMES, S.A., and MURRELL, J.V.: Smoke Release from Wall Linings in Full-Scale Compartment Fires, Fire Safety J., **2,** (1979), 61.

10.59 Katz, H.S., and Milewski, J.V.: *Handbook of Fillers and Reinforcements for Plastics,* Van Nostrand, New York, (1978).
10.60 Tesoro, G.C.: *Chemical Modification of Polymers with Flame-Retardant Compounds,* J. Polym. Sci: Macromol. Rev., **D13,** (1978), 283.
10.61 Buist, J., Grayson, S.J., and Woolley, W.D. (Eds.), *Fire and Cellular Polymers,* Applied Science, London, (1987).
10.62 Hilado, C.J.: Ed., *Flammability Handbook for Plastics, 3rd Ed.,* Technomic Publishers, Westport, USA, (1982).
10.63 Briggs, P.J.: *Fire Behaviour of PVC,* Presented at 3rd International PRI Conference "PVC '87", p.37/1, Brighton, UK, (1987).
10.64 *PVC in Fires,* Customer Service Note, British Plastics Federation, London, (1987).
10.65 Cullis, C.F., and Hirschler, M.M.: *The Combustion of Organic Polymers,* Clarendon Press, Oxford, (1981).
10.66 Hirschler, M.M.: *Flammability and Combustion Toxicity Parameters of Poly(Vinyl Chloride) in Perspective,* Presented at 3rd International PRI Conference "PVC '87", p.38/1, Brighton, UK, (1987).

Subject Index